The Local Structure of Algebraic K-Theory

For further volumes:
www.springer.com/series/6253

Algebra and Applications

Volume 18

Algebra and Applications aims to publish well written and carefully refereed monographs with up-to-date information about progress in all fields of algebra, its classical impact on commutative and noncommutative algebraic and differential geometry, K-theory and algebraic topology, as well as applications in related domains, such as number theory, homotopy and (co)homology theory, physics and discrete mathematics.

Particular emphasis will be put on state-of-the-art topics such as rings of differential operators, Lie algebras and super-algebras, group rings and algebras, C^*-algebras, Kac-Moody theory, arithmetic algebraic geometry, Hopf algebras and quantum groups, as well as their applications. In addition, Algebra and Applications will also publish monographs dedicated to computational aspects of these topics as well as algebraic and geometric methods in computer science.

Bjørn Ian Dundas • Thomas G. Goodwillie •
Randy McCarthy

The Local Structure of Algebraic K-Theory

Springer

Bjørn Ian Dundas
Department of Mathematics
University of Bergen
Bergen, Norway

Randy McCarthy
Department of Mathematics
University of Illinois
Urbana, IL, USA

Thomas G. Goodwillie
Mathematics Department
Brown University
Providence, USA

ISSN 1572-5553 ISSN 2192-2950 (electronic)
Algebra and Applications
ISBN 978-1-4471-4392-5 ISBN 978-1-4471-4393-2 (eBook)
DOI 10.1007/978-1-4471-4393-2
Springer London Heidelberg New York Dordrecht

Library of Congress Control Number: 2012947846

Mathematics Subject Classification: 19-02, 19D55, 55P43, 18G55, 16E40

Printed on acid-free paper

Springer is part of Springer Science+Business Media (www.springer.com)

Preface

Algebraic K-theory draws its importance from its effective codification of a mathematical phenomenon which occurs in as separate parts of mathematics as number theory, geometric topology, operator algebras, homotopy theory and algebraic geometry. In reductionistic language the phenomenon can be phrased as

there is no canonical choice of coordinates,

or, as so elegantly expressed by Hermann Weyl [312, p. 49]:

The introduction of numbers as coordinates ... is an act of violence whose only practical vindication is the special calculatory manageability of the ordinary number continuum with its four basic operations.

As such, algebraic K-theory is a meta-theme for mathematics, but the successful codification of this phenomenon in homotopy-theoretic terms is what has made algebraic K-theory a valuable part of mathematics. For a further discussion of algebraic K-theory we refer the reader to Chap. 1 below.

Calculations of algebraic K-theory are very rare and hard to come by. So any device that allows you to obtain new results is exciting. These notes describe one way to produce such results.

Assume for the moment that we know what algebraic K-theory is; how does it vary with its input?

The idea is that algebraic K-theory is like an analytic function, and we have this other analytic function called *topological cyclic homology* (*TC*) invented by Bökstedt, Hsiang and Madsen [27], and

the difference between K and TC is locally constant.

This statement will be proven below, and in its integral form it has not appeared elsewhere before.

The good thing about this, is that TC is occasionally possible to calculate. So whenever you have a calculation of K-theory you have the possibility of calculating all the K-values of input "close" to your original calculation.

So, for instance, if somebody (please) can calculate K-theory of the integers, many "nearby" applications in geometric topology (simply connected spaces) are

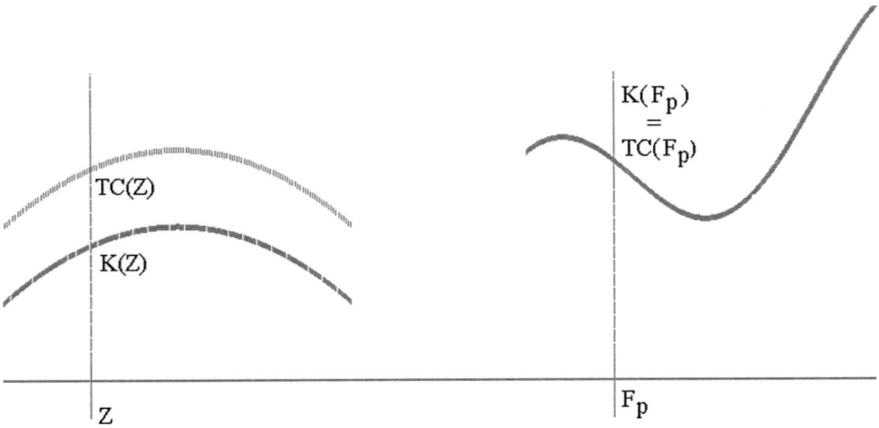

Fig. 0.1 The difference between K and TC is locally constant. The *left part of the figure* illustrates the difference between $K(\mathbf{Z})$ and $TC(\mathbf{Z})$ is quite substantial, but once you know this difference you know that it does not change in a "neighborhood" of $\mathbf{Z}$. In this neighborhood lies for instance all applications of algebraic K-theory of simply connected spaces, so here TC-calculations ultimately should lead to results in geometric topology as demonstrated by Rognes. On the *right hand side of the figure* you see that close to the finite field with p elements, K-theory and TC agree (this is a connective and p-adic statement: away from the characteristic there are other methods that are more convenient). In this neighborhood you find many interesting rings, ultimately resulting in Hesselholt and Madsen's calculations of the K-theory of local fields

available through TC-calculations (see e.g., [242, 243]). This means that calculations in motivic cohomology (giving K-groups of e.g., the integers) will actually have bearing on our understanding of diffeomorphisms of manifolds!

On a different end of the scale, Quillen's calculation of the K-theory of finite fields gives us access to "nearby" rings, ultimately leading to calculations of the K-theory of local fields [131]. One should notice that the illustration offered by Fig. 0.1 is not totally misleading: the difference between $K(\mathbf{Z})$ and $TC(\mathbf{Z})$ is substantial (though locally constant), whereas around the field $\mathbf{F}_p$ with p elements it is negligible.

Taking K-theory for granted (we'll spend quite some time developing it later), we should say some words about TC. Since K-theory and TC differ only by some locally constant term, they must have the same differential: $D_1K = D_1TC$ (see Fig. 0.2). For ordinary rings A this differential is quite easy to describe: it is the *homology* of the category $\mathcal{P}_A$ of finitely generated projective modules.

The homology of a category is like Hochschild homology, and as Connes observed, certain models of Hochschild homology carry a circle action which is useful when comparing with K-theory. Only, in the case of the homology of categories it turns out that the ground ring over which to take Hochschild homology is not an ordinary ring, but the so-called sphere spectrum. Taking this idea seriously, we end up with Bökstedt's *topological Hochschild homology THH*.

One way to motivate the construction of TC from *THH* is as follows. There is a transformation $K \to THH$ which we will call the *Dennis trace map*, and there is

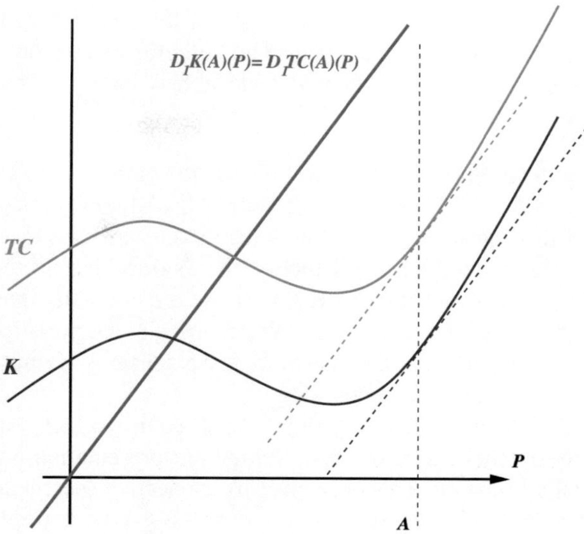

Fig. 0.2 The differentials "at an **S**-algebra A in the direction of the A-bimodule P" of K and TC are equal. For discrete rings the differential is the homology of the category of finitely generated projective modules. In this illustration the differential is the *magenta straight line* through the origin, K-theory is the *red curve* and TC is the shifted curve in *cyan* (Colon figure online)

a model for *THH* for which the Dennis trace map is just the *inclusion of the fixed points under the circle action*. That is, the Dennis trace can be viewed as a composite

$$K \cong THH^{\mathbb{T}} \subseteq THH$$

where $\mathbb{T}$ is the circle group.

The unfortunate thing about this statement is that it is *model dependent* in that fixed points do not preserve weak equivalences: if $X \to Y$ is a map of $\mathbb{T}$-spaces which is a weak equivalence of underlying spaces, normally the induced map $X^{\mathbb{T}} \to Y^{\mathbb{T}}$ will not be a weak equivalence. So, TC is an attempt to construct the $\mathbb{T}$-fixed points through techniques that **do** preserve weak equivalences.

It turns out that there is more to the story than this: *THH* possesses something called an *epicyclic structure* (which is not the case for all $\mathbb{T}$-spaces), and this allows us to approximate the $\mathbb{T}$-fixed points even better.

So in the end, the *cyclotomic trace* is a factorization

$$K \to TC \to THH$$

of the Dennis trace map.

The cyclotomic trace is the theme for this book. There is another paper devoted to this transformation, namely Madsen's eminent survey [192]. If you can get hold of a copy it is a great supplement to the current text.

It was originally an intention that readers who were only interested in discrete rings would have a path leading far into the material with minimal contact with ring spectra. This idea has to a great extent been abandoned since ring spectra and the techniques around them have become much more mainstream while these notes have matured. Some traces of this earlier approach can still be seen in that Chap. 1 does not depend at all on ring spectra, leading to the proof that stable K-theory of rings corresponds to homology of the category of finitely generated projective modules. Topological Hochschild homology is, however, interpreted as a functor of ring spectra, so the statement that stable K-theory is *THH* requires some background on ring spectra.

General Plan The general plan of the book is as follows.

In Sect. 1.1 we give some general background on algebraic K-theory. The length of this introductory section is justified by the fact that this book is primarily concerned with algebraic K-theory; the theories that fill the last chapters are just there in order to shed light on K-theory, we are not really interested in them for any other reason. In Sect. 1.2 we give Waldhausen's interpretation of algebraic K-theory and study in particular the case of radical extensions of rings. Finally, Sect. 1.3 compares stable K-theory and homology.

Chapter 2 aims at giving a crash course on ring spectra. In order to keep the presentation short we have limited our presentation only the simplest version: Segal's Γ-spaces. This only gives us connective spectra and the behavior with respect to commutativity issues leaves something to be desired. However, for our purposes Γ-spaces suffice and also fit well with Segal's version of algebraic K-theory, which we are using heavily later in the book.

Chapter 3 can (and perhaps should) be skipped on a first reading. It only asserts that various reductions are possible. In particular, K-theory of simplicial rings can be calculated degreewise "locally" (i.e., in terms of the K-theory of the rings appearing in each degree), simplicial rings are "dense" in the category of (connective) ring spectra, and all definitions of algebraic K-theory we encounter give the same result.

In Chap. 4, topological Hochschild homology is at long last introduced, first for ring spectra, and then in a generality suitable for studying the correspondence with algebraic K-theory. The equivalence between the topological Hochschild homology of a ring and the homology of the category of finitely generated projective modules is established in Sect. 4.3, which together with the results in Sect. 1.3 settle the equivalence between stable K-theory and topological Hochschild homology of rings.

In order to push the theory further we need an effective comparison between K-theory and *THH*, and this is provided by the Dennis trace map $K \to THH$ in the following chapter. We have here chosen a model which "localizes at the weak equivalences", and so conforms nicely with the algebraic case. For our purposes this works very well, but the reader should be aware that other models are more appropriate for proving structural theorems about the trace. The comparison between stable K-theory and topological Hochschild homology is finalized Sect. 5.3, using the trace. As a more streamlined alternative, we also offer a new and more direct trace construction in Sect. 5.4.

In Chap. 6 topological cyclic homology is introduced. This is the most involved of the chapters in the book, since there are so many different aspects of the theory that have to be set in order. However, when the machinery is set up properly, and the trace has been lifted to topological cyclic homology, the local correspondence between K-theory and topological cyclic homology is proved in a couple of pages in Chap. 7.

Chapter 7 ends with a quick and inadequate review of the various calculations of algebraic K-theory that have resulted from trace methods. We first review the general framework set up by Bökstedt and Madsen for calculating topological cyclic homology, and follow this through for three important examples: the prime field $\mathbf{F}_p$, the (p-adic) integers $\mathbf{Z}_p$ and the Adams summand ℓ_p. These are all close enough to $\mathbf{F}_p$ so that the local correspondence between K-theory and topological cyclic homology make these calculations into actual calculations of algebraic K-theory. We also discuss very briefly the Lichtenbaum-Quillen conjecture as seen from a homotopy theoretical viewpoint, which is made especially attractive through the comparison with topological cyclic homology. The inner equivariant workings of topological Hochschild homology display a rich and beautiful algebraic structure, with deep intersections with log geometry through the de Rham-Witt complex. This is prominent in Hesselholt and Madsen's calculation of the K-theory of local fields, but facets are found in almost all the calculations discussed in Sect. 7.3. We also briefly touch upon the first problem tackled through trace methods: the algebraic K-theory Novikov conjecture.

The Appendix collects some material that is used freely throughout the notes. Much of the material is available elsewhere in the literature, but for the convenience of the reader we have given the precise formulations we actually need and set them in a common framework. The reason for pushing this material to an appendix, and not working it into the text, is that an integration would have produced a serious eddy in the flow of ideas when only the most diligent readers will need the extra details. In addition, some of the results are used at places that are meant to be fairly independent of each other.

The rather detailed index is meant as an aid through the plethora of symbols and complex terminology, and we have allowed ourselves to make the unorthodox twist of adding hopefully helpful hints in the index itself, where this has not taken too much space, so that in many cases a brief glance at the index makes checking up the item itself unnecessary.

Displayed diagrams commute, unless otherwise noted. The ending of proofs that are just sketched or referred away and of statements whose verification is embedded in the preceding text are marked with a ☺.

Acknowledgments This book owes a lot to many people. The first author especially wants to thank Marcel Bökstedt, Bjørn Jahren, Ib Madsen and Friedhelm Waldhausen for their early and decisive influence on his view on mathematics. The third author would like to thank Marcel Bökstedt, Dan Grayson, John Klein, Jean-Louis Loday and Friedhelm Waldhausen whose support has made all the difference.

These notes have existed for quite a while on the net, and we are grateful for the helpful comments we have received from a number of people, in particular from Morten Brun, Lars Hesselholt, Harald Kittang, Birgit Richter, John Rognes, Stefan Schwede and Paul Arne Østvær.

A significant portion of the notes were written while visiting Stanford University, and the first author is grateful to Gunnar Carlsson and Ralph Cohen for the invitation and the request to give a course based on these notes, which gave the impetus to try to finish the project.

The manuscript has benefited immensely from some very thorough and helpful referee reports. We are very grateful for the time and consideration offered by the anonymous referees towards making the text as useful as possible. We are also grateful for the patience and encouragement offered by Springer Verlag and John Greenlees during the laborious years marking the end of the project.

Finally, the first author wants to thank his wife Siv and daughters Karen and Vår for their patience with him and apologize for all the time spent thinking, writing and generally not paying attention to the important things.

Leitfaden For the convenience of the reader we provide the following Leitfaden. It should not be taken too seriously, some minor dependencies are not shown, and many sections that are noted to depend on previous chapters are actually manageable if one is willing to retrace some cross referencing. In particular, Chap. 3 should be postponed upon a first reading.

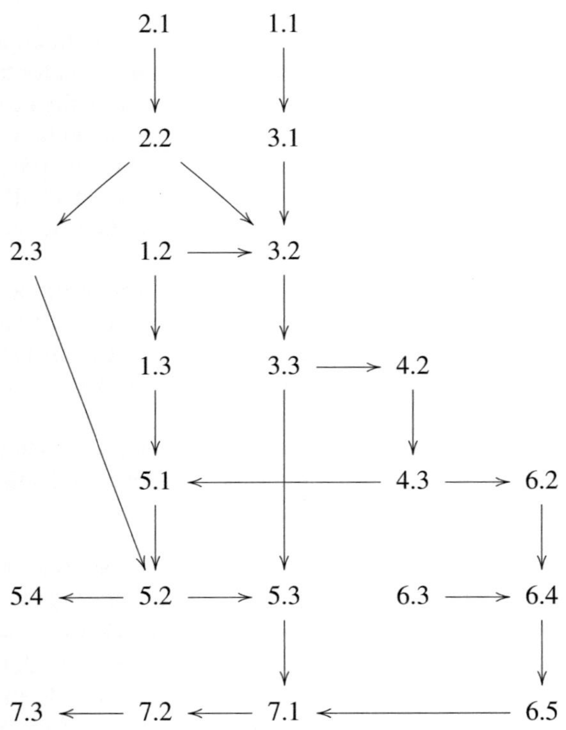

Contents

Chapter 1
Algebraic K-Theory

In this chapter we define and discuss the algebraic K-theory functor. This chapter will mainly be concerned with the algebraic K-theory of rings, but we will extend this notion at the end of the chapter. There are various possible extensions, but we will mostly focus on a class of objects that are close to rings. In later chapters this will be extended again to include ring spectra and even more exotic objects.

In the first section we give a quick nontechnical overview of K-theory. Many of the examples are but lightly touched upon and not needed later, but are included to give an idea of the scope of the theory. Some of the examples in the introduction may refer to concepts or ideas that are unfamiliar to the reader. If this is the case, the reader may consult the index to check whether this is a topic that will be touched upon again (and perhaps even explained), or if it is something that can be left for later investigations. In any case, the reader is encouraged to ignore such problems at a first reading. Although it only treats the first three groups, Milnor's book [213] is still one of the best elementary introductions to algebraic K-theory with Bass' book [13] providing the necessary support for more involved questions. For a more modern exposition one may consult Rosenberg's book [244]. For a fuller historical account, the reader may want to consult for instance [310] or [14].

In the second section we introduce Waldhausen's S-construction of algebraic K-theory and prove some of its basic properties.

The third section concerns itself with comparisons between K-theory and various homology theories, giving our first identification of the differential of algebraic K-theory, as discussed in the preface.

1.1 Introduction

The first appearance of what we now would call truly K-theoretic questions are the investigations of J.H.C. Whitehead (for instance [314, 315] or the later [316]), and Higman [133]. The name "K-theory" is much younger (said to be derived from the German word "Klassen"), and first appears in Grothendieck's work [1] in 1957 on

B.I. Dundas et al., *The Local Structure of Algebraic K-Theory*,
Algebra and Applications 18,
DOI 10.1007/978-1-4471-4393-2_1, © Springer-Verlag London 2013

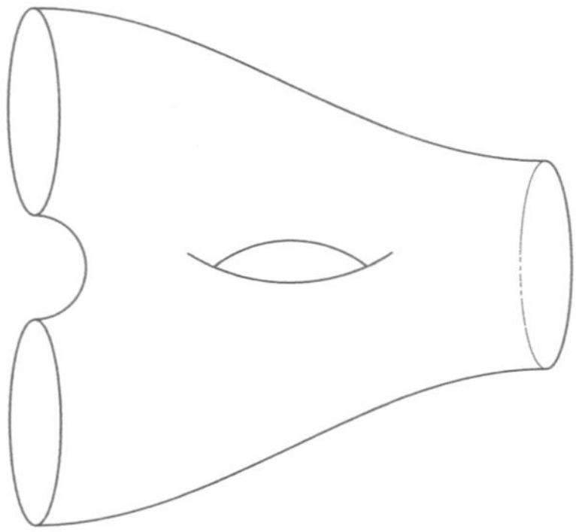

Fig. 1.1 A cobordism W between a disjoint union M of two circles and a single circle N

the Riemann-Roch theorem, see also [35]. But, even though it was not called K-theory, we can get some motivation by studying the early examples.

1.1.1 Motivating Example from Geometry: Whitehead Torsion

The "Hauptvermutung" states that two homeomorphic finite simplicial complexes have isomorphic subdivisions. The conjecture was formulated by Steinitz and Tietze in 1908, see [236] for references and a deeper discussion.

Unfortunately, the Hauptvermutung is not true: already in 1961 Milnor [212] gave concrete counterexamples built from lens spaces in all dimensions greater than six. To distinguish the simplicial structures he used an invariant of the associated chain complexes in what he called the *Whitehead group*. In the decade that followed, the Whitehead group proved to be an essential tool in topology, and especially in connection with problems related to "cobordisms". For a more thorough treatment of the following example, see Milnor's very readable article [210].

Let M and N be two smooth n-dimensional closed manifolds. A *cobordism* between M and N is an $n+1$-dimensional smooth compact manifold W with boundary the disjoint union of M and N, see Fig. 1.1 (in the oriented case we assume that M and N are oriented, and W is an oriented cobordism from M to N if it is oriented so that the orientation agrees with that on N and is the opposite of that on M).

Here we are interested in a situation where M and N are deformation retracts of W. Obvious examples are cylinders $M \times I$, where $I = [0, 1]$ is the closed unit interval.

More precisely: Let M be a closed, connected, smooth manifold of dimension $n > 5$. Suppose we are given an *h-cobordism* $(W; M, N)$, that is, a compact smooth $n + 1$ dimensional manifold W, with boundary the disjoint union of M and N, such that both the inclusions $M \subset W$ and $N \subset W$ are homotopy equivalences. The obvious examples are cylinders, i.e. diffeomorphic to $M \times I$ (see Fig. 1.2).

Question 1.1.1.1 *Is W diffeomorphic to $M \times I$?*

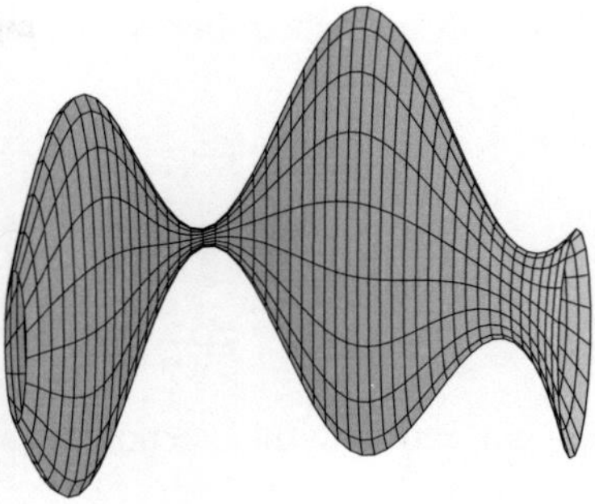

Fig. 1.2 An h-cobordism $(W; M, N)$. This one is a cylinder

It requires some imagination to realize that the answer to this question can be "no". In particular, in the low dimensions of the illustrations all h-cobordisms **are** cylinders.

However, this is not true in high dimensions, and the h-cobordism Theorem 1.1.1.2 below gives a precise answer to the question.

To fix ideas, let $M = L$ be a *lens space* of dimension, say, $n = 7$. That is, the cyclic group of order l, $\pi = \mu_l = \{1, e^{2\pi i/l}, \dots, e^{2\pi i(l-1)/l}\} \subseteq \mathbf{C}$, acts on the seven-dimensional sphere $S^7 = \{\mathbf{x} \in \mathbf{C}^4 \text{ s.t. } |\mathbf{x}| = 1\}$ by complex multiplication

$$\pi \times S^7 \to S^7 \quad (t, \mathbf{x}) \mapsto (t \cdot \mathbf{x})$$

and we let the lens space M be the quotient space $S^7/\pi = S^7/(\mathbf{x} \sim t \cdot \mathbf{x})$. Then M is a smooth manifold with fundamental group π.

Let

$$\cdots \xrightarrow{\partial} C_{i+1} \xrightarrow{\partial} C_i \xrightarrow{\partial} \cdots \longrightarrow C_0 \longrightarrow 0$$

be the relative cellular complex of the universal cover, calculating the homology $H_* = H_*(\tilde{W}, \tilde{M})$ (see Sects. 7 and 9 in [210] for details). Each C_i is a finitely generated free $\mathbf{Z}[\pi]$-module, and, up to orientation and translation by elements in π, has a preferred basis over $\mathbf{Z}[\pi]$ coming from the i-simplices added to get from M to W in some triangulation of the universal covering spaces. As always, the groups Z_i and B_i of i-cycles and i-boundaries are the kernel of $\partial\colon C_i \to C_{i-1}$ and image of $\partial\colon C_{i+1} \to C_i$. Since $M \subset W$ is a deformation retract, we have by homotopy invariance of homology that $H_* = 0$, and so $B_* = Z_*$.

By induction on i, we see that the exact sequence

$$0 \longrightarrow B_i \longrightarrow C_i \longrightarrow B_{i-1} \longrightarrow 0$$

is split. For each i we choose a splitting and consider the resulting isomorphism

$$C_i \underset{\cong}{\xrightarrow{\alpha_i}} B_i \oplus B_{i-1}.$$

This leads us to the following isomorphism

$$\begin{array}{ccc} \bigoplus_{i \text{ even}} C_i & \xrightarrow{\bigoplus_{i \text{ even}} \alpha_i} & \bigoplus_{i \text{ even}} B_i \oplus B_{i-1} \\ & & \cong \big\downarrow \text{can. rearrangement} \\ \bigoplus_{i \text{ odd}} C_i & \xrightarrow{\bigoplus_{i \text{ odd}} \alpha_i} & \bigoplus_{i \text{ odd}} B_i \oplus B_{i-1}. \end{array} \tag{1.1}$$

We will return to this isomorphism shortly in order to define the obstruction to the answer to Question 1.1.1.1 being "yes" (see Sect. 1.1.1.2), but first we need some basic definitions from linear algebra.

1.1.1.1 K_1 and the Whitehead Group

For any ring A (all the rings we consider are associative and unital) we may consider the ring $M_k(A)$ of $k \times k$ matrices with entries in A, as a monoid under multiplication (recall that a *monoid* satisfies all the axioms of a group except for the requirement that inverses must exist). The *general linear group* is the subgroup of invertible elements $GL_k(A)$. Take the colimit (or more concretely, the union) $GL(A) = \lim_{k\to\infty} GL_k(A) = \bigcup_{k\to\infty} GL_k(A)$ with respect to the stabilization

$$GL_k(A) \xrightarrow{g \mapsto g \oplus 1} GL_{k+1}(A)$$

(thus every element $g \in GL(A)$ can be thought of as an infinite matrix

$$\begin{bmatrix} g' & 0 & 0 & \dots \\ 0 & 1 & 0 & \dots \\ 0 & 0 & 1 & \dots \\ \vdots & \vdots & \vdots & \ddots \end{bmatrix}$$

with $g' \in GL_k(A)$ for some $k < \infty$). Let $E(A)$ be the subgroup of *elementary matrices* (i.e., $E_k(A) \subset GL_k(A)$ is the subgroup generated by the matrices e^a_{ij} with ones on the diagonal and a single nontrivial off-diagonal entry $a \in A$ in the ij position). The "Whitehead lemma" (see Lemma 1.1.2.2 below) implies that the quotient

$$K_1(A) = GL(A)/E(A)$$

is an abelian group. In the particular case where A is an integral group ring $\mathbf{Z}[\pi]$ we define the *Whitehead group* as the quotient

$$Wh(\pi) = K_1\big(\mathbf{Z}[\pi]\big)/\{\pm\pi\}$$

via $\{\pm\pi\} \subseteq GL_1(\mathbf{Z}[\pi]) \to K_1(\mathbf{Z}[\pi])$.

1.1.1.2 Classifying Cobordisms

Let $(W; M, N)$ be an h-cobordism, and consider the isomorphism $\bigoplus_{i \text{ even}} C_i \to \bigoplus_{i \text{ odd}} C_i$ given in Eq. (1.1) for the lens spaces, and similarly in general. This depended on several choices and in the preferred basis for the C_i it gives a matrix with coefficients in $\mathbf{Z}[\pi_1(M)]$. Stabilizing we get an element $\tau(W, M)_{\text{choices}} \in GL(\mathbf{Z}[\pi_1(M)])$ and a class $\tau(W, M) = [\tau(W, M)_{\text{choices}}] \in Wh(\pi_1(M))$.

The class $\tau(W, M)$ is independent of our preferred basis and choices of splittings and is called the *Whitehead torsion*.

The Whitehead torsion turns out to be a vital ingredient in Barden (Thesis, 1963), Mazur [202] and Stallings' [272] extension of the famous results of Smale [264] (where he proves the high dimensional Poincaré conjecture) beyond the simply connected case (for a proof, see also [163]):

Theorem 1.1.1.2 (Barden, Mazur, Stallings) *Let M be a compact, connected, smooth manifold of dimension ≥ 5 and let $(W; M, N)$ be an h-cobordism. The Whitehead torsion $\tau(W, M) \in Wh(\pi_1(M))$ is well defined, and τ induces a bijection*

$$\left\{\begin{matrix}\textit{diffeomorphism classes (rel. } M)\\ \textit{of } h\textit{-cobordisms } (W; M, N)\end{matrix}\right\} \longleftrightarrow Wh\big(\pi_1(M)\big)$$

In particular, $(W; M, N) \cong (M \times I; M, M)$ if and only if $\tau(W, M) = 0$.

Example 1.1.1.3 The Whitehead group, $Wh(\pi)$, has been calculated for only a very limited set of groups π. We list a few of them; for a detailed study of Wh of finite groups, see [220]. The first three refer to the lens spaces discussed above (see p. 375 in [210] for references).

1. $l = 1$, $M = S^7$. "Exercise": show that $K_1\mathbf{Z} = \{\pm 1\}$, and so $Wh(0) = 0$. Thus any h-cobordism of S^7 is diffeomorphic to $S^7 \times I$.
2. $l = 2$. $M = P^7$, the real projective 7-space. "Exercise": show that $K_1\mathbf{Z}[\mu_2] = \{\pm\mu_2\}$, and so $Wh(\mu_2) = 0$. Thus any h-cobordism of P^7 is diffeomorphic to $P^7 \times I$.
3. $l = 5$. $Wh(\mu_5) \cong \mathbf{Z}$ generated by the invertible element $t + t^{-1} - 1 \in \mathbf{Z}[\mu_5]$ (where t is a chosen fifth root of unity)—the inverse is $t^2 + t^{-2} - 1$. That is, there exist countably infinitely many non-diffeomorphic h-cobordisms with incoming boundary component S^7/μ_5.
4. Waldhausen [297]: If π is a free group, free abelian group, or the fundamental group of a submanifold of the three-sphere, then $Wh(\pi) = 0$.
5. Farrell and Jones [81]: If M is a closed Riemannian manifold with non-positive sectional curvature, then $Wh(\pi_1 M) = 0$.

Remark 1.1.1.4 The presentation of the Whitehead torsion differs slightly from that of [210]. It is easy to see that they are the same in the case where the B_i are free $\mathbf{Z}[\pi]$-modules (the splittings ensure that each B_i is "stably free" which is sufficient, but the argument is slightly more involved). Choosing bases we get matrices

$M_i \in GL(\mathbf{Z}[\pi])$ representing the isomorphisms $\alpha_i : C_i \cong B_i \oplus B_{i-1}$, and from the definition of $\tau(W, M)_{\text{choices}}$ we see that

$$\tau(W, M) = \Big(\sum_{i \text{ even}} [M_i]\Big) - \Big(\sum_{i \text{ odd}} [M_i]\Big) = \sum (-1)^i [M_i] \in Wh\big(\pi_1(M)\big).$$

1.1.2 K_1 of Other Rings

1. Commutative rings: The map from the units in A
$$A^* = GL_1(A) \to GL(A)/E(A) = K_1(A)$$
is split by the determinant map, and so the units of A is a split summand in $K_1(A)$. In certain cases (e.g., if A is *local* (A has a unique maximal ideal), or the integers in a number field, see next example) this is all of $K_1(A)$. We may say that the rest of $K_1(A)$ measures to what extent we can do Gauss elimination, in that $\ker\{\det: K_1(A) \to A^*\}$ is the group of equivalence classes of matrices up to stabilization in the number of variables and elementary row operations (i.e., multiplication by elementary matrices and multiplication of a row by an invertible element).
2. Let F be a number field (i.e., a finite extension of the rational numbers), and let $A \subseteq F$ be the ring of integers in F (i.e., the integral closure of $\mathbf{Z}$ in F). A result of Dirichlet asserts that A^* is finitely generated of rank $r_1 + r_2 - 1$ where r_1 (resp. $2r_2$) is the number of distinct real (resp. complex) embeddings of F, and in this case $K_1(A) \cong A^*$, see [213, Corollary 18.3] or the arguments on pp. 160–163.
3. Let $B \to A$ be an epimorphism of rings with kernel $I \subseteq rad(B)$—the Jacobson radical of B (that is, if $x \in I$, then $1 + x$ is invertible in B). Then
$$(1+I)^\times \longrightarrow K_1(B) \longrightarrow K_1(A) \longrightarrow 0$$
is exact, where $(1+I)^\times \subset GL_1(B)$ is the group $\{1 + x | x \in I\}$ under multiplication (see e.g., p. 449 in [13]). Moreover, if B is commutative and $B \to A$ is split, then
$$0 \longrightarrow (1+I)^\times \longrightarrow K_1(B) \longrightarrow K_1(A) \longrightarrow 0$$
is exact.

For later reference, we record the Whitehead lemma mentioned above. For this we need some definitions.

Definition 1.1.2.1 The *commutator* $[G, G]$ of a group G is the (normal) subgroup generated by all commutators $[g, h] = ghg^{-1}h^{-1}$. A group G is called *perfect* if it is equal to its commutator, or in other words, if its first homology group $H_1(G) = G/[G, G]$ vanishes. Any group G has a *maximal perfect subgroup*, which we call PG, and which is automatically normal. We say that G is *quasi-perfect* if $PG = [G, G]$.

The symmetric group Σ_n on $n \geq 5$ letters is quasi-perfect, since its commutator subgroup is the alternating group A_n, which in turn is a simple group. Further examples are provided by the

Lemma 1.1.2.2 (The Whitehead Lemma) *Let A be a unital ring. Then $GL(A)$ is quasi-perfect with maximal perfect subgroup $E(A)$, i.e.,*

$$\big[GL(A), GL(A)\big] = \big[E(A), GL(A)\big] = \big[E(A), E(A)\big] = E(A).$$

Proof See e.g., p. 226 in [13]. □

1.1.3 The Grothendieck Group K_0

Definition 1.1.3.1 Let $\mathfrak{C}$ be a small category and let $\mathcal{E}$ be a collection of diagrams $c' \to c \to c''$ in $\mathfrak{C}$. Then the Grothendieck group $K_0(\mathfrak{C}, \mathcal{E})$ is the abelian group, defined (up to unique isomorphism) by the following universal property. Any function f from the set of isomorphism classes of objects in $\mathfrak{C}$ to an abelian group A such that $f(c) = f(c') + f(c'')$ for all sequences $c' \to c \to c''$ in $\mathcal{E}$, factors uniquely through $K_0(\mathfrak{C})$.

If there is a final object $0 \in ob\mathfrak{C}$ such that for any isomorphism $c' \cong c \in \mathfrak{C}$ the sequence $c' \cong c \to 0$ is in $\mathcal{E}$, then $K_0(\mathfrak{C}, \mathcal{E})$ can be given as the free abelian group on the set of isomorphism classes $[c]$, of $\mathfrak{C}$, modulo the relations $[c] = [c'] + [c'']$ for $c' \to c \to c''$ in $\mathcal{E}$. Notice that $[0] = [0] + [0]$, so that $[0] = 0$.

Most often the pair $(\mathfrak{C}, \mathcal{E})$ will be an *exact category* in the sense that $\mathfrak{C}$ is an *additive category* (i.e., a category with all finite coproducts where the morphism sets are abelian groups and where composition is bilinear) such that there exists a full embedding of $\mathfrak{C}$ in an abelian category $\mathfrak{A}$, such that $\mathfrak{C}$ is closed under extensions in $\mathfrak{A}$ and $\mathcal{E}$ consists of the sequences in $\mathfrak{C}$ that are short exact in $\mathfrak{A}$.

Any additive category is an exact category if we choose the exact sequences to be the split exact sequences, but there may be other exact categories with the same underlying additive category. For instance, the category of abelian groups is an abelian category, and hence an exact category in the natural way, choosing $\mathcal{E}$ to consist of the short exact sequences. These are not necessary split, e.g., $\mathbf{Z} \xrightarrow{2} \mathbf{Z} \longrightarrow \mathbf{Z}/2\mathbf{Z}$ is a short exact sequence which does not split.

The definition of K_0 is a case of "additivity": K_0 is a (or perhaps, *the*) functor to abelian groups insensitive to extension issues. We will dwell more on this issue later, when we introduce the higher K-theories. Higher K-theory plays exactly the same rôle as K_0, except that the receiving category has a much richer structure than the category of abelian groups.

The choice of $\mathcal{E}$ will always be clear from the context, and we drop it from the notation and write $K_0(\mathfrak{C})$.

Example 1.1.3.2

1. Let A be a unital ring. An *A-module* is an abelian group M, together with a homomorphism $A \to \mathrm{End}(M)$ of rings, or otherwise said, a homomorphism $A \otimes M \to M$ of abelian groups, sending $a \otimes m$ to am with the property that $1m = m$ and $a(bm) = (ab)m$. Recall that an A-module M is *finitely generated* if there is a surjective homomorphism $A^n = A \oplus \cdots \oplus A \twoheadrightarrow M$ (n summands) of A-modules. An A-module P is *projective* if for all (solid) diagrams

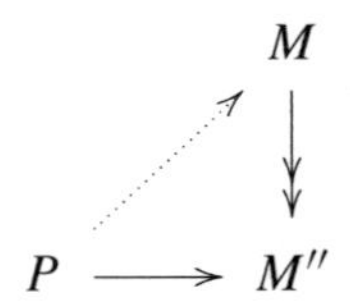

of A-modules where the vertical homomorphism is a surjection, there is a (dotted) homomorphism $P \to M$ making the resulting diagram commute. It is a consequence that an A-module P is finitely generated and projective precisely when there is an n and an A-module Q such that $A^n \cong P \oplus Q$. Note that Q is automatically finitely generated and projective.

   If, in a given subcategory of the category of A-modules we say that a certain sequence is exact, we usually mean that the sequence is exact when considered as a sequence of A-modules.

   If $\mathfrak{C} = \mathcal{P}_A$, the category of finitely generated projective A-modules, with the usual notion of (short) exact sequences, we often write $K_0(A)$ for $K_0(\mathcal{P}_A)$. Note that $\mathcal{P}_A$ is *split exact*, that is, all short exact sequences in $\mathcal{P}_A$ split. Thus we see that we could have defined $K_0(A)$ as the quotient of the free abelian group on the isomorphism classes in $\mathcal{P}_A$ by the relation $[P \oplus Q] \sim [P] + [Q]$. It follows that all elements in $K_0(A)$ can be represented as a difference $[P] - [F]$ where F is a finitely generated free A-module.
2. Inside $\mathcal{P}_A$ sits the category $\mathcal{F}_A$ of finitely generated free A-modules, and we let $K_0^f(A) = K_0(\mathcal{F}_A)$. If A is a principal ideal domain, then every submodule of a free module is free, and so $\mathcal{F}_A = \mathcal{P}_A$. This is so, e.g., for the integers, and we see that $K_0(\mathbf{Z}) = K_0^f(\mathbf{Z}) \cong \mathbf{Z}$, generated by the module of rank one. Generally, $K_0^f(A) \to K_0(A)$ is an isomorphism if and only if every finitely generated projective module is *stably free* (P and P' are said to be *stably isomorphic* if there is a finitely generated free A-module Q such that $P \oplus Q \cong P' \oplus Q$, and P is stably free if it is stably isomorphic to a free module). Whereas $K_0(A \times B) \cong K_0(A) \times K_0(B)$, the functor K_0^f does not preserve products: e.g., $\mathbf{Z} \cong K_0^f(\mathbf{Z} \times \mathbf{Z})$, while $K_0(\mathbf{Z} \times \mathbf{Z}) \cong \mathbf{Z} \times \mathbf{Z}$ giving an easy example of a ring where not all projectives are free.

3. Note that K_0 does not distinguish between stably isomorphic modules. This is not important in some special cases. For instance, if A is a commutative Noetherian ring of Krull dimension d, then every stably free module of rank $> d$ is free ([13, p. 239]).
4. The initial map $\mathbf{Z} \to A$ defines a map $\mathbf{Z} \cong K_0^f(\mathbf{Z}) \to K_0^f(A)$ which is always surjective, and in most practical circumstances, an isomorphism. If A has the *invariance of basis property*, that is, if $A^m \cong A^n$ if and only if $m = n$, then $K_0^f(A) \cong \mathbf{Z}$. Otherwise, $A = 0$, or there is an $h > 0$ and a $k > 0$ such that $A^m \cong A^n$ if and only if either $m = n$ or $m, n > h$ and $m \equiv n \mod k$. There are examples of rings with such h and k for all $h, k > 0$ (see [171] or [54]): let $A_{h,k}$ be the quotient of the free ring on the set $\{x_{ij}, y_{ji} | 1 \leq i \leq h, 1 \leq j \leq h+k\}$ by the matrix relations
$$[x_{ij}] \cdot [y_{ji}] = I_h, \quad \text{and} \quad [y_{ji}] \cdot [x_{ij}] = I_{h+k}.$$
Commutative (non-trivial) rings always have the invariance of basis property.
5. Let X be a compact Hausdorff topological space, and let $\mathfrak{C} = \text{Vect}(X)$ be the category of finite rank complex vector bundles on X, with exact sequences meaning the usual thing. Then $K_0(\text{Vect}(X))$ is the complex K-theory $K(X)$ of Atiyah and Hirzebruch [9]. Note that the possibility of constructing normal complements assures that $\text{Vect}(X)$ is a split exact category. Swan's theorem [280] states that the category $\text{Vect}(X)$ is equivalent to the category of finitely generated projective modules over the ring $C(X)$ of complex valued continuous functions on X. The equivalence is given by sending a bundle to its $C(X)$-module of sections. Furthermore, Bott periodicity (see the survey [36] or the neat proof [119]) states that there is a canonical isomorphism $K(S^2) \otimes K(X) \cong K(S^2 \times X)$. A direct calculation shows that $K(S^2) \simeq \mathbf{Z} \oplus \mathbf{Z}$ where it is customary to let the first factor be generated by the trivial bundle 1 and the second by $\xi - 1$ where ξ is the tautological line bundle on $S^2 = \mathbf{CP}^1$.
6. Let X be a scheme, and let $\mathfrak{C} = \text{Vect}(X)$ be the category of finite rank vector bundles on X. Then $K_0(\text{Vect}(X))$ is the $K(X)$ of Grothendieck. This is an example of K_0 of an exact category which is not split exact. The analogous statement to Swan's theorem above is that of Serre [258].

1.1.3.1 Example of Applications to Homotopy Theory

As an illustration we review Loday's [178] early application of the functors K_0 and K_1 to establishing a result about polynomial functions.

Let $T^n = \{(x_1, x_2, \ldots, x_{2n-1}, x_{2n}) \in \mathbf{R}^{2n} | x_{2i-1}^2 + x_{2i}^2 = 1, i = 1, \ldots, n\}$ be the n-dimensional torus and $S^n = \{(y_0, \ldots, y_n) \in \mathbf{R}^{n+1} | y_0^2 + \cdots + y_n^2 = 1\}$ the n-dimensional sphere. A polynomial function $T^n \to S^n$ is a polynomial function $f: \mathbf{R}^{2n} \to \mathbf{R}^{n+1}$ such that $f(T^n) \subseteq S^n$.

Proposition 1.1.3.3 (Loday [178]) *Let $n > 1$. Any polynomial function $f: T^n \to S^n$ is homotopic to a constant map.*

Sketch proof We only sketch the case $n = 2$. The other even dimensional cases are similar, whereas the odd cases uses K_1 instead of K_0. The heart of the matter is the following commutative diagram

$$\begin{array}{ccc} \mathbf{C}[y_0, y_1, y_2]/(y_0^2 + y_1^2 + y_2^2 - 1) & \longrightarrow & C(S^2) \\ f^* \downarrow & & f^* \downarrow \\ \mathbf{C}[x_1, x_2, x_3, x_4]/(x_1^2 + x_2^2 - 1, x_3^2 + x_4^2 - 1) & \longrightarrow & C(T^2) \end{array}$$

of $\mathbf{C}$-algebras, where the vertical maps are induced by the polynomial function f and the horizontal maps are defined as follows. If $X \subseteq \mathbf{R}^m$ is the zero set of some polynomial function $p = (p_1, \ldots, p_k)\colon \mathbf{R}^m \to \mathbf{R}^k$ there is a preferred map of $\mathbf{C}$-algebras $\mathbf{C}[x_1, \ldots, x_m]/(p_1, \ldots, p_k) \to C(X)$ given by sending the generator x_l to the composite function $X \subseteq \mathbf{R}^m \subseteq \mathbf{C}^m \twoheadrightarrow \mathbf{C}$ where the last map is projection onto the lth factor.

Let $\tilde{K}_0$ be the functor from rings to abelian groups whose value at A is the cokernel of the canonical map $K_0(\mathbf{Z}) \to K_0(A)$. Considering the resulting diagram

$$\begin{array}{ccc} \tilde{K}_0(\mathbf{C}[y_0, y_1, y_2]/(y_0^2 + y_1^2 + y_2^2 - 1)) & \longrightarrow & \tilde{K}_0(C(S^2)) \\ f^* \downarrow & & f^* \downarrow \\ \tilde{K}_0(\mathbf{C}[x_1, x_2, x_3, x_4]/(x_1^2 + x_2^2 - 1, x_3^2 + x_4^2 - 1)) & \longrightarrow & \tilde{K}_0(C(T^2)) \end{array}.$$

By Swan's theorem Example 1.1.3.2(5) we may identify the right hand vertical map with $f^*\colon \tilde{K}(S^2) \to \tilde{K}(T^2)$ (where $\tilde{K}(X)$ is the cokernel of the canonical map $K(*) \to K(X)$). Hence we are done if we can show

1. The top horizontal map is a surjection,
2. the lower left hand group is trivial and
3. a polynomial function $T^2 \to S^2$ is homotopic to a constant map if it induces the trivial map $\tilde{K}(S^2) \to \tilde{K}(T^2)$.

By the statements about complex K-theory in Example 1.1.3.2(5), $\tilde{K}(S^2)$ is a copy of the integers (generated by $\xi - 1$), so to see that the top horizontal map is a surjection it is enough to see that a generator is hit (i.e., the canonical line bundle is algebraic), and this is done explicitly in [178, Lemme 2].

The substitution $t_k = x_{2k-1} + i x_{2k}$ induces an isomorphism

$$\mathbf{C}[x_1, x_2, x_3, x_4]/(x_1^2 + x_2^2 - 1, x_3^2 + x_4^2 - 1) \cong \mathbf{C}[t_1, t_1^{-1}, t_2, t_2^{-1}],$$

and by [13, p. 636] $\tilde{K}_0(\mathbf{C}[t_1, t_1^{-1}, t_2, t_2^{-1}]) = 0$. This vanishing of a K-group is part of a more general statement about algebraic K-theory's behavior with respect to localizations and about polynomial rings over regular rings.

To see the last statement, one has to know that the Chern class is natural: the diagram

$$\begin{array}{ccc} \tilde{K}(S^2) & \xrightarrow{c_1} & H^2(S^2;\mathbf{Q}) \cong \mathbf{Q} \\ {\scriptstyle f^*}\downarrow & & {\scriptstyle f^*}\downarrow \\ \tilde{K}(T^2) & \xrightarrow{c_1} & H^2(T^2;\mathbf{Q}) \cong \mathbf{Q} \end{array}$$

commutes. Since $c_1(\xi_1 - 1) \neq 0$ we get that if the left vertical map is trivial, so is the right vertical map (which is multiplication by the degree). However, a map $f\colon T^2 \to S^2$ is homotopic to a constant map exactly if its degree is trivial. ☺

1.1.3.2 Geometric Example: Wall's Finiteness Obstruction

Let A be a space which is dominated by a finite CW-complex X (dominated means that there are maps $A \xrightarrow{i} X \xrightarrow{r} A$ such that $ri \simeq id_A$).

Question *is A homotopy equivalent to a finite CW-complex?*

The answer is *yes* if and only if a certain finiteness obstruction in the abelian group $\tilde{K}_0(\mathbf{Z}[\pi_1 A]) = \ker\{K_0(\mathbf{Z}[\pi_1 A]) \to K_0(\mathbf{Z})\}$ vanishes. So, for instance, if we know that $\tilde{K}_0(\mathbf{Z}[\pi_1 A])$ vanishes for algebraic reasons, we can always conclude that A is homotopy equivalent to a finite CW-complex. As for K_1, calculations of $K_0(\mathbf{Z}[\pi])$ are very hard, but we give a short list.

1.1.3.3 K_0 of Group Rings

1. If C_p is a cyclic group of prime order p less than 23, then $\tilde{K}_0(\mathbf{Z}[\pi])$ vanishes. The first nontrivial group is $\tilde{K}_0(\mathbf{Z}[C_{23}]) \cong \mathbf{Z}/3\mathbf{Z}$ (Kummer, see [213, p. 30]).
2. Waldhausen [297]: If π is a free group, free abelian group, or the fundamental group of a submanifold of the three-sphere, then $\tilde{K}_0(\mathbf{Z}[\pi]) = 0$.
3. Farrell and Jones [81]: If M is a closed Riemannian manifold with non-positive sectional curvature, then $\tilde{K}_0(\mathbf{Z}[\pi_1 M]) = 0$.

1.1.3.4 Facts About K_0 of Rings

1. If A is a commutative ring, then $K_0(A)$ has a ring structure. The additive structure comes from the direct sum of modules, and the multiplication from the tensor product.
2. If A is local, then $K_0(A) = \mathbf{Z}$.

3. Let A be a commutative ring. Define $rk_0(A)$ to be the split summand of $K_0(A)$ of classes of rank 0, cf. [13, p. 459]. The modules P for which there exists a Q such that $P \otimes_A Q \cong A$ form a category. The isomorphism classes form a group under tensor product. This group is called the Picard group, and is denoted $Pic(A)$. There is a "determinant" map $rk_0(A) \to Pic(A)$ which is always surjective. If A is a Dedekind domain (see [13, pp. 458–468]) the determinant map is an isomorphism, and $Pic(A)$ is isomorphic to the ideal class group $Cl(A)$.
4. Let A be the integers in a number field. Then Dirichlet tells us that $rk_0(A) \cong Pic(A) \cong Cl(A)$ is finite. For instance, if $A = \mathbf{Z}[e^{2\pi i/p}] = \mathbf{Z}[t]/\sum_{i=0}^{p-1} t^i$, the integers in the cyclotomic field $\mathbf{Q}(e^{2\pi i/p})$, then $K_0(A) \cong K_0(\mathbf{Z}[C_p])$ (Sect. 1.1.3.31).
5. If $f\colon B \to A$ is a surjection of rings with kernel I contained in the Jacobson radical, $rad(B)$, then $K_0(B) \to K_0(A)$ is injective ([13, p. 449]). It is an isomorphism if
   (a) B is complete in the I-adic topology ([13]),
   (b) (B, I) is a Hensel pair ([88]) or
   (c) f is split (as K_0 is a functor).
   That (B, I) is a *Hensel pair* means that if $f \in B[t]$ has image $\bar{f} \in A[t]$ and $a \in A = B/I$ satisfies $\bar{f}(a) = 0$ and $f'(a)$ is a unit in B/I, then there is a $b \in B$ mapping to a, and such that $f(b) = 0$. It implies that $I \subseteq rad(B)$.

1.1.3.5 An Example from Algebraic Geometry

Algebraic K-theory appears in Grothendieck's proof of the Riemann–Roch theorem, see Borel and Serre [35], where Bott's entry in Mathematical Reviews can serve as the missing introduction. Let X be a non-singular quasi-projective variety (i.e., a locally closed subvariety of some projective variety) over an algebraically closed field. Let $CH(X)$ be the Chow ring of cycles under linear equivalence (called $A(X)$ in [35, Sect. 6]) with product defined by intersection. Tensor product gives a ring structure on $K_0(X)$, and Grothendieck defines a natural ring homomorphism

$$ch\colon K_0(X) \to CH(X) \otimes \mathbf{Q},$$

similar to the Chern character for vector bundles, cf. [214]. This map has good functoriality properties with respect to pullback, i.e., if $f\colon X \to Y$, then

$$\begin{array}{ccc} K_0(X) & \xrightarrow{ch} & CH(X) \otimes \mathbf{Q} \\ {\scriptstyle f^!}\uparrow & & \uparrow{\scriptstyle f^*} \\ K_0(Y) & \xrightarrow{ch} & CH(Y) \otimes \mathbf{Q} \end{array}$$

commutes, where $f^!$ and f^* are given by pulling back along f. For proper morphisms $f\colon X \to Y$ [35, p. 100] there are "transfer maps" (defined as a sort of Euler characteristic) $f_!\colon K_0(X) \to K_0(Y)$ [35, p. 110] and direct image maps

$f_*\colon CH(X) \to CH(Y)$. The Riemann–Roch theorem is nothing but a quantitative measure of the fact that

$$\begin{array}{ccc} K_0(X) & \xrightarrow{ch} & CH(X) \otimes \mathbf{Q} \\ {\scriptstyle f_!}\downarrow & & {\scriptstyle f_*}\downarrow \\ K_0(Y) & \xrightarrow{ch} & CH(Y) \otimes \mathbf{Q} \end{array}$$

fails to commute: $ch(f_!(x)) \cdot Td(Y) = f_*(ch(x) \cdot Td(X))$ where $Td(X)$ is the value of the "Todd class" [35, p. 112] on the tangent bundle of X.

1.1.3.6 A Number-Theoretic Example

Let F be a number field and A its ring of integers. Then there is an exact sequence connecting K_1 and K_0:

$$0 \longrightarrow K_1(A) \longrightarrow K_1(F) \longrightarrow \bigoplus_{\mathfrak{m} \in Max(A)} K_0(A/\mathfrak{m}) \longrightarrow K_0(A) \longrightarrow K_0(F) \longrightarrow 0$$

(cf. [13, pp. 323, 702], or better [232, Corollary to Theorem 5] plus the fact that $K_1(A) \to K_1(F)$ is injective). The *zeta function* $\zeta_F(s)$ of F is defined as the meromorphic function on the complex plane $\mathbf{C}$ we get as the analytic continuation of

$$\zeta_F(s) = \sum_{I \text{ non-zero ideal in } A} |A/I|^{-s}.$$

This series converges for $Re(s) > 1$. The zeta function has a zero of order $r = \operatorname{rank}(K_1(A))$ (see Sect. 1.1.2.(2)) at $s = 0$, and the class number formula says that

$$\lim_{s \to 0} \frac{\zeta_F(s)}{s^r} = -\frac{R|K_0(A)_{tor}|}{|K_1(A)_{tor}|},$$

where $|-_{tor}|$ denotes the cardinality of the torsion subgroup, and the regulator R is a number that depends on the map δ above, see [175].

This is related to the Lichtenbaum-Quillen conjecture, which is now confirmed due to work of among many others Voevodsky, Suslin, Rost, Grayson (see Sect. 1.1.7 and Sect. 7.3.2 for references and a deeper discussion).

1.1.4 The Mayer–Vietoris Sequence

The reader may wonder why one chooses to regard the functors K_0 and K_1 as related. Example 1.1.3.6 provides one motivation, but that is cheating. Historically, it was an insight of Bass that K_1 could be obtained from K_0 in analogy with the definition of $K^1(X)$ as $K^0(S^1 \wedge X)$ (cf. Example 1.1.3.2(5)). This manifests itself in exact sequences connecting the two theories. As an example: if

$$\begin{array}{ccc} A & \longrightarrow & B \\ \downarrow & & \downarrow f \\ C & \xrightarrow{g} & D \end{array}$$

is a cartesian square of rings and g (or f) is surjective, then we have a long exact "Mayer–Vietoris" sequence

$$K_1(A) \longrightarrow K_1(B) \oplus K_1(C) \longrightarrow K_1(D) \longrightarrow K_0(A) \longrightarrow K_0(B) \oplus K_0(C) \longrightarrow K_0(D).$$

However, it is not true that this continues to the left. For one thing there is no simple analogy to the Bott periodicity $K^0(S^2 \wedge X) \cong K^0(X)$. Milnor proposed in [213] a definition of K_2 (see below) which would extend the Mayer–Vietoris sequence if **both** f and g are surjective, i.e., we have a long exact sequence

$$K_2(A) \longrightarrow K_2(B) \oplus K_2(C) \longrightarrow K_2(D) \longrightarrow K_1(A) \longrightarrow K_1(B) \oplus K_1(C) \longrightarrow K_1(D) \longrightarrow \cdots.$$

However, this was the best one could hope for:

Example 1.1.4.1 Swan [281] gave the following example showing that there exists no functor K_2 giving such a sequence if only g is surjective. Let A be commutative, and consider the pullback diagram

$$\begin{array}{ccc} A[t]/t^2 & \xrightarrow{t \mapsto 0} & A \\ {\scriptstyle a+bt \mapsto \begin{pmatrix} a & b \\ 0 & a \end{pmatrix}} \downarrow & & \downarrow \Delta \\ T_2(A) & \xrightarrow{g} & A \times A \end{array}$$

where $T_2(A)$ is the ring of upper triangular 2×2 matrices, g is the projection onto the diagonal, while Δ is the diagonal inclusion. As g splits $K_2(T_2(A)) \oplus K_2(A) \to K_2(A \times A)$ must be surjective, but, as we shall see below, $K_1(A[t]/t^2) \to K_1(T_2(A)) \oplus K_1(A)$ is not injective.

Recall that, since A is commutative, $GL_1(A[t]/t^2)$ is a direct summand of $K_1(A[t]/t^2)$. The element $1+t \in A[t]/t^2$ is invertible (and not the identity), but $[1+t] \neq [1] \in K_1(A[t]/t^2)$ is sent onto $[1]$ in $K_1(A)$, and onto

$$\left[\begin{pmatrix} 1 & 1 \\ 0 & 1 \end{pmatrix}\right] \sim \left[\begin{pmatrix} \begin{pmatrix} 1 & 1 \\ 0 & 1 \end{pmatrix} & 0 \\ 0 & 1 \end{pmatrix}\right] = [[e_{12}^{\begin{pmatrix} 1 & 0 \\ 0 & 0 \end{pmatrix}}, e_{21}^{\begin{pmatrix} 0 & 1 \\ 0 & 0 \end{pmatrix}}]] \sim [1] \in K_1\big(T_2(A)\big)$$

where the inner brackets are the commutator $[g,h] = ghg^{-1}h^{-1}$, as in Definition 1.1.2.1, of two elementary matrices (which by definition is trivial in K_1).

Using trace methods, one can measure the failure of excision and do concrete calculation, see Sect. 7.3.9.

1.1.5 Milnor's $K_2(A)$

Milnor's definition of $K_2(A)$ is given in terms of the Steinberg group, and turns out to be isomorphic to the second homology group $H_2(E(A))$ of the group of elementary matrices. Another, and more instructive way to say this is the following. The group $E(A)$ is generated by the matrices e_{ij}^a, $a \in A$ and $i \neq j$, and generally these generators are subject to lots of relations. There are, however, some relations which are more important than others, and furthermore are universal in the sense that they are valid for any ring: the so-called *Steinberg relations*. One defines the *Steinberg group* $St(A)$ to be exactly the group generated by symbols x_{ij}^a for every $a \in A$ and $i \neq j$ subject to these relations. Explicitly:

$$x_{ij}^a x_{ij}^b = x_{ij}^{a+b}$$

and

$$\left[x_{ij}^a, x_{kl}^b\right] = \begin{cases} 1 & \text{if } i \neq l \text{ and } j \neq k, \\ x_{il}^{ab} & \text{if } i \neq l \text{ and } j = k, \\ x_{kj}^{-ba} & \text{if } i = l \text{ and } j \neq k. \end{cases}$$

One defines $K_2(A)$ as the kernel of the surjection

$$St(A) \xrightarrow{x_{ij}^a \mapsto e_{ij}^a} E(A).$$

In fact,

$$0 \longrightarrow K_2(A) \longrightarrow St(A) \longrightarrow E(A) \longrightarrow 0$$

is a central extension of $E(A)$ (hence $K_2(A)$ is abelian), and $H_2(St(A)) = 0$, which makes it the "universal central extension" (see e.g., [165]).

The best references for K_i $i \leq 2$ are still Bass' [13] and Milnor's [213] books. Swan's paper [281] is recommended for an exposition of what optimistic hopes one might have to extend these ideas, and why some of these could not be realized (for instance, there is **no** functor K_3 such that the Mayer–Vietoris sequence extends, even if all maps are split surjective).

1.1.6 Higher K-Theory

At the beginning of the seventies there appeared suddenly a plethora of competing theories pretending to extend these ideas into a sequence of theories, $K_i(A)$ for $i \geq 0$. Some theories were more interesting than others, and many were equal. The one we are going to discuss in this paper is the Quillen K-theory, later extended by Waldhausen to a larger class of rings and categories.

As Quillen defines it, the K-groups are really the homotopy groups of a space. He gave three equivalent definitions, one by the "plus" construction discussed in Sect. 1.1.6.1 below (we also use it in Sect. 3.1.1), one via "group completion" and one by what he called the Q-construction. The group completion line of idea circulated as a preprint for a very long time, but in 1994 finally made it into the appendix of [87], while the Q-construction appears already in 1973 in [232]. That the definitions agree appeared in [108]. For a ring A, the homology of (a component of) the space $K(A)$ is nothing but the group homology of $GL(A)$. Using the plus construction and homotopy theoretic methods, Quillen calculated in [228] $K(\mathbf{F}_q)$, where $\mathbf{F}_q$ is the field with q elements. See Sect. 1.1.7(1) below for more details.

The advantage of the Q-construction is that it is more accessible to structural considerations. In the foundational article [232] Quillen uses the Q-construction to extend to the higher K-groups most of the general statements that were known to be true for K_0 and K_1.

However, given these fundamental theorems, of Quillen's definitions it is the plus construction that has proven most directly accessible to calculations (this said, very few groups were in the end calculated directly from the definitions, and by now indirect methods such as motivic cohomology and the trace methods that are the topic of this book have extended our knowledge far beyond the limitations of direct calculations).

1.1.6.1 Quillen's Plus Construction

We will now describe a variant of Quillen's definition of (a component of) the algebraic K-theory space of an associative ring A with unit via the plus construction. For more background, the reader may consult [16, 122], or [87].

We will be working in the category of simplicial sets (as opposed to topological spaces). The readers who are uncomfortable with this can think of simplicial sets (often referred to as simply "spaces") as topological spaces for the moment and consult Sect. 3.1.1 for further details. Later in the text simplicial techniques will become essential, so we have collected some basic facts about simplicial sets that are particularly useful for our applications in the Appendix.

If X is a simplicial set, $H_*(X) = H(X; \mathbf{Z})$ will denote the homology of X with trivial integral coefficients, and $\tilde{H}_*(X) = \ker\{H_*(X) \to H_*(pt) = \mathbf{Z}\}$ is the reduced homology.

Definition 1.1.6.1 Let $f\colon X \to Y$ be a map of connected simplicial sets with connected homotopy fiber F. We say that f is acyclic if $\tilde{H}_*(F) = 0$.

We see that the homotopy fiber of an acyclic map must have perfect fundamental group (i.e., $0 = \tilde{H}_1(F) \cong H_1(F) \cong \pi_1 F/[\pi_1 F, \pi_1 F]$). Recall from Definition 1.1.2.1 that any group π has a maximal perfect subgroup, which we call $P\pi$, and which is automatically normal.

1.1.6.2 Remarks on the Construction

There are various models for X^+, and the most usual is Quillen's original (originally used by Kervaire [164] on homology spheres, see also [179]). That is, regard X as a CW-complex, add 2-cells to X to kill $P\pi_1(X)$, and then kill the noise created in homology by adding 3-cells. See e.g., [122] for details on this and related issues. This process is also performed in details for the particular case $X = BA_5$ in Sect. 3.1.2.1.

In our simplicial setting, we will use a slightly different model, giving us strict functoriality (not just in the homotopy category), namely the partial integral completion of [40, p. 219]. Just as K_0 was defined by a universal property for functions into abelian groups, the integral completion constructs a universal element over simplicial abelian groups (the "partial" is there just to take care of pathologies such as spaces where the fundamental group is not quasi-perfect). For the present purposes we only have need for the following properties of the partial integral completion, and we defer the actual construction to Definition 3.1.1.5.

Proposition 1.1.6.2

1. *The assignment $X \mapsto X^+$ is an endofunctor of pointed simplicial sets, and there is a natural cofibration $q_X\colon X \to X^+$,*
2. *if X is connected, then q_X is acyclic, and*
3. *if X is connected then $\pi_1(q_X)$ is the projection killing the maximal perfect subgroup of $\pi_1 X$*

Then Quillen provides the theorem we need (for a proof and a precise simplicial formulation, see Theorem 3.1.1.7):

Theorem 1.1.6.3 *For X connected, Proposition* 1.1.6.2(2) *and* 1.1.6.2(3) *characterizes X^+ up to homotopy under X.*

The integral completion will reappear as an important technical tool in a totally different setting in Sect. 3.3.

Recall that the general linear group $GL(A)$ was defined as the union of the $GL_n(A)$. Form the *classifying space* (see Sect. A.2.6) of this group, $BGL(A)$. Whether you form the classifying space before or after taking the union is without consequence. Now, Quillen defines the connected cover of algebraic K-theory to be the realization $|BGL(A)^+|$ or rather, the homotopy groups,

$$K_i(A) = \begin{cases} \pi_i(BGL(A)^+) & \text{if } i > 0, \\ K_0(A) & \text{if } i = 0, \end{cases}$$

to be the K-groups of the ring A. We will use the following notation:

Definition 1.1.6.4 If A is a ring, then the *algebraic K-theory space* is

$$K(A) = BGL(A)^+.$$

Now, the Whitehead Lemma 1.1.2.2 tells us that $GL(A)$ is quasi-perfect with commutator $E(A)$, so

$$\pi_1 K(A) \cong GL(A)/PGL(A) = GL(A)/E(A) = K_1(A),$$

as expected. Furthermore, using the definition of $K_2(A)$ via the universal central extension, Sect. 1.1.5, it is not too difficult to prove that the K_2's of Milnor and Quillen agree: $K_2(A) = \pi_2(BGL(A)^+) \cong H_2(E(A))$ (and even $K_3(A) \cong H_3(St(A))$, see [96]).

One might regret that this space $K(A)$ has no homotopy in dimension zero, and this will be amended later. The reason we choose this definition is that the alternatives available to us at present all have their disadvantages. We might take $K_0(A)$ copies of this space, and although this would be a nice functor with the right homotopy groups, it will not agree with a more natural definition to come. Alternatively we could choose to multiply by $K_0^f(A)$ of Example 1.1.3.2(2) or $\mathbf{Z}$ as is more usual, but this has the shortcoming of not respecting products.

1.1.6.3 Other Examples of Use of the Plus Construction

1. Let $\Sigma_n \subset GL_n(\mathbf{Z})$ be the *symmetric group* of all permutations on n letters, and let $\Sigma_\infty = \lim_{n\to\infty} \Sigma_n$. Then the theorem of Barratt–Priddy–Quillen (e.g., [12]) states that $\mathbf{Z} \times B\Sigma_\infty^+ \simeq \lim_{k\to\infty} \Omega^k S^k$, so the groups $\pi_*(B\Sigma_\infty^+)$ are the "stable homotopy groups of spheres".
2. Let X be a connected space with abelian fundamental group. Then Kan and Thurston [154] have proved that X is homotopy equivalent to a BG^+ for some strange group G. With a slight modification, the theorem can be extended to arbitrary connected X.

3. Consider the *mapping class group* Γ_g of (isotopy classes of) diffeomorphisms of a surface of genus g (we are suppressing boundary issues). It is known that the colimit $B\Gamma_\infty$ of the classifying spaces as the genus goes to infinity has the same rational cohomology as $\mathcal{M}$, the *stable moduli space of Riemann surfaces*, and Mumford conjectured in [218] that the rational cohomology of $\mathcal{M}$ is a polynomial algebra generated by certain classes—the "Mumford classes"—κ_i with dimension $|\kappa_i| = 2i$. Since $B\Gamma_\infty$ and $B\Gamma_\infty^+$ have isomorphic cohomology groups, the Mumford conjecture follows by Madsen and Weiss' identification [193] of $\mathbf{Z} \times B\Gamma_\infty^+$ as the infinite loop space of a certain spectrum called $\mathbf{CP}_{-1}^\infty$ which (for badly understood reasons) will resurface in Sect. 7.3.8.1 (see also [91]). One should notice that prior to this, Tillmann [285] had identified $\mathbf{Z} \times B\Gamma_\infty^+$ with the infinite loop space associated to a category of cobordisms of one-dimensional manifolds.

1.1.6.4 Alternative Definitions of $K(A)$

In case the partial integral completion bothers you, for the space $BGL(A)$ it can be replaced by the following construction: choose an acyclic cofibration $BGL(\mathbf{Z}) \to BGL(\mathbf{Z})^+$ once and for all (by adding particular 2- and 3-cells), and define algebraic K-theory by means of the pushout square

$$\begin{array}{ccc} BGL(\mathbf{Z}) & \longrightarrow & BGL(A) \\ \downarrow & & \downarrow \\ BGL(\mathbf{Z})^+ & \longrightarrow & BGL(A)^+ \end{array} .$$

This will of course be functorial in A, and it can be verified that it has the right homotopy properties. However, at one point (e.g., in Chap. 3) we will need functoriality of the plus construction for more general spaces. All the spaces which we will need in these notes can be reached by choosing to do our handicrafted plus not on $BGL(\mathbf{Z})$, but on the space BA_5. See Sect. 3.1.2.1 for more details.

Another construction is due to Christian Schlichtkrull, [247, 2.2], who observed that the assignment $n \to BGL_n(A)$ can be extended to a functor from the category of finite sets and injective maps with $\{1, \dots, n\} \mapsto BGL_n(A)$, and that the homotopy colimit (see Definition A.7.0.1) is naturally equivalent to $BGL(A)^+$.

1.1.6.5 Comparison with Topological K-Theory

Quillen's definition of the algebraic K-theory of a ring fits nicely with the topological counterpart, as discussed in Example 1.1.3.2(5). If one considers the (topological) field $\mathbf{C}$, then the general linear group $GL_n(\mathbf{C})$ becomes a topological group. The classifying space construction applies equally well to topological groups, and we get the classifying space $B^{\mathrm{top}}GL_n(\mathbf{C})$. Vector bundles of rank n over a compact

Hausdorff topological space X are classified by unbased homotopy classes of maps into $B^{\mathrm{top}}GL_n(\mathbf{C})$, giving us the topological K-theory of Atiyah and Hirzebruch as the unbased homotopy classes of maps from X to $\mathbf{Z} \times B^{\mathrm{top}}GL(\mathbf{C})$. If X is based, reduced K-theory is given by based homotopy classes:

$$K^i(X) \cong \left[S^i \wedge X, \mathbf{Z} \times B^{\mathrm{top}}GL(\mathbf{C})\right].$$

The fundamental group of $B^{\mathrm{top}}GL(\mathbf{C})$ is trivial, and so the map

$$B^{\mathrm{top}}GL(\mathbf{C}) \to B^{\mathrm{top}}GL(\mathbf{C})^+$$

is an equivalence. To avoid the cumbersome notation, we notice that the Gram-Schmidt procedure guarantees that the inclusion of the unitary group $U(n) \subseteq GL_n(\mathbf{C})$ is an equivalence, and in the future we can use the convenient notation BU to denote any space with the homotopy type of $B^{\mathrm{top}}GL(\mathbf{C})$. The space $\mathbf{Z} \times BU$ is amazingly simple from a homotopy group point of view: $\pi_*(\mathbf{Z} \times BU)$ is the polynomial ring $\mathbf{Z}[u]$, where u is of degree 2 and is represented by the difference between the trivial and the tautological line bundle on $\mathbf{C}P^1 = S^2$. That multiplication by u gives an isomorphism $\pi_k BU \to \pi_{k+2} BU$ for $k > 0$ is a reflection of *Bott periodicity* $\Omega^2(\mathbf{Z} \times BU) \simeq \mathbf{Z} \times BU$) (for a cool proof, see [119]).

Similar considerations apply to the real case, with $\mathbf{Z} \times BO$ classifying real bundles. Its homotopy groups are 8-periodic.

1.1.7 Some Results and Calculations

In this section we will collect some results and calculations of algebraic K-theory that have been obtained by methods *different* from those that will be discussed in the chapters to come. The collection is somewhat idiosyncratic and often just picks out a piece of a more general result, but the reader is encouraged to pursue the references for further information.

For a discussion of results and calculations that do use trace methods and comparison to topological cyclic homology, see Sect. 7.3.

1. Quillen [228]: If $\mathbf{F}_q$ is the field with q elements, then

$$K_i(\mathbf{F}_q) \cong \begin{cases} \mathbf{Z} & \text{if } i = 0, \\ \mathbf{Z}/(q^j - 1)\mathbf{Z} & \text{if } i = 2j - 1, \\ 0 & \text{if } i = 2j > 0. \end{cases}$$

If $\bar{\mathbf{F}}_p$ is the algebraic closure of the prime field $\mathbf{F}_p$, then

$$K_i(\bar{\mathbf{F}}_p) \cong \begin{cases} \mathbf{Z} & \text{if } i = 0, \\ \mathbf{Q}/\mathbf{Z}[1/p] & \text{if } i = 2j - 1, \\ 0 & \text{if } i = 2j > 0. \end{cases}$$

The Frobenius automorphism $\Phi(a) = a^p$ induces multiplication by p^j on $K_{2j-1}(\bar{\mathbf{F}}_p)$, and the subgroup fixed by Φ^k is $K_{2j-1}(\mathbf{F}_{p^k})$.

A different way of phrasing this is to say that (the connected cover of) the algebraic K-theory space of $\mathbf{F}_q$ is equivalent to the homotopy fiber of a certain map $\psi^q - 1\colon BU \to BU$, where BU is the classifying space of the infinite unitary group (see Sect. 1.1.6.5) and ψ^q is the so-called qth Adams operation. The homotopy groups of BU are a copy of the integers in even positive dimensions and zero otherwise, and the qth Adams operation acts as q^j on $\pi_{2j}BU$.

2. Suslin [273]: "The algebraic K-theory of algebraically closed fields only depends on the characteristic, and away from the characteristic it always agrees with topological K-theory". More precisely:

   Let F be an algebraically closed field. The group $K_i(F)$ is divisible for $i \geq 1$. The torsion subgroup of $K_i(F)$ is zero if i is even, and it is isomorphic to

$$\begin{cases} \mathbf{Q}/\mathbf{Z}[1/p] & \text{if } char(F) = p > 0, \\ \mathbf{Q}/\mathbf{Z} & \text{if } char(F) = 0 \end{cases}$$

   if i is odd (see [277] for references).

   On the space level (not including K_0) Suslin's results are: If p is a prime different from the characteristic of the algebraically closed field F, then

$$K(F)^{\hat{}}_p \simeq BU^{\hat{}}_p$$

   where $^{\hat{}}_p$ is p-completion.

   If F is of characteristic $p > 0$, then $K(F)^{\hat{}}_p$ is contractible.

   Note in particular the pleasing formulation saying that $BGL(\mathbf{C})^+ \to B^{\mathrm{top}}GL(\mathbf{C})^+ \simeq B^{\mathrm{top}}GL(\mathbf{C})$ is an equivalence after p-completion. Even though $\mathbf{R}$ is not algebraically closed, the analogous result holds in the real case.

3. Naturally, the algebraic K-theory of the integers has been a key prize, and currently a complete calculation of the groups of degree divisible by 4 appears out of reach (relying on the so-called Vandiver's conjecture in number theory, which at present is known to hold for all prime numbers less than 12 million). We list here a few concrete results.

   - $K_0(\mathbf{Z}) = \mathbf{Z}$,
   - $K_1(\mathbf{Z}) = \mathbf{Z}/2\mathbf{Z}$,
   - $K_2(\mathbf{Z}) = \mathbf{Z}/2\mathbf{Z}$,
   - $K_3(\mathbf{Z}) = \mathbf{Z}/48\mathbf{Z}$ (Lee-Szczarba, 1976, [172]),
   - $K_4(\mathbf{Z}) = 0$ (Rognes, 2000, [241])
   - $K_5(\mathbf{Z}) = \mathbf{Z}$ (Elbaz-Vincent, Gangl and Soulé, 2002, [79]).

   We note the long time span from the identification of $K_3(\mathbf{Z})$ to that of $K_4(\mathbf{Z})$. In this period things did not stand still; there was much work on the so-called Lichtenbaum-Quillen conjecture, and other closely associated conjectures in motivic cohomology by a cohort of mathematicians including Voevodsky, Rost, Kahn, Suslin, Beilinson, Dwyer, Friedlander, Grayson, Mitchell, Levine, Soulé,

Table 1.1 The K-groups of the integers. The validity of the odd primary information assumes Vandiver's conjecture. Here k is the integer part of $1+\frac{n}{4}$, c_k is the numerator and w_{2k} the denominator of $(-1)^k\frac{1}{2}\zeta_{\mathbf{Q}}(1-2k)/2 = B_k/4k$ (where B_k is the kth Bernoulli number—numbered so that $B_1=\frac{1}{6}$, $B_2=\frac{1}{30},\ldots$), so that $w_2=24$, $w_4=240$ etc.

n mod 8	1	2	3	4	5	6	7	8
$K_n(\mathbf{Z})$	$\mathbf{Z}\oplus\mathbf{Z}/2$	$\mathbf{Z}/2c_k$	$\mathbf{Z}/2w_{2k}$	0	$\mathbf{Z}$	$\mathbf{Z}/c_k$	$\mathbf{Z}/w_{2k}$	0

Thomason, Wiles, Weibel, and many, many others. See Sect. 7.3.2 for some further information, or perhaps better, some more specialized and detailed source like Weibel's paper [309].

In 2000 Rognes and Weibel published a complete account [239] of the 2-torsion piece of $K_*(\mathbf{Z})$ following Voevodsky's proof of the Milnor conjecture [293]. The result can be stated in terms of a homotopy commutative square

$$\begin{array}{ccc} K(\mathbf{Z}[1/2]) & \longrightarrow & BO \\ \downarrow & & \downarrow \\ K(\mathbf{F}_3) & \longrightarrow & BU \end{array}$$

becoming homotopy cartesian after completion at 2, or in terms of the 2-primary information in the table one paragraph down.

For a more thorough discussion of the situation at odd primes we refer the reader to Weibel's survey [309], from which we have lifted Table 1.1 for the K-groups $K_n(\mathbf{Z})$ for $n>1$.

4. Quite early Borel [34] proved the following result. Let $\mathcal{O}_F$ be the integers in a number field F and n_j the order of vanishing of the zeta function

$$\zeta_F(s) = \sum_{0\neq I \text{ ideal in } \mathcal{O}_F} |\mathcal{O}_F/I|^{-s}$$

at $s=1-j$. Then

$$\operatorname{rank} K_i(\mathcal{O}_F) = \begin{cases} 0 & \text{if } i=2j>0, \\ n_j & \text{if } i=2j-1. \end{cases}$$

Example: If $F=\mathbf{Q}$, then

$$n_j = \begin{cases} 1 & \text{if } j=2k-1>1, \\ 0 & \text{otherwise.} \end{cases}$$

Furthermore, Quillen [231] proved that the groups $K_i(\mathcal{O}_F)$ are finitely generated.

Again, for a more thorough discussion we refer the reader to Weibel's survey [309] where the K-groups are expressed in similar terms as that of $K_*(\mathbf{Z})$ in the table above.

5. If A is a commutative ring, then $K_*(A) = \bigoplus_i K_i(A)$ is a graded commutative ring [179] (graded commutative means that $a \cdot b = (-1)^{|a||b|} b \cdot a$). Even more is true: $K_*(A)$ is a λ-ring [134]/[167] (the definition of a λ ring is most conveniently given by saying that it is a "coalgebra over the big Witt ring" [318], but the formulas are spelled out in the references). The λ-operations arise from the exterior powers $\Lambda^k M$ for A-modules M, and were used directly on this form by Grothendieck. However, for higher K-groups the fact that the exterior operations are non-linear (if $M' \rightarrowtail M \twoheadrightarrow M''$ is a short exact sequence we get an equality $[\Lambda^k M] = \sum_{i+j=k} [\Lambda^i M'][\Lambda^j M'']$ in K_0) means that one has to take quite indirect routes to get the operations. In [134]/[167] this is obtained by exploiting a certain universality over the representations (where the λ-operations are defined more readily) by the plus construction of $BGL(A)$. In [98] Gillet and Grayson showed how one could construct the loop space of the S-construction Definition 1.2.2.1 to get a combinatorial construction of the λ-operations [111] and the related Adams operations [113]. The Adams operations are ring homomorphisms.
6. Let A be a perfect ring of characteristic $p > 0$ (in characteristic $p > 0$ "perfect" means that the Frobenius endomorphism $a \mapsto a^p$ is an automorphism. For instance, all finite fields are perfect). Then the K-groups $K_i(A)$ are uniquely p-divisible for $i > 0$ (see [134] or [167]: the map induced by the Frobenius automorphism coincides with the pth Adams operation).
7. Gersten [97]/Waldhausen [297]: If A is a free associative ring, then the canonical ring homomorphism $\mathbf{Z} \to A$ induces an equivalence $K(\mathbf{Z}) \xrightarrow{\sim} K(A)$.
8. Waldhausen [297]: If G is a free group, free abelian group, or the fundamental group of a submanifold of the three-sphere, then there is a spectral sequence

$$E^2_{p,q} = H_p\big(G; K_q(\mathbf{Z})\big) \Rightarrow K_{p+q}\big(\mathbf{Z}[G]\big).$$

   This result is related to the algebraic K-theory *Novikov conjecture* about the so-called *assembly map*, which is also discussed briefly in Sect. 7.3.6.
9. Waldhausen [302]: The K-theory (in his sense) of the category of retractive spaces over a given space X, is equivalent to the product of the unreduced suspension spectrum of X (e.g., add a disjoint basepoint to get a based space X_+ and take its suspension spectrum $n \mapsto S^n \wedge X_+$) and the differentiable Whitehead spectrum of X. See also Sect. 3.2.3.2 and Sect. 7.3.8.1.
10. Goodwillie [102]: If $A \to B$ is a surjective map of rings such that the kernel is nilpotent, then the relative K-theory and the relative cyclic homology agree rationally.
11. Suslin/Panin:

$$K(\hat{\mathbf{Z}_p})^\wedge \simeq \operatorname*{holim}_{\overleftarrow{n}} K\big(\mathbf{Z}/p^n\mathbf{Z}\big)^\wedge$$

   where $^\wedge$ denotes profinite completion.

1.1.8 Where to Read

The Handbook of algebraic K-theory [86] contains many good surveys on the state of affairs in algebraic K-theory. Of older sources, one might mention the two very readable surveys [112] and [277] on the K-theory of fields and related issues. The article [216] is also recommended. For the K-theory of spaces see [301]. Some introductory books about higher K-theory exist: [16, 244, 270] and [147], and a "new" one (which looks very promising) is currently being written by Weibel [306]. The "Reviews in K-theory 1940–84" [194], is also helpful (although with both Mathematical Reviews and Zentralblatt on the web it naturally has lost some of its glory).

1.2 The Algebraic K-Theory Spectrum

Ideally, the so-called "higher K-theory" is nothing but a reformulation of the idea behind K_0: the difference is that whereas K_0 had values in abelian groups, K-theory has values in spectra, Sect. A.3.2. For convenience, we will follow Waldhausen and work with categories with cofibrations (see Sect. 1.2.1 below). When interested in the K-theory of rings we should, of course, apply our K-functor to the category $\mathcal{P}_A$ of finitely generated projective modules. The finitely generated projective modules form an exact category (see Sect. 1.1.3), which again is an example of a category with cofibrations.

There are many definitions of K-theory, each with its own advantages and disadvantages. Quillen began the subject with no less than three: the plus construction, the group completion approach and the "Q"-construction. Soon more versions appeared, but luckily most turned out to be equivalent to Quillen's whenever given the same input. We will eventually meet three: Waldhausen's "S"-construction which we will discuss in just a moment, Segal's Γ-space approach (see Sect. 2.3), and Quillen's plus construction (see Sect. 1.1.6.1 and Sect. 3.1.1).

1.2.1 Categories with Cofibrations

The source for these facts is Waldhausen's [301] from which we steal indiscriminately. That a category is *pointed* means that it has a chosen "zero object" 0 that is both initial and final.

Definition 1.2.1.1 A *category with cofibrations* is a pointed category $\mathcal{C}$ together with a subcategory $co\mathcal{C}$ satisfying

1. all isomorphisms are in $co\mathcal{C}$
2. all maps from the zero object are in $co\mathcal{C}$

3. if $A \to B \in co\mathcal{C}$ and $A \to C \in \mathcal{C}$, then the pushout

$$\begin{array}{ccc} A & \longrightarrow & B \\ \downarrow & & \downarrow \\ C & \longrightarrow & C \coprod_A B \end{array}$$

exists in $\mathcal{C}$, and the lower horizontal map is in $co\mathcal{C}$.

We will call the maps in $co\mathcal{C}$ simply *cofibrations*. Cofibrations may occasionally be written $\rightarrowtail$. A functor between categories with cofibrations is *exact* if it is pointed, takes cofibrations to cofibrations, and preserves the pushout diagrams of item (3).

Exact categories, as described in Sect. 1.1.3, are important examples. In these cases the monomorphisms in the short exact sequences are the cofibrations. In particular the category of finitely generated projective modules over a ring is a category with cofibrations:

Example 1.2.1.2 (The Category of Finitely Generated Projective Modules) Let A be a ring (unital and associative as always) and let $\mathcal{M}_A$ be the category of all A-modules. Conforming with the notation used elsewhere in the book, where $\mathcal{C}(c, c')$ denotes the set of maps $c \to c'$ in some category $\mathcal{C}$, we write $\mathcal{M}_A(M, N)$ for the group of A-module homomorphisms $M \to N$ instead of $\operatorname{Hom}_A(M, N)$.

We will eventually let the K-theory of the ring A be the K-theory of the category $\mathcal{P}_A$ of finitely generated projective right A-modules. The interesting structure of $\mathcal{P}_A$ as a category with cofibrations is to let the cofibrations be the injections $P' \rightarrowtail P$ in $\mathcal{P}_A$ such that the quotient P/P' is also in $\mathcal{P}_A$. That is, a homomorphism $P' \rightarrowtail P \in \mathcal{P}_A$ is a cofibration if it is the first part of a short exact sequence

$$0 \to P' \rightarrowtail P \twoheadrightarrow P'' \to 0$$

of projective modules. In this case the cofibrations are *split*, i.e., for any cofibration $j\colon P' \to P$ there exists a homomorphism $s\colon P \to P'$ in $\mathcal{P}_A$ such that $sj = id_{P'}$. Note that no choice of splitting is assumed in saying that j is split; some authors use the term "splittable".

A ring homomorphism $f\colon B \to A$ induces a pair of adjoint functors

$$\mathcal{M}_B \underset{f^*}{\overset{-\otimes_B A}{\rightleftarrows}} \mathcal{M}_A$$

where f^* is restriction of scalars. The adjunction isomorphism

$$\mathcal{M}_A(Q \otimes_B A, Q') \cong \mathcal{M}_B(Q, f^* Q')$$

is given by sending $L\colon Q \otimes_B A \to Q'$ to $q \mapsto L(q \otimes 1)$.

When restricted to finitely generated projective modules $- \otimes_B A$ induces a map $K_0(B) \to K_0(A)$ making K_0 into a functor.

Usually authors are not too specific about their choice of $\mathcal{P}_A$, but unfortunately this may not always be good enough. For one thing the assignment $A \mapsto \mathcal{P}_A$ should be functorial, and the problem is the annoying fact that if

$$C \xrightarrow{g} B \xrightarrow{f} A$$

are maps of rings, then $(M \otimes_C B) \otimes_B A$ and $M \otimes_C A$ are generally only naturally isomorphic (not equal).

So whenever pressed, $\mathcal{P}_A$ is the following category.

Definition 1.2.1.3 Let A be a ring. The category of *finitely generated projective A-modules* $\mathcal{P}_A$ is the following category with cofibrations. Its objects are the pairs (m, p), where m is a non-negative integer and $p = p^2 \in M_m(A)$. A morphism $(m, p) \to (n, q)$ is an A-module homomorphism of images $\mathrm{im}(p) \to \mathrm{im}(q)$. A cofibration is a split monomorphism (remember, a splitting is *not* part of the data).

Since $p^2 = p$ we get that $\mathrm{im}(p) \subseteq A^m \xrightarrow{p} \mathrm{im}(p)$ is the identity, and $\mathrm{im}(p)$ is a finitely generated projective module. Any finitely generated projective module in $\mathcal{M}_A$ is isomorphic to some such image. The full and faithful functor (i.e., bijective on morphism groups) $\mathcal{P}_A \to \mathcal{M}_A$ sending (m, p) to $\mathrm{im}(p)$ displays $\mathcal{P}_A$ as a category equivalent to the category of finitely generated projective objects in $\mathcal{M}_A$. With this definition $\mathcal{P}_A$ becomes a category with cofibrations, where $(m, p) \to (n, q)$ is a cofibration exactly when $\mathrm{im}(p) \to \mathrm{im}(q)$ is. The coproduct is given by $(m, p) \oplus (n, q) = (m + n, p \oplus q)$ where $p \oplus q$ is block sum of matrices.

Note that for any morphism $a\colon (m, p) \to (n, q)$ we may define

$$x_a\colon A^m \twoheadrightarrow \mathrm{im}(p) \xrightarrow{a} \mathrm{im}(q) \subseteq A^n,$$

and we get that $x_a = x_a p = q x_a$. In fact, when $(m, p) = (n, q)$, you get an isomorphism of rings

$$\mathcal{P}_A\big((m, p), (m, p)\big) \cong \big\{y \in M_m(A) \mid y = yp = py\big\}$$

via $a \mapsto x_a$, with inverse

$$y \mapsto \big\{\mathrm{im}(p) \subseteq A^m \xrightarrow{y} A^m \xrightarrow{p} \mathrm{im}(p)\big\}.$$

Note that the unit in the ring on the right hand side is the matrix p.

If $f\colon A \to B$ is a ring homomorphism, then $f_*\colon \mathcal{P}_A \to \mathcal{P}_B$ is given on objects by $f_*(m, p) = (m, f(p))$ ($f(p) \in M_m(B)$ is the matrix you get by using f on each entry in p), and on morphisms $a\colon (m, p) \to (n, q)$ by $f_*(a) = f(x_a)|_{\mathrm{im}(f(p))}$, which is well defined as $f(x_a) = f(q)f(x_a) = f(x_a)f(p)$. There is a natural isomorphism between

$$\mathcal{P}_A \longrightarrow \mathcal{M}_A \xrightarrow{M \mapsto M \otimes_A B} \mathcal{M}_B$$

and

$$\mathcal{P}_A \xrightarrow{f_*} \mathcal{P}_B \longrightarrow \mathcal{M}_B.$$

The assignment $A \mapsto \mathcal{P}_A$ is a functor from rings to exact categories.

Example 1.2.1.4 (The Category of Finitely Generated Free Modules) Let A be a ring. To conform with the strict definition of $\mathcal{P}_A$ in Definition 1.2.1.3, we define the category $\mathcal{F}_A$ of finitely generated free A-modules as the full subcategory of $\mathcal{P}_A$ with objects of the form $(n, 1)$, where 1 is the identity $A^n = A^n$. The inclusion $\mathcal{F}_A \subseteq \mathcal{P}_A$ is "cofinal" in the sense that given any object (m, p) in $\mathcal{P}_A$ there exists another object (n, q) in $\mathcal{P}_A$ such that $(n, q) \oplus (m, p) = (n + m, q \oplus p)$ is isomorphic to a free module. This will have the consequence that the K-theories of $\mathcal{F}_A$ and $\mathcal{P}_A$ only differ at K_0.

1.2.1.1 K_0 of Categories with Cofibrations

If $\mathcal{C}$ is a category with cofibrations, we let the "short exact sequences" be the cofiber sequences $c' \rightarrowtail c \twoheadrightarrow c''$, meaning that $c' \rightarrowtail c$ is a cofibration and the sequence fits in a pushout square

$$\begin{array}{ccc} c' & \longrightarrow & c \\ \downarrow & & \downarrow \\ 0 & \longrightarrow & c'' \end{array}.$$

This class is the class of objects of a category which we will call $S_2\mathcal{C}$. The maps are commutative diagrams

$$\begin{array}{ccccc} c' & \rightarrowtail & c & \twoheadrightarrow & c'' \\ \downarrow & & \downarrow & & \downarrow \\ d' & \rightarrowtail & d & \twoheadrightarrow & d'' \end{array}$$

Note that we can define cofibrations in $S_2\mathcal{C}$ too: a map like the one above is a cofibration if the vertical maps are cofibrations and the map from $c \coprod_{c'} d'$ to d is a cofibration.

Lemma 1.2.1.5 *With these definitions $S_2\mathcal{C}$ is a category with cofibrations.*

Proof Firstly, we have to prove that a composite of two cofibrations

$$\begin{array}{ccccc} c' & \longrightarrow & c & \longrightarrow & c'' \\ \downarrow & & \downarrow & & \downarrow \\ d' & \longrightarrow & d & \longrightarrow & d'' \\ \downarrow & & \downarrow & & \downarrow \\ e' & \longrightarrow & e & \longrightarrow & e'' \end{array}$$

again is a cofibration. The only thing to be checked is that the map from $c \coprod_{c'} e'$ to e is a cofibration, but this follows by Definition 1.2.1.1(1) and 1.2.1.1(3) since

$$c \coprod_{c'} e' \cong c \coprod_{c'} d' \coprod_{d'} e' \rightarrowtail d \coprod_{d'} e' \rightarrowtail e.$$

The axioms in Definition 1.2.1.1(1) and 1.2.1.1(2) are clear, and for 1.2.1.1(3) we reason as follows. Consider the diagram

$$\begin{array}{ccccc} d' & \longrightarrow & d & \longrightarrow & d'' \\ \uparrow & & \uparrow & & \uparrow \\ c' & \longrightarrow & c & \longrightarrow & c' \\ \downarrow & & \downarrow & & \downarrow \\ e' & \longrightarrow & e & \longrightarrow & e'' \end{array}$$

where the rows are objects of $S_2\mathcal{C}$ and the downwards pointing maps constitute a cofibration in $S_2\mathcal{C}$. Taking the pushout (which you get by taking the pushout of each column) the only nontrivial part of Definition 1.2.1.1(3) is that we have to check that $(e' \coprod_{c'} d') \coprod_{d'} d \to e \coprod_c d$ is a cofibration. But this is so since it is the composite

$$\left(e' \coprod_{c'} d'\right) \coprod_{d'} d \cong \left(e' \coprod_{c'} c\right) \coprod_{c} d \to e \coprod_{c} d$$

and the last map is a cofibration since $e' \coprod_{c'} c \to e$ is. □

There are three important functors

$$d_0, d_1, d_2 \colon S_2\mathcal{C} \to \mathcal{C}$$

sending a sequence $\mathbf{c} = \{c' \rightarrowtail c \twoheadrightarrow c''\}$ to $d_0(\mathbf{c}) = c''$, $d_1(\mathbf{c}) = c$ and $d_2(\mathbf{c}) = c'$.

Lemma 1.2.1.6 *The functors $d_i \colon S_2\mathcal{C} \to \mathcal{C}$ for $i = 0, 1, 2$, are all exact.*

Proof See [301, p. 323]. □

If $\mathcal{C}$ is a small category with cofibrations, we may define its zeroth algebraic K-group $K_0(\mathcal{C}) = K_0(\mathcal{C}, \mathcal{E})$ as in Definition 1.1.3.1, with $\mathcal{E} = obS_2\mathcal{C}$.

We now give a reformulation of the definition of K_0. We let $\pi_0(i\mathcal{C})$ be the set of isomorphism classes of $\mathcal{C}$. That a functor F from categories with cofibrations to abelian groups is "under $\pi_0 i$" then means that it comes equipped with a natural map $\pi_0(i\mathcal{C}) \to F(\mathcal{C})$, and a map between such functors must respect this structure.

Lemma 1.2.1.7 *The functor K_0 is the universal functor F under $\pi_0 i$ to abelian groups satisfying additivity, i.e., such that the natural map*

$$F(S_2\mathcal{C}) \xrightarrow{(d_0,d_2)} F(\mathcal{C}) \times F(\mathcal{C})$$

is an isomorphism.

Proof First one shows that K_0 satisfies additivity. For objects a and b in $\mathcal{C}$ let $a \vee b$ be their coproduct (under 0). Consider the splitting $K_0(\mathcal{C}) \times K_0(\mathcal{C}) \to K_0(S_2\mathcal{C})$ which sends $([a],[b])$ to $[a \rightarrowtail a \vee b \twoheadrightarrow b]$. We have to show that the composite

$$K_0(S_2\mathcal{C}) \xrightarrow{(d_0,d_2)} K_0(\mathcal{C}) \times K_0(\mathcal{C}) \longrightarrow K_0(S_2\mathcal{C})$$

sending $[a' \rightarrowtail a \twoheadrightarrow a'']$ to $[a' \rightarrowtail a' \vee a'' \twoheadrightarrow a''] = [a' = a' \to 0] + [0 \rightarrowtail a'' = a'']$ is the identity. But this is clear from the diagram

$$\begin{array}{ccccc} a' & = & a' & \longrightarrow & 0 \\ \| & & \downarrow & & \downarrow \\ a' & \longrightarrow & a & \longrightarrow & a'' \\ \downarrow & & \downarrow & & \| \\ 0 & \longrightarrow & a'' & = & a'' \end{array}$$

in $S_2S_2\mathcal{C}$. Let F be any other functor under $\pi_0 i$ satisfying additivity. By additivity the function $\pi_0(i\mathcal{C}) \to F(\mathcal{C})$ satisfies the additivity condition used in the definition of K_0 in Definition 1.1.3.1; so there is a unique factorization $\pi_0(i\mathcal{C}) \to K_0(\mathcal{C}) \to F(\mathcal{C})$ which for the same reason must be functorial. □

The question is: can we obtain deeper information about the category $\mathcal{C}$ if we allow ourselves a more fascinating target category than abelian groups? The answer is yes. If we use a category of spectra instead we get a theory—K-theory—whose homotopy groups are the K-groups introduced earlier.

1.2.2 Waldhausen's S-Construction

We now give Waldhausen's definition of the K-theory of a category with (isomorphisms and) cofibrations. (According to Waldhausen, the "S" is for "Segal" as in

Graeme B. Segal. According to Segal his construction was close to the "block-triangular" version given for additive categories in Definition 1.2.2.3 below. Apparently, Segal and Quillen were aware of this construction even before Quillen discovered his Q-construction, but it was not before Waldhausen reinvented it that it became apparent that the S-construction was truly useful. In fact, in a letter to Segal [229], Quillen comments: "... But it was only this spring that I succeeded in freeing myself from the shackles of the simplicial way of thinking and found the category $Q(\underline{\underline{B}})$".)

For any category $\mathcal{C}$, the *arrow category* $\mathcal{A}r\mathcal{C}$ (not to be confused with the *twisted* arrow category), is the category whose objects are the morphisms in $\mathcal{C}$, and where a morphism from $f: a \to b$ to $g: c \to d$ is a commutative diagram in $\mathcal{C}$

$$\begin{array}{ccc} a & \longrightarrow & c \\ {\scriptstyle f}\downarrow & & {\scriptstyle g}\downarrow . \\ b & \longrightarrow & d \end{array}$$

If $\mathcal{C} \to \mathcal{D}$ is a functor, we get an induced functor $\mathcal{A}r\mathcal{C} \to \mathcal{A}r\mathcal{D}$, and a quick check reveals that $\mathcal{A}r$ is itself a functor.

Consider the ordered set $[n] = \{0 < 1 < \cdots < n\}$ as a category and its arrow category $\mathcal{A}r[n]$.

Actually, since orientation differs in varying sources, let us be precise about this. The simplicial category Δ may be considered as a full subcategory of the category of small categories, by identifying $[n]$ with the category $\{0 \leftarrow 1 \leftarrow \cdots \leftarrow n\}$ (the idea is that you just insert a horizontal line to make $<$ into $<-$). Many authors consider instead the opposite category $[n]^o = \{0 \to 1 \to \cdots \to n\}$. Since we want to keep Waldhausen's *notation*, but still be consistent with our chosen convention we consider the arrow category $\mathcal{A}r([n]^o)$. So, in $\mathcal{A}r([11]^o)$ there is a unique morphism from the object $(2 \leq 4)$ to $(3 \leq 7)$ and no morphism the other way.

Definition 1.2.2.1 Let $\mathcal{C}$ be a category with cofibrations. Then $S\mathcal{C} = \{[n] \mapsto S_n\mathcal{C}\}$ is the simplicial category which in degree n is the category $S_n\mathcal{C}$ of functors $C: \mathcal{A}r([n]^o) \to \mathcal{C}$ satisfying the following properties

1. For all $j \geq 0$ we have that $C(j = j) = 0$ (the preferred null object in $\mathcal{C}$) and
2. if $i \leq j \leq k$, then $C(i \leq j) \rightarrowtail C(i \leq k)$ is a cofibration, and

$$\begin{array}{ccc} C(i \leq j) & \longrightarrow & C(i \leq k) \\ \downarrow & & \downarrow \\ C(j = j) & \longrightarrow & C(j \leq k) \end{array}$$

is a pushout.

The simplicial structure is induced by the cosimplicial category $[n] \mapsto \mathcal{A}r([n]^o)$.

To get one's hands on each individual category $S_n\mathcal{C}$, think of the objects as strings of cofibrations

$$C_{01} \rightarrowtail C_{02} \rightarrowtail \cdots \rightarrowtail C_{0n}$$

with compatible choices of cofibers $C_{ij} = C_{0j}/C_{0i}$, or equivalently as triangles

$$\begin{array}{ccccccccccc}
C_{01} & \rightarrowtail & C_{02} & \rightarrowtail & C_{03} & \rightarrowtail & \cdots & \rightarrowtail & C_{0,n-1} & \rightarrowtail & C_{0n} \\
 & & \downarrow & & \downarrow & & & & \downarrow & & \downarrow \\
 & & C_{12} & \rightarrowtail & C_{13} & \rightarrowtail & \cdots & \rightarrowtail & C_{1,n-1} & \rightarrowtail & C_{1n} \\
 & & & & \downarrow & & & & \downarrow & & \downarrow \\
 & & & & C_{23} & \rightarrowtail & \cdots & \rightarrowtail & C_{2,n-1} & \rightarrowtail & C_{2,n} \\
 & & & & & & & & \downarrow & & \downarrow \\
 & & & & & & \ddots & & \vdots & & \vdots \\
 & & & & & & & & \downarrow & & \downarrow \\
 & & & & & & & & C_{n-2,n-1} & \rightarrowtail & C_{n-2,n} \\
 & & & & & & & & & & \downarrow \\
 & & & & & & & & & & C_{n-1,n}
\end{array}$$

with horizontal arrows cofibrations and every square a pushout (the null object is placed in the corners below the diagonal).

If $\mathcal{C}$ is a category, we will let $ob\mathcal{C}$ be the class of objects in $\mathcal{C}$.

Since $obS_0\mathcal{C}$ is trivial, the fundamental group $\pi_1(obS\mathcal{C})$ is the quotient of the free group on the pointed set $ob\mathcal{C} = obS_1\mathcal{C}$ by the relation that $[c'] = [c'']^{-1}[c]$ for every $c' \rightarrowtail c \twoheadrightarrow c'' \in obS_2\mathcal{C}$ (this is the "edge loop" description of the fundamental group, and can be seen alternatively by using the Kan loop group description of the fundamental group of a space with only one zero simplex, see Appendix A.2.6.2). Hence, the difference between $K_0(\mathcal{C})$ and $\pi_1(obS\mathcal{C})$ is that the former is described as the quotient of the free *abelian* group on the *isomorphism classes* of objects in $\mathcal{C}$, modulo the same relations as $\pi_1(obS\mathcal{C})$.

Lemma 1.2.2.2 *Let $\mathcal{C}$ be a small category with cofibrations. Then there is a natural isomorphism $K_0(\mathcal{C}) \cong \pi_1(obS\mathcal{C})$.*

Proof An isomorphism $c' \xrightarrow{\cong} c$ in $\mathcal{C}$ can be considered as an element $c' \xrightarrow{\cong} c \to 0 \in obS_2\mathcal{C}$, and so $[c'] = [c]$ in $\pi_1(obS\mathcal{C})$. Since we then have that

$$[c'][c''] = [c'' \vee c'] = [c' \vee c''] = [c''][c']$$

we get that the fundamental group is an abelian group, and so $\pi_1 obS\mathcal{C}$ is the quotient of the free abelian group on the set of isomorphism classes of $\mathcal{C}$ by the relation $[c'] + [c''] = [c]$, which is just the formula for $K_0(\mathcal{C})$ arrived at in Sect. 1.1.3 □

Thus we have that $K_0(A) = K_0(\mathcal{P}_A)$ is the fundamental group of $obS\mathcal{P}_A$ if we choose the cofibrations to be the split monomorphisms, and it can be shown that $K_i(A)$ is $\pi_{i+1}(obS\mathcal{P}_A)$ for the other groups we discussed in the introduction (namely the algebraic cases $i = 1$ and $i = 2$, and also for the definition of the higher groups via the plus construction, see Sect. 3.2).

1.2.2.1 Additive Categories

Recall that an *Ab-category* [191] is a category where the morphism sets are abelian groups and where composition is bilinear (also called *linear category*). An *additive category* is an *Ab*-category with all finite products.

Let $\mathfrak{C}$ be an additive category, regarded as a category with cofibrations by letting the cofibrations be the split monomorphisms. With this choice we call $\mathfrak{C}$ a *split exact category*.

In these cases it is easier to see how the S-construction works. Note that if

$$c = (c_{0,1}, \dots, c_{i-1,i}, \dots, c_{n-1,n})$$

is a sequence of objects, then the sum diagram $\psi_n c$ with

$$(\psi_n c)_{ij} = \bigoplus_{i \le k \le j} c_{k-1,k}$$

and maps the obvious inclusions and projections, is an element in $S_n\mathfrak{C}$. Since $\mathfrak{C}$ is split exact every element of $S_n\mathfrak{C}$ is isomorphic to such a diagram. Maps between two such sum diagrams can be thought of as upper triangular matrices:

Definition 1.2.2.3 Let $\mathcal{C}$ be an *Ab*-category. For every $n > 0$, we define $T_n\mathcal{C}$—the *$n \times n$ upper triangular matrices* on $\mathcal{C}$—to be the category with objects $ob\mathcal{C}^n$, and morphisms

$$T_n\mathcal{C}\big((c_1, \dots, c_n), (d_1, \dots, d_n)\big) = \bigoplus_{1 \le j \le i \le n} \mathcal{C}(c_i, d_j)$$

with composition given by matrix multiplication.

Lemma 1.2.2.4 *Let $\mathfrak{C}$ be additive. Then the assignment ψ_q given in the discussion above defines a full and faithful functor*

$$\psi_q : T_q\mathfrak{C} \to S_q\mathfrak{C}$$

which is an equivalence of categories since $\mathfrak{C}$ is split exact. ☺

1.2.3 The Equivalence $obSC \to BiSC$

Lemma 1.2.3.1 below displays an amazing—and very useful—property about the simplicial set of objects of the S-construction: Considered as a functor from small categories with cofibrations to simplicial sets, it transforms natural isomorphisms to homotopies, and so sends equivalences of categories to homotopy equivalences.

This is reminiscent to the classifying space construction B discussed below (see also Appendix A.2.4), but is slightly weaker in that the classifying space takes all natural transformations to homotopies, whereas obS only takes the natural *iso*morphisms to homotopies.

All categories in this section are assumed to be small. For every $n \geq 0$, regard $[n] = \{0 < 1 < \cdots < n\}$ as a category (if $a \leq b$ there is a unique map $a \leftarrow b$), and maps in Δ as functors (hence we regard Δ as a full subcategory of the category of small categories). The *classifying space* (or *nerve*) of a small category $\mathcal{C}$ is the space (simplicial set) $B\mathcal{C}$ defined by

$$[q] \mapsto B_q\mathcal{C} = \{c_0 \leftarrow c_1 \leftarrow \cdots \leftarrow c_q \in \mathcal{C}\} = \big\{\text{functors } [q] \to \mathcal{C}\big\}.$$

Note that the standard $[q]$-simplex $\Delta[q] = \{[n] \mapsto \Delta([n], [q])\}$ is the nerve of the category $[n]$: $B[q] = \Delta[q]$. The standard fact that natural transformations induce homotopies comes from the fact that a natural transformation is the same as a functor $\mathcal{C} \times [1] \to \mathcal{D}$, and $B(\mathcal{C} \times [1]) \cong B\mathcal{C} \times B[1] = B\mathcal{C} \times \Delta[1]$. See also Appendix A.2.4.

Lemma 1.2.3.1 *If*

$$f_0, f_1 : \mathcal{C} \to \mathcal{D}$$

are naturally isomorphic exact functors, then they induce simplicially homotopic maps

$$obS\mathcal{C} \to obS\mathcal{D}.$$

Hence $\mathcal{C} \mapsto obS\mathcal{C}$ sends equivalences of categories to homotopy equivalences of spaces.

Proof (the same proof as in [301, 1.4.1]). We define a homotopy

$$H : obS\mathcal{C} \times B[1] \longrightarrow obS\mathcal{D}$$

from Sf_0 to Sf_1 as follows. The natural isomorphism $\eta\colon f_1 \cong f_0$ gives rise to a functor $F\colon \mathcal{C} \times [1] \to \mathcal{D}$ with $F(c,i) = f_i(c)$, and $F(c \to c', i \leq i')$ equal to the obvious composite $f_{i'}(c) \cong f_i(c) \to f_i(c')$. Let $c\colon \mathcal{A}r[n] \to \mathcal{C}$ be an object of $S_n\mathcal{C}$, and $\phi \in B_n[1] = \Delta([n],[1])$. Then $H(c,\phi)$ is the composite

$$\mathcal{A}r\big([n]^o\big) \longrightarrow \mathcal{A}r\big([n]^o\big) \times [n]^o \xrightarrow{(c,\phi)} \mathcal{C} \times [1]^o \cong \mathcal{C} \times [1] \xrightarrow{F} \mathcal{D}$$

where the first map sends $i \leq j$ to $(i \leq j, j)$, and where we have used the isomorphism $[1]^o \cong [1]$. This is an object in $S_n\mathcal{D}$ since $f \cong g$ is an isomorphism. □

We will use *bisimplicial sets* (functors from $\Delta^o \times \Delta^o$ to sets) quite freely, and may consider a simplicial set as a bisimplicial set which is constant in one simplicial direction. The *diagonal* of a bisimplicial set is the simplicial set you get by precomposing with the diagonal functor $\Delta^o \to \Delta^o \times \Delta^o$. We will simply say that a map of bisimplicial sets is an *equivalence* if its diagonal is a weak equivalence of simplicial sets. For this and related technicalities, the reader is invited to consult Appendix A.6.

If $\mathcal{C}$ is a category, then $i\mathcal{C} \subseteq \mathcal{C}$ is the subcategory with all objects, but only isomorphisms as morphisms.

Corollary 1.2.3.2 *If $t\mathcal{C} \subset i\mathcal{C}$ is a subcategory of the isomorphisms containing all objects, then the inclusion of the zero skeleton is an equivalence*

$$obS\mathcal{C} \xrightarrow{\simeq} BtS\mathcal{C}$$

where $tS_q\mathcal{C} \subseteq S_q\mathcal{C}$ is the subcategory whose morphisms are natural transformations coming from $t\mathcal{C}$.

Proof This follows by regarding the bisimplicial object

$$\big\{[p],[q] \mapsto B_p t S_q \mathcal{C}\big\}$$

as $obS_q\mathbf{N}_p(\mathcal{C}, t\mathcal{C})$, where $\mathbf{N}_p(\mathcal{C}, t\mathcal{C})$ is a full subcategory of the **category** $\mathbf{N}_p\mathcal{C}$ (see Appendix A.2.4) of functors $[p] \to \mathcal{C}$ and natural transformations between these. The objects of $\mathbf{N}_p(\mathcal{C}, t\mathcal{C})$ are the chains of maps in $t\mathcal{C}$, i.e., $ob\mathbf{N}_p(\mathcal{C}, t\mathcal{C}) = B_p t\mathcal{C}$.

Consider the functor $\mathcal{C} \to \mathbf{N}_p(\mathcal{C}, t\mathcal{C})$ given by sending c to the chain of identities on c (here we need that all identity maps are in $t\mathcal{C}$). It is an equivalence of categories. A splitting being given by e.g., sending $c_0 \leftarrow \cdots \leftarrow c_p$ to c_0: the natural isomorphism to the identity on $\mathbf{N}_p(\mathcal{C}, t\mathcal{C})$ is given by

$$\begin{array}{ccccccccc}
c_0 & \xleftarrow{\alpha_1} & c_1 & \xleftarrow{\alpha_2} & c_2 & \xleftarrow{\alpha_3} & \cdots & \xleftarrow{\alpha_p} & c_p \\
\Vert & & \downarrow{\scriptstyle \alpha_1} & & \downarrow{\scriptstyle \alpha_1\alpha_2} & & & & \downarrow{\scriptstyle \alpha_1\alpha_2\ldots\alpha_p} \\
c_0 & = & c_0 & = & c_0 & = & \cdots & = & c_0
\end{array}.$$

A morphism

$$\begin{array}{ccccccc} c'_0 & \leftarrow & c'_1 & \leftarrow & \dots & \leftarrow & c'p \\ \downarrow & & \downarrow & & & & \downarrow \\ c_0 & \leftarrow & c_1 & \leftarrow & \dots & \leftarrow & c_p \end{array}$$

in $\mathbf{N}_p(\mathcal{C}, t\mathcal{C})$ is said to be a cofibration if each of the vertical maps are cofibrations, giving $\mathbf{N}(\mathcal{C}, t\mathcal{C})$ the structure of a simplicial category with cofibrations.

Considering $obS\mathcal{C} \to BtS\mathcal{C}$ as a map of bisimplicial sets, we see that by Lemma 1.2.3.1 it is a homotopy equivalence

$$obS\mathcal{C} = obS\mathbf{N}_0(\mathcal{C}, t\mathcal{C}) \to obS\mathbf{N}_p(\mathcal{C}, t\mathcal{C}) = B_p tS\mathcal{C}$$

in every degree, and so by Theorem A.6.0.1 we obtain a weak equivalence of diagonals. □

1.2.3.1 Additivity

The fundamental theorem of the S-construction is the additivity theorem. For proofs we refer the reader to [301] or [203]. This result is actually not used explicitly anywhere in these notes, but it is our guiding theorem for all of K-theory. In fact, it shows that the S-construction is a true generalization of K_0, giving the same sort of universality for K-theory considered as a functor into spectra (see below).

Theorem 1.2.3.3 *Let $\mathcal{C}$ be a category with cofibrations. The natural map*

$$obS(S_2\mathcal{C}) \to obS(\mathcal{C}) \times obS(\mathcal{C})$$

is a weak equivalence. ☺

See also the more general formulation in Theorem 1.2.7.1.

1.2.4 The Spectrum

Continuing where Lemmas 1.2.1.5 and 1.2.1.6 left off, one checks that the definition of $S\mathcal{C}$ guarantees that it is in fact a simplicial category with cofibrations.

To be precise,

Definition 1.2.4.1 Let $\mathcal{C}$ be a category with cofibrations. A cofibration $c \rightarrowtail d \in S_q\mathcal{C}$ is a map such that for $0 < i \leq q$ the maps

$$c_{0i} \rightarrowtail d_{0i}$$

and

$$d_{0,i-1} \coprod_{c_{0,i-1}} c_{0i} \rightarrowtail d_{0i}$$

are all cofibrations.

Note that if $c \rightarrowtail d$ is a cofibration then it follows that all the maps $c_{ij} \rightarrowtail d_{ij}$ are cofibrations.

This means that we may take S of each $S_n\mathcal{C}$, and in this way obtain a bisimplicial object $SS\mathcal{C}$, and by iteration, a sequence of (multi)-simplicial objects $S^{(m+1)}\mathcal{C} = SS^{(m)}\mathcal{C}$.

Recall that a *spectrum* is a sequence of pointed spaces, $m \mapsto X^m$, $m \geq 0$, together with maps $S^1 \wedge X^m \to X^{m+1}$. See Appendix A.3.2 for further development of the basic properties of spectra, but recall that given a spectrum X, we define its homotopy groups as

$$\pi_q X = \varinjlim_k \pi_{k+q} X^k$$

(where the colimit is taken along the adjoint of the structure maps). A map of spectra $f: X \to Y$ is a *levelwise equivalence* if $f^n: X^n \to Y^n$ is a weak equivalence for every n, and a *stable equivalence* if it induces an isomorphism $\pi_*(f): \pi_* X \to \pi_* Y$. What we call levelwise equivalences are often called pointwise equivalences.

We will study another model for spectra in more detail in Chap. 2. Morally, spectra are beefed up versions of chain complexes, but in reality they give you much more.

If Z is any space (i.e., simplicial set) and $m, n \geq 0$, we have a natural map $\Delta([m],[n]) \times Z_n \to Z_m$ sending (ϕ, z) to $\phi^* z$, and varying m we get a simplicial map $\Delta[n] \times Z_n \to Z$ (an instance of the Yoneda map). In particular we get a map $\Delta[1] \times obS_1\mathcal{C} \to obS\mathcal{C}$, which, given that $S_0\mathcal{C} = *$ (i.e., $S\mathcal{C}$ is *reduced*) factors uniquely through a map $S^1 \wedge obS_1\mathcal{C} \to obS\mathcal{C}$ (since $S^1 = \Delta[1]/\partial\Delta[1]$). Upon identifying $\mathcal{C}$ and $S_1\mathcal{C}$ this gives a map

$$S^1 \wedge ob\mathcal{C} \to obS\mathcal{C}.$$

This means that the multi-simplicial sets

$$m \mapsto obS^{(m)}\mathcal{C} = ob\underbrace{S\ldots S}_{m \text{ times}}\mathcal{C}$$

form a spectrum after taking the diagonal.

Since $obS^{(m)}\mathcal{C}$ is connected in all of its m simplicial directions, the diagonal will be $m-1$-connected by Corollary A.6.0.8. A consequence of the additivity Theorem 1.2.3.3 is that this spectrum is almost an "Ω-spectrum" (see Sect. A.3.2): more precisely the adjoint maps $obS^{(m)}\mathcal{C} \to \Omega obS^{(m+1)}\mathcal{C}$ are equivalences for all $m > 0$. This is sometimes summarized by saying that $m \mapsto S^{(m)}\mathcal{C}$ is a *positive Ω-spectrum*. We won't need this fact.

For any category $\mathcal{D}$, let $i\mathcal{D} \subseteq \mathcal{D}$ be the subcategory with the same objects, but with only the isomorphisms as morphisms. As before, we get a map $S^1 \wedge Bi\mathcal{C} \to BiS\mathcal{C}$, and hence another spectrum $m \mapsto BiS^{(m)}\mathcal{C}$.

For each n, the degeneracies (i.e., the maps given by the unique maps to $[0]$ in Δ) induce an inclusion

$$obS^{(n)}\mathcal{C} = B_0 iS^{(n)}\mathcal{C} \to BiS^{(n)}\mathcal{C}$$

giving a map of spectra. That the two spectra are levelwise equivalent (that is, the maps $obS^{(n)}\mathcal{C} = B_0 iS^{(n)}\mathcal{C} \to BiS^{(n)}\mathcal{C}$ are all weak equivalences of spaces after taking diagonals) follows from Corollary 1.2.3.2.

Definition 1.2.4.2 Let $\mathcal{C}$ be a category with cofibrations. Then

$$\mathbf{K}(\mathcal{C}) = \{m \mapsto obS^{(m)}\mathcal{C}\}$$

is the K-theory spectrum of $\mathcal{C}$ (with respect to the isomorphisms).

In these notes we will only use this definition for categories with cofibrations which are *Ab*-categories. Exact categories are particular examples of *Ab*-categories with cofibrations.

The additivity Theorem 1.2.3.3 can be restated as a property of the K-theory spectrum: The natural map

$$obS^{(n)}(S_2\mathcal{C}) \longrightarrow obS^{(n)}(\mathcal{C}) \times obS^{(n)}(\mathcal{C})$$

is a weak equivalence for all $n \geq 0$. One should note that the claim that the map $\mathbf{K}(S_2\mathcal{C}) \to \mathbf{K}(\mathcal{C}) \times \mathbf{K}(\mathcal{C})$ is a *stable* equivalence follows almost automatically by the construction (see [301, 1.3.5]).

Scholium 1.2.4.3 *The following fact has been brought to our attention by Lars Hesselholt: If $\mathfrak{C}$ is an additive category, then the category $S_2\mathfrak{C}$ becomes an exact category if we define the sequence $c' \to c \to c''$ to be exact if $c' \to c$ is a cofibration in $S_2\mathfrak{C}$ and if for each $0 \leq i < j \leq 2$ the sequence $c'_{ij} \to c_{ij} \to c''_{ij}$ is split exact. However, all such short exact sequences split, making $S_2\mathfrak{C}$ itself no more than an additive category: given a cofibration $c' \rightarrowtail c$ in $S_2\mathfrak{C}$, we may choose retractions $c_{ij} \to c'_{ij}$ of $c'_{ij} \rightarrowtail c_{ij}$ in such a way that they form a retraction $c \to c'$ in $S_2\mathfrak{C}$ of $c' \rightarrowtail c$ (choose retractions of $c'_{01} \rightarrowtail c_{01}$ and $c_{01} \coprod_{c'_{01}} c'_{02} \rightarrowtail c_{02}$).*

Definition 1.2.4.4 (K-Theory of Rings) Let A be a ring (unital and associative as always). Then we define the K-theory spectrum of A, $\mathbf{K}(A)$, to be $\mathbf{K}(\mathcal{P}_A)$; the K-theory of the category of finitely generated projective right A-modules.

K-theory behaves nicely with respect to "cofinal" inclusions, see e.g., [271], and we cite the only case we need: Let $\mathcal{F}_A$ be the category of finitely generated free A-modules. The inclusion $\mathcal{F}_A \subseteq \mathcal{P}_A$ induces a homotopy fiber sequence (cf. Sect. A.5) of spectra

$$\mathbf{K}(\mathcal{F}_A) \longrightarrow \mathbf{K}(\mathcal{P}_A) \longrightarrow H\big(K_0(A)/K_0^f(A)\big)$$

where $H(M)$ is the Eilenberg–Mac Lane spectrum of an abelian group M (a spectrum whose only nonzero homotopy group is M in dimension zero. See Sect. A.3.2 for a construction). Hence the homotopy groups of $\mathbf{K}(\mathcal{F}_A)$ and $\mathbf{K}(A) = \mathbf{K}(\mathcal{P}_A)$ coincide in positive dimensions.

1.2.5 K-Theory of Split Radical Extensions

Recall that if B is a ring, the Jacobson radical $rad(M)$ of a B-module M is the intersection of all the kernels of maps from M to simple modules [13, p. 83]. Of particular importance to us is the case of a nilpotent ideal $I \subseteq B$. Then $I \subseteq rad(B)$ since $1 + I$ consists of units.

We now turn to the very special task of giving a suitable model for $\mathbf{K}(B)$ when $f \colon B \to A$ is a split surjection with kernel I contained in the Jacobson radical $rad(B) \subseteq B$. We have some low dimensional knowledge about this situation, namely Sects. 1.1.2(3) and 1.1.3.4(5) which tell us that $K_0(B) \cong K_0(A)$ and that the multiplicative group $(1+I)^\times$ maps surjectively onto the kernel of the surjection $K_1(B) \twoheadrightarrow K_1(A)$. Some knowledge of K_2 was also available already in the seventies (see e.g., [60, 292] and [189]).

We use the strictly functorial model explained in Definition 1.2.1.3 for the category of finitely generated projective modules $\mathcal{P}_A$ where an object is a pair (m, p) where m is a natural number and $p \in M_m A$ satisfies $p^2 = p$. If $j \colon A \to B$, then $j_*(m, p) = (m, j(p))$.

Lemma 1.2.5.1 *Let $f \colon B \to A$ be a split surjective k-algebra map with kernel I, and let $j \colon A \to B$ be a splitting. Let $c = (m, p) \in \mathcal{P}_A$ and $P = \mathrm{im}(p)$, and consider $\mathcal{P}_B(j_*c, j_*c)$ as a monoid under composition. The kernel of the monoid map*

$$f_* \colon \mathcal{P}_B(j_*c, j_*c) \to \mathcal{P}_A(c, c)$$

is isomorphic to the monoid of matrices $x = 1 + y \in M_m(B)$ such that $y \in M_m I$ and $y = yj(p) = j(p)y$. This is also naturally isomorphic to the set $\mathcal{M}_A(P, P \otimes_A j^*I)$. The monoid structure induced on $\mathcal{M}_A(P, P \otimes_A j^*I)$ is given by

$$\alpha \cdot \beta = (1+\alpha) \circ (1+\beta) - 1 = \alpha + \beta + \alpha \circ \beta$$

for $\alpha, \beta \in \mathcal{M}_A(P, P \otimes_A I)$ where $\alpha \circ \beta$ is the composite

$$P \xrightarrow{\beta} P \otimes_A I \xrightarrow{\alpha \otimes 1} P \otimes_A I \otimes_A I \xrightarrow{\text{multiplication in } I} P \otimes_A I.$$

Proof As in Definition 1.2.1.3, we identify $\mathcal{P}_B(j_*c, j_*c)$ with the set of matrices $x \in M_m(B)$ such that $x = xj(p) = j(p)x$ and likewise for $\mathcal{P}_A(c, c)$. The kernel consists of the matrices x for which $f(x) = p$ (the identity!), that is, the matrices of the form $j(p) + y$ with $y \in M_m(I)$ such that $y = yj(p) = j(p)y$. As a set, this is isomorphic to the claimed monoid, and the map $j(p) + y \mapsto 1 + y$ is a monoid isomorphism since $(j(p)+y)(j(p)+z) = j(p)^2 + yj(p) + j(p)z + yz = j(p) + y + z + yz \mapsto 1 + y + z + yz = (1 + y)(1 + z)$. The identification with $\mathcal{M}_A(P, P \otimes_A j^*I)$ is through the composite

$$\begin{aligned} \mathrm{Hom}_A(P, P \otimes_A j^*I) &\xrightarrow{\cong} \mathrm{Hom}_B(P \otimes_A B, P \otimes_A I) \\ &\xrightarrow{\phi \mapsto 1+\phi} \mathrm{Hom}_B(P \otimes_A B, P \otimes_A B) \xrightarrow{\cong} \mathcal{P}_B(j_*c, j_*c) \end{aligned}$$

where the first isomorphism is the adjunction isomorphism and the last isomorphism is the natural isomorphism between

$$\mathcal{P}_A \xrightarrow{j_*} \mathcal{P}_B \longrightarrow \mathcal{M}_B$$

and

$$\mathcal{P}_A \longrightarrow \mathcal{M}_A \xrightarrow{-\otimes_A B} \mathcal{M}_B.$$ □

Lemma 1.2.5.2 *In the same situation as the preceding lemma, if $I \subset Rad(B)$, then the kernel of*

$$f_* \colon \mathcal{P}_B(j_*c, j_*c) \longrightarrow \mathcal{P}_A(c, c)$$

is a group.

Proof To see this, assume first that $P \cong A^n$. Then

$$\mathcal{M}_A(P, P \otimes_A I) \cong M_n I \subseteq M_n\big(rad(B)\big) = rad\big(M_n(B)\big)$$

(we have that $M_n(rad(B)) = rad(M_n(B))$ since $\mathcal{M}_B(B^n, -)$ is an equivalence from B-modules to $M_n(B)$-modules, [13, p. 86]), and so $(1 + M_n(I))^\times$ is a group. If P is a direct summand of A^n, say $A^n = P \oplus Q$, and $\alpha \in \mathcal{M}_A(P, P \otimes_A I)$, then we have a diagram

$$\begin{array}{ccc} P \otimes_A B & \xrightarrow{1+\alpha} & P \otimes_A B \\ \downarrow & & \downarrow \\ A^n \otimes_A B & \xrightarrow{1+(\alpha,0)} & A^n \otimes_A B \end{array}$$

where the vertical maps are split injections. By the discussion above $1 + (\alpha, 0)$ must be an isomorphism, forcing $1 + \alpha$ to be one too. □

All of the above holds true if instead of considering module categories, we consider the S-construction of Waldhausen applied n times to the projective modules.

More precisely, let now c be some object in $S_p^{(n)}\mathcal{P}_A$. Then the set of morphisms $S_p^{(n)}\mathcal{M}_A(c, c\otimes_A j^*I)$ is still isomorphic to the monoid of elements sent to the identity under

$$S_p^{(n)}\mathcal{P}_B(j_*c, j_*c) \xrightarrow{f_*} S_p^{(n)}\mathcal{P}_A(c, c)$$

and, if I is radical, this is a group. We will usually suppress the simplicial indices and speak of elements in some unspecified dimension. We will also usually suppress the j^* that should be inserted whenever I is considered as an A-module.

We need a few technical definitions.

Definition 1.2.5.3 Let

$$0 \longrightarrow I \longrightarrow B \xrightarrow{f} A \longrightarrow 0$$

be a split extension of k-algebras with $I \subset Rad(B)$, and choose a splitting $j\colon A \to B$ of f. Let $t\mathcal{P}_B \subseteq \mathcal{P}_B$ be the subcategory with all objects, but with morphisms only the endomorphisms taken to the identity by f_*. Note that, since $I \subseteq rad(B)$, all morphisms in $t\mathcal{P}_B$ are automorphisms.

Let

$$tS_q^{(n)}\mathcal{P}_B \subseteq iS_q^{(n)}\mathcal{P}_B$$

be the subcategory with the same objects, but with morphisms transformations of diagrams in $S_q^{(n)}\mathcal{P}_B$ consisting of morphisms in $t\mathcal{P}_B$.

Consider the sequence of (multi) simplicial exact categories $n \mapsto \mathcal{D}_A^n B$ given by

$$ob\mathcal{D}_A^n B = obS^{(n)}\mathcal{P}_A \quad \text{and} \quad \mathcal{D}_A^n B(c, d) = S^{(n)}\mathcal{P}_B(j_*c, j_*d).$$

Let $t\mathcal{D}_A^n B \subset \mathcal{D}_A^n B$ be the subcategory containing all objects, but whose only morphisms are the automorphisms $S^{(n)}\mathcal{M}_A(c, c\otimes_A I)$ considered as the subset $\{b \in S^{(n)}\mathcal{P}_B(j_*c, j_*c) | f_*b = 1\} \subseteq \mathcal{D}_A^n B(c, c)$.

We set

$$\mathbf{K}_A B = \left\{ n \mapsto Bt\mathcal{D}_A^n B = \coprod_{m\in S^{(n)}\mathcal{P}_A} B\big(S^{(n)}\mathcal{M}_A(m, m\otimes_A I)\big) \right\}$$

where the bar construction is taken with respect to the group structure.

Recall that in the eyes of K-theory there really is no difference between the special type of automorphisms coming from t and all isomorphisms since by Corollary 1.2.3.2 the inclusions

$$obS^{(n)}\mathcal{P}_B \subseteq BtS^{(n)}\mathcal{P}_B \subseteq BiS^{(n)}\mathcal{P}_B$$

are both weak equivalences.

Note that $\mathcal{D}^n_A B$ depends not only on I as an A-bimodule (i.e., $A^o \otimes A$-module), but also on the multiplicative structure it inherits as an ideal in B. We have a factorization

$$S^{(n)}\mathcal{P}_A \xrightarrow{j_!} \mathcal{D}^n_A B \xrightarrow{j_\#} S^{(n)}\mathcal{P}_B$$

where $j_!$ is the identity on object, and j_* on morphisms, and $j_\#$ is the fully faithful functor sending $c \in obt\mathcal{D}^n_A B = obS^{(n)}\mathcal{P}_A$ to $j_*c \in obS^{(n)}\mathcal{P}_B$ (and the identity on morphisms). We see that $\mathbf{K}_A B$ is a subspectrum of $\{n \mapsto BiS^{(n)}\mathcal{P}_B\}$ via

$$t\mathcal{D}^n_A B \longrightarrow tS^{(n)}\mathcal{P}_B \subseteq iS^{(n)}\mathcal{P}_B.$$

Theorem 1.2.5.4 *Let $f : B \to A$ be a split surjection of k-algebras with splitting j and kernel $I \subset rad(B)$. Then*

$$\mathcal{D}^n_A B \xrightarrow{j_\#} S^{(n)}\mathcal{P}_B, \textit{and its restriction } t\mathcal{D}^n_A B \xrightarrow{j_\#} tS^{(n)}\mathcal{P}_B$$

are (degreewise) equivalences of simplicial exact categories, and so the chain

$$\mathbf{K}_A B(n) = Bt\mathcal{D}^n_A B \subseteq BtS^{(n)}\mathcal{P}_B \supseteq obS^{(n)}\mathcal{P}_B = \mathbf{K}(B)(n)$$

consists of weak equivalences.

Proof To show that

$$\mathcal{D}^n_A B \xrightarrow{j_\#} S^{(n)}\mathcal{P}_B$$

is an equivalence, all we have to show is that every object in $S^{(n)}\mathcal{P}_B$ is isomorphic to something in the image of $j_\#$. We will show that $c \in S^{(n)}\mathcal{P}_B$ is isomorphic to $j_* f_* c = j_\#(j_! f_* c)$.

Let $c = (m, p) \in ob\mathcal{P}_B$, $P = \mathrm{im}(p)$. Consider the diagram with short exact columns

$$\begin{array}{ccc}
\mathrm{im}(p)\cdot I & \dashrightarrow & \mathrm{im}(jf(p))\cdot I \\
\big\downarrow & & \big\downarrow \\
\mathrm{im}(p) & \overset{\eta_p}{\dashrightarrow} & \mathrm{im}(jf(p)) \\
\big\downarrow{\scriptstyle \pi} & & \big\downarrow{\scriptstyle \pi'} \\
f^*\mathrm{im}(f(p)) & = & f^*\mathrm{im}(fjf(p))
\end{array}$$

Since $\mathrm{im}(p)$ is projective there exist a (not necessarily natural) lifting η_p. Let C be the cokernel of η_p. A quick diagram chase shows that $C \cdot I = C$. Since $\mathrm{im}(jf(p))$, and hence C, is finitely generated, Nakayama's Lemma 3.1.4.1 tells us that C is trivial. This implies that η_p is surjective, but $\mathrm{im}(jf(p))$ is also projective, so η_p must be split surjective. Call the splitting ϵ. Since $\pi\epsilon = \pi'\eta_p\epsilon = \pi'$ the argument above applied to ϵ shows that ϵ is also surjective. Hence η_p is an isomorphism. Thus, every object $c \in ob\mathcal{P}_B$ is isomorphic to $j_*(f_*c)$.

Let $c \in obS^{(n)}\mathcal{P}_B$. Then c and j_*f_*c are splittable diagrams with isomorphic vertices. Choosing isomorphisms on the "diagonal" we can extend these to the entire diagram, and so c and j_*f_*c are indeed isomorphic as claimed, proving the first assertion.

To show that

$$t\mathcal{D}^n_A B \xrightarrow{j_\#} tS^{(n)}\mathcal{P}_B$$

is an equivalence, note first that this functor is also fully faithful. We know that any $c \in ob\, tS^{(n)}\mathcal{P}_B = obS^{(n)}\mathcal{P}_B$ is isomorphic in $S^{(n)}\mathcal{P}_B$ to j_*f_*c, and the only thing we need to show is that we can choose this isomorphism in t. Let $\iota\colon c \to j_*f_*c \in iS^{(n)}\mathcal{P}_B$ be any isomorphism. Consider

$$c \xrightarrow{\iota} j_*f_*c = j_*f_*j_*f_*c \xrightarrow{j_*f_*(\iota^{-1})} j_*f_*c.$$

Since $f_*(j_*f_*(\iota^{-1}) \circ \iota) = f_*(\iota^{-1}) \circ f_*(\iota) = 1_{f_*c}$ the composite $j_*f_*(\iota^{-1}) \circ \iota$ is an isomorphism in $tS^n_q\mathcal{P}$ from c to $j_\#(j_!f_*c)$. □

Definition 1.2.5.5 In the context of Definition 1.2.5.3, let

$$\tilde{\mathbf{K}}_A B = \mathbf{K}_A B/\mathbf{K}(A) = \left\{ n \mapsto \bigvee_{m \in S^{(n)}\mathcal{P}_A} B\big(S^{(n)}\mathcal{M}_A(m, m \otimes_A I)\big) \right\}.$$

This is a model for the relative algebraic K-theory of $B \to A$ in that Theorem 1.2.5.4 says that

$$\tilde{\mathbf{K}}_A B \xrightarrow{\sim} \mathbf{K}(B)/\mathbf{K}(A)$$

is a (levelwise) equivalence of spectra. The latter spectrum is stably equivalent to the homotopy fiber of $\mathbf{K}(B) \to \mathbf{K}(A)$. To see this, consider the square

$$\begin{array}{ccc} \mathbf{K}(B) & \longrightarrow & \mathbf{K}(A) \\ \downarrow & & \downarrow \\ \mathbf{K}(B)/\mathbf{K}(A) & \longrightarrow & * \end{array}$$

It is a (homotopy) cocartesian square of spectra, and hence homotopy cartesian. (In spectrum dimension n this is a cocartesian square, and the spaces involved are at least $n-1$-connected, and so all maps are $n-1$-connected. Then Blakers–Massey Corollary A.8.2.2 tells us that the square is $(n-1)+(n-1)-1 = 2n-3$ homotopy cartesian.) This means that the homotopy fiber of the upper horizontal map maps by a weak equivalence to the homotopy fiber of the lower horizontal map.

1.2.5.1 Analyticity Properties of $\mathbf{K}_A(B)$

The following result will not be called for until Lemma 7.2.1.3, and may be safely skipped at a first reading until the result is eventually referred back to, but is placed here since it uses notation that is better kept local.

Although we are not using the notion of calculus of functors in these notes, we will in many cases come quite close. The next lemma, which shows how $\mathbf{K}_A(B)$ behaves under certain inverse limits, can be viewed as an example of this. A twist which will reappear later is that we do not ask whether the functor turns "cocartesianness" into "cartesianness", but rather to what extent the functor preserves inverse limits. The reason for this is that in many cases the coproduct structure of the source category can be rather messy, whereas some forgetful functor tells us exactly what the limits should be.

For the basics on cubes see Appendix A.8. In particular, a *strongly cocartesian n-cube* is an n-cube where each two-dimensional face is cocartesian.

Let Split be the category of split radical extensions over a given ring A. Note that if $B \twoheadrightarrow A$ is an object in Split with kernel the A-bimodule P, then $B \twoheadrightarrow A$ is isomorphic to $A \ltimes P \twoheadrightarrow A$ where $A \ltimes P$ is $A \oplus P$ as an abelian group with multiplication given by $(a', p') \cdot (a, p) = (a'a, a'p + p'a + p'p)$.

The category sSplit of simplicial objects in Split then inherits the notion of k-cartesian cubes via the forgetful functor down to simplicial sets. By "final maps" in an n-cube we mean the maps induced from the n inclusions of the subsets of cardinality $n - 1$ in $\{1, \dots, n\}$.

If $A \ltimes P \in s\mathrm{Split}$ it makes sense to talk about $\mathbf{K}(A \ltimes P)$ by applying the functor in every degree, and diagonalizing.

Lemma 1.2.5.6 *Let $A \ltimes \mathcal{P}$ be a strongly cartesian n-cube in s*Split *such all the final maps are k-connected. Then $\mathbf{K}(A \ltimes \mathcal{P})$ is $(1 + k)n$-cartesian.*

Proof Fix the non-negative integer q, the tuple $p = (p_1, \dots, p_q)$ and the object $c \in obS^{(q)}_p \mathcal{P}_A$. The cube $S^{(q)}_p \mathcal{M}_A(c, c \otimes_A \mathcal{P})$ is also strongly cartesian (it is so as a simplicial set, and so as a simplicial group), and the final maps are still k-connected. Taking the bar of this gives us a strongly cartesian cube $BS^{(q)}_p \mathcal{M}_A(c, c \otimes_A \mathcal{P})$, but whose final maps will be $k+1$-connected. By the Blakers–Massey Corollary A.8.2.2 this means that $BS^{(q)}_p \mathcal{M}_A(c, c \otimes_A \mathcal{P})$ will be $(k + 2)n - 1$-cocartesian. The same will be true for

$$\coprod_{c \in obS^{(q)}_p \mathcal{P}_A} BS^{(q)}_p \mathcal{M}_A(c, c \otimes_A \mathcal{P}).$$

Varying p and remembering that each multi-simplicial space is $(q - 1)$-connected in the p direction, we see that the resulting cube is $q + (k + 2)n - 1$-cocartesian, cf. Corollary A.6.0.8. Varying also q, we see that this means that the cube of spectra $\mathbf{K}(A \ltimes \mathcal{P})$ is $(k + 2)n - 1$ cocartesian, or equivalently $(k + 2)n - 1 - (n - 1) = (k + 1)n$-cartesian. □

The importance of this lemma will become apparent as we will approximate elements in Split by means of cubical diagrams in sSplit where all but the initial node will be "reduced" in the sense that the zero skeletons will be exactly the trivial extension $A = A$.

1.2.6 Categories with Cofibrations and Weak Equivalences

Definition 1.2.4.2 of the K-theory of a category with cofibrations does not immediately cover more general situations where we are interested in incorporating some structure of weak equivalences, e.g., simplicial rings. Waldhausen [301] covers this case also, and demands only that the category of weak equivalences $w\mathcal{C} \subseteq \mathcal{C}$ contains all isomorphisms and satisfies the *gluing lemma*, that is, if the left horizontal maps in the commutative diagram

$$\begin{array}{ccccc} d & \leftarrowtail & c & \rightarrow & e \\ \downarrow & & \downarrow & & \downarrow \\ d' & \leftarrowtail & c' & \rightarrow & e' \end{array}$$

are cofibrations and the vertical maps are weak equivalences, then the induced map

$$d \coprod_c e \to d' \coprod_{c'} e'$$

is also a weak equivalence. Cf. Lemma 5.2.6 of [139] which proves the gluing lemma in most cases of interest.

In this case $S\mathcal{C}$ inherits a subcategory of weak equivalences, $wS\mathcal{C}$ satisfying the same conditions by declaring that a map is a weak equivalence if it is on all nodes. We iterate this construction and define

$$\mathbf{K}(\mathcal{C}, w) = \{m \mapsto BwS^{(m)}\mathcal{C}\}. \tag{1.2}$$

Corollary 1.2.3.2 then says that

$$\mathbf{K}(\mathcal{C}) \xrightarrow{\simeq} \mathbf{K}(\mathcal{C}, i)$$

is an equivalence of spectra.

One should note that there really is no need for the new definition, since the old covers all situations by the following observation. If we let $\mathbf{N}_q\mathcal{C}$ be the category of functors $[q] \to \mathcal{C}$ and natural transformations between these, we can let $\mathbf{N}_q(\mathcal{C}, w)$ be the full subcategory of $\mathbf{N}_q\mathcal{C}$ with $ob\mathbf{N}_q(\mathcal{C}, w) = B_q w\mathcal{C}$. Letting q vary, this is a simplicial category with cofibrations, and we have an canonical isomorphism

$$\mathbf{K}(\mathcal{C}, w)(m) = BwS^{(m)}\mathcal{C} \cong obS^{(m)}\mathbf{N}(\mathcal{C}, w) = \mathbf{K}\big(\mathbf{N}(\mathcal{C}, w)\big).$$

Some authors use the word "Waldhausen category" to signify a category with cofibrations and weak equivalences.

1.2.7 Other Important Facts About the K-Theory Spectrum

The following theorems are important for the general framework of algebraic K-theory and we include them for the reader's convenience. We will neither need them for the development of the theory nor prove them, but we still want to use them in some examples and draw the reader's attention to them. The papers [110] and [271] of Grayson and Staffeldt give very concrete and nice proofs of Quillen's original statements in the context of Waldhausen's construction. In addition, Waldhausen [301], Thomason [284] and Schlichting's [245] papers are good sources.

Theorem 1.2.7.1 (Additivity Theorem: [301, Sect. 1.4] and [203]) *Let $\mathcal{C}$ be a category with cofibrations and weak equivalences $w\mathcal{C}$. Then*

$$BwSS_2\mathcal{C} \to BwS\mathcal{C} \times BwS\mathcal{C}$$

is an equivalence, and the structure map $BwS^{(m)}\mathcal{C} \to \Omega BwS^{(m+1)}\mathcal{C}$ is an equivalence for $m > 0$. ☺

In order to state the next theorems, we need to define some important notions about categories with cofibrations and weak equivalences

Definition 1.2.7.2 A subcategory of a category is said to satisfy the *two-out-of-three property* if given two composable morphisms

$$c \xleftarrow{g} b \xleftarrow{f} a$$

in the ambient category, one has that, if two of f, g and gf are in the smaller category, then all three are.

The category of weak equivalences in a category with cofibrations and weak equivalences is said to satisfy the *extension axiom* if for any given map

$$\begin{array}{ccc} c & \rightarrowtail & c' \\ \downarrow & & \downarrow \\ d & \rightarrowtail & d' \end{array}$$

of cofibrations where $c \to d$ and the induced map of cofibers $c'/c \to d'/d$ are weak equivalences it follows that the map $c' \to d'$ is a weak equivalence too.

Let $\mathcal{C}$ and $\mathcal{D}$ be categories with weak equivalences and cofibrations. A functor $F: \mathcal{C} \to \mathcal{D}$ is *exact* if it preserves weak equivalences and is exact as a functor of categories with cofibrations (i.e., preserves cofibrations and pushout squares along cofibrations, cf. Definition 1.2.1.1).

An exact functor of categories with weak equivalences and cofibrations $F: \mathcal{C} \to \mathcal{D}$ has the *approximation property* if

1. $w\mathcal{C} = F^{-1}(w\mathcal{D})$
2. given $F(c) \to d \in \mathcal{D}$, there is a cofibration $c \rightarrowtail c'$ in $\mathcal{C}$ and a weak equivalence $F(c') \xrightarrow{\sim} d$ in $\mathcal{D}$ such that the induced diagram

$$\begin{array}{ccc} F(c) & \longrightarrow & d \\ & \searrow \quad \nearrow_{\sim} & \\ & F(c') & \end{array}$$

commutes

Waldhausen refers to the two-out-of-three property by saying that the smaller category satisfies the *saturation axiom*. One notices how subcategories characterized by some induced invariant (like homotopy groups) being isomorphisms will always satisfy the two-out-of-three property. On the other hand, there are interesting applications (like simple maps) where the two-out-of-three property does not hold.

We state Schlichting's version [245, Theorem 10] of the approximation theorem.

Theorem 1.2.7.3 (Approximation Theorem 1.6.7 [301] and [245]) *Let $\mathcal{C}$ and $\mathcal{D}$ be categories with cofibrations and weak equivalences, where the weak equivalences have the two-out-of-three property, and suppose that every morphism in $\mathcal{C}$ may be factored as a cofibration followed by a weak equivalence. If the exact functor $F: \mathcal{C} \to \mathcal{D}$ has the approximation property, then the induced maps $Bw\mathcal{C} \to Bw\mathcal{D}$ and $BwS\mathcal{C} \to BwS\mathcal{D}$ are weak equivalences.* ☺

It does not escape notice that with Schlichting's formulation, even Ravel's "Fool's morning song" (cf. axiom Cyl 3 on p. 348 of [301]) is redundant.

The next theorem, the so-called fibration theorem is of importance in localization situations, and is often referred to as the localization theorem, and again we use Schlichting's formulation [245, Theorem 11]

Theorem 1.2.7.4 (Fibration Theorem [108, 245, 301] and [232]) *Let $\mathcal{C}$ be a category with cofibrations and* two *subcategories $v\mathcal{C} \subseteq w\mathcal{C}$ of weak equivalences. Let $\mathcal{C}^w \subseteq \mathcal{C}$ be the full subcategory with cofibrations of objects c such that the unique map $* \to c$ is in $w\mathcal{C}$. This inherits two subcategories of weak equivalences $v\mathcal{C}^w = \mathcal{C}^w \cap v\mathrm{C}$ and $w\mathcal{C}^w = \mathcal{C}^w \cap w\mathcal{C}$. Assume every map in $\mathcal{C}$ may be factored as a cofibration followed by a map in $w\mathcal{C}$ and that $w\mathcal{C}$ has the two-out-of-three property and satisfies the extension axiom Definition* 1.2.7.2. *Then the square*

$$\begin{array}{ccc} BwS\mathcal{C}^w & \longrightarrow & BwS\mathcal{C}^w \\ \downarrow & & \downarrow \\ BvS\mathcal{C} & \longrightarrow & BwS\mathcal{C} \end{array}$$

is homotopy cartesian, and the upper right term, $BwSC^w$ is contractible. ☺

In his foundational paper [232], Quillen states a dévissage theorem for the K-theory of abelian categories. Although there has been serious effort, this theorem has still not found a satisfactory formulation in Waldhausen's setup, see e.g., [317], [110, 5.1], [271, 4.1] (the last two with short proofs).

Still it is a very important theorem and we state it with the original conditions.

Theorem 1.2.7.5 (Dévissage Theorem [232, Theorem 4]) *Let $\mathcal{A}$ be an essentially small abelian category and $\mathcal{B}$ a full additive subcategory closed under taking subobjects and quotient objects. If for each object a of $\mathcal{A}$ there is a finite filtration $0 = a_0 \subseteq a_1 \subseteq \cdots \subseteq a_n = a$ such that each subquotient a_j/a_{j-1} is in $\mathcal{B}$, then the map $BiS\mathcal{B} \to BiS\mathcal{A}$ induced by the inclusion is a weak equivalence.* ☺

We state the resolution theorem (see also [110, 4.1] and [271, 3.1]).

Theorem 1.2.7.6 (Resolution Theorem [232, Theorem 3]) *Assume $\mathcal{P} \subseteq \mathcal{M}$ is a full exact subcategory of an exact category $\mathcal{M}$, closed under exact sequences, extensions and cokernels. Assume that for any $M \in \mathcal{M}$ there is a short exact sequence $0 \to M \to P \to P'' \to 0$ in $\mathcal{M}$ with P, P'' in $\mathcal{P}$, then the map $BiS\mathcal{P} \to BiS\mathcal{M}$ induced by the inclusion is a weak equivalence.* ☺

Some of the theory Waldhausen develops for his construction will appear later in other contexts. The stable approach we will eventually adopt, avoids the use of machinery such as spherical objects and cell filtrations.

1.3 Stable K-Theory is Homology

In this section we will try to connect K-theory to homology. This is done by considering "small perturbations" in input in K-theory, giving a "linear" theory: the "directional derivative" of K-theory. This is then compared with the classical concept of homology, and the two are shown to be equal.

This theme will be revisited later, most noticeably in Sect. 5.3 where stable K-theory is compared with topological Hochschild homology through the trace map. This map is proved to be an equivalence, using that—in view of Theorem 4.3.4.1 and Theorem 4.3.5.16—the homology of the category of finitely generated modules may be identified with topological Hochschild homology.

1.3.1 Split Surjections with Square-Zero Kernels

If A is a unital ring, and P is any A-bimodule (i.e., an $A^o \otimes A$ module, but with no multiplicative structure as part of the data), we define the ring $A \ltimes P$ simply to be

$A \oplus P$ as an A-bimodule, and with multiplication $(a', p')(a, p) = (a'a, a'p + p'a)$. That is, $P^2 = 0$ when P is considered as the kernel of the projection $A \ltimes P \twoheadrightarrow A$ sending (a, p) to a.

Algebraically, this is considered to be a "small" deformation of A (the elements of P are so small that their products vanish!). And the difference between $K(A \ltimes P)$ and $K(A)$ reflects the local structure of K-theory. The goal is to measure this difference.

Considered as a functor from A-bimodules, $P \mapsto \mathbf{K}(A \ltimes P)$ is not *additive*, even if we remove the part coming from $\mathbf{K}(A)$. That is, if we let

$$\tilde{\mathbf{K}}(A \ltimes P) = \mathrm{hofib}\big\{\mathbf{K}(A \ltimes P) \longrightarrow \mathbf{K}(A)\big\}$$

then the map $\tilde{\mathbf{K}}(A \ltimes (P \oplus Q)) \to \tilde{\mathbf{K}}(A \ltimes P) \times \tilde{\mathbf{K}}(A \ltimes Q)$ induced by the projections is not an equivalence in general. For instance, we have by [159] that $\pi_2\tilde{\mathbf{K}}(\mathbf{Z} \ltimes P) \cong \bigwedge^2 P \oplus P/2P$ for all abelian groups P. Hence

$$\begin{aligned}
\pi_2\tilde{\mathbf{K}}\big(\mathbf{Z} \ltimes (P \oplus Q)\big) &\cong \bigwedge^2(P \oplus Q) \oplus (P \oplus Q)/2(P \oplus Q) \\
&\cong \Big(\bigwedge^2 P \oplus P/2P\Big) \oplus \Big(\bigwedge^2 Q \oplus Q/2Q\Big) \oplus P \otimes Q \\
&\cong \pi_2\tilde{\mathbf{K}}(\mathbf{Z} \ltimes P) \oplus \pi_2\tilde{\mathbf{K}}(\mathbf{Z} \ltimes Q) \oplus (P \otimes Q)
\end{aligned}$$

where the tensor product expresses the non-linearity.

There are means of forcing linearity upon a functor, which will eventually give stable K-theory, and the aim of this section is to prove that this linear theory is equivalent to the homology of the category of finitely generated projective A-modules.

1.3.2 The Homology of a Category

Let $\mathcal{C}$ be an *Ab*-category (that is: a category enriched in *Ab*, the category of abelian groups, see Appendix A.10.2.2. *Ab*-categories are also known as "linear categories", "ringoids", "rings with many objects" and unfortunately, some call them "additive categories", a term we reserve for pointed *Ab*-categories with sum). The important thing to remember is that the homomorphism sets are equipped with an abelian group structure, such that composition is bilinear.

We say that $\mathcal{C}$ is *flat* if the morphism sets are flat as abelian groups. A $\mathcal{C}$-bimodule is an *Ab*-functor (linear functor) $\mathcal{C}^o \otimes \mathcal{C} \to Ab$ (see also Appendix A.10.4). The category $Ab^{\mathcal{C}^o \otimes \mathcal{C}}$ of $\mathcal{C}$-bimodules forms an abelian category with "enough projectives", so we are free to do homological algebra. If $\mathcal{C}$ is flat, the *Hochschild homology* of $\mathcal{C}$ with coefficients in $M \in Ab^{\mathcal{C}^o \otimes \mathcal{C}}$ is customarily defined as

$$\mathrm{Tor}_*^{Ab^{\mathcal{C}^o \otimes \mathcal{C}}}(M, \mathcal{C})$$

(see [215]). There is a standard simplicial abelian group (complex) whose homotopy groups calculate the Hochschild homology groups, namely

$$[q] \mapsto HH(\mathcal{C}, M)_q = \bigoplus_{c_0,\dots,c_q \in ob\mathcal{C}} M(c_0, c_q) \otimes \bigotimes_{1 \le i \le q} \mathcal{C}(c_i, c_{i-1})$$

with face and degeneracies as in Hochschild homology (see [215], and also below).

Let $\mathcal{C}$ be any category (that is, not necessarily an *Ab*-category). It is not uncommon to to call functors $\mathcal{C}^o \times \mathcal{C} \to Ab$ "bifunctors". We note immediately that, by adjointness of the free and forgetful functors

$$Ens \overset{\mathbf{Z}}{\rightleftarrows} Ab$$

connecting abelian groups to sets, a "bifunctor" is nothing but a $\mathbf{Z}\mathcal{C}$-bimodule in the *Ab*-enriched world; that is, an *Ab*-functor $\mathbf{Z}\mathcal{C}^o \otimes \mathbf{Z}\mathcal{C} \cong \mathbf{Z}(\mathcal{C}^o \times \mathcal{C}) \to Ab$. So, for any "bifunctor" (i.e., $\mathbf{Z}\mathcal{C}$-bimodule) M we may define the homology of $\mathcal{C}$ with respect to M as

$$H_*(\mathcal{C}, M) = \pi_* HH(\mathbf{Z}\mathcal{C}, M)$$

(notice that $\mathbf{Z}\mathcal{C}$ is flat). The standard complex $HH(\mathbf{Z}\mathcal{C}, M)$ calculating this homology, is naturally isomorphic to the following complex $F(\mathcal{C}, M)$:

Definition 1.3.2.1 Let $\mathcal{C}$ be a category and M a $\mathbf{Z}\mathcal{C}$-bimodule. Then the *homology of $\mathcal{C}$ with coefficients in M*, is the simplicial abelian group $F(\mathcal{C}, M)$ which in degree q is given by

$$F_q(\mathcal{C}, M) = \bigoplus_{c_0 \leftarrow \cdots \leftarrow c_q \in B_q\mathcal{C}} M(c_0, c_q) \cong \bigoplus_{c_0,\dots,c_q \in ob\mathcal{C}} M(c_0, c_q) \otimes \bigotimes_{1 \le i \le q} \mathbf{Z}\mathcal{C}(c_i, c_{i-1})$$

and with simplicial structure defined as follows. We write elements of $F_q(\mathcal{C}, M)$ as sums of elements of the form (x, α) where $x \in M(c_0, c_q)$ and

$$\alpha = c_0 \xleftarrow{\alpha_1} \cdots \xleftarrow{\alpha_q} c_q \in B_q\mathcal{C}.$$

Then

$$d_i(x, \alpha) = \begin{cases} (M(\alpha_1, 1)x, d_0\alpha) & \text{if } i = 0, \\ (x, d_i\alpha) & \text{if } 0 < i < q, \\ (M(1, \alpha_q)x, d_q\alpha) & \text{if } i = q \end{cases}$$

and $s_i(x, \alpha) = (x, s_i\alpha)$.

Remark 1.3.2.2 The homology of $\mathcal{C}$, or rather $F(\mathcal{C}, -) : Ab^{\mathbf{Z}\mathcal{C}^o \otimes \mathbf{Z}\mathcal{C}} \to sAb$, is characterized up to equivalence by the properties

1. If $M \in obAb^{\mathbf{Z}\mathcal{C}^o \otimes \mathbf{Z}\mathcal{C}}$ is projective, then $F(\mathcal{C}, M) \to H_0(\mathcal{C}, M)$ is an equivalence,

2. the functor $F(\mathcal{C},-)\colon Ab^{\mathbf{Z}\mathcal{C}^o\otimes\mathbf{Z}\mathcal{C}}\to sAb$ takes short exact sequences to fiber sequences, together with
3. the values of $H_0(\mathcal{C},M)$.

In particular, this means that if we have a map to or from some other theory satisfying (1) and (2), and inducing an isomorphism on π_0, then this map is an equivalence.

1.3.3 Incorporating the S-Construction

In order to compare with K-theory, we will incorporate the S-construction into the source of the homology functor.

Let $\mathcal{C}$ be a small category, and M a $\mathbf{Z}\mathcal{C}$-bimodule (i.e., a functor from $\mathcal{C}^o\times\mathcal{C}$ to abelian groups). Recall how bimodules are extended to diagram categories (see Appendix A.10.4 for the general situation).

If $\mathfrak{C}$ is an exact category, consider the full subcategory $S_q\mathfrak{C}$ of the category $[\mathcal{A}r([q]^o),\mathfrak{C}]$ of functors from $\mathcal{A}r([q]^o)$ to $\mathfrak{C}$. Let M be a $\mathfrak{C}$-bimodule. Then S_qM is defined, and is given by

$$S_qM(c,d)$$
$$=\left\{\{m_{ij}\}\in\prod_{0\le i\le j\le q}M(c_{ij},d_{ij})\;\middle|\;M(1,d_{ij}\to d_{kl})m_{ij}=M(c_{ij}\to c_{kl},1)m_{kl}\right\}$$

(if you like ends, this has the compact and pleasing notation $S_qM(c,d)=\int_{i\le j}M(c_{ij},d_{ij})$). Note that, if M is not pointed (i.e., a $\tilde{\mathbf{Z}}\mathfrak{C}$-bimodule) we may have elements in the groups $M(c_{ii},d_{ii})=M(0,0)$, but these are uniquely determined by the values in the other groups. (In fact, if $\mathfrak{C}$ is split exact, then the projection $S_qM(c,d)\to M(c_{0q},d_{0,q})$ is a split monomorphism—a retract is constructed using a choice of splitting.)

The construction $[q]\mapsto S_qM$ is functorial in q, in the sense that for every map $\phi\colon[p]\to[q]\in\Delta$ there are natural maps $\phi^*\colon S_pM\to\phi^*S_qM$. Let $\mathfrak{C}$ be an exact category, and M a pointed $\mathfrak{C}$-bimodule. Note that, due to the bimodule maps $\phi^*\colon S_pM\to\phi^*S_qM$

$$F(S\mathfrak{C},SM)=\big\{[p],[q]\mapsto F_p(S_q\mathfrak{C},S_qM)\big\}$$

is a bisimplicial abelian group.

Again we get a map $S^1\wedge F(\mathfrak{C},M)\to F(S\mathfrak{C},SM)$ making

$$\mathbf{F}(\mathfrak{C},M)=\big\{n\mapsto F\big(S^{(n)}\mathfrak{C},S^{(n)}M\big)\big\}$$

a spectrum. In the special case $\mathfrak{C}=\mathcal{P}_A$, and $M(c,d)=\mathrm{Hom}_A(c,d\otimes_A P)$ for some A-bimodule P, we define

$$\mathbf{F}(A,P)=\mathbf{F}\big(\mathcal{P}_A,\mathrm{Hom}_A(-,-\otimes_A P)\big).\tag{1.3}$$

Note that this can not cause any confusion as the spectrum $\mathbf{F}(\mathfrak{C}, M)$ was before only defined for additive categories (and not for nontrivial rings). We will also consider the associated spectra $\mathbf{F}_q$ for $q \geq 0$ (with the obvious definition using $\mathbf{F}_q$ instead of $\mathbf{F}$).

Lemma 1.3.3.1 *Let $\mathfrak{C}$ be an additive category and let M be a pointed $\mathbf{Z}\mathfrak{C}$-bimodule. Let*

$$\eta\colon F_q(\mathfrak{C}, M) \to \Omega F_q(S\mathfrak{C}, SM)$$

denote the (adjoint of the) structure map. Then the two composites in the non-commutative diagram

$$\begin{array}{ccc} F_q(\mathfrak{C}, M) & \xrightarrow{d_0^q} & F_0(\mathfrak{C}, M) \\ {\scriptstyle\eta}\downarrow & & \downarrow{\scriptstyle s_0^q} \\ \Omega F_q(S\mathfrak{C}, SM) & \xleftarrow{\eta} & F_q(\mathfrak{C}, M) \end{array}$$

are homotopic.

Proof There are three maps $d_0, d_1, d_2\colon F_q(S_2\mathcal{C}, S_2M) \to F_q(\mathcal{C}, M)$ induced by the structure maps $S_2\mathcal{C} \to S_1\mathcal{C} = \mathcal{C}$, see Lemma 1.2.1.6. The two maps

$$\eta d_1 \text{ and } \eta d_0 * \eta d_2\colon F_q(S_2\mathcal{C}, S_2M) \to F_q(\mathcal{C}, M) \to \Omega F_q(S\mathcal{C}, SM)$$

are homotopic, where $\eta d_0 * \eta d_2$ denotes the loop product (remember: the simplicial loop space $\Omega X = \mathcal{S}_*(S^1, \sin|X|)$ is isomorphic to the singular complex of the space of based loops in $|X|$). This is so for general reasons: if X is a reduced simplicial set, then the two maps ηd_1 and $\eta d_0 * \eta d_2$ are homotopic as maps

$$X_2 \to X_1 \xrightarrow{\eta} \Omega X$$

where the latter map is induced by the adjoint of the canonical map $S^1 \wedge X_1 \to X$ (composed with $X \to \sin|X|$).

In the diagrams below we use the following notation:

1. i_1 is the inclusion into the first summand, pr_2 the second projection, Δ the diagonal and $\nabla\colon c \oplus c \to c$ the difference $(a, b) \mapsto a - b$,
2. $\beta_i = \alpha_i \dots \alpha_q\colon c_q \to c_{i-1}$, $\Delta_i = (1 \oplus \beta_i)\Delta$, and $\nabla_i = \nabla(1 \oplus \beta_i)$

(exercise: check that the claimed elements of $S_2M(-,-)$ are well defined). We define two maps

$$E, D\colon F_q(\mathfrak{C}, M) \to F_q(S_2\mathfrak{C}, S_2M)$$

by sending $(\alpha_0, \{\alpha_i\}) = (\alpha_0 \in M(c_0, c_q), \{c_{i-1} \xleftarrow{\alpha_i} c_i\})$ to

$$
\begin{aligned}
&E\big(\alpha_0,\{\alpha_i\}\big)\\
&=\left(\begin{pmatrix} 0 \\ M(pr_2,\Delta)\alpha_0 \\ \alpha_0 \end{pmatrix}\in S_2M\left(\begin{array}{cc} c_q & c_q \\ {\scriptstyle i_1}\big\downarrow & {\scriptstyle i_1}\big\downarrow \\ c_q\oplus c_0, & c_q\oplus c_q \\ {\scriptstyle pr_2}\big\downarrow & {\scriptstyle pr_2}\big\downarrow \\ c_0 & c_q \end{array}\right),\left\{\begin{array}{ccc} c_q & = & c_q \\ {\scriptstyle i_1}\big\downarrow & & {\scriptstyle i_1}\big\downarrow \\ c_q\oplus c_{i-1} & \xleftarrow{1\oplus\alpha_i} & c_q\oplus c_i \\ {\scriptstyle pr_2}\big\downarrow & & {\scriptstyle pr_2}\big\downarrow \\ c_{i-1} & \xleftarrow{\alpha_i} & c_i \end{array}\right\}\right)
\end{aligned}
$$

and

$$
\begin{aligned}
&D\big(\alpha_0,\{\alpha_i\}\big)\\
&=\left(\begin{pmatrix} M(\beta_1,1)\alpha_0 \\ M(pr_2,\Delta)\alpha_0 \\ 0 \end{pmatrix}\in S_2M\left(\begin{array}{cc} c_q & c_q \\ {\scriptstyle \Delta_1}\big\downarrow & {\scriptstyle \Delta}\big\downarrow \\ c_q\oplus c_0, & c_q\oplus c_q \\ {\scriptstyle \nabla_1}\big\downarrow & {\scriptstyle \nabla}\big\downarrow \\ c_0 & c_q \end{array}\right),\left\{\begin{array}{ccc} c_q & = & c_q \\ {\scriptstyle \Delta_i}\big\downarrow & & {\scriptstyle \Delta_{i+1}}\big\downarrow \\ c_q\oplus c_{i-1} & \xleftarrow{1\oplus\alpha_i} & c_q\oplus c_i \\ {\scriptstyle \nabla_i}\big\downarrow & & {\scriptstyle \nabla_{i+1}}\big\downarrow \\ c_{i-1} & \xleftarrow{\alpha_i} & c_i \end{array}\right\}\right)
\end{aligned}
$$

Since $d_2E = d_0D = 0$ we get that

$$\eta = \eta d_0E \simeq \eta d_1E = \eta d_1D \simeq \eta d_2D = \eta s_0^q d_0^q. \qquad \square$$

Corollary 1.3.3.2 *In the situation of the lemma, the inclusion of degeneracies induces a stable equivalence of spectra*

$$\mathbf{F}_0(\mathfrak{C}, M) \xrightarrow{\sim} \mathbf{F}(\mathfrak{C}, M)$$

and in particular, if A is a ring and P an A-bimodule, then

$$\mathbf{F}_0(A, P) \xrightarrow{\sim} \mathbf{F}(A, P).$$

Proof It is enough to show that for every q the map $\mathbf{F}_0(\mathfrak{C}, M) \to \mathbf{F}_q(\mathfrak{C}, M)$ induced by the degeneracy is a stable equivalence (since loops of simplicial connected spaces may be performed in each degree, see Corollary A.6.0.4, and since a degreewise equivalence of simplicial spaces induces an equivalence on the diagonal, see Theorem A.6.0.1). In other words, we must show that for every q and k

$$\pi_0 \lim_{m\to\infty} \Omega^{m+k} F_0\big(S^{(m)}\mathfrak{C}, S^{(m)}M\big) \xrightarrow{s_0^q} \pi_0 \lim_{m\to\infty} \Omega^{m+k} F_q\big(S^{(m)}\mathfrak{C}, S^{(m)}M\big)$$

is an isomorphism. It is a split injection by definition, and a surjection by Lemma 1.3.3.1. $\square$

1.3.4 K-Theory as a Theory of Bimodules

Let A be a ring and let $A \ltimes P \to A$ be any split radical extension. Recall the $\tilde{\mathbf{K}}_A$ construction of Definition 1.2.5.5. The last part of Theorem 1.2.5.4 says that

$$\mathbf{K}(A \ltimes P)/\mathbf{K}(A) \simeq \tilde{\mathbf{K}}_A(A \ltimes P) = \left\{ n \mapsto \bigvee_{m \in obS^{(n)}\mathcal{P}_A} B\big(S^{(n)}\mathcal{M}_A(m, m \otimes_A P)\big)\right\}.$$

Notice the striking similarity with

$$\mathbf{F}_0(\mathcal{P}_A, M) = \left\{ n \mapsto \bigoplus_{m \in obS^{(n)}\mathcal{P}_A} M(m,m)\right\}.$$

In the special case where $P^2 = 0$ the group structure on $\mathrm{Hom}_A(c, c \otimes_A P)$ for $c \in S_q^{(n)}\mathcal{P}_A$ is just the summation of maps: let $f, g \in \mathrm{Hom}_A(c, c \otimes_A P)$, then $f \cdot g = (1+f)(1+g) - 1 = f + g + f \circ g$, where $f \circ g$ is the composite

$$c \xrightarrow{\ g\ } c \otimes_A P \xrightarrow{\ f \otimes 1\ } c \otimes_A P \otimes_A P \to c \otimes_A P$$

where the last map is induced by the multiplication in $P \subseteq A \ltimes P$, which is trivial. So $f \cdot g = f + g$. This means that the natural isomorphism

$$\begin{aligned} B_q \mathrm{Hom}_A(c, c \otimes_A P) = \mathrm{Hom}_A(c, c \otimes_A P)^{\times q} &\cong \mathrm{Hom}_A\big(c, (c \otimes_A P)^{\times q}\big) \\ &\cong \mathrm{Hom}_A\big(c, c \otimes_A P^{\times q}\big) = \mathrm{Hom}_A(c, c \otimes_A B_q P) \end{aligned}$$

induces a simplicial isomorphism. Hence

$$M = B\big(S^{(n)}\mathcal{M}_A(-, - \otimes_A P)\big) \cong S^{(n)}\mathcal{M}_A(-, - \otimes_A BP)$$

is a (simplicial) $\mathcal{P}_A$-bimodule, and the only difference between $\tilde{\mathbf{K}}_A(A \ltimes P)$ and $\mathbf{F}_0(\mathcal{P}_A, M)$ is that the first is built up of wedge summands, whereas the second is built up of direct sums.

Here stable homotopy enters. Recall that a space X is 0-*connected* (or just *connected*, since in our simplicial setting there is no danger of confusion between notions like connected and path connected) if $\pi_0 X$ is a point, and if X is connected it is k-*connected* for a $k > 0$ if for all vertices $x \in X_0$ we have that $\pi_q(X, x) = 0$ for $0 < q \leq k$. A space is -1-connected by definition if it is nonempty. A map $X \to Y$ is k-*connected* if its homotopy fiber is $(k-1)$-connected. We use the same convention for simplicial rings and modules.

Note 1.3.4.1 If A is a ring and P is a simplicial A-bimodule, we let $A \ltimes P$ be the simplicial ring with q-vertices $A \ltimes P_q$ (the square zero extension). When we write $\tilde{\mathbf{K}}_A(A \ltimes P)$ in the statement below, we mean simply $\{[q] \mapsto \tilde{\mathbf{K}}_A(A \ltimes P_q)\}$ (or the associated diagonal)—the "degreewise K-theory". Taking K-theory degreewise in this sense is quite rarely the right thing to do, but in the case of radical extensions we will see in Sect. 3.1.4 that it agrees with the more common definitions the reader will find elsewhere, both in the literature and in Sect. 3.1. Likewise $\mathbf{F}_0(A, BP) = \{[q] \mapsto \mathbf{F}_0(A, B_q P_q) = \mathbf{F}_0(A, P_q^{\times q})\}$.

The difference between wedge and direct sum vanishes stably, which accounts for

Theorem 1.3.4.2 *Let A be a ring and P an m-connected simplicial A-bimodule, the inclusion $\bigvee \subseteq \bigoplus$ induces a $2m+2$-connected map*

$$\tilde{\mathbf{K}}_A(A \ltimes P) \to \mathbf{F}_0(A, BP) \cong B\mathbf{F}_0(A, P).$$

Proof Corollary A.8.2.4 says that if X is n-connected and Y is m-connected, then the inclusion $X \vee Y \to X \times Y$ is $m+n$-connected, and so the same goes for finitely many factors. Now, finite sums of modules are the same as products of underlying sets, and infinite sums are filtered colimits of the finite sub-sums. Since the functors in question commute with filtered colimits, and the filtered colimit of k-connected maps is k-connected, the result follows. □

1.3.4.1 Removing the Bar

What is the rôle of the bar construction in Theorem 1.3.4.2? Removing it on the K-theory side, that is in $\mathbf{K}_A(A \ltimes P)$, we are invited to look at

$$\left\{n \mapsto \coprod_{c \in obS^{(n)}\mathcal{P}_A} \mathrm{Hom}_A(c, c \otimes_A P)\right\}. \tag{1.4}$$

We identify this as follows. Let $\mathcal{E}_A P$ be the exact category with objects pairs (c, f) with $c \in ob\mathcal{P}_A$ and $f \in \mathrm{Hom}_A(c, c \otimes_A P)$, and morphisms $(c, f) \to (d, g)$ commutative diagrams of A-modules

$$\begin{array}{ccc} c & \xrightarrow{h} & d \\ {\scriptstyle f}\downarrow & & \downarrow{\scriptstyle g} \\ c \otimes_A P & \xrightarrow{h \otimes 1} & d \otimes_A P \end{array}$$

We have a functor $\mathcal{E}_A P \to \mathcal{P}_A$ given by $(c, f) \mapsto c$, and a sequence in $\mathcal{E}_A P$ is exact if it is sent to an exact sequence in $\mathcal{P}_A$. As examples we have that $\mathcal{E}_A 0$ is isomorphic to $\mathcal{P}_A$, and $\mathcal{E}_A A$ is what is usually called the category of endomorphisms on $\mathcal{P}_A$. We see that the expression (1.4) is just the K-theory spectrum $\mathbf{K}(\mathcal{E}_A P) = \{x \mapsto obS^{(n)}\mathcal{E}_A P\}$.

Definition 1.3.4.3 Let A be a unital ring. Set C_A to be the functor from the category of A-bimodules to the category of spectra given by

$$\mathsf{C}_A(P)=\mathbf{K}(\mathcal{E}_A P)/\mathbf{K}(A)=\left\{n\mapsto \bigvee_{c\in S^{(n)}\mathcal{P}_A} S^{(n)}\mathcal{M}_A(c,c\otimes_A P)\right\}$$

(the homomorphism groups $S^{(n)}\mathcal{M}_A(c,c\otimes_A P)$ are pointed in the zero map).

With this definition we can restate Theorem 1.2.5.4 for the square zero case as

$$\mathsf{C}_A(BP)\simeq \operatorname{hofib}\left\{\mathbf{K}(A\ltimes P)\to\mathbf{K}(A)\right\}.$$

Note that, in the language of Definition 1.2.5.3, yet another way of writing $\mathsf{C}_A P$ is as the spectrum $\{n\mapsto N_0^{cy}t\mathcal{D}_A^n(A\ltimes P)/N_0^{cy}t\mathcal{D}_A^n(A)=N_0^{cy}t\mathcal{D}_A^n(A\ltimes P)/obS^{(n)}\mathcal{P}_A\}$.

We are free to introduce yet another spectrum direction in $\mathsf{C}_A P$ by observing that we have natural maps $S^1\wedge\mathsf{C}_A P\to\mathsf{C}_A(BP)$ given by

$$S^1\wedge\bigvee M\cong\bigvee(S^1\wedge M)\to\bigvee(\tilde{\mathbf{Z}}[S^1]\otimes M)\cong\bigvee BM.$$

Here $\tilde{\mathbf{Z}}$ is the left adjoint of the forgetful functor from abelian groups to pointed sets, $\tilde{\mathbf{Z}}[X]=\mathbf{Z}[X]/\mathbf{Z}[*]$, see Sect. A.3.1, which has the property that $\tilde{\mathbf{Z}}[X\wedge Y]\cong\tilde{\mathbf{Z}}[X]\otimes\tilde{\mathbf{Z}}[Y]$, and we get a canonical isomorphism between the classifying space BM of an abelian group M and $\tilde{\mathbf{Z}}[S^1]\otimes M$, see Sect. A.3.1.2.

Aside 1.3.4.4 There are **two** natural maps $\mathbf{K}(A)\to\mathbf{K}(\mathcal{E}_A A)$, given by sending $c\in obS^{(n)}\mathcal{P}_A$ to either $(c,0)$ or $(c,1)$ in $obS^{(n)}\mathcal{E}_A A$. The first is used when forming $\mathsf{C}_A P$, and the latter gives rise to a map

$$\mathbf{K}(A)\to\mathsf{C}_A A.$$

Composing this with $\mathsf{C}_A A\to\Omega\mathsf{C}_A(BA)=\Omega\tilde{\mathbf{K}}_A(A[t]/t^2)\to\Omega\mathbf{K}(A[t]/t^2)/\mathbf{K}(A)$, we get a weak map (i.e., a chain of maps where the arrows pointing the wrong way are weak equivalences)

$$\mathbf{K}(A)\to\Omega\mathbf{K}\big(A[t]/t^2\big)/\mathbf{K}(A)\xleftarrow{\sim}\operatorname{hofib}\left\{\mathbf{K}\big(A[t]/t^2\big)\to\mathbf{K}(A)\right\}$$

(cf. [160] or [299]) where *hofib* is (a functorial choice representing) the homotopy fiber.

The considerations above are related to the results of Grayson in [109]. Let A be commutative and $R=S^{-1}A[t]$ where $S=1+tA[t]$. The theorem above says that $\tilde{\mathbf{K}}_A(A[t]/t^2)=\mathbf{K}(\mathcal{E}_A(BA))/\mathbf{K}(A)$ is equivalent to $\mathbf{K}(A[t]/t^2)/\mathbf{K}(A)$, whereas Grayson's theorem tells us that the "one-simplices" of this, i.e., $\mathsf{C}_A A=\mathbf{K}(\mathcal{E}_A A)/\mathbf{K}(A)$ is equivalent to the loop of $\tilde{\mathbf{K}}_A(R)\simeq\mathbf{K}(R)/\mathbf{K}(A)$.

1.3.4.2 More General Bimodules

Before we go on to reformulate Theorem 1.3.4.2 in the more fashionable form "stable K-theory is homology" we will allow our K-functor more general bimodules so that we have symmetry in the input.

Definition 1.3.4.5 Let $\mathfrak{C}$ be an exact category and M a pointed $\mathbf{Z}\mathfrak{C}$-bimodule. Then we define the spectrum

$$\mathsf{C}_{\mathfrak{C}}(M) = \left\{ n \mapsto \bigvee_{c \in obS^{(n)}\mathfrak{C}} S^{(n)} M(c,c) \right\}.$$

The structure maps

$$S^1 \wedge \bigvee_{c \in ob\mathfrak{C}} M(c,c) \to \bigvee_{c \in obS\mathfrak{C}} SM(c,c)$$

are well defined, because $\bigvee_{c \in obS_0\mathfrak{C}} S_0 M(c,c) = M(0,0) = 0$ since we have demanded that M is pointed.

The notation should not cause confusion, although $\mathsf{C}_A P = \mathsf{C}_{\mathcal{P}_A} \mathrm{Hom}_A(-, c \otimes_A P)$, since the ring A is not an exact category (except when $A = 0$, and then it doesn't matter).

If M is bilinear, this is the K-theory spectrum of the following category, which we will call $\mathcal{E}_{\mathfrak{C}}(M)$. The objects are pairs (c, f) with $c \in ob\mathfrak{C}$ and $f \in M(c,c)$ and a morphism from (c,m) to (c',m') is an $f \in \mathfrak{C}(c,c')$ such that $M(f,1)m' = M(1,f)m$. A sequence $(c',m') \to (c,m) \to (c'',m'')$ is exact if the underlying sequence $c' \to c \to c''$ is exact.

1.3.5 Stable K-Theory

Recall from Sect. 1.3.1 that the functor $P \mapsto \tilde{\mathbf{K}}(A \ltimes P)$ is not additive when considered as a functor from A-bimodules. If F is a pointed (simplicial) functor from A-bimodules to spectra, we define its *first differential*, $D_1 F$, by

$$D_1 F(P) = \lim_{\overrightarrow{k}} \Omega^k F\big(B^k P\big),$$

where F is applied in each degree to the k-fold bar construction. We have a transformation $F \to D_1 F$. If F already were additive, then $F \to D_1 F$ would be a weak equivalence. This means that $D_1 F$ is initial (in the homotopy category) among additive functors under F, and is a left adjoint (in the homotopy categories) to the inclusion of the additive functors into all functors from A-bimodules to spectra.

Definition 1.3.5.1 Let A be a simplicial ring and P an A-bimodule. Then

$$\mathbf{K}^S(A,P)=D_1\mathsf{C}_A(P)=\varinjlim_k \Omega^k\mathsf{C}_A\big(B^kP\big).$$

If $\mathfrak{C}$ is an exact category and M a $\mathfrak{C}$-bimodule, then

$$\mathbf{K}^S(\mathfrak{C},M)=D_1\mathsf{C}_{\mathfrak{C}}(M)=\varinjlim_k \Omega^k\mathsf{C}_{\mathfrak{C}}\big(M\otimes S^k\big),$$

where for a finite pointed set X, $M\otimes X$ is the bimodule sending c,d to $M(c,d)\otimes \tilde{\mathbf{Z}}X$.

Again, the isomorphism $\mathbf{K}^S(A,P)\cong\mathbf{K}^S(\mathcal{P}_A,\mathrm{Hom}_A(-,-\otimes_A P))$ should cause no confusion. If M is a pointed simplicial $\mathfrak{C}$-bimodule, we apply $\mathsf{C}_{\mathfrak{C}}$ degreewise.

We note that there is a chain of equivalences

$$\begin{array}{c}
\mathbf{K}^S(A,P)=D_1\Omega\mathsf{C}_A(BP)\\
\simeq\big\downarrow\\
D_1\Omega\big(\mathbf{K}(\mathcal{P}_{A\ltimes P},i)/\mathbf{K}(A)\big)\\
\big\uparrow\simeq\\
D_1\Omega\big(\mathbf{K}(A\ltimes P)/\mathbf{K}(A)\big)\\
\simeq\big\downarrow\\
D_1\Omega\mathrm{hofib}\big\{\mathbf{K}(A\ltimes P)\to\mathbf{K}(A)\big\}=\operatorname{holim}_{\overrightarrow{k}}\Omega^k\mathrm{hofib}\big\{\mathbf{K}\big(A\ltimes B^{k-1}P\big)\to\mathbf{K}(A)\big\}
\end{array}$$

and the target spectrum is the (spectrum version of the) usual definition of *stable K-theory*, cf. [160] and [299].

In the rational case Goodwillie proved in [101] that stable K-theory is equivalent to Hochschild homology (see later). In general this is not true, and we now turn to the necessary modification.

Theorem 1.3.5.2 *Let $\mathfrak{C}$ be an exact category and M an m-connected pointed simplicial $\mathfrak{C}$-bimodule. The inclusion $\bigvee\subseteq\bigoplus$ induces a $2m$-connected map*

$$\mathsf{C}_{\mathfrak{C}}M\to\mathbf{F}_0(\mathfrak{C},M)$$

and

$$D_1\mathsf{C}_{\mathfrak{C}}\xrightarrow{\simeq}D_1\mathbf{F}_0(\mathfrak{C},-)\xleftarrow{\simeq}\mathbf{F}_0(\mathfrak{C},-)$$

are equivalences. Hence

$$\mathbf{K}^S(\mathfrak{C},M)\simeq\mathbf{F}_0(\mathfrak{C},M)\xrightarrow{\sim}\mathbf{F}(\mathfrak{C},M).$$

In particular, for A a ring and P an A-bimodule, the map $\mathsf{C}_A P \to \mathbf{F}_0(A, P)$ gives rise to natural equivalences

$$\mathbf{K}^S(A, P) = D_1\mathsf{C}_A \xrightarrow{\simeq} D_1\mathbf{F}_0(A, -) \xleftarrow{\simeq} \mathbf{F}_0(A, -) \xrightarrow{\simeq} \mathbf{F}(A, P).$$

Proof The equivalence

$$D_1\mathbf{F}_0(\mathfrak{C}, -) \xleftarrow{\simeq} \mathbf{F}_0(\mathfrak{C}, -)$$

follows since by Corollary 1.3.3.2 the inclusion by the degeneracies $\mathbf{F}_0(\mathfrak{C}, -) \to \mathbf{F}(\mathfrak{C}, -)$ is an equivalence, and the fact that $\mathbf{F}(\mathfrak{C}, -)$ is additive, and so unaffected by the differential. The rest of the argument follows as before. □

Adding up the results, we get the announced theorem:

Corollary 1.3.5.3 *Let $\mathfrak{C}$ be an additive category, and M a bilinear $\mathfrak{C}$ bimodule. Then we have natural isomorphisms*

$$\pi_*\mathbf{K}^S(\mathfrak{C}, M) \cong H_*(\mathfrak{C}, M)$$

and in particular

$$\pi_*\mathbf{K}^S(A, P) \cong H_*\big(\mathcal{P}_A, \mathcal{M}_A(-, - \otimes_A P)\big).$$

Proof The calculations of homotopy groups follows from the fact that $\mathbf{F}(\mathfrak{C}, M)$ is an Ω-spectrum (and so $\pi_*\mathbf{F}(\mathfrak{C}, M) \cong \pi_* F(\mathfrak{C}, M) = H_*(\mathfrak{C}, M)$). This follows from the equivalence

$$F(\mathfrak{C}, M) \sim THH(\mathfrak{C}, M)$$

and results on topological Hochschild homology in Chap. 4. However, for the readers who do not plan to cover this material, we provide a proof showing that $\mathbf{F}$ is an Ω-spectrum directly without use of stabilizations at the end of this section, see Proposition 1.3.6.5. □

1.3.6 A Direct Proof of "F is an Ω-Spectrum"

Much of what is to follow makes sense in an *Ab*-category setting. For convenience, we work in the setting of additive categories, and we choose zero objects which we always denote 0.

Definition 1.3.6.1 Let $G\colon \mathcal{A} \to \mathcal{B}$ be an additive functor. We let the *twisted product category* $\mathcal{A} \times_G \mathcal{B}$ be the linear category with objects $\mathrm{ob}(\mathcal{A}) \times \mathrm{ob}(\mathcal{B})$ and

$$\mathcal{A} \times_G \mathcal{B}\big((a, b), (a', b')\big) = \mathcal{A}(a, a') \oplus \mathcal{B}(b, b') \oplus \mathcal{B}\big(G(a), b'\big)$$

with composition given by

$$(f,g,h)\circ(f',g',h') = (f\circ f', g\circ g', h\circ G(f') + g\circ h').$$

If M is an $\mathcal{A}$-bimodule and N is a $\mathcal{B}$-bimodule, with an $\mathcal{A}$-bimodule map $G^*\colon M \to G^*N$ we define the $\mathcal{A}\times_G \mathcal{B}$-bimodule $M\times_G N$ by

$$M\times_G N((a,b),(a',b')) = M(a,a')\oplus N(b,b')\oplus N(G(a),b')$$

with bimodule action defined by

$$\begin{aligned}(M\times_G N)&((f,g,h),(f',g',h'))(m,n,n_G)\\ &= (M(f,f')m, N(g,g'), N(Gf,h')G^*m + N(h,g')n + N(Gf,g')n_G).\end{aligned}$$

From now on, we assume for convenience that M and N are pointed (i.e., take zero in either coordinate to zero). Note the following structure.

1. An inclusion
$$\mathcal{A} \xrightarrow{in_{\mathcal{A}}} \mathcal{A}\times_G \mathcal{B}$$
sending $f\colon a\to a'$ $ob\mathcal{A}$ to $(f,0,0)\colon (a,0)\to(a',0)$,
2. an $\mathcal{A}$-bimodule map $M\to in_{\mathcal{A}}^*(M\times_G N)$,
3. a projection
$$\mathcal{A}\times_G \mathcal{B} \xrightarrow{pr_{\mathcal{A}}} \mathcal{A},$$
4. and an $\mathcal{A}\times_G\mathcal{B}$-bimodule map $M\times_G N\to pr_{\mathcal{A}}^*M$.

Likewise for $\mathcal{B}$. The composite

$$\begin{aligned}F(\mathcal{A},M)\oplus F(\mathcal{B},N) &\xrightarrow{in_{\mathcal{A}}+in_{\mathcal{B}}} F(\mathcal{A}\times_G\mathcal{B}, M\times_G N)\\ &\xrightarrow{(pr_{\mathcal{A}}\oplus pr_{\mathcal{B}})\Delta} F(\mathcal{A},M)\oplus F(\mathcal{B},N)\end{aligned}$$

is the identity.

Lemma 1.3.6.2 (F is "Additive") *With the notation as above*

$$F(\mathcal{A},M)\oplus F(\mathcal{B},N) \xrightarrow{in_{\mathcal{A}}+in_{\mathcal{B}}} F(\mathcal{A}\times_G\mathcal{B}, M\times_G N)$$

is an equivalence.

Proof We will show that the "other" composite

$$\begin{aligned}F(\mathcal{A}\times_G\mathcal{B}, M\times_G N) &\xrightarrow{(pr_{\mathcal{A}}\oplus pr_{\mathcal{B}})\Delta} F(\mathcal{A},M)\oplus F(\mathcal{B},N)\\ &\xrightarrow{in_{\mathcal{A}}+in_{\mathcal{B}}} F(\mathcal{A}\times_G\mathcal{B}, M\times_G N)\end{aligned}$$

is homotopic to the identity. Let $\mathbf{x} = (x_0; x_1, \dots, x_q) \in F_q(\mathcal{A} \times_G \mathcal{B}, M \times_G N)$, where

$$x_0 = (m, n, n_G) \in M \times_G N\big((a_0, b_0), (a_q, b_q)\big), \quad \text{and}$$

$$x_i = (f_i, g_i, h_i) \in \mathcal{A} \times_G \mathcal{B}\big((a_i, b_i), (a_{i-1}, b_{i-1})\big), \quad \text{for } i > 0.$$

Then x is sent to

$$\begin{aligned} J(x) = &\big((m, 0, 0); in_{\mathcal{A}} pr_{\mathcal{A}} x_1, \dots, in_{\mathcal{A}} pr_{\mathcal{A}} x_q\big) \\ &+ \big((0, n, 0); in_{\mathcal{B}} pr_{\mathcal{B}} x_1, \dots, in_{\mathcal{B}} pr_{\mathcal{B}} x_q\big). \end{aligned}$$

We define a homotopy between the identity and J as follows. Let $x_i^1 = (f_i, 0, 0) \in (\mathcal{A} \times_G \mathcal{B})((a_i, b_i), (a_{i-1}, 0))$ and $x_i^2 = (0, g_i, 0) \in (\mathcal{A} \times_G \mathcal{B})((0, b_i), (a_{i-1}, b_{i-1}))$. If $\phi_i \in \Delta([q], [1])$ is the map with inverse image of 0 of cardinality i, we define

$$H \colon F(\mathcal{A} \times_G \mathcal{B}, M \times_G N) \times \Delta \to F(\mathcal{A} \times_G \mathcal{B}, M \times_G N)$$

by the formula

$$\begin{aligned} H(\mathbf{x}, \phi_i) = &\big((m, 0, 0); in_{\mathcal{A}} pr_{\mathcal{A}} x_1, \dots, in_{\mathcal{A}} pr_{\mathcal{A}} x_{i-1}, x_i^1, x_{i+1}, \dots, x_q\big) \\ &- \big((0, n, n_G); x_1, \dots, x_{i-1}, x_i^2, in_{\mathcal{B}} pr_{\mathcal{B}} x_{i+1}, \dots, in_{\mathcal{B}} pr_{\mathcal{B}} x_q\big) \\ &+ \big((0, n, 0); in_{\mathcal{B}} pr_{\mathcal{B}} x_1, \dots, in_{\mathcal{B}} pr_{\mathcal{B}} x_q\big) \\ &+ \big((0, n, n_G); x_1, \dots, x_q\big) \end{aligned}$$

(note that in the negative summand, it is implicit that n_G is taken away when $i = 0$). □

Lemma 1.3.6.3 *Let $\mathfrak{C}$ be an additive category and M a bilinear bimodule. Then the natural map*

$$S_q\mathfrak{C} \xrightarrow{c \mapsto (c_{0,1}, \dots, c_{q-1,q})} \mathfrak{C}^{\times q}$$

induces an equivalence

$$F(S_q\mathfrak{C}, S_q M) \xrightarrow{\sim} F(\mathfrak{C}^{\times q}, M^{\times q}) \xrightarrow{\sim} F(\mathfrak{C}, M)^{\times q}$$

Proof Recall the equivalence $\psi_q \colon T_q\mathfrak{C} \to S_q\mathfrak{C}$ of Lemma 1.2.2.4, and note that if $G_q \colon \mathfrak{C} \to T_q\mathfrak{C}$ is defined by $c \mapsto G_q(c) = (0, \dots, 0, c)$, then we have an isomorphism $T_{q+1}\mathfrak{C} \cong \mathfrak{C} \times_{G_q} T_q\mathfrak{C}$. Furthermore, if M is a linear bimodule, then we define $T_q M = \psi_q^* S_q M$, and we have that $T_{q+1} M \cong M \times_G T_q M$.

Hence

$$\begin{aligned} F(S_q\mathfrak{C}, S_q M) &\xleftarrow{\sim} F(T_q\mathfrak{C}, T_q M) \cong F(\mathfrak{C} \times_G T_{q-1}\mathfrak{C}, M \times_G T_{q-1} M) \\ &\xleftarrow{\sim} F(\mathfrak{C}, M) \oplus F(T_{q-1}\mathfrak{C}, T_{q-1} M) \end{aligned}$$

and by induction we get that

$$F(S_q\mathfrak{C}, S_qM) \xleftarrow{\sim} F(\mathfrak{C}, M)^{\times q}$$

and this map is a right inverse to the map in the statement. □

Definition 1.3.6.4 (cf. Sect. A.2.7) For any simplicial category $\mathcal{D}$ we may define the path category $P\mathcal{D}$ by setting $P_q\mathcal{D} = \mathcal{D}_{q+1}$ and letting the face and degeneracy functors be given by raising all indices by one. The unused d_0 defines a functor $P\mathcal{D} \to \mathcal{D}$, and we have a map $\mathcal{D}_1 = P_0\mathcal{D} \to P\mathcal{D}$ given by the degeneracies.

Then $\mathcal{D}_0 \to P\mathcal{D}$ (given by degeneracies in $\mathcal{D}$) defines a simplicial homotopy equivalence, see Lemma A.2.7.2, with inverse given by $\prod_{1\le i\le q+1} d_i : P_q\mathcal{D} \to \mathcal{D}_0$.

Proposition 1.3.6.5 *Let $\mathfrak{C}$ be an additive category, and M a bilinear bimodule, then*

$$F(\mathfrak{C}, M) \to \Omega F(S\mathfrak{C}, SM)$$

is an equivalences.

Proof Consider

$$F(\mathfrak{C}, M) \to F(PS\mathfrak{C}, PSM) \to F(S\mathfrak{C}, SM). \tag{1.5}$$

For every q we have equivalences

$$\begin{array}{ccccc} F(\mathfrak{C}, M) & \longrightarrow & F(P_qS\mathfrak{C}, P_qSM) & \longrightarrow & F(S_q\mathfrak{C}, S_qM) \\ \| & & \downarrow\wr & & \downarrow\wr \\ F(\mathfrak{C}, M) & \longrightarrow & F(\mathfrak{C}, M)^{\times q+1} & \longrightarrow & F(\mathfrak{C}, M)^{\times q} \end{array}$$

where the lower sequence is the trivial split fibration. As all terms are bisimplicial abelian groups the sequence (1.5) must be a fiber sequence (see Theorem A.6.0.3) where the total space is contractible. □

Chapter 2
Gamma-Spaces and S-Algebras

In this chapter we will introduce the so-called Γ-spaces. The reader can think of these as (very slight) generalizations of (simplicial) abelian groups. The surprising fact is that this minor generalization is big enough to encompass a wide and exotic variety of new examples.

The use of Γ-spaces also fixes another disparity. Quillen defined algebraic K-theory to be a functor from things with abelian group structure (such as rings or exact categories) to abelian groups. We have taken the view that K-theory takes values in spectra, and although spectra are almost as good as abelian groups, this is somehow unsatisfactory. The introduction of Γ-spaces evens this out, in that K-theory now takes things with a Γ-space structure (such as **S**-algebras, or the Γ-space analog of exact categories) to Γ-spaces.

Furthermore, this generalization enables us to include new fields of study, such as the K-theory of spaces, into serious consideration. It is also an aid—almost a prerequisite—when trying to understand the theories to be introduced in later chapters.

To be quite honest, Γ-spaces should not be thought of as a generalization of simplicial abelian groups, but rather of simplicial abelian (symmetric) monoids, since there need not be anything resembling inverses in the setting we use (as opposed to Segal's original approach). On the other hand, it is very easy to "group complete": it is a stabilization process.

The reader should be aware that Γ-spaces give us just one of several solutions to an old and important problem in stable homotopy theory. After having been put on sound foundations by Boardman in the mid 1960's (see [295] or [4, III]), the smash product played a central rôle in stable homotopy theory for decades, but until the 1990's one only knew the construction in the "stable homotopy category", and did not know how to realize the smash products in any category of spectra without inverting the stable equivalences.

Several solutions to this problem came more or less at the same time. In the summer of 1993 Elmendorf, Kriz and May's built the categorical foundations for a point

B.I. Dundas et al., *The Local Structure of Algebraic K-Theory*,
Algebra and Applications 18,
DOI 10.1007/978-1-4471-4393-2_2, © Springer-Verlag London 2013

set level smash product, influenced by an observation of Hopkins [138] on coequalizers, making their original intended approach using "monadic bar constructions" more or less obsolete. The team was later joined by Mandell and the construction underwent some changes (for instance, some problems with the unit were solved) before it appeared in the book [80]. Around the same time, Jeff Smith gave talks where he offered another solution which he called "symmetric spectra". Together with Hovey and Shipley he documented this approach in [141] (see also the unpublished notes on symmetric spectra by Schwede [251]).

The Γ-space approach to the problem of having a point set level construction of the smash product appeared in 1999 [188] in a paper by Lydakis, and has the advantage of being by far the simplest, but the disadvantage of only giving connective spectra. The solution is simple, and the techniques were well known in the 1970's, and the authors have come to understand that the construction of the smash product in Γ-spaces was known, but dismissed as "much too simple" to have the right homotopy properties, see also Sect. 2.1.2.3 below.

Since then many variants have been introduced (most notably, orthogonal spectra) and there has been some reconciliation between the different setups (see in particular [196]). In retrospect, it turns out that Bökstedt in his investigations [30] in the 1980's on topological Hochschild homology had struck upon the smash product for simplicial functors [187] in the sense that he gave the correct definition for what it means for a simplicial functor to be an algebra over the sphere spectrum, see also Gunnarsson's preprint [117]. Bökstedt called what was to become **S**-algebras in simplicial functors (with some connectivity hypotheses) "FSP", short for "functors with smash products". Also, orthogonal ring spectra had appeared in [200], although not recognized as monoids in a monoidal structure.

The Γ-spaces have one serious shortcoming, and that is that they do not model strictly commutative ring spectra in the same manner as their competitors (see Lawson [170]). Although this mars the otherwise beautiful structure, it will not affect anything of what we will be doing, and we use Γ-spaces because of their superior concreteness and simplicity.

2.1 Algebraic Structure

2.1.1 Γ-Objects

A Γ-object is a functor from the category of finite sets. We need to be quite precise about this, and the details follow.

2.1.1.1 The Category Γ^o

Roughly, Γ^o is the category of pointed finite sets—a fundamental building block for much of mathematics. To be more precise, we choose a skeleton, and let Γ^o be the category with one object $k_+ = \{0, 1, \ldots, k\}$ for every natural number k, and with morphism sets $\Gamma^o(m_+, n_+)$ the set of functions $f\colon \{0, 1, \ldots, m\} \to \{0, 1, \ldots, n\}$ such that $f(0) = 0$. The notation k_+ is chosen to distinguish it from the *ordered* set $[k] = \{0 < 1 < \cdots < k\}$.

In [257] Segal considered the opposite category and called it Γ, and this accounts for the awkward situation where we call the most fundamental object in mathematics the opposite of something. Some people object to this so strongly that they write Γ when Segal writes Γ^o. We follow Segal's convention.

2.1.1.2 Motivation

A *symmetric monoid* is a set M together with a multiplication and a unit element so that any two maps $M^{\times j} \to M$ obtained by composing maps in the diagram

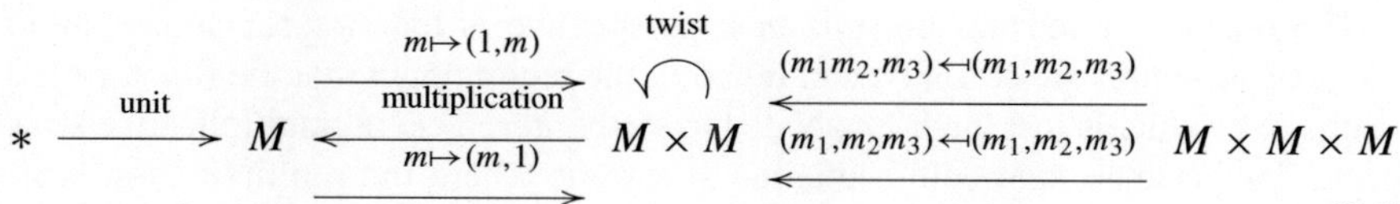

(2.1)

are equal. Thinking of multiplication as "two things coming together" as in the map $2_+ \to 1_+$ given by

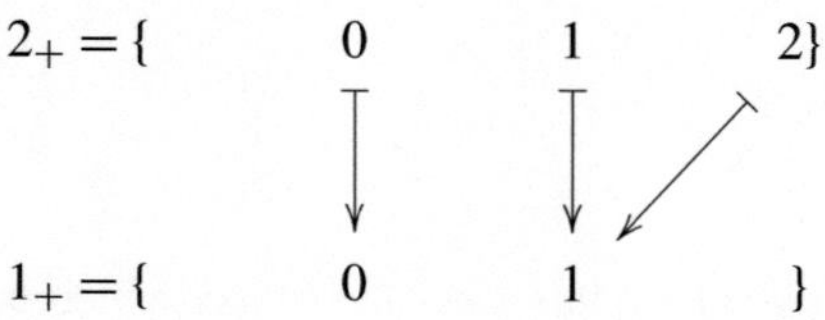

we see that the diagram (2.1) is mirrored by the diagram

$$0_+ \longrightarrow 1_+ \rightleftarrows 2_+ \leftleftarrows 3_+$$

in Γ^o, where the two arrows $1_+ \to 2_+$ are given by

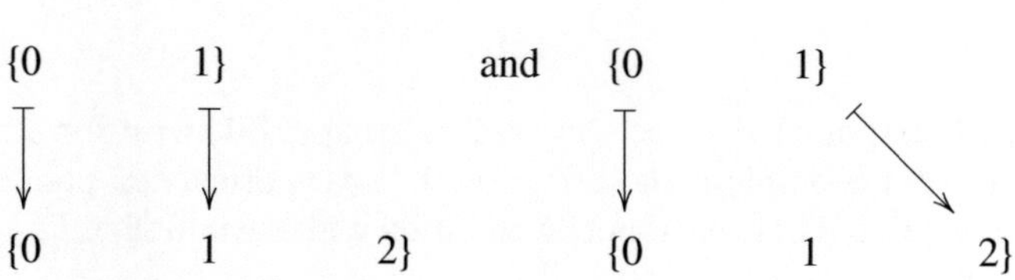

and the maps $3_+ \to 2_+$ are

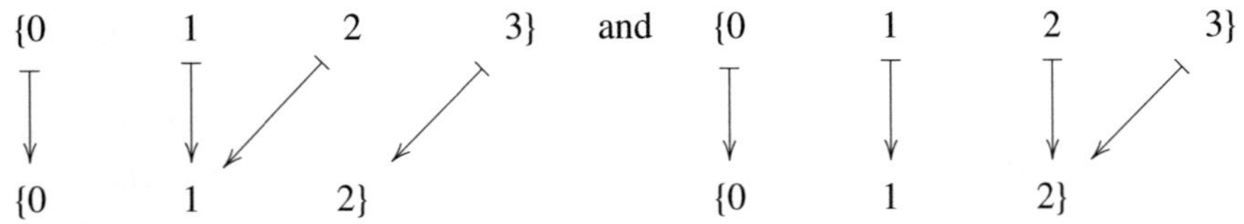

(there are more maps in Γ^o, but these suffice for the moment). So we could say that a symmetric monoid is a functor from this part of Γ^o to pointed sets sending 0_+ to the one-point set and sending wedge sum to product (e.g., $3_+ = 2_+ \vee 1_+$ must be sent to the product of the values at 2_+ and 1_+, i.e., the triple product of the value at 1_+).

This doesn't seem very helpful until one notices that this extends to all of Γ^o, and the requirement of sending 0_+ to the one-point set and wedge sum to product fixes the behavior in the sense that there is a one-to-one correspondence between such functors from Γ^o to sets and symmetric monoids; see Example 2.1.2.1(1) below for more details.

The reason for introducing this new perspective is that we can model multiplicative structures functorially, and relaxing the requirement that the functor sends wedge to product is just the trick needed to study more general multiplicative structures. For instance, one could imagine situations where the multiplication is not naturally defined on $M \times M$, but on some bigger space like $M \times M \times X$, giving an entire family of multiplications varying over the space X. This is exactly what we need when we are going to study objects that are, say, commutative only up to homotopy. Variants of this idea are Lawvere's algebraic theories, operads and multicategories.

2.1.1.3 Γ-Objects

If $\mathcal{C}$ is a pointed category (i.e., it has a chosen object which is both initial and final) one may consider pointed functors $\Gamma^o \to \mathcal{C}$ (often called a Γ-*object* in $\mathcal{C}$) and natural transformations between such functors. This defines a category we call $\Gamma\mathcal{C}$. Most notably we have the category

$$\Gamma\mathcal{S}_*$$

of Γ-spaces, that is pointed functors from Γ^o to pointed simplicial sets, or equivalently, of simplicial Γ-objects in the category of pointed sets. If $\mathcal{A} = sAb$ is the category of simplicial abelian groups, we may define

$$\Gamma\mathcal{A},$$

the category of simplicial Γ-objects in abelian groups. Likewise for other module categories. Another example is the category of Γ-categories, i.e., pointed functors from Γ^o to categories. These must not be confused with the notion of $\Gamma\mathcal{S}_*$-categories (see Sect. 2.1.6).

2.1.2 The Category $\Gamma\mathcal{S}_*$ of Γ-Spaces

We start with some examples of Γ-spaces.

Example 2.1.2.1

1. Let M be an abelian group. If we consider M as a mere pointed set, we can not reconstruct the abelian group structure. However, if we consider M as a Γ-pointed set, HM, as follows, there is no loss of structure. Send k_+ to the *set*

$$HM(k_+) = M \otimes \tilde{\mathbf{Z}}[k_+] \cong M^{\times k},$$

where $\tilde{\mathbf{Z}}[k_+]$ is the free abelian group on the pointed set k_+ (and so is the sum of k copies of $\mathbf{Z}$). A function $f \in \Gamma^o(k_+, n_+)$ gives rise to the homomorphism $f_*\colon HM(k_+) \to HM(n_+)$ sending the k-tuple $(m_1, \ldots, m_k) \in M^{\times k}$ to the n-tuple

$$\left(\left(\sum_{j\in f^{-1}(1)} m_j\right), \ldots, \left(\sum_{j\in f^{-1}(n)} m_j\right)\right)$$

(where $m_0 = 0$).

Alternative description: $HM(X) = Ens_*(X, M)$, and if $f\colon X \to Y \in \Gamma^o$, then $f_*\colon HM(X) \to HM(Y)$ sends ϕ to $y \mapsto f_*\phi(y) = \sum_{x\in f^{-1}(y)} \phi(x)$.

In effect, this defines a functor

$$\bar{H}\colon sAb = \mathcal{A} \to \Gamma\mathcal{A},$$

and we follow by the forgetful functor $U\colon \Gamma\mathcal{A} \to \Gamma\mathcal{S}_*$, so that

$$H = U\bar{H}.$$

Both HM and $\bar{H}M$ will be referred to as the *Eilenberg–Mac Lane* objects associated with M. The reason is that, through the functor from Γ-spaces to spectra defined in Definition 2.2.1.8, these Γ-objects naturally give rise to the so-called Eilenberg-Mac Lane spectra Sect. A.3.2.

2. The inclusion $\Gamma^o \subset Ens_* \subset \mathcal{S}_*$ is called in varying sources, $\mathbf{S}$ (for "sphere spectrum"), Id (for "identity"), etc. We will call it $\mathbf{S}$.

Curiously, the Barratt–Priddy–Quillen theorem (see e.g., [257] or [12]) states that $\mathbf{S}$ is "stably equivalent" (defined in Definition 2.2.1.5) to the K-theory of the category Γ^o (in the interpretation of Sect. 2.3).

3. If X is a pointed simplicial set and M is a Γ-space, then $M \wedge X$ is the Γ-space sending $Y \in ob\Gamma^o$ to $M(Y) \wedge X$. Dually, we let $\underline{\mathcal{S}_*}(X, M)$ be the Γ-space

$$Y \mapsto \underline{\mathcal{S}_*}\big(X, M(Y)\big) = \big\{[q] \mapsto \mathcal{S}_*\big(X \wedge \Delta[q]_+, M(Y)\big)\big\}$$

(see Sect. A.2.3 for facts on function spaces). Note that $\Gamma\mathcal{S}_*(M \wedge X, N)$ is naturally isomorphic to $\Gamma\mathcal{S}_*(M, \underline{\mathcal{S}_*}(X, N))$: the former set consists of natural maps $M(Y)\wedge X \to N(Y)$, whereas the latter consists of natural maps $M(Y) \to \underline{\mathcal{S}_*}(X, N(Y))$. The natural isomorphism is then induced by the adjunction between $\wedge$ and the internal mapping space in $\mathcal{S}_*$.

For any simplicial set X, we let $\mathbf{S}[X] = \mathbf{S}\wedge X_+$, and we see that this is a left adjoint to the functor $R\colon \Gamma\mathcal{S}_* \to \mathcal{S}_*$ evaluating at 1_+.

4. For $X \in ob\Gamma^o$, let $\Gamma^X \in ob\Gamma\mathcal{S}_*$ be given by

$$\Gamma^X(Y) = \Gamma^o(X, Y)$$

Note that $\mathbf{S} = \Gamma^{1_+}$.

The notion of Γ-spaces we are working with is slightly more general than Segal's [257]. It is usual to call Segal's Γ-spaces *special*:

Definition 2.1.2.2 A Γ-space M is said to be *special* if the canonical maps

$$M(k_+) \to \prod_k M(1_+)$$

(induced by the k maps $k_+ \to 1_+$ with support a single element) are weak equivalences for all $k_+ \in ob\Gamma$. This induces a symmetric monoid structure on $\pi_0 M(1_+)$ via

$$\pi_0 M(1_+) \times \pi_0 M(1_+) \xleftarrow{\cong} \pi_0 M(2_+) \longrightarrow \pi_0 M(1_+),$$

induced by the function $\phi\colon 2_+ \to 1_+$ with $\phi^{-1}(1) = \{1, 2\}$, and we say that M is *very special* if this is an abelian group structure.

The difference between Γ-spaces and very special Γ-spaces is not really important. Any Γ-space M gives rise to a very special Γ-space, say QM, in one of many functorial ways, such that there is a "stable equivalence" $M \xrightarrow{\sim} QM$ (see Definition 2.2.1.5). However, the larger category of all Γ-spaces is nicer for formal reasons, and the very special Γ-spaces are just nice representatives in each stable homotopy class.

2.1.2.1 The Smash Product

There is a close connection between Γ-spaces and spectra (there is a functor defined in Definition 2.2.1.8 that induces an equivalence on homotopy categories), and so the question of what the smash product of two Γ-spaces should be could be expected to be a complicated issue. M. Lydakis [187, 188] realized that this was not the case: the simplest candidate works just beautifully.

If we have two Γ-spaces M and N, we may consider the "external smash", i.e., the functor $\Gamma^o \times \Gamma^o \to \mathcal{S}_*$ which sends (X, Y) to $M(X)\wedge N(Y)$. The category Γ^o

has its own smash product, and we want some "universal filler" in

$$\begin{array}{ccc} \Gamma^o \times \Gamma^o & \xrightarrow{(X,Y)\mapsto M(X)\wedge N(Y)} & \mathcal{S}_* \\ {\scriptstyle\wedge}\downarrow & & , \\ \Gamma^o & & \end{array}$$

where $\wedge\colon \Gamma^o \times \Gamma^o \to \Gamma^o$ is the smash (sending (X,Y) to $X\wedge Y$). The solutions to these kinds of questions are called "left Kan extensions" [191], and in our case it takes the following form:

Let $Z \in \Gamma^o$ and let $\wedge/Z$ be the over category (cf. Sect. A.2.4.2), i.e., the category whose objects are tuples (X, Y, v) where $(X,Y) \in \Gamma^o \times \Gamma^o$ and $v\colon X\wedge Y \to Z \in \Gamma^o$, and where a morphism $(X,Y,v) \to (X',Y',v')$ is a pair of functions $f\colon X \to X'$ and $g\colon Y \to Y'$ in Γ^o such that $v = v' \circ (f\wedge g)$.

Then the smash product $(M\wedge N)(Z)$ is defined as the colimit of the composite

$$\wedge/Z \xrightarrow{(X,Y,v)\mapsto(X,Y)} \Gamma^o \times \Gamma^o \xrightarrow{(X,Y)\mapsto M(X)\wedge N(Y)} \mathcal{S}_*,$$

that is

$$(M\wedge N)(Z) = \varinjlim_{(X,Y,v)\in\wedge/Z} M(X)\wedge N(Y).$$

In the language of coends, this becomes particularly perceptive:

$$(M\wedge N)(Z) = \int^{(X,Y)} \big(M(X)\wedge N(Y)\big)\wedge\Gamma^o(X\wedge Y, Z)$$

the "weighted average of all the handicrafted smash products $M(X)\wedge N(Y)$"; the weight being the number of functions $X\wedge Y \to Z$.

Remark 2.1.2.3 Note that a map from a smash product $M\wedge M' \to N \in \Gamma\mathcal{S}_*$ is uniquely described by giving a map $M(X)\wedge M'(Y) \to N(X\wedge Y)$ which is natural in $X, Y \in ob\Gamma^o$.

2.1.2.2 The Closed Structure

Theorem 2.1.2.4 below states that the smash product endows the category of Γ-spaces with a structure of a *closed* category (which is short for closed symmetric monoidal category). For a thorough discussion see Definition A.10.1.1, but for now recall that *symmetric monoidal* means that the functor $\wedge\colon \Gamma\mathcal{S}_* \times \Gamma\mathcal{S}_* \to \Gamma\mathcal{S}_*$ is associative, symmetric and unital (**S** is the unit) up to coherent isomorphisms, and that it is *closed* means that in addition there is an "internal morphism object" with reasonable behavior.

The Γ-space of morphisms from M to N is defined by setting

$$\underline{\Gamma\mathcal{S}_*}(M,N) = \big\{(k_+,[q]) \mapsto \underline{\Gamma\mathcal{S}_*}(M,N)(k_+)_q = \Gamma\mathcal{S}_*\big(M\wedge\Delta[q]_+, N(k_+\wedge -)\big)\big\}.$$

Theorem 2.1.2.4 (Lydakis) *With these definitions of smash and morphism object, there are choices of coherency isomorphisms such that* $(\Gamma\mathcal{S}_*, \wedge, \mathbf{S})$ *becomes a closed category* (*cf. Definition* A.10.1.1).

Proof (See [188] for further details) First one uses the definitions to show that there is a natural isomorphism $\underline{\Gamma\mathcal{S}_*}(M\wedge N, P) \cong \underline{\Gamma\mathcal{S}_*}(N, \underline{\Gamma\mathcal{S}_*}(M, P))$ (using the (co)end-descriptions of smash and internal morphism objects this can be written as follows

$$\begin{aligned}
\underline{\Gamma\mathcal{S}_*}(M\wedge N, P)(V) &= \int_Z \underline{\mathcal{S}_*}\Big(\int^{X,Y} M(X)\wedge N(Y)\wedge\Gamma^o(X\wedge Y, Z), P(V\wedge Z)\Big)\\
&\cong \int_{X,Y} \underline{\mathcal{S}_*}\Big(M(X)\wedge N(Y), \int_Z \underline{\mathcal{S}_*}\big(\Gamma^o(X\wedge Y, Z), P(V\wedge Z)\big)\Big)\\
&\cong \int_{X,Y} \underline{\mathcal{S}_*}\big(M(X)\wedge N(Y), P\big(V\wedge(X\wedge Y)\big)\big)\\
&\cong \int_X \underline{\mathcal{S}_*}\Big(N(Y), \int_Y \underline{\mathcal{S}_*}\big(M(X), P\big((V\wedge X)\wedge Y\big)\big)\Big)\\
&= \underline{\Gamma\mathcal{S}_*}\big(N, \underline{\Gamma\mathcal{S}_*}(M, P)\big)(V),
\end{aligned}$$

with $X, Y, Z, V \in ob\Gamma^o$ and where the isomorphisms are induced by the (dual) Yoneda lemma, associativity and the closed structure of $\underline{\mathcal{S}_*}$).

The symmetry $M\wedge N \cong N\wedge M$ follows from the construction of the smash product, and associativity follows by comparing with

$$M\wedge N\wedge P = \Big\{V \mapsto \varinjlim_{X\wedge Y\wedge Z\to V} M(X)\wedge N(Y)\wedge P(Z)\Big\}.$$

Recall from Example 2.1.2.1(4) that $\Gamma^X(Y) = \Gamma^o(X, Y)$ and note that $\mathbf{S} = \Gamma^{1_+}$, $\underline{\Gamma\mathcal{S}_*}(\Gamma^X, M) \cong M(X\wedge -)$ and $\Gamma^X\wedge\Gamma^Y \cong \Gamma^{X\wedge Y}$. We get that $M\wedge\mathbf{S} = M\wedge\Gamma^{1_+} \cong M$ since $\underline{\Gamma\mathcal{S}_*}(M\wedge\mathbf{S}, N) \cong \underline{\Gamma\mathcal{S}_*}(M, \underline{\Gamma\mathcal{S}_*}(\mathbf{S}, N)) \cong \underline{\Gamma\mathcal{S}_*}(M, N)$ for any N.

That all diagrams that must commute actually do so follows from the crucial observation Lemma 2.1.2.5 below (with the obvious definition of the multiple smash product). □

Lemma 2.1.2.5 *Any natural automorphism* ϕ *of expressions of the form*

$$M_1\wedge M_2\wedge\cdots\wedge M_n$$

must be the identity (*i.e.*, $\mathrm{Aut}(\bigwedge^n : \Gamma\mathcal{S}_*{}^{\times n} \to \Gamma\mathcal{S}_*)$ *is the trivial group*).

Proof The analogous statement is true in Γ^o, since any element in $X_1\wedge X_2\wedge\cdots\wedge X_n$ is in the image of a map from $1_+\wedge 1_+\wedge\cdots\wedge 1_+$, and so any natural automorphism must fix this element.

Fixing a dimension, we may assume that the M_i are discrete, and we must show that $\phi(z) = z$ for any $z \in \bigwedge M_i(Z)$. By construction, z is an equivalence class represented, say, by an element $(x_1, \dots, x_m) \in \bigwedge M_i(X_i)$ in the $f\colon \bigwedge X_i \to Z$ summand of the colimit. Represent each $x_i \in M_i(X_i)$ by a map $f_i\colon \Gamma^{X_i} \to M_i$ (so that $f_i(X_i = X_i) = x_i$). Then z is the image of $\wedge id_{X_i}$ in the f-summand of the composite

$$\Bigl(\bigwedge \Gamma^{X_i}\Bigr)(Z) \xrightarrow{\wedge f_i} \Bigl(\bigwedge M_i\Bigr)(Z).$$

Hence it is enough to prove the lemma for $M_i = \Gamma^{X_i}$ for varying X_i. But $\bigwedge \Gamma^{X_i} \cong \Gamma^{\bigwedge X_i}$ and

$$\Gamma\mathcal{S}_*\bigl(\Gamma^{\bigwedge X_i}, \Gamma^{\bigwedge X_i}\bigr) \cong \Gamma^o\Bigl(\bigwedge X_i, \bigwedge X_i\Bigr),$$

and we are done. □

The crucial word in Lemma 2.1.2.5 is "natural". There is just one automorphism of the **functor** $\bigwedge^n\colon \Gamma\mathcal{S}_*{}^{\times n} \to \Gamma\mathcal{S}_*$ whereas there are, of course, nontrivial actions on individual expressions $M_1 \wedge \cdots \wedge M_n$. One should note that the functor in **one** variable $M \mapsto M \wedge \cdots \wedge M$ has full Σ_n-symmetry.

2.1.2.3 Day's Product

Theorem 2.1.2.4 also follows from a much more general theorem of Day [59], not relying on the special situation in Lemma 2.1.2.5.

In hindsight it may appear as a mystery that the smash product took so long to appear on the stage, given that the problem was well publicized and the technical construction had been known since 1970. Rainer Vogt had considered this briefly, and commented in an email in 2009: "I did not know of Day's product but discovered it myself (later than Day in the 80's). Then Roland [Schwänzl] and I thought a little about it. Since we considered special Γ-spaces only and the product did not preserve those we lost interest, in particular after we realised that we would get an associative and commutative smash product for connective spectra which we did not believe exists. When many years later Lydakis exploited this construction we could have kicked ourselves."

2.1.3 Variants

The proof that $\Gamma\mathcal{S}_*$ is a closed category works if $\mathcal{S}_*$ is exchanged for other suitable closed categories with colimits. In particular $\Gamma\mathcal{A}$, the category of Γ-objects in the category $\mathcal{A} = sAb$ of simplicial abelian groups, is a closed category. The unit is $\bar{H}\mathbf{Z} = \{X \mapsto \tilde{\mathbf{Z}}[X]\}$ (it is $H\mathbf{Z}$ as a set, but we remember the group structure, see

Example 2.1.2.1(1)), the tensor is given by

$$(M \otimes N)(Z) = \varinjlim_{X \wedge Y \to Z} M(X) \otimes N(Y)$$

and the internal function object is given by

$$\underline{\Gamma\mathcal{A}}(M, N) = \big\{X, [q] \mapsto \Gamma\mathcal{A}\big(M \otimes \mathbf{Z}\big[\Delta[q]\big], N(-\wedge X)\big)\big\}.$$

2.1.3.1 $\Gamma\mathcal{S}_*$ vs. $\Gamma\mathcal{A}$

The adjoint functor pair between abelian groups and pointed sets

$$Ens_* \underset{U}{\overset{\tilde{\mathbf{Z}}}{\rightleftarrows}} Ab,$$

where U is the forgetful functor, induces an adjoint functor pair

$$\Gamma\mathcal{S}_* \underset{U}{\overset{\tilde{\mathbf{Z}}}{\rightleftarrows}} \Gamma\mathcal{A}.$$

The homomorphisms $\mathbf{Z} \to \tilde{\mathbf{Z}}(1_+)$ (sending n to $n \cdot 1$) and $\tilde{\mathbf{Z}}(X) \wedge \tilde{\mathbf{Z}}(Y) \to \tilde{\mathbf{Z}}(X \wedge Y)$ (sending the generator $x \otimes y$ to the generator $x \wedge y$) are isomorphisms and $\tilde{\mathbf{Z}}\colon (Ens_*, \wedge, 1_+) \to (Ab, \otimes, \mathbf{Z})$ is a *strong symmetric monoidal functor* (see Definition A.10.1.3 for details, but briefly a strong symmetric monoidal functor is a symmetric monoidal functor F such that the structure maps $F(a) \otimes F(b) \to F(a \otimes b)$ and $1 \to F(1)$ are isomorphisms). It follows immediately that $\tilde{\mathbf{Z}}\colon (\Gamma\mathcal{S}_*, \wedge, \mathbf{S}) \to (\Gamma\mathcal{A}, \otimes, \bar{H}\mathbf{Z})$ is strong symmetric monoidal. In particular $\tilde{\mathbf{Z}}\mathbf{S} \cong \bar{H}\mathbf{Z}$,

$$\tilde{\mathbf{Z}}(M \wedge N) \cong \tilde{\mathbf{Z}}M \otimes \tilde{\mathbf{Z}}N$$

and

$$\underline{\Gamma\mathcal{S}_*}(M, UP) \cong U\underline{\Gamma\mathcal{A}}(\tilde{\mathbf{Z}}M, P)$$

satisfying the necessary associativity, commutativity and unit conditions.

Later, we will see that the category $\Gamma\mathcal{A}$, for all practical (homotopical) purposes can be exchanged for $sAb = \mathcal{A}$. The comparison functors come from the adjoint pair

$$\mathcal{A} \underset{R}{\overset{\bar{H}}{\rightleftarrows}} \Gamma\mathcal{A}$$

where $\bar{H}P(X) = P \otimes \tilde{\mathbf{Z}}[X]$ and $RM = M(1_+)$. We see that $R\bar{H} = id_{\mathcal{A}}$. The other adjunction, $\bar{H}R \to id_{\Gamma\mathcal{A}}$, is discussed in Lemma 2.1.3.1 below. Both $\bar{H}$ and R are symmetric monoidal functors.

2.1.3.2 Special Objects

We say that $M \in ob\Gamma\mathcal{A}$ is *special* if its underlying Γ-space $UM \in ob\Gamma\mathcal{S}_*$ is special, i.e., if for all finite pointed sets X and Y the canonical map

$$UM(X \vee Y) \xrightarrow{\sim} UM(X) \times UM(Y)$$

is a weak equivalence in $\mathcal{S}_*$. The following lemma has the consequence that all special objects in $\Gamma\mathcal{A}$ can be considered to be in the image of $\bar{H}: sAb = \mathcal{A} \to \Gamma\mathcal{A}$:

Lemma 2.1.3.1 *Let $M \in ob\Gamma\mathcal{A}$ be special. Then the unit of adjunction $(\bar{H}RM)(k_+) \to M(k_+)$ is an equivalence.*

Proof Since M is special, we have that $M(k_+) \to \prod_k M(1_+)$ is an equivalence. On the other hand, if we precompose this map with the unit of adjunction

$$(\bar{H}RM)(k_+) = M(1_+) \otimes \tilde{\mathbf{Z}}[k_+] \to M(k_+)$$

we get an isomorphism. □

2.1.3.3 Additivization

There is also a Dold-Puppe-type construction: $L: \Gamma\mathcal{A} \to \mathcal{A}$ which is **left** adjoint to $\bar{H}$: Consider the three pointed functions $pr_1, pr_2, \nabla: 2_+ \to 1_+$ with nonzero value $pr_1(1) = \nabla(1) = \nabla(2) = pr_2(2) = 1$. Then L is given by

$$LM = \operatorname{coker}\{M(pr_1) - M(\nabla) + M(pr_2): M(2_+) \to M(1_+)\}.$$

This functor is intimately connected with the subcategory of $\Gamma\mathcal{A}$ consisting of "additive", or coproduct preserving functors $\Gamma^o \to \mathcal{A}$.

The additive objects are uniquely defined by their value at 1_+, and we get an isomorphism $M \cong \bar{H}(M(1_+)) = \bar{H}RM$. Using this we may identify $\mathcal{A}$ with the full subcategory of additive objects in $\Gamma\mathcal{A}$, and the inclusion into $\Gamma\mathcal{A}$ has a left adjoint given by $\bar{H}L$.

Note that all the functors L, R and $\bar{H}$ between A and $\Gamma\mathcal{A}$ are strong symmetric monoidal.

Just the same considerations could be made with Ab exchanged for the category of k-modules for any commutative ring k.

2.1.4 S-*Algebras*

In any monoidal category there is a notion of a *monoid* (see Definition A.10.1.4). The reason for the name is that a monoid in the usual sense is a monoid in

$(Ens, \times, *)$. Furthermore, the axioms for a ring is nothing but the statement that it is a monoid in $(Ab, \otimes, \mathbf{Z})$. For a commutative ring k, a k-algebra is no more than a monoid in (k-mod, $\otimes_k$, k), and so it is natural to define **S**-algebras the same way:

Definition 2.1.4.1 An **S**-algebra A is a monoid in $(\Gamma\mathcal{S}_*, \wedge, \mathbf{S})$.

This means that A is a Γ-space together with maps $\mu = \mu^A\colon A\wedge A \to A$ and $1\colon \mathbf{S} \to A$ such that the diagrams

$$\begin{array}{ccccc} A\wedge(A\wedge A) & \xrightarrow{\cong} & (A\wedge A)\wedge A & \xrightarrow{\mu\wedge id} & A\wedge A \\ \downarrow{\scriptstyle id\wedge\mu} & & & & \downarrow{\scriptstyle \mu} \\ A\wedge A & & \xrightarrow{\mu} & & A \end{array}$$

(the isomorphism is the associativity isomorphism of the smash product) and

$$\begin{array}{ccccc} \mathbf{S}\wedge A & \xrightarrow{1\wedge id} & A\wedge A & \xleftarrow{id\wedge 1} & A\wedge\mathbf{S} \\ & {\scriptstyle\cong}\searrow & \downarrow{\scriptstyle\mu} & \swarrow{\scriptstyle\cong} & \\ & & A & & \end{array}$$

commute, where the diagonal maps are the natural unit isomorphisms.

We say that an **S**-algebra is *commutative* if $\mu = \mu \circ tw$ where

$$A\wedge A \xrightarrow[\cong]{tw} A\wedge A$$

is the twist isomorphism.

Remark 2.1.4.2 In the definition of an **S**-algebra, any knowledge of the symmetric monoidal category structure is actually never needed, since maps $M\wedge N \to P$ **out of** the smash products is uniquely characterized by a map $M(X)\wedge N(Y) \to P(X\wedge Y)$ natural in $X, Y \in ob\Gamma^o$. So, since the multiplication is a map from the smash $A\wedge A \to A$, it can alternatively be defined as a map $A(X)\wedge A(Y) \to A(X\wedge Y)$ natural in both X and Y.

This was the approach of Bökstedt [30] when he defined FSP's. These are simplicial functors from finite spaces to spaces with multiplication and unit, such that the natural diagrams commute, plus some stability conditions. These stability conditions are automatically satisfied if we start out with functors from Γ^o (and then apply degreewise and diagonalize if we want $X \in s\Gamma^o$ as input), see Lemma 2.2.1.3. On the other hand, we shall later see that there is no loss of generality to consider only **S**-algebras.

2.1.4.1 Variants

An $\bar{H}\mathbf{Z}$-algebra is a monoid in $(\Gamma\mathcal{A}, \otimes, \bar{H}\mathbf{Z})$. (This is, for all practical purposes, equivalent to the more sophisticated notion of $H\mathbf{Z} = U\bar{H}\mathbf{Z}$-algebras arising from the fact that there is a closed category $(H\mathbf{Z}\text{-mod}, \wedge_{H\mathbf{Z}}, H\mathbf{Z})$, see Example 2.1.5.6 below.) Since the functors

$$\Gamma\mathcal{S}_* \underset{U}{\overset{\tilde{\mathbf{Z}}}{\rightleftarrows}} \Gamma\mathcal{A} \quad \begin{matrix} \xrightarrow{L} \\ \xleftarrow{\bar{H}} \\ \xrightarrow{R} \end{matrix} \quad \mathcal{A}$$

all are monoidal they send monoids to monoids. For instance, if A is a simplicial ring, then $\bar{H}A$ is an $\bar{H}\mathbf{Z}$-algebra and HA is an $\mathbf{S}$-algebra (it is even an $H\mathbf{Z}$-algebra):

Example 2.1.4.3

1. Let A be a simplicial ring, then HA is an $\mathbf{S}$-algebra with multiplication

$$HA \wedge HA \to H(A \otimes A) \to HA$$

and unit $\mathbf{S} \to \tilde{\mathbf{Z}}\mathbf{S} \cong H\mathbf{Z} \to HA$.

In particular, note the $\mathbf{S}$-algebra $H\mathbf{Z}$. It is given by $X \mapsto \tilde{\mathbf{Z}}[X]$, the "integral homology", and the unit map $X = \mathbf{S}(X) \to H\mathbf{Z}(X) = \tilde{\mathbf{Z}}[X]$ is the Hurewicz map of Appendix A.3.1.

2. Of course, $\mathbf{S}$ is the initial $\mathbf{S}$-algebra. If M is a simplicial monoid, the *spherical monoid algebra* $\mathbf{S}[M]$ is given by

$$\mathbf{S}[M](X) = M_+ \wedge X$$

with obvious unit and with multiplication coming from the monoid structure. Note that $R\tilde{\mathbf{Z}}\mathbf{S}[M] \cong \mathbf{Z}[M]$.

3. If A is an $\mathbf{S}$-algebra, then A^o, the *opposite* of A, is the $\mathbf{S}$-algebra given by the Γ-space A with the same unit $\mathbf{S} \to A$, but with the twisted multiplication

$$A \wedge A \underset{\cong}{\xrightarrow{tw}} A \wedge A \xrightarrow{\mu} A.$$

4. If A and B are $\mathbf{S}$-algebras, their smash $A \wedge B$ is a new $\mathbf{S}$-algebra with multiplication

$$(A \wedge B) \wedge (A \wedge B) \xrightarrow{id \wedge tw \wedge id} (A \wedge A) \wedge (B \wedge B) \to A \wedge B,$$

and unit $\mathbf{S} \cong \mathbf{S} \wedge \mathbf{S} \to A \wedge B$.

5. If A and B are two $\mathbf{S}$-algebras, the product $A \times B$ is formed pointwise: $(A \times B)(X) = A(X) \times B(X)$ and with componentwise multiplication and diagonal unit. The coproduct also exists, but is more involved.

6. Matrices: If A is an **S**-algebra, we define the **S**-algebra of $n \times n$ matrices $Mat_n A$ by

$$Mat_n A(X) = \underline{\mathcal{S}_*}\big(n_+, n_+ \wedge A(X)\big) \cong \prod_n \bigvee_n A(X)$$

—the matrices with only "one entry in each column". The unit is the diagonal, whereas the multiplication is determined by

$$\begin{array}{c} Mat_n A(X) \wedge Mat_n A(Y) = \underline{\mathcal{S}_*}\big(n_+, n_+ \wedge A(X)\big) \wedge \underline{\mathcal{S}_*}\big(n_+, n_+ \wedge A(Y)\big) \\ \big\downarrow {\scriptstyle id \wedge (\text{smashing with } id_{A(X)})} \\ \underline{\mathcal{S}_*}\big(n_+, n_+ \wedge A(X)\big) \wedge \underline{\mathcal{S}_*}\big(n_+ \wedge A(X), n_+ \wedge A(X) \wedge A(Y)\big) \\ \big\downarrow {\scriptstyle \text{composition}} \\ \underline{\mathcal{S}_*}\big(n_+, n_+ \wedge A(X) \wedge A(Y)\big) \\ \big\downarrow {\scriptstyle \text{multiplication}} \\ \underline{\mathcal{S}_*}\big(n_+, n_+ \wedge A(X \wedge Y)\big) = Mat_n A(X \wedge Y). \end{array}$$

We note that for a simplicial ring B, there is a natural map of **S**-algebras (sending some wedges to products, and rearranging the order)

$$Mat_n HB \to HM_n B$$

where $M_n B$ is the ordinary matrix ring. This map is a stable equivalence as defined in Definition 2.2.1.5. We also have a "Whitehead sum"

$$Mat_n(A) \times Mat_m(A) \xrightarrow{\ \vee\ } Mat_{n+m}(A)$$

which is the block sum listing the first matrix in the upper left hand corner and the second matrix in the lower right hand corner. This sum is sent to the ordinary Whitehead sum under the map $Mat_n HB \to HM_n B$.

2.1.5 A-Modules

If A is a ring, we define a left A-module to be an abelian group M together with a map $A \otimes M \to M$ satisfying certain properties. In other words, it is a "$(A \otimes -)$-algebra" where $(A \otimes -)$ is the triple on abelian groups sending P to $A \otimes P$. Likewise

Definition 2.1.5.1 Let A be an **S**-algebra. A (left) A-module is an $(A \wedge -)$-algebra.

To be more explicit, a left A-module is a pair (M, μ^M) where $M \in ob\Gamma\mathcal{S}_*$ and

$$A \wedge M \xrightarrow{\mu^M} M \in \Gamma\mathcal{S}_*$$

such that

$$\begin{array}{ccc} A\wedge A\wedge M & \xrightarrow{id\wedge\mu^M} & A\wedge M \\ {\scriptstyle \mu^A\wedge id}\downarrow & & {\scriptstyle \mu^M}\downarrow \\ A\wedge M & \xrightarrow{\mu^M} & M \end{array}$$

commutes and such that the composite

$$M \cong \mathbf{S}\wedge M \xrightarrow{1\wedge id} A\wedge M \xrightarrow{\mu^M} M$$

is the identity.

If M and N are A-modules, an A-module map $M\to N$ is a map of Γ-spaces compatible with the A-module structure (an "$(A\wedge -)$-algebra morphism").

Remark 2.1.5.2

1. Note that, as remarked for $\mathbf{S}$-algebras in Remark 2.1.4.2, the structure maps defining A-modules could again be defined directly without reference to the internal smash in $\Gamma\mathcal{S}_*$.
2. One defines *right A-modules and A-bimodules* as A^o-modules and $A^o\wedge A$-modules.
3. Note that an $\mathbf{S}$-module is no more than a Γ-space. In general, if A is a commutative $\mathbf{S}$-algebra, then the concepts of left or right modules agree.
4. If A is a simplicial ring, then an HA-module does not need to be of the sort HP for an A-module P, but we shall see that the difference between A-modules and HA-modules is for most applications irrelevant.

Definition 2.1.5.3 Let A be an $\mathbf{S}$-algebra. Let M be an A-module and M' an A^o-module. The smash product $M'\wedge_A M$ is the Γ-space given by the coequalizer

$$M'\wedge_A M = \varinjlim \left\{ M'\wedge A\wedge M \rightrightarrows M'\wedge M \right\}$$

where the two maps represent the two actions.

Definition 2.1.5.4 Let A be an $\mathbf{S}$-algebra and let M, N be A-modules. The Γ-space of A-module maps is defined as the equalizer

$$\underline{\mathcal{M}}_A(M,N) = \varprojlim \left\{ \underline{\Gamma\mathcal{S}_*}(M,N) \rightrightarrows \underline{\Gamma\mathcal{S}_*}(A\wedge M, N) \right\}$$

where the first map is induced by the action of A on M, and the second is

$$\underline{\Gamma\mathcal{S}_*}(M,N) \to \underline{\Gamma\mathcal{S}_*}(A\wedge M, A\wedge N) \to \underline{\Gamma\mathcal{S}_*}(A\wedge M, N)$$

induced by the action of A on N.

From these definitions, the following proposition is immediate.

Proposition 2.1.5.5 *Let k be a commutative* **S***-algebra. Then the smash product and morphism object over k endow the category $\mathcal{M}_k$ of k-modules with the structure of a closed category.* ☺

Example 2.1.5.6 (k-Algebras) If k is a commutative **S**-algebra, the monoids in the closed monoidal category (k-mod, $\wedge_k$, k) are called k-algebras. The most important example to us is the category of $H\mathbf{Z}$-algebras. A crucial point we shall return to later is that the homotopy categories of $H\mathbf{Z}$-algebras and simplicial rings are equivalent.

2.1.6 $\Gamma\mathcal{S}_*$-Categories

Since ($\Gamma\mathcal{S}_*$, $\wedge$, $\mathbf{S}$) is a (symmetric monoidal) closed category it makes sense to talk of a *$\Gamma\mathcal{S}_*$-category*, i.e., a collection of objects $ob\mathcal{C}$ and for each pair of objects $c, d \in ob\mathcal{C}$ a Γ-space $\underline{\mathcal{C}}(c,d)$ of "morphisms", with multiplication

$$\underline{\mathcal{C}}(c,d)\wedge\underline{\mathcal{C}}(b,c) \longrightarrow \underline{\mathcal{C}}(b,d)$$

and unit

$$\mathbf{S} \longrightarrow \underline{\mathcal{C}}(c,c)$$

satisfying the usual identities analogous to the notion of an **S**-algebra (as a matter of fact: an **S**-algebra is precisely (the Γ-space of morphisms in) a $\Gamma\mathcal{S}_*$-category with one object). See Sect. A.10.2 for more details on enriched category theory.

In particular, $\Gamma\mathcal{S}_*$ is itself a $\Gamma\mathcal{S}_*$-category. As another example; from Definition 2.1.5.4 of the Γ-space of A-module morphisms, the following fact follows immediately.

Proposition 2.1.6.1 *Let A be an* **S***-algebra. Then the category of A-modules is a $\Gamma\mathcal{S}_*$-category.* ☺

Further examples of $\Gamma\mathcal{S}_*$-categories:

Example 2.1.6.2

1. Any $\Gamma\mathcal{S}_*$-category $\mathcal{C}$ has an underlying $\mathcal{S}_*$-category $R\mathcal{C}$, or just $\mathcal{C}$ again for short, with function spaces $(R\mathcal{C})(c,d) = R(\underline{\mathcal{C}}(c,d)) = \underline{\mathcal{C}}(c,d)(1_+)$ (see Example 2.1.2.1(3)). The prime example being $\Gamma\mathcal{S}_*$ itself, where we always drop the R from the notation.

   A $\Gamma\mathcal{S}_*$-category with only one object is what we call an **S**-algebra (just as a k-*mod*-category with only one object is a k-algebra), and this is closely connected to Bökstedt's notion of an FSP. In fact, a "ring functor" in the sense of [70] is the same as a $\Gamma\mathcal{S}_*$-category when restricted to $\Gamma^o \subseteq \mathcal{S}_*$, and conversely, any $\Gamma\mathcal{S}_*$-category is a ring functor when extended degreewise.

2. Just as the Eilenberg–Mac Lane construction takes rings to **S**-algebras Example 2.1.4.3(1), it takes *Ab*-categories to $\Gamma\mathcal{S}_*$-categories. Let $\mathcal{E}$ be an *Ab*-category (i.e., enriched in abelian groups). Then using the Eilenberg–Mac Lane-construction of Example 2.1.4.3(1) on the morphism groups gives a $\Gamma\mathcal{S}_*$-category which we will call $\tilde{\mathcal{E}}$ (it could be argued that it ought to be called $H\mathcal{E}$, but somewhere there has got to be a conflict of notation, and we choose to sin here). To be explicit: if $c, d \in ob\mathcal{E}$, then $\tilde{\mathcal{E}}(c,d)$ is the Γ-space which sends $X \in ob\Gamma^o$ to $\mathcal{E}(c,d) \otimes \tilde{\mathbf{Z}}[X] = H(\mathcal{E}(c,d))(X)$.
3. Let $\mathcal{C}$ be a pointed $\mathcal{S}_*$-category. The category $\Gamma\mathcal{C}$ of pointed functors $\Gamma^o \to \mathcal{C}$ is a $\Gamma\mathcal{S}_*$-category by declaring that
$$\underline{\Gamma\mathcal{C}}(c,d)(X) = \Gamma\mathcal{C}\big(c, d(X\wedge -)\big) \in ob\mathcal{S}_*.$$
4. Let $(\mathcal{C}, \sqcup, e)$ be a symmetric monoidal category. An augmented symmetric monoid in $\mathcal{C}$ is an object c together with maps $c \sqcup c \to c$, $e \to c \to e$ satisfying the usual identities. A slick way of encoding all the identities of an augmented symmetric monoid c is to identify it with its bar complex (Eilenberg–Mac Lane object) $\bar{H}c \colon \Gamma^o \to \mathcal{C}$ where
$$\bar{H}c(k_+) = \sqcup^{k_+} c = \overbrace{c \sqcup \cdots \sqcup c}^{k \text{ times}} \quad \big(\sqcup^{0_+} c = e\big).$$
That is, an augmented symmetric monoid is a rigid kind of Γ-object in $\mathcal{C}$; it is an Eilenberg–Mac Lane object.
5. Adding (3) and (4) together we get a functor from symmetric monoidal categories to $\Gamma\mathcal{S}_*$-categories, sending $(\mathcal{C}, \sqcup, e)$ to the $\Gamma\mathcal{S}_*$-category with objects the augmented symmetric monoids, and with morphism objects
$$\Gamma\mathcal{C}\big(\bar{H}c, \bar{H}d(X\wedge -)\big).$$
6. Important special case: If $(\mathcal{C}, \vee, e)$ is a *category with sum* (i.e., e is both final and initial in $\mathcal{C}$, and $\vee$ is a coproduct), then **all** objects are augmented symmetric monoids and
$$\Gamma\mathcal{C}\big(\bar{H}c, \bar{H}d(k_+\wedge -)\big) \cong \mathcal{C}\left(c, \bigvee^{k_+} d\right),$$
where $\bigvee^{k_+} d = d \vee \cdots \vee d$ (k-summands).

2.1.6.1 The $\Gamma\mathcal{S}_*$-Category $\mathcal{C}^\vee$

The last Example 2.1.6.2(6) is so important that we introduce the following notation. Let $(\mathcal{C}, \vee, e)$ be a category with sum (i.e., e is both final and initial in $\mathcal{C}$, and $\vee$ is a coproduct), then $\mathcal{C}^{\vee}$ is the $\Gamma\mathcal{S}_*$-category with $ob\mathcal{C}^{\vee} = ob\mathcal{C}$ and

$$\mathcal{C}^{\vee}(c,d)(X) = \mathcal{C}\Big(c, \bigvee^{X} d\Big).$$

If $(\mathcal{E}, \oplus, 0)$ is an Ab-category with sum (what is often called an *additive category*), then the $\tilde{\mathcal{E}}$ of Example 2.1.6.2(2) and $\mathcal{E}^{\oplus}$ coincide:

$$\tilde{\mathcal{E}}(c,d)(n_+) \cong \mathcal{E}(c,d)^{\times n} \cong \mathcal{E}\big(c, d^{\oplus n}\big) = \mathcal{E}^{\oplus}(c,d)(n_+),$$

since finite sums and products coincide in an additive category, see [191, p. 194].

It is worth noting that the structure of Example 2.1.6.2(6) when applied to $(\Gamma\mathcal{S}_*, \vee, 0_+)$ is different from the $\Gamma\mathcal{S}_*$-enrichment we have given to $\Gamma\mathcal{S}_*$ when declaring it to be a symmetric monoidal **closed** category under the smash product. Then $\underline{\Gamma\mathcal{S}_*}(M,N)(X) = \Gamma\mathcal{S}_*(M, N(X\wedge -))$. However, $\vee^X N \cong X\wedge N \to N(X\wedge -)$ is a stable equivalence (see Definition 2.2.1.5), and in some cases this is enough to ensure that

$$\underline{\Gamma\mathcal{S}_*}^{\vee}(M,N)(X) \cong \Gamma\mathcal{S}_*(M, X\wedge N) \to \Gamma\mathcal{S}_*\big(M, N(X\wedge -)\big) = \underline{\Gamma\mathcal{S}_*}(M,N)(X)$$

is a stable equivalence.

2.1.6.2 A Reformulation

When talking in the language of Ab-categories (linear categories), a ring is just (the morphism group in) an Ab-category with one object, and an A-module M corresponds to a functor from A to Ab: the ring homomorphism $A \to \mathrm{End}(M)$ giving the abelian group M a structure of an A-module is exactly the data needed to give a functor from the category (corresponding to) A to Ab sending the single object to M.

In the setting of $\Gamma\mathcal{S}_*$-categories, we can similarly reinterpret **S**-algebras and their modules. An **S**-algebra A is simply a $\Gamma\mathcal{S}_*$-category with only one object, and an A-module is a $\Gamma\mathcal{S}_*$-functor from A to $\Gamma\mathcal{S}_*$.

Thinking of A-modules as $\Gamma\mathcal{S}_*$-functors $A \to \Gamma\mathcal{S}_*$ the definitions of smash and morphism objects can be elegantly expressed as

$$M'\wedge_A M = \int^A M'\wedge M$$

and

$$\underline{\mathrm{Hom}}_A(M,N) = \int_A \underline{\Gamma\mathcal{S}_*}(M,N).$$

If B is another **S**-algebra, M' a $B\wedge A^o$-module we get $\Gamma\mathcal{S}_*$-adjoint functors

$$\mathcal{M}_A \underset{\mathcal{M}_B(M',-)}{\overset{M'\wedge_A -}{\leftrightarrows}} \mathcal{M}_B$$

due to the canonical isomorphism

$$\underline{\mathcal{M}_B}(M' \wedge_A N, P) = \int_B \underline{\Gamma\mathcal{S}_*}\left(\int^A M' \wedge N, P\right)$$
$$\cong \int_A \underline{\Gamma\mathcal{S}_*}\left(M', \int_B \underline{\Gamma\mathcal{S}_*}(N, P)\right) = \underline{\mathcal{M}_A}(N, \underline{\mathcal{M}_B}(M', P))$$

which follows from the definitions, the Fubini theorem for ends and the fact that $\Gamma\mathcal{S}_*$ is closed symmetric monoidal ($P \in ob\mathcal{M}_B$).

2.2 Stable Structures

In this section we will discuss the homotopical properties of Γ-spaces and **S**-algebras. Historically Γ-spaces are nice representations of connective spectra and the choice of equivalences reflects this. That is, in addition to the obvious pointwise equivalences, we have the so-called stable equivalences. The functors of **S**-algebras we will define, such as K-theory, should respect stable equivalences. Any **S**-algebra can, up to a canonical stable equivalence, be replaced by a very special one.

2.2.1 The Homotopy Theory of $\mathbf{\Gamma}$*-Spaces*

To define the stable structure we need to take a different view to Γ-spaces.

2.2.1.1 Gamma-Spaces as Functors of Spaces

Let M be a Γ-space. It is a (pointed) functor $M\colon \Gamma^o \to \mathcal{S}_*$, and by extension by colimits and degreewise application followed by the diagonal we may think of it as a functor $\mathcal{S}_* \to \mathcal{S}_*$. To be explicit, we first extend from the skeletal category Γ^o to all finite pointed sets (in a chosen universe) by, for each finite pointed set S of cardinality $k+1$, choosing a pointed isomorphism $\alpha_S\colon S \cong k_+$ (α_{k_+} is chosen to be the identity), setting $M(S) = M(k_+)$ and if $f\colon S \to T$ is a pointed function of finite sets we define $M(f)$ to be $M(\alpha_T f \alpha_S^{-1})$. If X is a pointed set, we define

$$M(X) = \varinjlim_{Y \subseteq X} M(Y),$$

where the colimit varies over the finite pointed subsets $Y \subseteq X$, and so M is a (pointed) functor $\mathit{Ens}_* \to \mathcal{S}_*$. For this to be functorial, we—as always—assume that all colimits are actually chosen (and not something only defined up to unique isomorphism). Finally, if $X \in ob\mathcal{S}_*$, we set

$$M(X) = \mathrm{diag}^*\{[q] \mapsto M(X_q)\} = \{[q] \mapsto M(X_q)_q\}.$$

Aside 2.2.1.1 For those familiar with the language of coends, the extensions of a Γ-space M to an endofunctor on spaces can be done all at once: if X is a space, then

$$M(X) = \int^{k_+} X^{\times k} \wedge M(k_+).$$

In yet other words, we do the left Kan extension

$$\begin{array}{ccc} \Gamma^o & \xrightarrow{M} & \mathcal{S}_* \\ {\scriptstyle \mathbf{S}}\downarrow & \nearrow & \\ \mathcal{S}_*. & & \end{array}$$

2.2.1.2 Gamma-Spaces as Simplicial Functors

The fact that these functors come from degreewise applications of a functor on (discrete) sets make them "simplicial" (more precisely: they are $\mathcal{S}_*$-functors), i.e., they give rise to simplicial maps

$$\underline{\mathcal{S}_*}(X, Y) \to \underline{\mathcal{S}_*}\big(M(X), N(Y)\big)$$

which results in natural maps

$$M(X) \wedge Y \to M(X \wedge Y) \tag{2.2}$$

coming from the identity on $X \wedge Y$ through the composite

$$\begin{aligned} \mathcal{S}_*(X \wedge Y, X \wedge Y) &\cong \mathcal{S}_*\big(Y, \underline{\mathcal{S}_*}(X, X \wedge Y)\big) \\ &\to \mathcal{S}_*\big(Y, \underline{\mathcal{S}_*}\big(M(X), M(X \wedge Y)\big)\big) \cong \mathcal{S}_*\big(M(X) \wedge Y, M(X \wedge Y)\big) \end{aligned}$$

(where the isomorphisms are the adjunction isomorphisms of the smash/function space adjoint pair). In particular this means that Γ-spaces define spectra: the nth term is given by $M(S^n)$, and the structure map is $S^1 \wedge M(S^n) \to M(S^{n+1})$ where S^n is $S^1 = \Delta[1]/\partial\Delta[1]$ smashed with itself n times, see also Definition 2.2.1.8 below.

Definition 2.2.1.2 If $M \in ob\Gamma\mathcal{S}_*$, then the *homotopy groups* of M are defined as

$$\pi_q M = \varinjlim_{k} \pi_{k+q} M\big(S^k\big).$$

Note that $\pi_q M = 0$ for $q < 0$, by the following lemma.

Lemma 2.2.1.3 *Let* $M \in \Gamma\mathcal{S}_*$.

1. *If* $Y \xrightarrow{\sim} Y' \in \mathcal{S}_*$ *is a weak equivalence then* $M(Y) \xrightarrow{\sim} M(Y')$ *is a weak equivalence also.*

2. *If X is an n-connected pointed space, then $M(X)$ is n-connected also.*
3. *If X is an n-connected pointed space, then the canonical map of Eq.* (2.2) $M(X)\wedge Y \to M(X\wedge Y)$ *is* $2n$*-connected.*

Proof Let $\mathbf{L}M$ be the simplicial Γ-space given by

$$\mathbf{L}M(X)_p = \bigvee_{Z_0,\dots,Z_p\in(\Gamma^o)^{\times p+1}} M(Z_0)\wedge\Gamma^o(Z_0,Z_1)\wedge\cdots\wedge\Gamma^o(Z_{p-1},Z_p)\wedge\Gamma^o(Z_p,X)$$

with operators determined by

$$d_i(f\wedge\alpha_1\wedge\cdots\wedge\alpha_p\wedge\beta) = \begin{cases}(M(\alpha_1)(f)\wedge\alpha_2\wedge\cdots\wedge\alpha_p\wedge\beta) & \text{if } i=0,\\ (f\wedge\alpha_1\wedge\cdots\wedge\alpha_{i+1}\circ\alpha_i\wedge\cdots\wedge\beta) & \text{if } 1\le i\le p-1,\\ (f\wedge\alpha_1\wedge\cdots\wedge\alpha_{p-1}\wedge(\beta\circ\alpha_p)) & \text{if } i=p,\end{cases}$$

$$s_j(f\wedge\alpha_1\wedge\cdots\wedge\alpha_p\wedge\beta) = (f\wedge\cdots\alpha_j\wedge\mathrm{id}\wedge\alpha_{j+1}\cdots\wedge\beta)$$

($\mathbf{L}M$ is an example of a "homotopy coend", or a "bar construction"). Consider the natural transformation

$$\mathbf{L}M \xrightarrow{\ \eta\ } M$$

determined by

$$(f\wedge\alpha_1\wedge\cdots\wedge\beta)\mapsto M(\beta\circ\alpha_p\circ\cdots\circ\alpha_1)(f).$$

For each $Z\in ob\Gamma^o$ we obtain a simplicial homotopy inverse to η_Z by sending $f\in M(Z)$ to $(f\wedge\mathrm{id}_Z\wedge\cdots\wedge\mathrm{id}_Z)$. Since $\mathbf{L}M$ and M both commute with filtered colimits we see that η is an equivalence on all pointed sets and so by Theorem A.6.0.1, η is an equivalence for all pointed simplicial sets because $\mathbf{L}M$ and M are applied degreewise. Thus, for all pointed simplicial sets X the map η_X is a weak equivalence

$$\mathbf{L}M(X)\xrightarrow{\sim} M(X).$$

(1) If $Y\xrightarrow{\sim} Y'$ is a weak equivalence then $\mathcal{S}_*(k_+,Y)\cong Y^{\times k}\xrightarrow{\sim}(Y')^{\times k}\cong\mathcal{S}_*(k_+,Y')$ is a weak equivalence for all k. But this implies that $\mathbf{L}M(Y)_p\xrightarrow{\sim}\mathbf{L}M(Y')_p$ for all p and hence, by Theorem A.6.0.1, that $\mathbf{L}M(Y)\xrightarrow{\sim}\mathbf{L}M(Y')$.

(2) If X is n-connected for some $n\ge 0$, then $\mathcal{S}_*(k_+,X)\cong X^{\times k}$ is n-connected for all k and hence $\mathbf{L}M(X)_p$ is n-connected for all p. Thus, by Theorem A.6.0.5 we see that $\mathbf{L}M(X)$ is n-connected also.

(3) If X is n-connected and X' is m-connected then, by Corollary A.8.2.4, $X\vee X'\to X\times X'$ is $(m+n)$-connected and so $Y\wedge(X\times X')\to(Y\wedge X')\times(Y\wedge X')$ is

$(m+n)$-connected also by the commuting diagram

$$\begin{array}{ccc} Y \wedge (X \vee X') & \longrightarrow & Y \wedge (X \times X') \\ {\cong}\downarrow & & \downarrow \\ (Y \wedge X) \vee (Y \wedge X') & \longrightarrow & (Y \wedge X) \times (Y \wedge X') \end{array}$$

since both horizontal maps are $(m+n)$-connected. By induction we see that

$$Y \wedge \mathcal{S}_*(k_+, X) \to \mathcal{S}_*(k_+, Y \wedge X)$$

is $2n$-connected for all k and so $Y \wedge \mathbf{L}M(X)_p \to \mathbf{L}M(Y \wedge X)_p$ is $2n$-connected for all p. Since a simplicial space which is $k > 0$-connected in every degree has a k-connected diagonal (e.g., by Theorem A.6.0.5) we can conclude that $Y \wedge \mathbf{L}M(X) \to \mathbf{L}M(Y \wedge X)$ is $2n$-connected. □

Following Schwede [253] we now define two closed model category structures on $\Gamma\mathcal{S}_*$ (these differ very slightly from the structures considered by Bousfield and Friedlander [39] and Lydakis [188]). For basics on model categories see Appendix A.4. We will call these model structures the "pointwise" and the "stable" structures":

Definition 2.2.1.4 Pointwise structure: A map $M \to N \in \Gamma\mathcal{S}_*$ is a *pointwise fibration* (resp. *pointwise equivalence*) if $M(X) \to N(X) \in \mathcal{S}_*$ is a fibration (resp. weak equivalence) for every $X \in ob\Gamma$. The map is a (pointwise) *cofibration* if it has the lifting property with respect to maps that are both pointwise fibrations and pointwise equivalences, i.e., $i\colon A \to X \in \Gamma\mathcal{S}_*$ is a cofibration if for every pointwise fibration $f\colon E \to B \in \Gamma\mathcal{S}_*$ that is a pointwise equivalence and for every solid commutative diagram

$$\begin{array}{ccc} A & \longrightarrow & E \\ {\scriptstyle i}\downarrow & \overset{s}{\nearrow} & \downarrow{\scriptstyle \simeq}\ {\scriptstyle f} \\ X & \longrightarrow & B \end{array}$$

there exists a (dotted) map $s\colon X \to E$ making the resulting diagram commute.

From this one constructs the stable structure. Note that the cofibrations in the two structures are the same! Because of this we often omit the words "pointwise" and "stable" when referring to cofibrations.

Definition 2.2.1.5 Stable structure: A map of Γ-spaces is a stable equivalence if it induces an isomorphism on homotopy groups (defined in Definition 2.2.1.2). It is a (stable) cofibration if it is a (pointwise) cofibration, and it is a stable fibration if it has the lifting property with respect to all maps that are both stable equivalences and cofibrations.

As opposed to simplicial sets, not all Γ-spaces are cofibrant. Examples of cofibrant objects are the Γ-spaces Γ^X of Example 2.1.2.1(4) (and so the simplicial Γ-spaces $\mathbf{L}M$ defined in the proof of Lemma 2.2.1.3 are cofibrant in every degree, so that $\mathbf{L}M \to M$ can be thought of as a cofibrant resolution).

We shall see in Corollary 2.2.1.7 that the stably fibrant objects are the very special Γ-spaces which are pointwise fibrant.

2.2.1.3 Important Convention

The stable structure will by far be the most important to us, and so when we occasionally forget the qualification "stable", and say that a map of Γ-spaces is a fibration, a cofibration or an equivalence this is short for it being a **stable** fibration, cofibration or equivalence. We will say "pointwise" when appropriate.

Theorem 2.2.1.6 *Both the pointwise and the stable structures define closed model category structures* (*see Sect.* A.4.2) *on* $\Gamma\mathcal{S}_*$. *Furthermore, these structures are compatible with the* $\Gamma\mathcal{S}_*$*-category structure. More precisely*: *If* $M \overset{i}{\rightarrowtail} N$ *is a cofibration and* $P \overset{p}{\twoheadrightarrow} Q$ *is a pointwise* (*resp. stable*) *fibration, then the canonical map*

$$\underline{\Gamma\mathcal{S}_*}(N, P) \to \underline{\Gamma\mathcal{S}_*}(M, P) \prod_{\underline{\Gamma\mathcal{S}_*}(M,Q)} \underline{\Gamma\mathcal{S}_*}(N, Q) \tag{2.3}$$

is a pointwise (*resp. stable*) *fibration, and if in addition i or p is a pointwise* (*resp. stable*) *equivalence, then* (2.3) *is a pointwise* (*resp. stable*) *equivalence.*

Sketch proof (cf. Schwede [253]) That the pointwise structure is a closed simplicial model category is essentially an application of Quillen's basic theorem [235, II4] to the category of Γ-sets. The rest of the pointwise claim follows from the definition of $\underline{\Gamma\mathcal{S}_*}(-,-)$.

As to the stable structure, all the axioms but one follow from the pointwise structure. If $f: M \to N \in \Gamma\mathcal{S}_*$, one must show that there is a factorization $M \overset{\sim}{\rightarrowtail} X \twoheadrightarrow N$ of f as a cofibration which is a stable equivalence, followed by a stable fibration. We refer the reader to [253]. We refer the reader to the same source for compatibility of the stable structure with the $\Gamma\mathcal{S}_*$-enrichment. ☺

Note that, since the cofibrations are the same in the pointwise and the stable structure, a map is both a pointwise equivalence and a pointwise fibration if and only if it is both a stable equivalence and a stable fibration.

Corollary 2.2.1.7 *Let M be a* Γ*-space. Then M is stably fibrant* (*i.e.*, $M \to *$ *is a stable fibration*) *if and only if it is very special and pointwise fibrant.*

Proof If M is stably fibrant, $M \to *$ has the lifting property with respect to all maps that are stable equivalences and cofibrations, and hence also to the maps that are pointwise equivalences and cofibrations; that is, M is pointwise fibrant. Let $X, Y \in ob\Gamma^o$, then $\Gamma^X \vee \Gamma^Y \to \Gamma^{X\vee Y} \cong \Gamma^X \times \Gamma^Y$ is a cofibration and a (stable) equivalence. This means that if M is stably fibrant, then

$$\underline{\Gamma\mathcal{S}_*}\big(\Gamma^{X\vee Y}, M\big) \to \underline{\Gamma\mathcal{S}_*}\big(\Gamma^X \vee \Gamma^Y, M\big)$$

is a stable equivalence and a stable fibration, which is the same as saying that it is a pointwise equivalence and a pointwise fibration, which means that

$$M(X \vee Y) \cong \Gamma\mathcal{S}_*\big(\Gamma^{X\vee Y}, M\big) \to \Gamma\mathcal{S}_*\big(\Gamma^X \vee \Gamma^Y, M\big) \cong M(X) \times M(Y)$$

is an equivalence. Here, as elsewhere, we have written $\Gamma\mathcal{S}_*(-,-)$ for the underlying morphism space $R\underline{\Gamma\mathcal{S}_*}(-,-)$. Similarly, the map

$$\mathbf{S} \vee \mathbf{S} \xrightarrow{in_1+\Delta} \mathbf{S} \times \mathbf{S}$$

is a stable equivalence. When $\pi_0\Gamma\mathcal{S}_*(-, M)$ is applied to this map we get the assignment $(a,b) \mapsto (a, a+b)\colon \pi_0 M(1_+)^{\times 2} \to \pi_0 M(1_+)^{\times 2}$.

If M is fibrant this must be an isomorphism, and so $\pi_0 M(1_+)$ has inverses.

Conversely, suppose that M is pointwise fibrant and very special. Let $M \overset{i}{\underset{\sim}{\rightarrowtail}} N \twoheadrightarrow *$ be a factorization into a map that is a stable equivalence and cofibration followed by a stable fibration. Since both M and N are very special i must be a pointwise equivalence, and so has a section (from the pointwise structure), which means that M is a retract of a stably fibrant object since we have a lifting in the diagram in the pointwise structure

$$\begin{array}{ccc} M & = & M \\ {\scriptstyle\simeq}\big\downarrow{\scriptstyle i} & & \big\downarrow \\ N & \longrightarrow & * \end{array}.$$

□

2.2.1.4 A Simple Fibrant Replacement Functor

In the approach we will follow, it is a strange fact that we will never need to replace a Γ-space with a cofibrant one, but we will constantly need to replace them by stably fibrant ones. There is a particularly easy way to do this: let M be any Γ-space, and set

$$QM(X) = \lim_{\overrightarrow{k}} \Omega^k M\big(S^k \wedge X\big),$$

cf. the analogous construction for spectra in Sect. A.3.2.1. Obviously the map $M \to QM$ is a stable equivalence, and QM is pointwise fibrant and very special (use e.g., Lemma 2.2.1.3). For various purposes, this replacement Q will not be good enough. Its main deficiency is that it will not take **S**-algebras to **S**-algebras.

2.2.1.5 Comparison with Spectra

We have already observed that Γ-spaces give rise to spectra:

Definition 2.2.1.8 Let M be a Γ-space. Then the *spectrum associated* with M is the sequence

$$\underline{M} = \{k \mapsto M(S^k)\}$$

where S^k is $S^1 = \Delta[1]/\partial\Delta[1]$ smashed with itself k times, together with the structure maps $S^1 \wedge M(S^k) \to M(S^1 \wedge S^k) = M(S^{k+1})$ of Eq. (2.2).

The assignment $M \mapsto \underline{M}$ is a simplicial functor

$$\Gamma\mathcal{S}_* \xrightarrow{M \mapsto \underline{M}} \mathcal{S}pt$$

(where $\mathcal{S}pt$ is the category of spectra, see Appendix A.3.2 for details) and it follows from the considerations in [39] that it induces an equivalence between the stable homotopy categories of Γ-spaces and connective spectra.

Crucial for the general acceptance of Lydakis' definition of the smash product was the following (where $\text{conn}(X)$ is the connectivity of X):

Proposition 2.2.1.9 *Let M and N be Γ-spaces and X and Y spaces. If M is cofibrant, then the canonical map*

$$M(X)\wedge N(Y) \to (M\wedge N)(X\wedge Y)$$

is n-connected with $n = \text{conn}(X) + \text{conn}(Y) + \min(\text{conn}(X), \text{conn}(Y))$.

Sketch proof (see [188] for further details). The proof goes by induction, first treating the case $M = \Gamma^o(n_+, -)$, and observing that then $M(X)\wedge N(Y) \cong X^{\times n}\wedge N(Y)$ and $(M\wedge N)(X\wedge Y) \cong N((X\wedge Y)^{\times n})$. Hence, in this case the result follows from Lemma 2.2.1.3(3). ☺

Corollary 2.2.1.10 *Let M and N be Γ-spaces with M cofibrant. Then $\underline{M\wedge N}$ is stably equivalent to a handicrafted smash product of spectra, e.g.,*

$$n \mapsto \left\{\lim_{\overrightarrow{k,l}} \Omega^{k+l}\left(S^n \wedge M(S^k)\wedge N(S^l)\right)\right\}.$$ ☺

2.2.2 A Fibrant Replacement for S-Algebras

Note that if A is a simplicial ring, then the Eilenberg–Mac Lane object HA of Example 2.1.2.1(1) is a very special Γ-space, and so maps between simplicial rings induce maps that are stable equivalences if and only if they are pointwise equivalences. Hence any functor respecting pointwise equivalences of **S**-algebras will have good homotopy properties when restricted to simplicial rings.

When we want to apply functors to all **S**-algebras A, we frequently need to replace our **S**-algebras by a very special **S**-algebras before feeding them to our functor in order to ensure that the functor will preserve stable equivalences. This is a potential problem since the fibrant replacement functor Q presented in Sect. 2.2.1.4 does not take **S**-algebras to **S**-algebras.

For this we need a gadget explored by Breen [41] and Bökstedt [30]. Breen noted the need for a refined stabilization of the Eilenberg-Mac Lane spaces for rings and Bökstedt noted that when he wanted to extend Hochschild homology to **S**-algebras or rather FSPs (see Chap. 4) in general, the face maps were problematic as they involved the multiplication, and this was not well behaved with respect to naïve stabilization. Both mention Illusie [145] as a source of inspiration.

2.2.2.1 The Category $\mathcal{I}$

Let $\mathcal{I} \subset \Gamma^o$ be the subcategory with all objects, but only the injective maps. This has much more structure than the natural numbers considered as the subcategory where we only allow the standard inclusion $\{0, 1, \dots, n-1\} \subset \{0, 1, \dots, n\}$. Most importantly, the wedge sum of two sets $x_0, x_1 \mapsto x_0 \vee x_1$ induces a natural transformation $\mathcal{I} \times \mathcal{I} \to \mathcal{I}$. To be quite precise, the sum is given by $k_+ \vee l_+ = (k+l)_+$ with inclusion maps $k_+ \to (k+l)_+$ sending $i \in k_+$ to $i \in (k+l)_+$, and $l_+ \to (k+l)_+$ sending $j > 0 \in l_+$ to $k + j \in (k+l)_+$. Note that $\vee$ is strictly associative and unital: $(x \vee y) \vee z = x \vee (y \vee z)$ and $0_+ \vee x = x = x \vee 0_+$ (but symmetric only up to isomorphism).

This results in a simplicial category $\{[p] \mapsto \mathcal{I}^{p+1}\}$ with structure maps given by sending $\mathbf{x} = (x_0, \dots, x_q) \in \mathcal{I}^{q+1}$ to

$$d_i(\mathbf{x}) = \begin{cases} (x_0, \dots, x_i \vee x_{i+1}, \dots, x_q) & \text{for } 0 \le i < q, \\ (x_q \vee x_0, , x_1, \dots, x_{p-1}) & \text{for } i = q, \end{cases}$$

$$s_i(\mathbf{x}) = (x_0, \dots, x_i, 0_+, x_{i+1}, \dots, x_p) \quad \text{for } 0 \le i \le q.$$

Below, and many times later, we will use the symbol $Map_*(X, Y)$ to signify the homotopy-theoretically sensible mapping space $\underline{\mathcal{S}_*}(X, \sin|Y|)$ (which, in view of the geometric realization/singular complex adjunction is naturally isomorphic to the singular complex of the space $Top_*(|X|, |Y|)$ of pointed maps with the compact open topology) between pointed simplicial sets X and Y. If Y is fibrant, the map $\underline{\mathcal{S}_*}(X, Y) \to Map_*(X, Y)$ induced by the unit of adjunction $Y \to \sin|Y|$ is a weak equivalence. For more details on the geometric realization/singular complex adjoint pair, the reader may consult Appendix A.2.1. On several occasions we will need that smashing with a pointed space A induces a map $Map_*(X, Y) \to Map_*(A \wedge X, A \wedge Y)$, sending the q-simplex $\Delta[q]_+ \wedge X \to \sin|Y|$ to

$$\Delta[q]_+ \wedge (A \wedge X) \cong A \wedge \big(\Delta[q]_+ \wedge X\big) \to A \wedge \sin|Y| \to \sin|A \wedge Y|,$$

where the first isomorphism is the symmetry structure isomorphism of $\wedge$, the middle map is induced by the map in question and the last map is adjoint to the composite $A \to \underline{\mathcal{S}_*}(Y, A\wedge Y) \to \underline{\mathcal{S}_*}(\sin|Y|, \sin|A\wedge Y|)$ (where the first map is adjoint to the identity and the last map induced by the $\mathcal{S}_*$-functor $\sin|-|$).

Definition 2.2.2.1 If $x = k_+ \in ob\mathcal{I}$, we let $|x| = k$—the number of non-base points. We will often not distinguish notationally between x and $|x|$. For instance, an expression like S^x will mean S^1 smashed with itself $|x|$ times: $S^{0_+} = S^0$, $S^{(k+1)_+} = S^1\wedge S^{k_+}$. Likewise Ω^x will mean $Map_*(S^x, -)$. If $\phi\colon x \to y \in \mathcal{I}$, then $S^{(|y|-|x|)}\wedge S^x \to S^y$ is the isomorphism which inserts the jth factor of S^x as the $\phi(j)$th factor of S^y and distributes the factors of $S^{(|y|-|x|)}$ over the remaining factors of S^y, keeping the order. If M is a Γ-space and X is a finite pointed set, the assignment $x \mapsto \Omega^x M(S^x\wedge X)$ is a functor, where $\phi\colon x \to y$ is sent to

$$\Omega^x M\big(S^x\wedge X\big) \to \Omega^{(|y|-|x|)+|x|}\big(S^{(|y|-|x|)}\wedge M\big(S^x\wedge X\big)\big) \quad \to$$
$$\Omega^{(|y|-|x|)+|x|} M\big(S^{(|y|-|x|)}\wedge S^x\wedge X\big) \cong \Omega^y M\big(S^y\wedge X\big),$$

where the first map is the suspension, the second is induced by the structure map of M and the last isomorphism is conjugation by the isomorphism $S^{(|y|-|x|)}\wedge S^x \to S^y$ described above. Let $T_0 M$ be the Γ-space

$$T_0 M = \Big\{X \mapsto \underset{\overrightarrow{x\in\mathcal{I}}}{\operatorname{holim}}\, \Omega^x M\big(S^x\wedge X\big)\Big\}$$

The reason for the notation $T_0 M$ will become apparent in Chap. 4 (no, it is not because it is the tangent space of something).

We would like to know that this has the right homotopy properties, i.e., that $T_0 M$ is equivalent to

$$QM = \Big\{X \mapsto \underset{\overrightarrow{k}}{\lim}\, \Omega^k M\big(S^k\wedge X\big)\Big\}.$$

One should note that, as opposed to $\mathbf{N}$, the category $\mathcal{I}$ is not filtering, so we must stick with the homotopy colimits. However, $\mathcal{I}$ possesses certain good properties which overcome this difficulty. Bökstedt attributes the idea behind the following very important stabilization lemma to Illusie [145]. Still, we attach Bökstedt's name to the result to signify the importance his insight at this point was to the development of the cyclotomic trace. See [30, 1.5], but also [192, 2.3.7] or [42, 2.5.1] and compare with [145, VI, 4.6.12] and [41].

Lemma 2.2.2.2 (Bökstedt's Approximation Lemma) *Let $G\colon \mathcal{I}^{q+1} \to \mathcal{S}_*$ be a functor, $\mathbf{x} \in ob\mathcal{I}^{q+1}$, and consider the full subcategory $F_{\mathbf{x}} \subseteq \mathcal{I}^{q+1}$ of objects supporting maps from $\mathbf{x}$. Assume G sends maps in $F_{\mathbf{x}}$ to n-connected maps. Then the canonical map*

$$G(\mathbf{x}) \to \underset{\overrightarrow{\mathcal{I}^{q+1}}}{\operatorname{holim}}\, G$$

is n-connected.

Proof Since $\mathcal{I}^{q+1}$ has an initial object, Lemma A.7.4.1 tells us that we may work with unbased homotopy colimits. Consider the functor

$$\mu_{\mathbf{x}}\colon \mathcal{I}^{q+1} \xrightarrow{\mathbf{y}\mapsto\mu_{\mathbf{x}}(y)=\mathbf{x}\vee\mathbf{y}} \mathcal{I}^{q+1},$$

factoring over the inclusion $F_{\mathbf{x}} \subseteq \mathcal{I}^{q+1}$ The second inclusion $\mathbf{y} \subseteq \mathbf{x} \vee \mathbf{y}$ defines a natural transformation $\eta_{\mathbf{x}}$ from the identity to $\mu_{\mathbf{x}}$. This natural transformation translates to a homotopy from the identity to the map

$$\operatorname*{holim}_{\overrightarrow{\mathcal{I}^{q+1}}} G \xrightarrow{G\eta_{\mathbf{x}}} \operatorname*{holim}_{\overrightarrow{\mathcal{I}^{q+1}}} G\mu_{\mathbf{x}} \xrightarrow{(\mu_{\mathbf{x}})_*} \operatorname*{holim}_{\overrightarrow{\mathcal{I}^{q+1}}} G,$$

showing that $\operatorname{holim}_{\overrightarrow{F_{\mathbf{x}}}} G \to \operatorname{holim}_{\overrightarrow{\mathcal{I}^{q+1}}} G$ is a split surjection in the homotopy category. Likewise, the same natural transformation restricted to $F_{\mathbf{x}}$ gives a homotopy from the identity to

$$\operatorname*{holim}_{\overrightarrow{F_{\mathbf{x}}}} G \xrightarrow{G\eta_{\mathbf{x}}} \operatorname*{holim}_{\overrightarrow{F_{\mathbf{x}}}} G\mu_{\mathbf{x}} \xrightarrow{(\mu_{\mathbf{x}})_*} \operatorname*{holim}_{\overrightarrow{F_{\mathbf{x}}}} G,$$

showing that $\operatorname{holim}_{\overrightarrow{F_{\mathbf{x}}}} G \to \operatorname{holim}_{\overrightarrow{\mathcal{I}^{q+1}}} G$ is a split injection in the homotopy category. Together this shows that the map $\operatorname{holim}_{\overrightarrow{F_{\mathbf{x}}}} G \to \operatorname{holim}_{\overrightarrow{\mathcal{I}^{q+1}}} G$ is a weak equivalence. Hence it is enough to show that $G(\mathbf{x}) \to \operatorname{holim}_{\overrightarrow{F_{\mathbf{x}}}} G$ is n-connected.

Repeating the same argument as above with the constant functor $*$ instead of G, we see that $B(F_{\mathbf{x}}) \cong \operatorname{holim}_{\overrightarrow{F_{\mathbf{x}}}} * \to \operatorname{holim}_{\overrightarrow{\mathcal{I}^{q+1}}} * \cong B(\mathcal{I}^{q+1})$ is an equivalence, and the latter space is contractible since $\mathcal{I}^{q+1}$ has an initial object. Quillen's theorem B in the form of Lemma A.7.4.2 then states that $G(\mathbf{x}) \to \operatorname{holim}_{\overrightarrow{F_{\mathbf{x}}}} G$ is n-connected. □

Lemma 2.2.2.3 *Let M be a Γ-space. Then T_0M is very special and the natural transformation $M \to T_0M$ is a stable equivalence of Γ-spaces.*

Proof Let X and Y be pointed finite sets and $y = k_+ \in ob\mathcal{I}$. Consider the diagram

$$\begin{array}{ccc}
\Omega^y\big[M(S^y)\wedge(X\vee Y)\big] & \longrightarrow & \Omega^y\big(M(S^y)\wedge X\big) \times \Omega^y\big(M(S^y)\wedge Y\big) \\
\downarrow & & \downarrow \\
\Omega^y\big[M\big(S^y\wedge(X\vee Y)\big)\big] & \longrightarrow & \Omega^y M\big(S^y\wedge X\big) \times \Omega^y M\big(S^y\wedge Y\big) \\
\downarrow & & \downarrow \\
\operatorname*{holim}_{\overrightarrow{x\in\mathcal{I}}} \Omega^x\big[M\big(S^x\wedge(X\vee Y)\big)\big] & \longrightarrow & \operatorname*{holim}_{\overrightarrow{x\in\mathcal{I}}} \Omega^x M\big(S^x\wedge X\big) \times \operatorname*{holim}_{\overrightarrow{x\in\mathcal{I}}} M\big(S^x\wedge Y\big)
\end{array}$$

where the horizontal maps are induced by the projections from $X \vee Y$ to X and to Y and are $k-2$-connected by the Freudenthal suspension Theorem A.8.2.3; the top vertical maps are the maps defined in Eq. (2.2) and are $k-2$-connected by Lemma 2.2.1.3(3); and the bottom vertical maps are the canonical maps into the homotopy colimits and are k-connected by Bökstedt's approximation Lemma 2.2.2.2. The bottom map is the map $T_0M(X \vee Y) \to T_0M(X) \times T_0M(Y)$ induced by the projections. Since k can be chosen arbitrarily, this shows that the Γ-space T_0M is very special.

The same reasoning shows that the canonical map

$$\operatorname*{holim}_{\overrightarrow{k \in \mathbf{N}}} \Omega^k M\big(S^k \wedge X\big) \to \operatorname*{holim}_{\overrightarrow{x \in \mathcal{I}}} \Omega^x M\big(S^x \wedge X\big)$$

induced by the inclusion $\mathbf{N} \subseteq \mathcal{I}$ is a weak equivalence under $M(X)$. Since by Lemma A.7.3.2 the canonical map $\operatorname{holim}_{\overrightarrow{k \in \mathbf{N}}} \Omega^k M(S^k \wedge X) \to \lim_{\overrightarrow{k \in \mathbf{N}}} \Omega^k M(S^k \wedge X) = (QM)(X)$ is a weak equivalence under $M(X)$ we are done. □

Note that $T_0M(X)$ is usually not a fibrant space, and so T_0M is not stably fibrant either, but the lemma shows that e.g., $\sin|T_0M|$ *is* stably fibrant.

A *stable equivalence* of **S**-algebras is a map of **S**-algebras that is a stable equivalence when considered as a map of Γ-spaces.

Lemma 2.2.2.4 *The functor T_0 maps* **S***-algebras to* **S***-algebras, and the natural transformation $id \to T_0$ is a stable equivalence of* **S***-algebras.*

Proof Given Lemma 2.2.2.3, we only need to establish the multiplicative properties. Let A be an **S**-algebra. We have to define the multiplication and the unit of T_0A. The unit is obvious: $\mathbf{S} \to T_0\mathbf{S} \to T_0A$.

If $F, G\colon J \to \mathcal{S}_*$ are functors, distributivity of $\wedge$ over $\vee$ gives a natural isomorphism $(\operatorname{holim}_{\overrightarrow{J}} F) \wedge (\operatorname{holim}_{\overrightarrow{J}} G) \cong \operatorname{holim}_{\overrightarrow{J \times J}} F \wedge G$. Using the map $\Omega^x A(S^x \wedge X) \wedge \Omega^y A(S^y \wedge Y) \to \Omega^{x \vee y}(A(S^x \wedge X) \wedge A(S^y \wedge Y))$ that smashes maps together, the multiplication in A and the sum in $\mathcal{I}$, the multiplication in T_0A is given by the composite

$$\begin{aligned} T_0A(X) \wedge T_0A(Y) &\longrightarrow \operatorname*{holim}_{\overrightarrow{(x,y) \in \mathcal{I}^2}} \Omega^{x \vee y}\big(A\big(S^x \wedge X\big) \wedge A\big(S^y \wedge Y\big)\big) \\ &\xrightarrow{\text{mult. in } A} \operatorname*{holim}_{\overrightarrow{(x,y) \in \mathcal{I}^2}} \Omega^{x \vee y} A\big(S^{x \vee y} \wedge X \wedge Y\big) \\ &\xrightarrow{\vee \text{ in } \mathcal{I}} \operatorname*{holim}_{\overrightarrow{z \in \mathcal{I}}} \Omega^z A\big(S^z \wedge X \wedge Y\big) = T_0A(X \wedge Y). \end{aligned}$$

Checking that this gives a unital and associative structure on T_0A follows by using the same properties in $\mathcal{I}$ and A. That the map $A \to T_0A$ is a map of **S**-algebras is now immediate. □

Remark 2.2.2.5 It is noteworthy that the fibrant replacement Q is **not** monoidal and will not take **S**-algebras to **S**-algebras; the presence of nontrivial automorphisms in $\mathcal{I}$ is of vital importance. We discuss this further in Note 4.2.2.6 since it crucial to Bökstedt's definition of topological Hochschild homology. Notice that if A is an **S**-algebra, then the multiplication in A provides $T_0A(1_+)$ with the structure of a simplicial monoid. However, even if A is commutative, the automorphisms of $\mathcal{I}$ prevent $T_0M(1_+)$ from being commutative (unless $A = *$), thus saving us from Lewis' pitfalls [174].

Corollary 2.2.2.6 *Any $\bar{H}\mathbf{Z}$-algebra is functorially stably equivalent to $\bar{H}$ of a simplicial ring. In particular, if A is an $\mathbf{S}$-algebra, then $\tilde{\mathbf{Z}}A$ is functorially stably equivalent to H of a simplicial ring.*

Proof The T_0 construction can equally well be performed in $\bar{H}\mathbf{Z}$-modules: let $\Omega^1_{Ab}M$ be $\underline{\mathcal{S}_*}(S^1, M)$, which is an $\bar{H}\mathbf{Z}$-module if M is, and let the homotopy colimit be given by the usual formula except the wedges are replaced by sums (see Sect. A.7.4.1 for further details). Let $R_0A = \operatorname{holim}_{\overrightarrow{x\in\mathcal{I}}} \Omega^x_{Ab}A(S^x)$. This is an $\bar{H}\mathbf{Z}$-algebra if A is. There is a natural equivalence $R_0A \to R_0(\sin|A|)$ and a natural transformation $T_0UA \to UR_0(\sin|A|)$ (U is the forgetful functor). By Lemma A.7.4.6 and Lemma 2.2.1.3(2) you get that $T_0UA(S^n) \to UR_0(\sin|A|)(S^n)$ is $(2n-1)$-connected. But since both sides are special Γ-spaces, this means that $T_0UA \xrightarrow{\sim} UR_0\sin|A| \xleftarrow{\sim} UR_0A$ is a natural chain of weak equivalences. (Alternatively, we could have adapted Bökstedt's approximation theorem to prove directly that $A \to R_0A$ is a stable equivalence.)

Consequently, if A is a $\bar{H}\mathbf{Z}$-algebra, there is a functorial stable equivalence $A \to R_0A$ of $\bar{H}\mathbf{Z}$-algebras. But R_0A is special and for such algebras the unit of adjunction $\bar{H}R \to 1$ is an equivalence by Lemma 2.1.3.1. □

2.2.3 Homotopical Algebra in the Category of A-Modules

Although it is not necessary for the subsequent development, we list a few facts pertaining to the homotopy structure on categories of modules over **S**-algebras. The stable structure on A-modules is inherited in the usual way from the stable structure on Γ-spaces.

Definition 2.2.3.1 Let A be an **S**-algebra. We say that an A-module map is an *equivalence* (resp. *fibration*) if it is a stable equivalence (resp. stable fibration) of Γ-spaces. The *cofibrations* are defined by the lifting property.

Theorem 2.2.3.2 *With these definitions, the category of A-modules is a closed model category compatibly enriched in $\Gamma\mathcal{S}_*$: if $M \overset{i}{\rightarrowtail} N$ is a cofibration and $P \overset{p}{\twoheadrightarrow} Q$ is a fibration, then the canonical map*

$$\underline{\operatorname{Hom}}_A(N, P) \xrightarrow{(i^*, p_*)} \underline{\operatorname{Hom}}_A(M, P) \prod_{\underline{\operatorname{Hom}}_A(M,Q)} \underline{\operatorname{Hom}}_A(N, Q)$$

is a stable fibration, and if in addition i or p is an equivalence, then (i^*, p_*) is a stable equivalence.

Sketch proof (For a full proof, consult [253]). For the proof of the closed model category structure, see [255, 3.1.1]. For the proof of the compatibility with the enrichment, see the proof of [255, 3.1.2] where the commutative case is treated. ☺

The smash product behaves as expected (see [188] and [253] for proofs):

Proposition 2.2.3.3 *Let A be an $\mathbf{S}$-algebra, and let M be a cofibrant A^o-module. Then $M\wedge_A -\colon A\text{-mod} \to \Gamma\mathcal{S}_*$ sends stable equivalences to stable equivalences. If N is an A-module there are first quadrant spectral sequences*

$$\mathrm{Tor}_p^{\pi_* A}(\pi_* M, \pi_* N)_q \Rightarrow \pi_{p+q}(M\wedge_A N)$$

$$\pi_p\big(M\wedge_A(H\pi_q N)\big) \Rightarrow \pi_{p+q}(M\wedge_A N)$$

If $A \to B$ is a stable equivalence of $\mathbf{S}$-algebras, then the derived functor of $B\wedge_A$— induces an equivalence between the homotopy categories of A and B-modules. ☺

2.2.3.1 k-Algebras

Let k be a commutative $\mathbf{S}$-algebra. In the category of k-algebras, we call a map a fibration or a weak equivalence if it is a stable fibration or stable equivalence of Γ-spaces. The cofibrations are as usual the maps with the right (right meaning correct: in this case left is right) lifting property. With these definitions the category of k-algebras becomes a closed simplicial model category [253]. We will need the analogous result for $\Gamma\mathcal{S}_*$-categories.

2.2.4 Homotopical Algebra in the Category of $\Gamma\mathcal{S}_*$-Categories

Definition 2.2.4.1 A $\Gamma\mathcal{S}_*$-functor of $\Gamma\mathcal{S}_*$-categories $F\colon \mathcal{C} \to \mathcal{D}$ is a *stable equivalence* if for all $c, c' \in ob\mathcal{C}$ the map

$$\mathcal{C}\big(c, c'\big) \to \mathcal{D}\big(Fc, Fc'\big) \in \Gamma\mathcal{S}_*$$

is a stable equivalence, and for any $d \in ob\mathcal{D}$ there is a $c \in ob\mathcal{C}$ and an isomorphism $Fc \cong d$.

Likewise, an $\mathcal{S}$-functor of $\mathcal{S}$-categories $F\colon \mathcal{C} \to \mathcal{D}$ is a *weak equivalence* if for all $c, c' \in ob\mathcal{C}$ the map $\mathcal{C}(c, c') \to \mathcal{D}(Fc, Fc') \in \mathcal{S}$ is a weak equivalence, and for any $d \in ob\mathcal{D}$ there is a $c \in ob\mathcal{C}$ and an isomorphism $Fc \cong d$.

A functor which is surjective on isomorphism classes is sometimes called "essentially surjective".

Recall that a $\Gamma\mathcal{S}_*$-*equivalence* is a $\Gamma\mathcal{S}_*$-functor $\mathcal{C}\xrightarrow{F}\mathcal{D}$ for which there exists a $\Gamma\mathcal{S}_*$-functor $\mathcal{C}\xleftarrow{G}\mathcal{D}$ and $\Gamma\mathcal{S}_*$-natural isomorphisms $id_{\mathcal{C}}\cong GF$ and $id_{\mathcal{D}}\cong FG$.

Lemma 2.2.4.2 *Every stable equivalence of $\Gamma\mathcal{S}_*$-categories can be written as a composite of a stable equivalence inducing the identity on the objects and a $\Gamma\mathcal{S}_*$-equivalence.*

Proof Let $F\colon \mathcal{C}\to\mathcal{D}$ be a stable equivalence. Let $\underline{F}$ be the $\Gamma\mathcal{S}_*$-category with the same objects as $\mathcal{C}$, but with morphisms given by $\underline{F}(c,c')=\mathcal{D}(Fc,Fc')$. Then F factors as $\mathcal{C}\to\underline{F}\to\mathcal{D}$ where the first map is the identity on objects and a stable equivalence on morphisms, and the second is induced by F on objects, and is the identity on morphisms. The latter map is a $\Gamma\mathcal{S}_*$-equivalence: for every $d\in ob\mathcal{D}$ **choose** a $c_d\in ob\mathcal{C}$ and an isomorphism $d\cong Fc_d$. As one checks, the application $d\mapsto c_d$ defines the inverse $\Gamma\mathcal{S}_*$-equivalence. □

So stable equivalences are the more general, and may be characterized as composites of $\Gamma\mathcal{S}_*$-equivalences and stable equivalences that induce the identity on the set of objects. Likewise for weak equivalences of $\mathcal{S}$-categories.

2.3 Algebraic K-Theory

2.3.1 K-Theory of Symmetric Monoidal Categories

A symmetric monoid can be viewed as a symmetric monoidal category (an SMC) with just identity morphisms. A symmetric monoid M gives rise to a Γ-space HM via the formula $k_+\mapsto M^{\times k}$ (see Example 2.1.2.1(1) and [257]), the Eilenberg–Mac Lane object of M. Algebraic K-theory, as developed in Segal's paper [257], is an extension of this to symmetric monoidal categories (see also [260] or [282]), such that for every symmetric monoidal category $\mathcal{C}$ we have a Γ-category $\bar{H}\mathcal{C}$.

For a finite set X, let $\mathcal{P}X$ be the set of subsets of X. If S and T are two disjoint subsets of X, then $S\coprod T$ is again a subset of X. If all the coherence isomorphisms symmetric monoidal category $(\mathcal{C},\sqcup,e)$ were identities we could define the algebraic K-theory as the Γ-category which evaluated at $k_+\in\Gamma^o$ was the category whose objects were all functions $\mathcal{P}\{1,\dots,k\}\to ob\mathcal{C}$ sending $\coprod$ to $\sqcup$ and $\emptyset$ to e:

$$\begin{pmatrix}\mathcal{P}\{1,\dots,k\}\\ \coprod\\ \emptyset\end{pmatrix}\to\begin{pmatrix}ob\mathcal{C}\\ \sqcup\\ e\end{pmatrix}.$$

Such a function is uniquely given by declaring what its values are on all subsets $\{i\} \subset \{1, \ldots, k\}$ and so this is nothing but $\mathcal{C}$ times itself k times.

In the non-strict case this loosens up only a bit. If $(\mathcal{C}, \sqcup, e)$ is a symmetric monoidal category, $\bar{H}\mathcal{C}(k_+)$ is the symmetric monoidal category whose objects are the pointed functors $\mathcal{P}\{1, \ldots, k\} \to \mathcal{C}$ taking $\coprod$ to $\sqcup$ up to coherent isomorphisms. More precisely (remembering that displayed diagrams commute unless otherwise explicitly stated not to)

Definition 2.3.1.1 Let $(\mathcal{C}, \sqcup, e)$ be a symmetric monoidal category. Let $k_+ \in ob\Gamma^o$. An object of $\bar{H}\mathcal{C}(k_+)$ is a function $a\colon \mathcal{P}\{1, \ldots, k\} \to ob\mathcal{C}$ together with a choice of isomorphisms

$$\alpha_{S,T}\colon a_S \sqcup a_T \to a_{S\coprod T}$$

for every pair $S, T \subseteq \{1, \ldots, k\}$ such that $S \cap T = \emptyset$, satisfying the following conditions:

1. $a_\emptyset = e$,
2. the morphisms $a_{\emptyset,S}\colon e \sqcup a_S \to a_{\emptyset \coprod S} = a_S$ and $a_{S,\emptyset}\colon a_S \sqcup e \to a_{S\coprod\emptyset} = a_S$ are the structure isomorphisms in $\mathcal{C}$,
3. the diagram

$$\begin{array}{ccccc} (a_S \sqcup a_T)\sqcup a_U & & \xrightarrow{\text{associativity}} & & a_S\sqcup(a_T\sqcup a_U)\\ \downarrow{\scriptstyle \alpha_{S,T}\sqcup id} & & & & \downarrow{\scriptstyle id\sqcup\alpha_{T,U}} \\ a_{S\coprod T}\sqcup a_U & \xrightarrow[\alpha_{S\coprod T,U}]{} & a_{S\coprod T\coprod U} & \xleftarrow[\alpha_{S,T\coprod U}]{} & a_S\sqcup a_{T\coprod U}\end{array}$$

   commutes and
4. the diagram

$$\begin{array}{ccc} a_S \sqcup a_T & \longrightarrow & a_T \sqcup a_S \\ & \searrow{\scriptstyle \alpha_{S,T}} & \downarrow{\scriptstyle \alpha_{T,S}} \\ & & a_{S\coprod T} = a_{T\coprod S} \end{array}$$

   commutes, where the unlabelled arrow is the corresponding structure isomorphism in $\mathcal{C}$.

A morphism $f\colon (a, \alpha) \to (b, \beta) \in \bar{H}\mathcal{C}(X)$ is a collection of morphisms

$$f_S\colon a_S \to b_S \in \mathcal{C}$$

such that

1. $f_\emptyset = id_e$ and

2. the diagram

$$\begin{array}{ccc} a_S \sqcup a_T & \xrightarrow{f_S \sqcup f_T} & b_S \sqcup b_T \\ {\scriptstyle \alpha_{S,T}}\downarrow & & {\scriptstyle \beta_{S,T}}\downarrow \\ a_{S \coprod T} & \xrightarrow{f_{S \coprod T}} & b_{S \coprod T} \end{array}$$

commutes.

If $\phi\colon k_+ \to l_+ \in \Gamma^o$, then $\bar{H}\mathcal{C}(k_+) \to \bar{H}\mathcal{C}(l_+)$ is defined by sending $a\colon \mathcal{P}\{1,\dots,k\} \to \mathcal{C}$ to

$$\mathcal{P}\{1,\dots,l\} \xrightarrow{\phi^{-1}} \mathcal{P}\{1,\dots,k\} \xrightarrow{a} \mathcal{C}$$

(this makes sense as ϕ was pointed at 0), with corresponding isomorphism

$$a_{\phi^{-1}(S),\phi^{-1}(T)}\colon a_{\phi^{-1}S} \sqcup a_{\phi^{-1}T} \to a_{\phi^{-1}(S)\coprod\phi^{-1}(T)} = a_{\phi^{-1}(S\coprod T)}.$$

This defines the Γ-category $\bar{H}\mathcal{C}$, which again is obviously functorial in $\mathcal{C}$, giving the functor

$$\bar{H}\colon \text{symmetric monoidal categories} \to \Gamma\text{-categories}.$$

The classifying space $B\bar{H}\mathcal{C}$ forms a Γ-space which is often called the (direct sum) *algebraic K-theory of* $\mathcal{C}$.

If $\mathcal{C}$ is discrete, or in other words, $\mathcal{C} = ob\mathcal{C}$ is a symmetric monoid, then this is exactly the Eilenberg–Mac Lane spectrum of $ob\mathcal{C}$.

Note that $\bar{H}\mathcal{C}$ becomes a **special** Γ-category in the sense that

Lemma 2.3.1.2 *Let* $(\mathcal{C}, \sqcup, e)$ *be a symmetric monoidal category. The canonical map*

$$\bar{H}\mathcal{C}(k_+) \to \bar{H}\mathcal{C}(1_+) \times \cdots \times \bar{H}\mathcal{C}(1_+)$$

is an equivalence of categories.

Proof We do this by producing an equivalence $E_k\colon \mathcal{C}^{\times k} \to \bar{H}\mathcal{C}(k_+)$ such that

$$\begin{array}{ccc} \mathcal{C}^{\times k} & = & \mathcal{C}^{\times k} \\ {\scriptstyle E_k}\downarrow & & {\scriptstyle E_1^{\times k}}\downarrow \\ \bar{H}\mathcal{C}(k_+) & \longrightarrow & \bar{H}\mathcal{C}(1_+)^{\times k} \end{array}$$

commutes. The equivalence E_k is given by sending $(c_1,\dots,c_k) \in ob\mathcal{C}^{\times k}$ to $E_k(c_1,\dots,c_k) = \{(a_S, \alpha_{S,T})\}$ where

$$a_{\{i_1,\dots,i_j\}} = c_{i_1} \sqcup \big(c_{i_2} \sqcup \cdots \sqcup (c_{i_{k-1}} \sqcup c_{i_k}) \cdots\big)$$

and $\alpha_{S,T}$ is the unique isomorphism we can write up using only the structure isomorphisms in $\mathcal{C}$. Likewise for morphisms. A quick check reveals that this is an equivalence (check the case $k = 1$ first), and that the diagram commutes. □

2.3.1.1 Enrichment in $\Gamma\mathcal{S}_*$

The definitions above make perfect sense also in the $\Gamma\mathcal{S}_*$-enriched world, and we may speak about symmetric monoidal $\Gamma\mathcal{S}_*$-categories $\mathcal{C}$.

A bit more explicitly: a *symmetric monoidal $\Gamma\mathcal{S}_*$-category* is a tuple $(\mathcal{C}, \sqcup, e, \alpha, \lambda, \rho, \gamma)$ such that $\mathcal{C}$ is a $\Gamma\mathcal{S}_*$-category, $\sqcup\colon \mathcal{C} \times \mathcal{C} \to \mathcal{C}$ is a $\Gamma\mathcal{S}_*$-functor, $e \in ob\mathcal{C}$ and α, λ, ρ and γ are $\Gamma\mathcal{S}_*$-natural transformations satisfying the usual requirements listed in Definition A.10.1.1.

The definition of $\bar{H}\mathcal{C}$ at this generality is as follows: the objects in $\bar{H}\mathcal{C}(k_+)$ are the same as before (2.3.1.1), and the Γ-space $\bar{H}\mathcal{C}((a,\alpha),(b,\beta))$ is defined as the equalizer

$$\bar{H}\mathcal{C}\big((a,\alpha),(b,\beta)\big)(k_+) \longrightarrow \prod_{\emptyset\neq S\subseteq\{1,\dots,k\}} \mathcal{C}(a_S, b_S) \rightrightarrows \prod_{\substack{\emptyset\neq S,T\subseteq\{1,\dots,k\}\\ S\cap T=\emptyset}} \mathcal{C}(a_S \sqcup a_T, b_{S\coprod T}).$$

The (S, T)-components of the two maps in the equalizer are the two ways around

$$\begin{array}{ccccc}
\prod_U \mathcal{C}(a_U, b_U) & \xrightarrow{\mathrm{proj}_S\times\mathrm{proj}_T} & \mathcal{C}(a_S, b_S) \times \mathcal{C}(a_T, b_T) & \xrightarrow{\sqcup} & \mathcal{C}(a_S \sqcup a_T, b_S \sqcup b_T) \\
\Big\downarrow{\scriptstyle \mathrm{proj}_{S\coprod T}} & & & & \Big\downarrow{\scriptstyle (\beta_{S,T})_*} \\
\mathcal{C}(a_{S\coprod T}, b_{S\coprod T}) & & \xrightarrow{\quad(\alpha_{S,T})^*\quad} & & \mathcal{C}(a_S \sqcup a_T, b_{S\coprod T}).
\end{array}$$

2.3.1.2 Categories with Sum

The simplest example of symmetric monoidal $\Gamma\mathcal{S}_*$-categories comes from categories with sum (i.e., $\mathcal{C}$ is pointed and has a coproduct $\vee$). If $\mathcal{C}$ is a category with sum we consider it as a $\Gamma\mathcal{S}_*$-category via the enrichment

$$\mathcal{C}^{\vee}(c,d)(k_+) = \mathcal{C}\left(c, \bigvee^{k} d\right)$$

(see Sect. 2.1.6.1).

The sum structure survives to give $\mathcal{C}^{\vee}$ the structure of a symmetric $\Gamma\mathcal{S}_*$-monoidal category:

$$
\begin{aligned}
&(\mathcal{C}^\vee \times \mathcal{C}^\vee)\big((c_1, c_2), (d_1, d_2)\big)(k_+)\\
&= \mathcal{C}\Big(c_1, \bigvee^k d_1\Big) \times \mathcal{C}\Big(c_2, \bigvee^k d_2\Big)\\
&\overset{\vee}{\to} \mathcal{C}\Big(c_1 \vee c_2, \Big(\bigvee^k d_1\Big) \vee \Big(\bigvee^k d_2\Big)\Big) \cong \mathcal{C}\Big(c_1 \vee c_2, \bigvee^k (d_1 \vee d_2)\Big)\\
&= \mathcal{C}^\vee(c_1 \vee c_2, d_1 \vee d_2)(k_+).
\end{aligned}
$$

Categories with sum also have a particular transparent K-theory. The data for a symmetric monoidal category above simplifies in this case to $\bar{H}\mathcal{C}(k_+)$ having as objects functors from the pointed category of subsets and inclusions of $k_+ = \{0, 1, \dots, k\}$, sending 0_+ to 0 and pushout squares to pushout squares, see also Sect. 3.2.1.1.

2.3.2 Quite Special **Γ***-Objects*

Let $\mathcal{C}$ be a Γ-$\Gamma\mathcal{S}_*$-category, i.e., a pointed functor $\mathcal{C}\colon \Gamma^o \to \Gamma\mathcal{S}_*$-categories. We say that $\mathcal{C}$ is *special* if for each pair of finite pointed sets X and Y the canonical $\Gamma\mathcal{S}_*$-functor $\mathcal{C}(X \vee Y) \to \mathcal{C}(X) \times \mathcal{C}(Y)$ is a $\Gamma\mathcal{S}_*$-equivalence of $\Gamma\mathcal{S}_*$-categories. So, for instance, if $\mathcal{C}$ is a symmetric monoidal category, then $\bar{H}\mathcal{C}$ is special. We need a slightly weaker notion.

Definition 2.3.2.1 Let $\mathcal{D}$ be a Γ-$\Gamma\mathcal{S}_*$-category. We say that $\mathcal{D}$ is *quite special* if for each pair of finite pointed sets $X, Y \in ob\Gamma^o$ the canonical map $\mathcal{D}(X \vee Y) \to \mathcal{D}(X) \times \mathcal{D}(Y)$ is a stable equivalence of $\Gamma\mathcal{S}_*$-categories (see Definition 2.2.4.1 for definition).

Likewise, a functor $\mathcal{D}\colon \Gamma^o \to \mathcal{S}$-categories is *quite special* if $\mathcal{D}(X \vee Y) \to \mathcal{D}(X) \times \mathcal{D}(Y)$ is a weak equivalence of $\mathcal{S}$-categories Definition 2.2.4.1.

Typically, theorems about special $\mathcal{D}$ remain valid for quite special $\mathcal{D}$.

Lemma 2.3.2.2 *Let $\mathcal{D}\colon \Gamma^o \to \mathcal{S}$-categories be quite special. Then $B\mathcal{D}$ is special.*

Proof This follows since the classifying space functor B preserves products and by [75] takes weak equivalences of $\mathcal{S}$-categories to weak equivalences of simplicial sets. □

Recall the fibrant replacement functor T_0 of Definition 2.2.2.1. The same proof as in Lemma 2.2.2.4 gives that if we use T_0 on all the morphism objects in a $\Gamma\mathcal{S}_*$-category we get a new category where the morphism objects are stably fibrant.

Lemma 2.3.2.3 *Let $\mathcal{D}\colon \Gamma^o \to \Gamma\mathcal{S}_*$-categories be quite special. Then $T_0\mathcal{D}$ is quite special.*

Proof This follows since T_0 preserves stable equivalences, and since

$$T_0(M \times N) \xrightarrow{\sim} T_0 M \times T_0 N$$

is a stable equivalence for any $M, N \in ob\Gamma\mathcal{S}_*$. Both these facts follow from the definition of T_0 and Bökstedt's approximation Lemma 2.2.2.2. □

2.3.3 A Uniform Choice of Weak Equivalences

When considering a discrete ring A, the algebraic K-theory can be recovered from knowing only the *iso*morphisms of finitely generated projective A-modules. We will show in Sect. 3.2.1 that the algebraic K-theory of A, as defined through Waldhausen's S-construction in Chap. 1, is equivalent to what you get if you apply Segal's construction $\bar{H}$ to the groupoid $i\mathcal{P}_A$ of finitely generated A-modules and isomorphisms between them. So, non-invertible homomorphisms are not seen by algebraic K-theory.

This is not special for the algebraic K-theory of discrete rings. Most situations where you would be interested in applying Segal's $\bar{H}$ to an (ordinary) symmetric monoidal category, it turns out that only the isomorphisms matter.

This changes when one's attention turns to symmetric monoidal categories where the morphisms form Γ-spaces. Then one focuses on "weak equivalences" within the category rather than on isomorphisms. Luckily, the "universal choice" of weak equivalences, is the most useful one. This choice is good enough for our applications, but has to be modified in more complex situations where we must be free to choose our weak equivalences. For a trivial example of this, see Note 2.3.3.3 below.

Consider the path components functor π_0 as a functor from $\mathcal{S}$-categories (or $\Gamma\mathcal{S}_*$-categories) to categories by letting $ob(\pi_0\mathcal{C}) = ob\mathcal{C}$, and $(\pi_0\mathcal{C})(c,d) = \pi_0(\mathcal{C}(c,d))$. This works by the monoidality of π_0. Likewise, the evaluation at 1_+ (see Example 2.1.2.1(3)), $R\colon \Gamma\mathcal{S}_* \to \mathcal{S}$ induces a functor R from $\Gamma\mathcal{S}_*$-categories to $\mathcal{S}$-categories. Note that $\pi_0\mathcal{C} \cong \pi_0 R T_0 \mathcal{C}$.

Define the functor

$$\omega\colon \Gamma\mathcal{S}_*\text{-categories} \to \mathcal{S}\text{-categories}$$

by means of the (categorical) pullback

$$\begin{array}{ccc} \omega\mathcal{C} & \xrightarrow{w_\mathcal{C}} & RT_0\mathcal{C} \\ \downarrow & & \downarrow \\ i\pi_0\mathcal{C} & \longrightarrow & \pi_0\mathcal{C} \end{array}$$

where $i\pi_0\mathcal{C}$ is the subcategory of isomorphisms in $\pi_0\mathcal{C}$.

Lemma 2.3.3.1 *Let $\mathcal{C}$ be a quite special Γ-$\Gamma\mathcal{S}_*$-category. Then $\omega\mathcal{C}$ is a quite special Γ-$\mathcal{S}$-category.*

Proof That $\mathcal{C}$ is quite special implies that $RT_0\mathcal{C}$ is quite special, since stable equivalences of stably fibrant Γ-spaces are pointwise equivalences, and hence taken to weak equivalences by R. The map $RT_0\mathcal{C} \to \pi_0 RT_0\mathcal{C} \cong \pi_0\mathcal{C}$ is a (pointwise) fibration since R takes fibrant Γ-spaces to fibrant spaces.

Furthermore, $\pi_0\mathcal{C}$ is special since π_0 takes stable equivalences of $\Gamma\mathcal{S}_*$-spaces to isomorphisms. The subcategory of isomorphisms in a special Γ-category is always special (since the isomorphism category in a product category is the product of the isomorphism categories), so $i\pi_0\mathcal{C}$ is special too.

We have to know that the pullback behaves nicely with respect to this structure. The map $RT_0\mathcal{C}(X \vee Y) \to RT_0\mathcal{C}(X) \times RT_0\mathcal{C}(Y)$ is a weak equivalence. Hence it is enough to show that if $\mathcal{A} \to \mathcal{B}$ is a weak equivalence of $\mathcal{S}$-categories with fibrant morphism spaces, then $i\pi_0\mathcal{A} \times_{\pi_0\mathcal{A}} \mathcal{A} \to i\pi_0\mathcal{B} \times_{\pi_0\mathcal{B}} \mathcal{B}$ is a weak equivalence. Notice that $ob\mathcal{A} \cong ob(i\pi_0\mathcal{A} \times_{\pi_0\mathcal{A}} \mathcal{A})$ and that two objects in $\mathcal{A}$ are isomorphic if and only if they are isomorphic as objects of $i\pi_0\mathcal{A} \times_{\pi_0\mathcal{A}} \mathcal{A}$ (and likewise for $\mathcal{B}$). Hence we only have to show that the map induces a weak equivalence on morphism spaces, which is clear since pullbacks along fibrations are equivalent to homotopy pullbacks. □

Lemma 2.3.3.2 *Let $\mathcal{E}$ be an Ab-category with subcategory $i\mathcal{E}$ of isomorphisms, and let $\tilde{\mathcal{E}}$ be the associated $\Gamma\mathcal{S}_*$-category (see Example* 2.1.6.2(2)). *Then the natural map $i\mathcal{E} \to \omega\tilde{\mathcal{E}}$ is a stable equivalence.*

Proof Since $\tilde{\mathcal{E}}$ has stably fibrant morphism objects $T_0\tilde{\mathcal{E}} \overset{\sim}{\longleftarrow} \tilde{\mathcal{E}}$ and by construction $R\tilde{\mathcal{E}} = \mathcal{E}$ (considered as an $\mathcal{S}$-category). This means also that $\pi_0\tilde{\mathcal{E}} \cong \mathcal{E}$, and the result follows. □

Note 2.3.3.3 So, for *Ab*-categories our uniform choice of weak equivalences essentially just picks out the isomorphisms, which is fine since that is what we usually want. For modules over **S**-algebras they also give a choice which is suitable for K-theory (more about this later).

However, occasionally this construction will not pick out the weak equivalences you had in mind. As an example, consider the category Γ^o itself with its monoidal structure coming from the sum. It turns out that the category of isomorphisms $i\Gamma^o = \coprod_{n\geq 0} \Sigma_n$ is an extremely interesting category: its algebraic K-theory is equivalent to the sphere spectrum by the Barratt–Priddy–Quillen theorem (see e.g., [257, Proposition 3.5]).

However, since Γ^o is a category with sum, by Sect. 2.3.1.2 it comes with a natural enrichment $(\Gamma^o)^\vee$. We get that $(\Gamma^o)^\vee(m_+, n_+)(k_+) \cong \Gamma^o(m_+, k_+ \wedge n_+)$. But in the language of Example 2.1.4.3(6), this is nothing but the n by m matrices over the sphere spectrum. Hence $(\Gamma^o)^\vee$ is isomorphic to the $\Gamma\mathcal{S}_*$-category whose objects are the natural numbers, and where the Γ-space of morphisms from n to m

is $Mat_{n,m}\mathbf{S} = \prod_m \vee_n \mathbf{S}$. The associated uniform choice of weak equivalences are exactly the "homotopy invertible matrices" $\widehat{GL}_n(\mathbf{S})$ of Definition 3.2.3.1, and the associated algebraic K-theory is the algebraic K-theory of **S**—also known as Waldhausen's algebraic K-theory of a point $A(*)$, see Sect. 3.2.3.

That **S** and $A(*)$ are different can for instance be seen from the fact that the stable homotopy groups of spheres are finite in positive dimension, whereas $A(*)$ is rationally equivalent to the K-theory of the integers, which is infinite cyclic in degree 5. For a further discussion, giving partial calculations, see Sect. 7.3.

Chapter 3
Reductions

In this chapter we will perform two important reductions. We will also clean up some of the mess caused by our use of varying definitions for algebraic K-theory along the way.

The first reduction takes place in Sect. 3.1 and tells one that our handling of simplicial rings in Note 1.3.4.1 is not in conflict with the usual conventions of algebraic K-theory, and in particular the one we obtain from Sect. 2.3.3. This is of importance even if one is only interested in ordinary rings: there are certain points (in Chap. 5) where even the statements for ordinary rings relies on functoriality of algebraic K-theory for the category of simplicial rings.

Armed with Sect. 3.1 and with Sect. 3.2, which tells us that all the various definitions of K-theory agree, those only interested in applications to discrete rings are free to pass on to Chap. 4.

The second reduction, which you will find in Sect. 3.3, is the fact that for most practical purposes, theorems that are true for simplicial rings are true in general for **S**-algebras. One may think of this as a sort of denseness property, coupled with the fact that the requirement that a functor is "continuous" is rather weak.

3.1 Degreewise K-Theory

Algebraic K-theory is on one hand a group-completion device, which is apparent from the definition of K_0. When looking at K_1 we can also view it as an "abelianization" device. You kill off the commutator of the general linear group to get K_1. To get K_2 you "kill off" yet another piece where some homology group vanishes. The procedure of killing off stuff to which homology is blind ends in group-theory at this point, but if you are willing to go into spaces, you may continue, and that is just what Quillen's plus-construction is all about.

When studying stable K-theory in Sect. 1.3, we had to introduce simplicial rings into the picture, and it turned out that we could be really naïve about it: we just applied our constructions in every dimension. That this works is quite surprising.

B.I. Dundas et al., *The Local Structure of Algebraic K-Theory*,
Algebra and Applications 18,
DOI 10.1007/978-1-4471-4393-2_3,

When one wants to study K-theory of simplicial rings, the degreewise application of the K-functor only rarely gives anything interesting. One way to get an interesting K-theory would be to take the S-construction of some suitable category of modules, but instead of isomorphisms use weak equivalences. Another, and simpler way, is to use Quillen's plus construction on a nice space similar to the classifying space of the general linear group. This is what we will do in this section, but it will not be proven until the next section that the two approaches are equivalent (by means of a yet another approach to K-theory due to Segal, see Sect. 2.3). The plus construction has the advantage that the comparison between the "correct" and degreewise definitions is particularly simple.

3.1.1 The Plus Construction

In this section we collect the facts we need about Quillen's plus construction made into a functorial construction, using Bousfield and Kan's integral completion functor [40]. For a long time, the best sources on the plus construction were the papers by Loday [179] and Hausmann and Husemoller [122] and Berrick's textbook [16], but Quillen's original account which had circulated for a very long time finally made it into the appendix of [87].

3.1.1.1 Acyclic Maps

Recall from Definition 1.1.6.1 that a map of pointed connected spaces is called *acyclic* if the reduced integral homology of the homotopy fiber vanishes. We need some facts about acyclic maps.

If Y is a connected space, we may form its *universal cover* $\tilde{Y}$ as follows. From $\sin|Y|$, form the space B by identifying two simplices $u, v \in \sin|Y|_q$ whenever, considered as maps $\Delta[q] \to \sin|Y|$, they agree on the one-skeleton of $\Delta[q]$. Then $\sin|Y| \twoheadrightarrow B$ is a fibration of fibrant spaces [201, 8.2], and $\tilde{Y}$ is defined by the pull-back diagram

$$\begin{array}{ccc} \tilde{Y} & \longrightarrow & B^{\Delta[1]} \\ \downarrow & & \downarrow \\ Y & \longrightarrow & B. \end{array}$$

Note that $\tilde{Y} \to Y$ is a fibration with fibers equivalent to the discrete set $\pi_1 Y$.

Lemma 3.1.1.1 *Let $f\colon X \to Y$ be a map of connected spaces, and $\tilde{Y}$ the universal cover of Y. Then f is acyclic if and only if*

$$H_*(X \times_Y \tilde{Y}) \to H_*(\tilde{Y})$$

is an isomorphism.

Proof We may assume that $X \to Y$ is a fibration of pointed spaces with fiber F. Then the projection $X \times_Y \tilde{Y} \to \tilde{Y}$ is also a fibration with fiber F, and the Serre spectral sequence [99, IV.5.1]

$$H_p\bigl(\tilde{Y}; H_q(F)\bigr) \Rightarrow H_{p+q}(X \times_Y \tilde{Y})$$

gives that if $\tilde{H}_*(F) = 0$, then the edge homomorphism (which is induced by $X \times_Y \tilde{Y} \to \tilde{Y}$) is an isomorphism as claimed.

Conversely, assume $H_*(X \times_Y \tilde{Y}) \to H_*(\tilde{Y})$ is an isomorphism. Then it is easy to check directly that $\tilde{H}_q(F) = 0$ for $q \leq 1$. Assume we have shown that $\tilde{H}_q(F) = 0$ for $q < k$ for a $k \geq 2$. Then the spectral sequence gives an exact sequence

$$H_{k+1}(X \times_Y \tilde{Y}) \xrightarrow{\cong} H_{k+1}(\tilde{Y}) \longrightarrow H_k(F) \longrightarrow H_k(X \times_Y \tilde{Y}) \xrightarrow{\cong} H_k(\tilde{Y}) \longrightarrow 0$$

which implies that $H_k(F) = 0$ as well. □

The lemma can be reformulated using homology with local coefficients: $H_*(\tilde{Y}) = H_*(Y; \mathbf{Z}[\pi_1 Y])$ and

$$H_*(X \times_Y \tilde{Y}) \cong H_*\bigl(\mathbf{Z}[\tilde{X}] \otimes_{\mathbf{Z}[\pi_1 X]} \mathbf{Z}[\pi_1 Y]\bigr) = H_*\bigl(X; f^*\mathbf{Z}[\pi_1 Y]\bigr),$$

so f is acyclic if and only if it induces an isomorphism

$$H_*\bigl(X; f^*\mathbf{Z}[\pi_1 Y]\bigr) \cong H_*\bigl(Y; \mathbf{Z}[\pi_1 Y]\bigr).$$

This can be stated in more general coefficients:

Corollary 3.1.1.2 *A map $f\colon X \to Y$ of connected spaces is acyclic if and only if for any local coefficient system $\mathcal{G}$ on Y, f induces an isomorphism*

$$H_*\bigl(X; f^*\mathcal{G}\bigr) \cong H_*(Y; \mathcal{G}).$$

Proof By the lemma we only need to verify one implication. If $i\colon F \to Y$ is the fiber of f, then the Serre spectral sequence gives

$$H_p\bigl(Y; H_q\bigl(F; i^* f^*\mathcal{G}\bigr)\bigr) \Rightarrow H_{p+q}\bigl(X; f^*\mathcal{G}\bigr).$$

However, $i^* f^*\mathcal{G}$ is a trivial coefficient system, so if $\tilde{H}_*(F) = 0$, the edge homomorphism must be an isomorphism. □

This reformulation of acyclicity is useful, for instance when proving the following lemma.

Lemma 3.1.1.3 *Let*

$$\begin{array}{ccc} X & \xrightarrow{f} & Y \\ {\scriptstyle g}\downarrow & & \downarrow{\scriptstyle g'} \\ Z & \xrightarrow{f'} & S \end{array}$$

be a pushout square of connected spaces with f acyclic, and either f or g a cofibration. Then f' is acyclic.

Proof Let $\mathcal{G}$ be a local coefficient system on S. Using the characterization Corollary 3.1.1.2 of acyclic maps as maps inducing isomorphism in homology with arbitrary coefficients, we get by excision that

$$H_*(S, Z; \mathcal{G}) \cong H_*\big(Y, X; \big(g'\big)^*\mathcal{G}\big) = 0$$

implying that f' is acyclic. □

Lemma 3.1.1.4 *Let $f\colon X \to Y$ be a map of connected spaces. Then f is a weak equivalence if and only if it is acyclic and induces an isomorphism of the fundamental groups.*

Proof Let F be the homotopy fiber of f. If f induces an isomorphism $\pi_1 X \cong \pi_1 Y$ on fundamental groups, then $\pi_1 F$ is abelian. If f is acyclic, then $\pi_1 F$ is perfect. Only the trivial group is both abelian and perfect, so $\pi_1 F = 0$. As $\tilde{H}_* F = 0$ the Whitehead theorem tells us that F is contractible. □

3.1.1.2 The Functorial Construction

We now give a functorial construction of the plus construction, following the approach of Bousfield and Kan [40, p. 218].

If X is any set, we may consider the free abelian group generated by X, and call it $\mathbf{Z}[X]$. If X is pointed we let $\tilde{\mathbf{Z}}[X] = \mathbf{Z}[X]/\mathbf{Z}[*]$. This defines a functor $\mathit{Ens}_* \to \mathit{Ab}$ which is adjoint to the forgetful functor $U\colon \mathit{Ab} \to \mathit{Ens}_*$, and extends degreewise to all spaces. The transformation given by the inclusion of the generators $X \to \tilde{\mathbf{Z}}[X]$ (where we symptomatically have forgotten to write the forgetful functor) induces the Hurewicz homomorphism $\pi_*(X) \to \pi_*(\tilde{\mathbf{Z}}[X]) = \tilde{H}_*(X)$.

As explained in Appendix A.1.3, the fact that $\tilde{\mathbf{Z}}$ is a left adjoint implies that it gives rise to a cosimplicial space $\underset{\sim}{\mathbf{Z}}$ via

$$\underset{\sim}{\mathbf{Z}}[X] = \big\{[n] \mapsto \tilde{\mathbf{Z}}^{n+1}[X]\big\},$$

where the superscript $n+1$ means that we have used the functor $\tilde{\mathbf{Z}}$ $n+1$ times. The total space (see Sect. A.2.8) of this cosimplicial space is called the *integral completion* of X and is denoted $\mathbf{Z}_\infty X$.

Bousfield and Kan define the integral completion in a slightly different, but isomorphic, manner, which has the advantage of removing the seeming dependence on a base point. Let $e_X\colon \mathbf{Z}[X] \to \mathbf{Z}$ be the homomorphism that sends $\sum n_i x_i$ to $\sum n_i$. Instead of considering the abelian group $\tilde{\mathbf{Z}}[X]$, Bousfield and Kan consider the set $e_X^{-1}(1)$. The composite $e_X^{-1}(1) \subseteq \mathbf{Z}[X] \twoheadrightarrow \tilde{\mathbf{Z}}[X]$ is a bijection, and one may define the integral completion for non-based spaces using $X \mapsto e_X^{-1}(1)$ instead.

If $f: X \to Y$ is a function of sets, then we define the space $\dot{\mathbf{Z}}[X]$ and the map $\dot{\mathbf{Z}}[X] \to Y$ by

$$\dot{\mathbf{Z}}[X] = e_X^{-1}(1) \cap \coprod_{y \in Y} \mathbf{Z}[f^{-1}(y)]$$

$$= \left\{ \sum n_i x_i \;\middle|\; f(x_1) = \cdots = f(x_n), \sum n_i = 1 \right\} \xrightarrow{\sum n_i x_i \mapsto f(x_1)} Y.$$

Note that if Y is a one-point space, then $\dot{\mathbf{Z}}[X] = e_X^{-1}(1)$, but usually $\dot{\mathbf{Z}}[X]$ will not be an abelian group. This construction is natural in f, and we may extend it to spaces, giving a cosimplicial subspace of $\underset{\sim}{\mathbf{Z}}[X]$, whose total is called the *fiberwise integral completion* of X (or rather, of f).

If X is a space, there is a natural fibration $\sin|X| \to \sin|X|/P$ given by "killing, in each component, $\pi_i(X)$ for $i > 1$ and the maximal perfect subgroup $P\pi_1(X) \subseteq \pi_1(X)$". More precisely, let $\sin|X|/P$ be the space obtained from $\sin|X|$ by identifying two simplices $u, v \in \sin|X|_q$ whenever, for every injective map $\phi \in \Delta([1],[q])$, we have $d_i\phi^* u = d_i\phi^* v$ for $i = 0, 1$, and

$$[\phi^* u]^{-1} * [\phi^* v] = 0 \in \pi_1(X, d_0\phi^* u)/P\pi_1(X, d_0\phi^* u).$$

The projection $\sin|X| \to \sin|X|/P$ is a fibration.

Definition 3.1.1.5 The plus construction $X \mapsto X^+$ is the functor given by the fiberwise integral completion of $\sin|X| \to \sin|X|/P$ (called the *partial integral completion* in Sect. 1.1.6.1), and $q_X: X \to X^+$ is the natural transformation coming from the inclusion $X \subseteq \dot{\mathbf{Z}}[\sin|X|]$.

That this is the desired definition follows from [40, p. 219], where they use the alternative description of Corollary 3.1.1.2 for an acyclic map:

Proposition 3.1.1.6 *If X is a pointed connected space, then*

$$q_X: X \to X^+$$

is an acyclic map killing the maximal perfect subgroup of the fundamental group. ☺

We note that q_X is always a cofibration (= inclusion).

3.1.1.3 Uniqueness of the Plus Construction

Proposition 3.1.1.6 characterizes the plus construction X^+ up to homotopy under X:

Theorem 3.1.1.7 *Consider the (solid) diagram of connected spaces*

$$\begin{array}{ccc} X & \xrightarrow{\ f\ } & Y \\ {\scriptstyle q_X}\big\downarrow & \overset{h}{\nearrow} & \\ X^+ & & \end{array}$$

If Y is fibrant and $P\pi_1 X \subseteq \ker\{\pi_1 X \to \pi_1 Y\}$, then there exists a dotted map $h\colon X^+ \to Y$ making the resulting diagram commutative. Furthermore, the map is unique up to homotopy, and is a weak equivalence if f is acyclic.

Proof Let $S = X^+ \coprod_X Y$ and consider the solid diagram

$$\begin{array}{ccccc} X & \xrightarrow{\ f\ } & Y & = & Y \\ {\scriptstyle q_X}\big\downarrow & & {\scriptstyle g}\big\downarrow & \overset{H}{\nearrow} & \big\downarrow \\ X^+ & \xrightarrow{\ f'\ } & S & \longrightarrow & * \end{array}$$

By Lemma 3.1.1.3, we know that g is acyclic. The van Kampen theorem [99, III.1.4] tells us that $\pi_1 S$ is the "free product" $\pi_1 X^+ *_{\pi_1 X} \pi_1 Y$, and the hypotheses imply that $\pi_1 Y \to \pi_1 S$ must be an isomorphism.

By Lemma 3.1.1.4, this means that g is a weak equivalence. Furthermore, as q_X is a cofibration, so is g. Thus, as Y is fibrant, there exists a dotted H making the diagram commutative, and we may choose $h = Hf'$. By the universal property of S, any h must factor through f', and the uniqueness follows by the uniqueness of H.

If f is acyclic, then both $f = hq_X$ and q_X are acyclic, and so h must be acyclic. Furthermore, as f is acyclic $\ker\{\pi_1 X \to \pi_1 Y\}$ must be perfect, but as $P\pi_1 X \subseteq \ker\{\pi_1 X \to \pi_1 Y\}$ we must have $P\pi_1 X = \ker\{\pi_1 X \to \pi_1 Y\}$. So, h is acyclic and induces an isomorphism on the fundamental group, and by Lemma 3.1.1.4 h is an equivalence. □

Lemma 3.1.1.8 *Let $X \to Y$ be a k-connected map of connected spaces. Then $X^+ \to Y^+$ is also k-connected.*

Proof Either one uses the characterization of acyclic maps by homology with local coefficients, and checks by hand that the lemma is right in low dimensions, or one can use our choice of construction and refer it away: [40, p. 113 and p. 42]. □

3.1.1.4 The Plus Construction on Simplicial Spaces

The plus construction on the diagonal of a simplicial space (bisimplicial set) may be performed degreewise in the following sense. Remember, Definition 1.1.2.1, that a quasi-perfect group is a group G in which the maximal perfect subgroup is the commutator: $PG = [G, G]$. The *diagonal* $\mathrm{diag}^* X$ of a simplicial space $X = \{[s] \mapsto X_s\}$ is the space obtained by precomposing with the diagonal functor $\mathrm{diag}\colon \Delta^o \to \Delta^o \times \Delta^o$, so that $(\mathrm{diag}^* X)_s = (X_s)_s$.

Lemma 3.1.1.9 *Let $\{[s] \mapsto X_s\}$ be a simplicial space such that X_s is connected for every $s \geq 0$. Let $X^+ = \{[s] \mapsto X_s^+\}$ be the "degreewise plus-construction". Consider the diagram*

$$\begin{array}{ccc} \mathrm{diag}^* X & \longrightarrow & \mathrm{diag}^*(X^+) \\ \downarrow & & \downarrow \\ (\mathrm{diag}^* X)^+ & \longrightarrow & (\mathrm{diag}^*(X^+))^+ \end{array},$$

where the upper horizontal map is induced by the plus construction $q_{X_s}\colon X_s \to X_s^+$, and the lower horizontal map is plus of the upper horizontal map.

The lower horizontal map is always an equivalence, and the right vertical map is an equivalence if and only if $\pi_1 \mathrm{diag}^*(X^+)$ has no nontrivial perfect subgroup. This is true if, for instance, $\pi_1(X_0^+)$ is abelian, which follows if $\pi_1(X_0)$ is quasi-perfect.

Proof Let $A(X_s)$ be the homotopy fiber of $q_{X_s}\colon X_s \to X_s^+$, and consider the sequence

$$A(X) = \{[s] \mapsto A(X_s)\} \longrightarrow X \xrightarrow{q_X} X^+$$

of simplicial spaces. As X_s and X_s^+ are connected, Theorem A.6.0.3 gives that

$$\mathrm{diag}^* A(X) \longrightarrow \mathrm{diag}^* X \xrightarrow{\mathrm{diag}^* q_X} \mathrm{diag}^*(X^+)$$

is a fiber sequence. But as each $A(X_s)$ is acyclic, the spectral sequence Theorem A.6.0.5 calculating the homology of a bisimplicial set gives that $\tilde{H}_*(\mathrm{diag}^* A(X)) = 0$, and so the map $\mathrm{diag}^* q_X\colon \mathrm{diag}^* X \to \mathrm{diag}^*(X^+)$ is acyclic. The lower horizontal map $(\mathrm{diag}^* X)^+ \to (\mathrm{diag}^*(X^+))^+$ in the displayed square is thus the plus of an acyclic map, and hence acyclic itself. However, $P\pi_1((\mathrm{diag}^* X)^+)$ is trivial, so this map must be an equivalence.

The right vertical map is the plus construction applied to $\mathrm{diag}^*(X^+)$, and so is an equivalence if and only if it induces an equivalence on π_1, i.e., if $P\pi_1(\mathrm{diag}^*(X^+)) = *$. If $\pi_1(X_0)$ is quasi-perfect, then $\pi_1(X_0^+) = \pi_1(X_0)/P\pi_1(X_0) = H_1(X_0)$ is abelian, and so the quotient $\pi_1 \mathrm{diag}^*(X^+)$ is also abelian, and hence has no perfect (nontrivial) subgroups. □

Remark 3.1.1.10 Note that some condition is needed to ensure that $\pi_1 \mathrm{diag}^*(X^+)$ is without nontrivial perfect subgroups, for let $X_q = BF_q$ where $F \xrightarrow{\sim} P$ is a free resolution of a perfect group P. Then $\mathrm{diag}^*(X^+) \simeq BP \not\simeq BP^+$.

3.1.1.5 Nilpotent Fibrations and the Plus Construction

Let π and G be groups, and let π act on G (through group homomorphisms). The action is *nilpotent* if there exists a finite filtration

$$* = G_{n+1} \subseteq G_n \subseteq \cdots \subseteq G_2 \subseteq G_1 = G$$

respected by the action, such that each $G_{i+1} \subset G_i$ is a normal subgroup and such that the quotients G_i/G_{i+1} are abelian with induced trivial action.

A group G is said to be *nilpotent* if the self-action via inner automorphisms is nilpotent.

Definition 3.1.1.11 If $f: E \to B$ is a fibration of connected spaces with connected fiber F, then $\pi_1(E)$ acts on each $\pi_i(F)$ (see Sect. A.5.1), and we say that f is *nilpotent* if these actions are nilpotent. Generally, we will say that a map of connected spaces $X \to Y$ is nilpotent if the associated fibration is.

Lemma 3.1.1.12 *If $F \to E \to B$ is any fiber sequence of connected spaces where $\pi_1 E$ acts trivially on $\pi_* F$, then $E \to B$ is nilpotent.*

Proof Since $\pi_q F$ is abelian for $q > 1$, a trivial action is by definition nilpotent, and the only thing we have to show is that the action of $\pi_1 E$ on $\pi_1 F$ is nilpotent. Let $A' = \ker\{\pi_1 F \to \pi_1 E\}$ and $A'' = \ker\{\pi_1 E \to \pi_1 B\}$. Since $\pi_1 E$ acts trivially on A' and A'', and both are abelian (the former as it is the cokernel of $\pi_2 E \to \pi_2 B$, and the latter as it is in the center of $\pi_1 E$), $\pi_1 E$ acts nilpotently on $\pi_1 F$. □

Lemma 3.1.1.13 *Let $f: X \to Y$ be a map of connected spaces. If either*

1. *f fits in a fiber sequence $X \xrightarrow{f} Y \longrightarrow Z$ where Z is connected and $P\pi_1(Z)$ is trivial, or*
2. *f is nilpotent,*

then

$$\begin{array}{ccc} X & \xrightarrow{f} & Y \\ {\scriptstyle q_X}\downarrow & & \downarrow{\scriptstyle q_Y} \\ X^+ & \xrightarrow{f^+} & Y^+ \end{array}$$

is (homotopy) cartesian.

Proof Part 1. Consider the map of fiber sequences

$$\begin{array}{ccccc} X & \xrightarrow{f} & Y & \longrightarrow & Z \\ \downarrow & & {\scriptstyle q_Y}\downarrow & & \downarrow{\scriptstyle q_Z} \\ F & \longrightarrow & Y^+ & \longrightarrow & Z^+ \end{array}$$

in the homotopy category. Since $P\pi_1 Z$ is trivial, $q_Z\colon Z \to Z^+$ is an equivalence, and so the homotopy fibers of $X \to F$ and q_Y are equivalent. Hence the map $X \to F$ is acyclic. To see that $X \to F$ is equivalent to q_X, Theorem 3.1.1.7 gives that we must show that $\pi_1 X \twoheadrightarrow \pi_1 F$ is surjective and that $\pi_1 F$ is without nontrivial perfect subgroups. Surjectivity follows by chasing the map of long exact sequences of fibrations. That a perfect subgroup $P \subseteq \pi_1 F$ must be trivial follows since $\pi_1 Y^+$ is without nontrivial perfect subgroups, and so P must be a subgroup of the abelian subgroup $\ker\{\pi_1 F \to \pi_1 Y^+\} \cong \operatorname{coker}\{\pi_2 Y^+ \to \pi_2 Z^+\}$.

Part 2. That f is nilpotent is equivalent, up to homotopy, to the statement that f factors as a tower of fibrations

$$Y = Y_0 \xleftarrow{\ f_1\ } Y_1 \xleftarrow{\ f_2\ } \cdots \xleftarrow{\ f_k\ } Y_k = X$$

where each f_i fits in a fiber sequence

$$Y_i \xrightarrow{\ f_i\ } Y_{i-1} \longrightarrow K(G_i, n_i)$$

with $n_i > 1$ (see e.g., [40, p. 61]). But statement 1 tells us that this implies that

$$\begin{array}{ccc} Y_i & \longrightarrow & Y_i^+ \\ \downarrow & & \downarrow \\ Y_{i-1} & \longrightarrow & Y_{i-1}^+ \end{array}$$

is cartesian, and by induction on k, the statement follows. □

3.1.2 K-Theory of Simplicial Rings

A simplicial monoid M is called *group-like* if $\pi_0 M$ is a group. This has the nice consequence that we may form a good classifying space. That is, if BM is (the diagonal of) the space you get by taking the classifying space degreewise, then $\Omega BM \simeq M$ (see Corollary A.6.1.3).

If A is a simplicial (associative and unital) ring, Waldhausen [298] defined $\widehat{GL}_n(A)$ as the pullback of the diagram

$$\begin{array}{ccc} \widehat{GL}_n(A) & \longrightarrow & M_n(A) \\ \downarrow & & \downarrow \\ GL_n(\pi_0 A) & \longrightarrow & M_n(\pi_0 A) \end{array}$$

(since the rightmost vertical map is a surjection of simplicial abelian groups, it is a fibration, cf. Sect. A.4.1(3), which means that the square is also a *homotopy* pullback in the sense of Sect. A.8). Just as in the discrete case, $\widehat{GL}_n(A)$ sits inside $\widehat{GL}_{n+1}(A)$

via $m \mapsto m \oplus 1$, and we let $\widehat{GL}(A)$ be the union of the $\widehat{GL}_n(A)$. As $\pi_0\widehat{GL}_n(A) = GL_n(\pi_0(A))$ we get that $\widehat{GL}_n(A)$, and hence also $\widehat{GL}(A)$, is group-like.

In analogy with the definition of the algebraic K-theory space Definition 1.1.6.4 of a ring, Waldhausen suggested the following definition.

Definition 3.1.2.1 If A is a simplicial ring, then the *algebraic K-theory space* of A is

$$K(A) = B\widehat{GL}(A)^+.$$

We note that

$$\pi_1 K(A) = \pi_1 B\widehat{GL}(A)/P\big(\pi_1 B\widehat{GL}(A)\big) = GL(\pi_0 A)/P\big(GL(\pi_0 A)\big) = K_1(\pi_0 A)$$

(where $P()$ denotes the maximal perfect subgroup, Definition 1.1.2.1). This pattern does not continue, the homotopy fiber of the canonical map $K(A) \to K(\pi_0 A)$ has in general highly nontrivial homotopy groups. Waldhausen proves in [298, Proposition 1.2] that if k is the first **positive** number for which $\pi_k A$ is nonzero, the first nonvanishing homotopy group of the homotopy fiber of the map $K(A) \to K(\pi_0 A)$ sits in dimension $k+1$ and is isomorphic to the zeroth Hochschild homology group $HH_0(\pi_0 A, \pi_k A) = \pi_k A/[\pi_0 A, \pi_k A]$. We shall not prove this now, but settle for the weaker

Lemma 3.1.2.2 *If $B \to A$ is a $k > 0$-connected map of simplicial rings, then the induced map $K(B) \to K(A)$ is $(k+1)$-connected.*

Proof Obviously $M_n A \to M_n B$ is k-connected. As $k > 0$, we have $\pi_0 B \cong \pi_0 A$, and so $\widehat{GL}_n(B) \to \widehat{GL}_n(A)$ is also k-connected. Hence, the map of classifying spaces $B\widehat{GL}(B) \to B\widehat{GL}(A)$ is $(k+1)$-connected, and we are done as the plus construction preserves connectivity of maps by Lemma 3.1.1.8. □

3.1.2.1 Spaces Under BA_5

Let A_n be the alternating group on n letters. For $n \geq 5$ this is a perfect group with no nontrivial normal subgroups. We give a description of Quillen's plus for BA_5 by adding cells. Since A_5 is perfect, it is enough, by Theorem 3.1.1.7, to display a map $BA_5 \to Y$ inducing an isomorphism in integral homology, where Y is simply connected.

Let α be a nontrivial element in A_5. This can be thought of as a map $S^1 = \Delta[1]/\partial\Delta[1] \to BA_5$ (consider α as an element in $B_1 A_5$; since $B_0 A_5 = *$ this is a loop). Form the pushout

$$\begin{array}{ccc} S^1 & \longrightarrow & D^2 \\ {\scriptstyle\alpha}\downarrow & & \downarrow \\ BA_5 & \longrightarrow & X_1. \end{array}$$

Since A_5 has no nontrivial normal subgroups, the van Kampen theorem [99, III.1.4] tells us that X_1 is simply connected. The homology sequence of the pushout splits up into

$$0 \to H_2(A_5) \to H_2(X_1) \to H_1(S^1) \to 0, \quad \text{and} \quad H_q(A_5) \cong H_q(X_1), \quad \text{for } q \neq 1.$$

Since $H_1(S^1) \cong \mathbf{Z}$, we may choose a splitting $\mathbf{Z} \to H_2(X_1) \cong \pi_2(X_1)$, and we let $\beta\colon S^2 \to \sin|X_1|$ represent the image of a generator of $\mathbf{Z}$. Form the pushout

$$\begin{array}{ccc} S^2 & \longrightarrow & D^3 \\ {\scriptstyle\beta}\downarrow & & \downarrow \\ \sin|X_1| & \longrightarrow & X_2. \end{array}$$

We get isomorphisms $H_q(X_1) \cong H_q(X_2)$ for $q \neq 2, 3$, and an exact diagram

$$\begin{array}{ccccccccccc} & & & & & & & & H_2(A_5) & & \\ & & & & & & & & \downarrow & & \\ 0 & \to & H_3(X_1) & \to & H_3(X_2) & \to & H_2(S^2) & \to & H_2(X_1) & \to & H_2(X_2) & \to & 0. \\ & & & & & & & & \downarrow & & \\ & & & & & & & & H_1(S^1) & & \end{array}$$

However, by the definition of β, the composite $H_2(S^2) \to H_2(X_1) \to H_1(S^1)$ is an isomorphism. Hence $H_3(X_1) \cong H_3(X_2)$ and $H_2(A_5) \cong H_2(X_2)$. Collecting these observations, we get that the map $BA_5 \to X_2$ is an isomorphism in homology and $\pi_1 X_2 = 0$, and

$$BA_5 \to \text{“}BA_5^+\text{”} = X_2$$

is a model for the plus construction.

Proposition 3.1.2.3 *Let $\mathcal{C}$ be the category of spaces under BA_5 with the property that if $BA_5 \to Y \in ob\mathcal{C}$ then the image of A_5 normally generates $P\pi_1 Y$. Then the bottom arrow in the pushout diagram*

$$\begin{array}{ccc} BA_5 & \longrightarrow & \text{“}BA_5^+\text{”} \\ \downarrow & & \downarrow \\ Y & \longrightarrow & \text{“}Y^+\text{”} \end{array}$$

is a functorial model for the plus construction in $\mathcal{C}$.

Proof As it is clearly functorial, we only have to check the homotopy properties of $Y \to$ "Y^+" as given in Theorem 3.1.1.7. By Lemma 3.1.1.3, it is acyclic, and by van Kampen [99, III.1.4] π_1("Y^+") $= \pi_1 Y *_{A_5} \{1\}$. Using that the image of A_5 normally generates $P\pi_1 Y$ we get that π_1("Y^+") $= \pi_1 Y/P\pi_1 Y$, and we are done. □

Example 3.1.2.4 If R is some ring, then we get a map

$$A_5 \subseteq \Sigma_5 \subseteq \Sigma_\infty \subseteq GL(\mathbf{Z}) \to GL(R).$$

We will show that $E(R)$ is normally generated by

$$\alpha = \begin{bmatrix} 0 & 1 & 0 \\ 0 & 0 & 1 \\ 1 & 0 & 0 \end{bmatrix} \in A_3 \subseteq A_5.$$

The relation $e^1_{41} = [[\alpha, e^{-1}_{43}], e^{-1}_{21}]$ reduces the problem to showing that e^1_{41} normally generates $E(R)$, which follows from the Steinberg relations Sect. 1.1.5: if $r \in R$, then $e^r_{i1} = [e^r_{i4}, e^1_{41}]$ if $1 \neq i \neq 4$, $e^r_{41} = [e^r_{4i}, e^1_{i1}]$ if $i \neq 1$, $e^r_{4j} = [e^1_{41}, e^r_{1j}]$ if $1 \neq j \neq 4$ and finally $e^r_{ij} = [e^r_{i1}, e^1_{1j}]$ if $1 \neq j \neq i \neq 4$. Hence, all spaces under BA_5, satisfying the requirement that the map on fundamental groups is the inclusion $A_5 \subseteq GL(R)$ lie in $\mathcal{C}$. In particular, if A is an $\mathbf{S}$-algebra we get that $B\widehat{GL}(A)$, as defined in 3.2.3.1 below, is in this class, since $\pi_1 B\widehat{GL}(A) \cong GL(\pi_0 A)$, and the algebraic K-theory of $\mathbf{S}$-algebras *could* be defined as "$B\widehat{GL}(A)^+$".

3.1.3 Degreewise K-Theory

Let A be a simplicial ring (unital and associative as always). Waldhausen's construction $B\widehat{GL}(A)^+$ is very different from what we get if we apply Quillen's definition to A degreewise, i.e.,

$$K^{deg}(A) = \operatorname{diag}^*\big\{[q] \mapsto K(A_q)\big\}.$$

This is also a useful definition. For instance, we know by [97] that if A is a regular and right Noetherian ring, then $K(A)$ agrees with the Karoubi–Villamayor K-theory of A, which may be defined to be the degreewise K-theory of a simplicial ring $\Delta A = \{[q] \mapsto A[t_0, \ldots, t_q]/\sum t_i = 1\}$ with

$$d_i t_j = \begin{cases} t_j & \text{if } j < i, \\ 0 & \text{if } i = j, \\ t_{j-1} & \text{if } j > i. \end{cases}$$

That is, for regular right Noetherian rings the canonical map $K(A) \to K^{deg}(\Delta A)$ is a weak equivalence, and interestingly, it is $K^{deg}(\Delta A)$ which is the central actor in important theories like motivic homotopy theory, not $K(A)$. On the other hand, since $t_1 \in \Delta_1 A$ is a path between 0 and 1, and any connected unital simplicial ring is contractible ("multiplication by a path from 0 to 1" gives a contraction), we get by Lemma 3.1.2.2 that $K(\Delta A)$ is contractible, and so, in this case Waldhausen's functor gives very little information.

For ease of notation, let $GL(A)$ be the simplicial group $\{[t] \mapsto GL(A_t)\}$ obtained by applying GL to every degree of A and let $BGL(A)$ be the diagonal of the bisimplicial set $\{[s],[t] \mapsto B_s GL(A_t)\}$.

Lemma 3.1.3.1 *Let A be a simplicial ring. There is a natural chain of weak equivalences*

$$K^{deg}(A) \xrightarrow{\sim} K^{deg}(A)^+ \xleftarrow{\sim} BGL(A)^+.$$

Proof The Whitehead Lemma 1.1.2.2 states that $K_1(A_0)$ is abelian, and so Lemma 3.1.1.9 with $X = BGL(A)$ gives the desired equivalences. □

The inclusion $GL(A) \subset \widehat{GL}(A)$ induces a map

$$BGL(A)^+ \to B\widehat{GL}(A)^+ = K(A)$$

By Lemma 3.1.3.1, the first space is equivalent to $K^{deg}(A)$, and it is of interest to know what information is preserved by this map.

Example 3.1.3.2 The following example is rather degenerate, but still of great importance. For instance, it was the example we considered when talking about stable algebraic K-theory in Sect. 1.3.5.

Let A be a discrete ring, and let P be a *reduced* A-bimodule (in the sense that it is a simplicial bimodule, and $P_0 = 0$). Consider the square zero extension $A \ltimes P$ as in Sect. 1.3.1 (that is, $A \ltimes P$ is isomorphic to $A \oplus P$ as a simplicial abelian group, and the multiplication is given by $(a_1, p_1) \cdot (a_2, p_2) = (a_1 a_2, a_1 p_2 + p_1 a_2)$). Then one sees that $GL(A \ltimes P)$ is actually equal to $\widehat{GL}(A \ltimes P)$: as P is reduced and A discrete $GL(\pi_0(A \ltimes P)) \cong GL(\pi_0 A) = GL(A)$ and as P is square zero $\ker\{GL(A \ltimes P) \to GL(A)\} = (1 + M(P))^\times \cong M(P)$. Hence, all "homotopy invertible" matrices are actually invertible: $GL(A \ltimes P) = \widehat{GL}(A \ltimes P)$.

If you count the number of occurrences of the comparison of degreewise and ordinary K-theory in what is to come, it is this trivial example that will pop up most often. However, we have essential need of the more general cases too. We are content with only an equivalence, and even more so, only an equivalence in relative K-theory. In order to extend this example to cases where A might not be discrete and P not reduced, we have to do some preliminary work.

3.1.3.1 Degreewise vs. Ordinary K-Theory of Simplicial Rings

Recall the definition of the subgroup of elementary matrices $E \subseteq GL$. For this section, we reserve the symbol $K_1(A)$ for the quotient of simplicial groups $\{[q] \mapsto$

$K_1(A_q)\} = GL(A)/E(A)$, which must not be confused with $\pi_1 K(A) \cong K_1(\pi_0 A)$. Let $\widehat{E}(A) \subset \widehat{GL}(A)$ consist of the components of Waldhausen's grouplike monoid $\widehat{GL}(A)$ (see Sect. 3.1.2) belonging to the subgroup $E(\pi_0 A) \subseteq GL(\pi_0 A)$ of elementary matrices. Much of the material in this section is adapted from the paper [64].

Theorem 3.1.3.3 *Let A be a simplicial ring. Then*

$$\begin{array}{ccc} BGL(A) & \longrightarrow & BGL(A)^+ \\ \downarrow & & \downarrow \\ B\widehat{GL}(A) & \longrightarrow & B\widehat{GL}(A)^+ \end{array}$$

is (homotopy) cartesian.

Proof Note that both horizontal maps in the left square of

$$\begin{array}{ccccc} BE(A) & \longrightarrow & BGL(A) & \longrightarrow & BK_1(A) \\ \downarrow & & \downarrow & & \downarrow \\ B\widehat{E}(A) & \longrightarrow & B\widehat{GL}(A) & \longrightarrow & BK_1(\pi_0 A) \end{array}$$

satisfy the conditions in Lemma 3.1.1.13(1), since both rows are fiber sequences with base spaces simplicial abelian groups.

So we are left with proving that

$$\begin{array}{ccc} BE(A) & \longrightarrow & BE(A)^+ \\ \downarrow & & \downarrow \\ B\widehat{E}(A) & \longrightarrow & B\widehat{E}(A)^+ \end{array}$$

is cartesian, but by Lemma 3.1.1.13(2) this follows from the lemma below. □

Lemma 3.1.3.4 (cf. [83] or [276]) *The map $BE(A) \to B\widehat{E}(A)$ is nilpotent.*

Proof For $1 \leq k \leq \infty$ let $j_k \colon E_k(A) \to \widehat{E}_k(A)$ be the inclusions and let F_k be the homotopy fiber of $Bj_k \colon BE_k(A) \to B\widehat{E}_k(A)$. For convenience we abbreviate our notation for the colimits under stabilization $\widehat{E}_k(A) \to \widehat{E}_{k+1}(A)$, given by block sum $g \mapsto g \oplus 1$, and write $j = j_\infty \colon E(A) \subseteq \widehat{E}(A)$ and $F = F_\infty$.

Instead of showing that the action of $\pi_0 E(A) \cong \pi_1 BE(A)$ on $\pi_*(F)$ is nilpotent, we show that $\pi_0 E(A) \to \pi_0 Map_*(F, F) \to \mathrm{End}(\pi_*(F))$ is trivial. In view of Lemma 3.1.1.12 this is sufficient, and it is in fact an equivalent statement since $\pi_0 E(A)$ is perfect (being a quotient of $E(A_0)$) and any nilpotent action of a perfect group is trivial.

We have an isomorphism

$$Map_*(F, F) \cong \lim_{\overleftarrow{n}} Map_*(F_n, F)$$

and, since homotopy groups commute with filtered colimits,

$$\mathrm{End}\big(\pi_*(F)\big) \cong \lim_{\overleftarrow{n}} \mathrm{Hom}\big(\pi_*(F_n), \pi_*(F)\big).$$

Hence it is enough to show that for each k the composite

$$E(A_0) \twoheadrightarrow \pi_0 E(A) \to \lim_{\overleftarrow{n}} \pi_0 Map_*(F_n, F) \to \pi_0 Map_*(F_k, F)$$
$$\to \mathrm{Hom}\big(\pi_*(F_k), \pi_*(F)\big)$$

is trivial.

Now we fix a $k > 2$. To show that the homomorphism is trivial, it is enough to show that a set of normal generators is in the kernel. In Example 3.1.2.4 above, we saw that $e^1_{4,1}$ is a normal generator for $E(A_0)$, and by just the same argument $e^1_{k+1,1}$ will also normally generate $E(A_0)$, so it is enough to show that $e^1_{k+1,1}$ is killed by $E(A_0) \to \mathrm{Hom}(\pi_*(F_k), \pi_*(F))$.

Consider the simplicial category $j_k/1$ with objects $\widehat{E}_k(A)$ and where a morphism in degree q from m to n is a $g \in E_k(A_q)$ such that $m = n \cdot g$. The classifying space $B(j_k/1)$ is isomorphic to the bar construction $B(\widehat{E}_k(A), E_k(A), *) = \{[q] \mapsto \widehat{E}_k(A) \times E_k(A)^{\times q}\}$. The forgetful functor $j_k/1 \to E_k(A)$ (where $E_k(A)$ is considered to be a simplicial category with one object in each degree) induces an equivalence $B(j_k/1) \xrightarrow{\sim} F_k$ (see e.g., Corollary A.6.1.4) compatible with stabilization $t\colon E_k(A) \to E_{k+1}(A)$. By Theorem A.5.2.1, the action on the fiber

$$B(j_k/1) \times E_k(A) \xrightarrow{\sim} B(j_k/1) \times \Omega B E_k(A) \to B(j_k/1)$$

is induced by the simplicial functor

$$j_k/1 \times E_k(A) \xrightarrow{(m,g) \mapsto i_g(m)} j_k/1$$

(where $E_k(A)$ now is considered as a simplicial discrete category with one object for every element in $E_k(A)$ and only identity morphisms) sending (m, g) to $i_g(m) = gmg^{-1}$.

In order to prove that $e^1_{k+1,1}$ is killed, we consider the factorization

$$E(A_0) \to \pi_0 Map_*\big(B(j_k/1), B(j/1)\big) \to \mathrm{Hom}\big(\pi_*(F_k), \pi_*(F)\big)$$

and show that $e^1_{k+1,1}$ is killed already in $\pi_0 Map_*(B(j_k/1), B(j/1))$.

As natural transformations give rise to homotopies (cf. Sect. A.2.4.1), we are done if we display a natural simplicial isomorphism between t and $i_{e^1_{k+1,1}} \circ t$ in the category of pointed functors $[j_k/1, j/1]_*$, where $t(m) = m \oplus I$ and $i_{e^1_{k+1,1}}(m) = e^1_{k+1,1} m e^{-1}_{k+1,1}$. If $m = (m_{ij}) \in M_k(A)$ is any matrix, we have that $i_{e^1_{k+1,1}}(t(m)) = t(m) \cdot \tau(m)$ where

$$\tau(m) = e^{-1}_{k+1,1} \cdot \prod_{1 \le j \le k} e^{m_{1j}}_{k+1,j}.$$

It is easy to check that $\tau(m)$ is simplicial ($\psi^*\tau(m) = \tau(\psi^* m)$ for $\psi \in \Delta$) and natural in $m \in j_k/1$. Thus, $m \mapsto \tau(m)$ is the desired natural isomorphism between $i_{e^1_{k+1,1}} t$ and t in $[j_k/1, j/1]_*$. □

The outcome is that we are free to choose our model for the homotopy fiber of the plus construction applied to $B\widehat{GL}(A)$ among the known models for the homotopy fiber of the plus construction applied to $BGL(A)$:

Corollary 3.1.3.5 *If X is any functor from discrete rings to spaces with a natural transformation $X(-) \to BGL(-)$ such that*

$$X(A) \to BGL(A) \to BGL(A)^+$$

is a fiber sequence for any ring A, then X extends degreewise to a functor of simplicial rings with a natural transformation $X \to BGL \to B\widehat{GL}$ such that

$$X(A) \to B\widehat{GL}(A) \to B\widehat{GL}(A)^+$$

is a fiber sequence for any simplicial ring A.

Proof By Theorem 3.1.3.3 is enough to show that $[q] \mapsto X(A_q)$ is equivalent to the homotopy fiber of $BGL(A) \to BGL(A)^+$, but this will follow if $\{[q] \mapsto BGL(A_q)\}^+ \to \{[q] \mapsto BGL(A_q)^+\}$ is an equivalence. By Lemma 3.1.1.9 this is true since $GL(A_0)$ is quasi-perfect, which is part of the Whitehead Lemma 1.1.2.2. □

3.1.4 K-Theory of Simplicial Radical Extensions May Be Defined Degreewise

If $f: B \to A$ is a map of simplicial (associative and unital) rings, we will let $K(f)$ denote the homotopy fiber of the induced map $K(B) \to K(A)$.

If f is surjective and $I_q = \ker\{f_q : B_q \to A_q\}$ is inside the *Jacobson radical* $rad(B_q) \subseteq B_q$ (that is, $1 + x$ is invertible in B_q if $x \in I_q$) for every $q \geq 0$ we say that f is a *radical extension*. This situation has been treated previously also in Sects. 1.1.2 and 1.2.5.

We recall some basic properties of radical extensions. Notice that the Jacobson radical is a two-sided ideal, and that any nil-ideal (i.e., an ideal consisting of nilpotent elements) is contained in the Jacobson radical.

Recall following the slightly extended version of Nakayama's lemma from Bass [13, pp. 85–86].

Lemma 3.1.4.1 *Let $f: B \to A$ be a radical extension of discrete rings with kernel I. Then*

1. (*Nakayama's lemma*) *If M is a finitely generated B-module such that $MI = M$, then $M = 0$.*

2. *For every* n*, the two-sided ideal* $M_n(I)$ *is contained in the Jacobson radical of the matrix ring* $M_n(B)$.
3. *For every* n*, the induced group homomorphism* $GL_n(B) \to GL_n(A)$ *is a surjection with kernel* $(1+M_n(I))^\times$*, the subgroup of* $GL_n(B)$ *of matrices of the form* $1+m$ *where* $m \in M_n(I)$.

☺

We studied radical extensions of discrete rings in Sect. 1.2.5, and by the following proposition, this gives information about the simplicial case as well:

Proposition 3.1.4.2 *Let* $f\colon B \twoheadrightarrow A$ *be a radical extension of simplicial rings. Then the relative K-theory* $K(f)$ *is equivalent to* $\operatorname{diag}^*\{[q] \mapsto K(f_q)\}$.

Proof The proof follows closely the one given in [102] for the nilpotent case. Let $I = \ker\{f\colon B \to A\}$. Since all spaces are connected we may just as well consider

$$[q] \mapsto \operatorname{hofib}\{BGL(B_q)^+ \to BGL(A_q)^+\}.$$

As $\pi_1(BGL(-)^+)$ has values in abelian groups, we see by Lemma 3.1.1.9. that $\operatorname{diag}^*\{[q] \mapsto BGL(-)^+\}$ is equivalent to the plus of the diagonal $BGL(A)^+$. Hence to prove the proposition it is enough to prove that

$$\begin{array}{ccc} BGL(B)^+ & \longrightarrow & B\widehat{GL}(B)^+ \\ \downarrow & & \downarrow \\ BGL(A)^+ & \longrightarrow & B\widehat{GL}(A)^+ \end{array}$$

is homotopy cartesian.

Note that $GL_n(B_q) \to GL_n(A_q)$ is a group epimorphism with kernel $(1+M_n(I_q))^\times$, the multiplicative group of all $n \times n$ matrices of the form $1+m$ where m has entries in I. Hence $B(1+M(I))^\times$ is the (homotopy) fiber of $BGL(B) \to BGL(A)$.

Furthermore, $(1+M_n(I))^\times$ is also the fiber of the epimorphism of group-like simplicial monoids $\widehat{GL}_n(B) \to \widehat{GL}_n(A)$. This follows as $J = \ker\{\pi_0(B) \to \pi_0(A)\}$ is a radical ideal in $\pi_0(B)$ (for any $x \in J$, the sum $1+x$ is invertible in $\pi_0 B$ since there is a $y \in I_0$ mapping to x such that $1+y$ is invertible in B_0), which implies that

$$\begin{aligned}(1+M_n(J))^\times &= \ker\{GL_n(\pi_0(B)) \to GL_n(\pi_0(A))\} \\ &= \ker\{M_n(\pi_0(B)) \to M_n(\pi_0(A))\}.\end{aligned}$$

Consequently $B(1+M(I))^\times$ is also the (homotopy) fiber of $B\widehat{GL}(B) \to B\widehat{GL}(A)$, and so

$$\begin{array}{ccc} BGL(B) & \longrightarrow & BGL(A) \\ \downarrow & & \downarrow \\ B\widehat{GL}(B) & \longrightarrow & B\widehat{GL}(A) \end{array}$$

is homotopy cartesian. By Theorem 3.1.3.3 all vertical squares in the cube

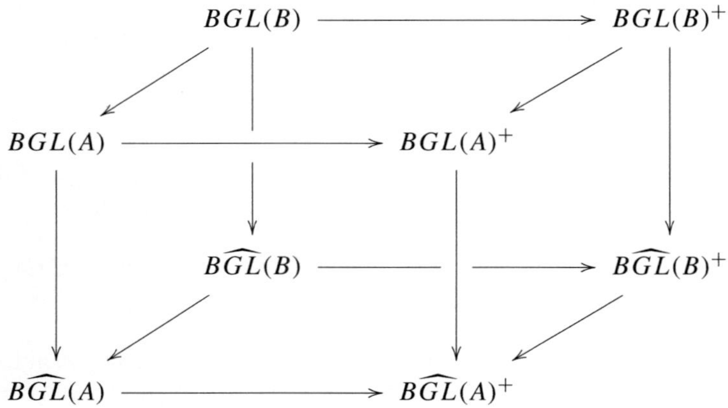

of reduced spaces and 1-connected maps, except possibly

$$\begin{array}{ccc} BGL(B)^+ & \longrightarrow & B\widehat{GL}(B)^+ \\ \downarrow & & \downarrow \\ BGL(A)^+ & \longrightarrow & B\widehat{GL}(A)^+ \end{array}$$

are homotopy cartesian, and so this square is also homotopy cartesian. □

Example 3.1.4.3 (The Resolving Complex and Stein Relativization) When we have an extension where the kernel is not in the radical, the difference between degreewise and ordinary K-theory is significant. However, there is a historical precedence for studying relative situations by means of degreewise K-theory. We have already seen in Example 1.1.4.1 that the most naïve kind of excision fails for algebraic K-theory. Related to this is the classical method of describing relative K-theory. In Bass' [13] and Milnor's [213] books on K-theory, the Stein relativization is used to describe relative K-theory. As is admitted in Milnor's book [213, p. 56], this is not a satisfactory description, and we will give the reason why it works in low dimensions, but fails higher up. See [281] to get further examples of the failure. The reader might also want to consult [180] and [166].

Let $f: A \to B$ be a surjection of associative rings with unit, and define the *Stein relativization* as the cokernel

$$K_i^{Stein}(f) = \operatorname{coker}\{K_i(A) \to K_i(A \times_B A)\}$$

of the map induced by the diagonal $A \to A \times_B A$. The question is: when do we have exact sequences

$$\cdots \to K_{i+1}(A) \to K_{i+1}(B) \to K_i^{Stein}(f) \to K_i(A) \to K_i(B) \to \cdots,$$

or more precisely, how far is

$$\begin{array}{ccc} K(A\times_B A) & \xrightarrow{pr_1} & K(A) \\ {\scriptstyle pr_2}\downarrow & & {\scriptstyle f}\downarrow \\ K(A) & \xrightarrow{f} & K(B) \end{array}$$

from being cartesian?

The failure turns up for $i=2$, but this oughtn't be considered as bad as was fashionable at the time: The Stein relativization can be viewed as a first approximation to the homotopy fiber as follows. Let S be the "resolving complex", i.e., the simplicial ring given in dimension q as the $q+1$ fold product of A over f with the various projections and diagonals as face and degeneracies

$$\cdots A\times_B A\times_B A \;\substack{\longrightarrow\\\longrightarrow\\\longrightarrow}\; A\times_B A \;\substack{\longrightarrow\\\longrightarrow}\; A.$$

This gives a factorization $A\to S\to B$ where the former map is inclusion of the zero skeleton, and the latter is a weak equivalence (since $A\to B$ was assumed to be a surjection). Now, as one may check directly, $GL(-)$ respects products, and

$$GL(S_q)\cong GL(A)\times_{GL(B)}\cdots\times_{GL(B)} GL(A)$$

($q+1$ $GL(A)$ factors). Just as for the simplicial ring S, this simplicial group is concentrated in degree zero, but as GL does not respect surjections we see that $\pi_0(GL(S))\cong im\{GL(A)\to GL(B)\}$ may be different from $GL(B)$. But this is fine, for as $E(-)$ respects surjections we get that $GL(B)/im\{GL(A)\to GL(B)\}\cong \bar{K}_1(B)=K_1(B)/im\{K_1(A)\to K_1(B)\}$, and we get a fiber sequence

$$BGL(S)\to B\widehat{GL}(S)\to B\bar{K}_1(B)$$

where the middle space is equivalent to $BGL(B)$. Applying Theorem 3.1.3.3 (overkill as $\bar{K}_1(B)$ is abelian) to $BGL(S)\to B\widehat{GL}(S)$ we get that there is a fiber sequence

$$K^{deg}(S)\to K(B)\to B\bar{K}_1(B)$$

which means that $\phi(f)=\text{hofib}\{K(A)\to K^{deg}(S)\}$ is the connected cover of the homotopy fiber of $K(A)\to K(B)$.

We may regard $\phi(f)$ as a simplicial space $[q]\mapsto \phi_q(f)=\text{hofib}\{K(A)\to K(S_q)\}$. Then $\phi_0(f)=0$ and $\pi_i(\phi_1(f))=K_{i+1}^{Stein}(f)$. An analysis shows that $d_0-d_1+d_2\colon \pi_0(\phi_2(f))\to\pi_0(\phi_1(f))$ is zero, whereas $d_0-d_1+d_2-d_3\colon \pi_0(\phi_3(f))\to \pi_0(\phi_2(f))$ is surjective, so the E_2 term of the spectral sequence associated to the simplicial space looks like

$$\begin{array}{lllll} 0 & K_3^{Stein}(f)/? & \dots & & \\ 0 & K_2^{Stein}(f)/? & ? & \dots & \\ 0 & K_1^{Stein}(f) & 0 & ? & \dots \end{array}$$

This gives that $K_1^{Stein}(f)$ is correct, whereas $K_2^{Stein}(f)$ surjects onto π_2 of relative K-theory.

3.2 Agreement of the Various K-Theories

This section aims at removing any uncertainty due to the many definitions of algebraic K-theory that we have used. In Sect. 3.2.1 we show that the approaches of Waldhausen and Segal agree, at least for additive categories. In Sect. 3.2.2 we show that Segal's machine is an infinite delooping of the plus-construction, and show how this is related to group-completion. In Sect. 3.2.3 we give the definition of the algebraic K-theory space of an **S**-algebra. For *spherical group rings* as in Example 2.1.4.3(2), i.e., **S**-algebras of the form $\mathbf{S}[G]$ for G a simplicial group, we show that the algebraic K-theory space of $\mathbf{S}[G]$ is the same as Waldhausen's algebraic K-theory of the classifying space BG. Lastly, we show that the definition of the algebraic K-theory of an **S**-algebra as defined in Chap. 2 is the infinite delooping of the plus-construction.

3.2.1 The Agreement of Waldhausen's and Segal's Approaches

We give a quick proof of the fact that the S-construction of Chap. 1 and the $\bar{H}$-construction of Chap. 2 coincide when applied to additive categories. This fact is much more general, and applies to a large class of categories with cofibrations and weak equivalences where the cofibrations are "splittable up to weak equivalences", see Waldhausen's [301, Sect. 1.8].

3.2.1.1 Segal's Construction Applied to Categories with Cofibrations

Let $\mathfrak{C}$ be a category with cofibrations. By forgetting structure we may consider it as a category with sum and apply Segal's Γ-space machine Sect. 2.3 to it, or we may apply Waldhausen's S-construction Definition 1.2.2.1.

Note that Segal's Γ-space machine could be reinterpreted as the functor $\bar{H}\mathfrak{C}$ from the category Γ^o of finite pointed sets to the category of categories with sum, whose value at $k_+ = \{0, \dots, k\}$ is the category $\bar{H}\mathfrak{C}(k_+)$ described as follows. Its objects are functors to $\mathfrak{C}$ from the pointed category of subsets and inclusions of $k_+ = \{0, 1, \dots, k\}$, sending $0_+ = \{0\}$ to the zero object $0 \in \mathfrak{C}$ and pushout squares to pushout squares in $\mathfrak{C}$. The morphisms are simply natural transformations of such diagrams. For instance, $\bar{H}\mathfrak{C}(1_+)$ is isomorphic to $\mathfrak{C}$, whereas $\bar{H}\mathfrak{C}(2_+)$ consists of pushout diagrams

$$\begin{array}{ccc} 0 & \longrightarrow & c_{\{0,1\}} \\ \downarrow & & \downarrow \\ c_{\{0,2\}} & \longrightarrow & c_{\{0,1,2\}} \end{array}.$$

We see that $\bar{H}\mathfrak{C}(k_+)$ is equivalent as a category to $\mathfrak{C}^{\times k}$ via the map sending a functor $c \in ob\bar{H}\mathfrak{C}(k_+)$ to $c_{\{0,1\}}, \ldots, c_{\{0,k\}}$. However, $\mathfrak{C}^{\times k}$ is not necessarily functorial in k, making $\bar{H}\mathfrak{C}$ the preferred model for the bar construction of $\mathfrak{C}$.

Also, this formulation of $\bar{H}\mathfrak{C}$ is naturally isomorphic to the one we gave in Sect. 2.3, but the advantage is that it is easier to compare with Waldhausen's construction.

Any functor from Γ^o is naturally a simplicial object by precomposing with the circle $S^1 \colon \Delta^o \to \Gamma^o$ (after all, the circle is a simplicial **finite pointed** set). We could of course precompose with any other simplicial finite pointed set, and part of the point about Γ-spaces was that if M was a functor from Γ^o to sets, then $\{m \mapsto M(S^m)\}$ is a spectrum.

3.2.1.2 The Relative $\bar{H}$-Construction

In order to compare Segal's and Waldhausen's constructions it will be convenient to have a concrete model for the homotopy fiber of $\bar{H}$ applied to an exact functor $\mathcal{C} \to \mathcal{D}$ of categories with sum (or more generally, a symmetric monoidal functor of symmetric monoidal categories). To this end we define the simplicial Γ-category $C_{\mathcal{C}\to\mathcal{D}}$ by the (categorical) pullback

$$\begin{array}{ccc} C_{\mathcal{C}\to\mathcal{D}}(X) & \longrightarrow & \bar{H}\mathcal{D}(PS^1 \wedge X) \\ \downarrow & & \downarrow \\ \bar{H}\mathcal{C}(S^1 \wedge X) & \longrightarrow & \bar{H}\mathcal{D}(S^1 \wedge X) \end{array}$$

(as always, for this to be a functor we need to have made choices of pullbacks in the category of sets). Here PS^1 is the "path space" on S^1 as defined in Appendix A.2.7: $(PS^1)_q = S^1_{q+1}$ where the ith face map is the $i+1$st face map in S^1, and where the zeroth face map of S^1 induces a weak equivalence $PS^1 \to S^1_0 = *$. The point of this construction is Lemma 3.2.1.3 which displays it as a relative version of the $\bar{H}$-construction, much like the usual construction involving the path space in topological spaces.

Usually categorical pullbacks are of little value, but in this case it turns out that the categorical pullback is equivalent to the *fiber product*.

Definition 3.2.1.1 Let $\mathcal{C}_1 \xrightarrow{f_1} \mathcal{C}_0 \xleftarrow{f_2} \mathcal{C}_2$ be a diagram of categories. The *fiber product* $\prod(f_1, f_2)$ is the category whose objects are tuples (c_1, c_2, α) where $c_i \in$

obC_i for $i = 1,2$ and α is an isomorphism in $\mathcal{C}_0$ from f_1c_1 to f_2c_2; and where a morphism from (c_1, c_2, α) to (d_1, d_2, β) is a pair of morphisms $g_i : c_i \to d_i$ for $i = 1,2$ such that

$$\begin{array}{ccc} f_1c_1 & \xrightarrow{\alpha} & f_2c_2 \\ {\scriptstyle f_1g_1}\downarrow & & \downarrow{\scriptstyle f_2g_2} \\ f_1d_1 & \xrightarrow{\beta} & f_2d_2 \end{array}$$

commutes.

Fiber products (like homotopy pullbacks) are good because of their invariance: if you have a diagram

$$\begin{array}{ccccc} \mathcal{C}_1 & \xrightarrow{f_1} & \mathcal{C}_0 & \xleftarrow{f_2} & \mathcal{C}_2 \\ \downarrow\simeq & & \downarrow\simeq & & \downarrow\simeq \\ \mathcal{C}_1' & \xrightarrow{f_1'} & \mathcal{C}_0' & \xleftarrow{f_2'} & \mathcal{C}_2' \end{array}$$

where the vertical maps are equivalences, you get an equivalence $\prod(f_1, f_2) \to \prod(f_1', f_2')$. Note also the natural map $F : \mathcal{C}_1 \times_{\mathcal{C}_0} \mathcal{C}_2 \to \prod(f_1, f_2)$ sending (c_1, c_2) to $(c_1, c_2, 1_{f_1c_1})$.

This map is occasionally an equivalence, as is exemplified in the following lemma. If $\mathcal{C}$ is a category, then $Iso\,\mathcal{C}$ is the class of isomorphisms, and if f is a morphism, let sf be its source and tf be its target.

Lemma 3.2.1.2 *Let $\mathcal{C}_1 \xrightarrow{f_1} \mathcal{C}_0 \xleftarrow{f_2} \mathcal{C}_2$ be a diagram of categories, and assume that the map of classes*

$$Iso\,\mathcal{C}_1 \xrightarrow{g \mapsto (f_1g, sg)} Iso\,\mathcal{C}_0 \times_{obC_0} obC_1$$

has a section (the pullback is taken along source and f_1). Then the natural map

$$F : \mathcal{C}_1 \times_{\mathcal{C}_0} \mathcal{C}_2 \to \prod(f_1, f_2)$$

is an equivalence.

Proof Let $\sigma : Iso\,\mathcal{C}_0 \times_{obC_0} obC_1 \to Iso\,\mathcal{C}_1$ be a section, and define

$$G : \prod(f_1, f_2) \to \mathcal{C}_1 \times_{\mathcal{C}_0} \mathcal{C}_2$$

by $G(c_1, c_2, \alpha) = (t\sigma(\alpha, c_1), c_2)$ and $G(g_1, g_2) = (\sigma(d_1, \beta)g_1\sigma(c_1, \alpha)^{-1}, g_2)$. Checking the diagrams proves that F and G are inverses up to natural isomorphisms built out of σ. □

One should think about the condition in Lemma 3.2.1.2 as a categorical equivalent of the Kan-condition in simplicial sets. This being one of the very few places you can find an error (even tiny and in the end totally irrelevant) in Waldhausen's papers, it is cherished by his fans, since in [301] he seems to claim that the pullback is equivalent to the fiber products if f_1 has a section (which is false). At this point there is even a small error in [118, p. 257], where it seems that they claim that the map in Lemma 3.2.1.2 factors through f_1.

Now, $Iso\, \bar{H}\mathcal{D}(PS^1 \wedge X) \to Iso\, \bar{H}\mathcal{D}(S^1 \wedge X) \times_{ob\bar{H}\mathcal{D}(S^1 \wedge X)} ob\bar{H}\mathcal{D}(PS^1 \wedge X)$ has a section given by pushouts in the relevant diagrams. Hence $C_{\mathcal{C}\to\mathcal{D}}(X)$ is equivalent to the fiber product category, and as such is invariant under equivalences. Consequently the natural map

$$C_{\mathcal{C}\to\mathcal{D}}(k_+)_q \longrightarrow \mathcal{C}^{\times qk} \times_{\mathcal{D}^{\times qk}} \mathcal{D}^{\times(q+1)k} \cong \mathcal{C}^{\times qk} \times \mathcal{D}^{\times k}$$

is an equivalence. If we consider categories with sum and weak equivalences, we get a structure of sum and weak equivalence on $C_{\mathcal{C}\to\mathcal{D}}$ as well, with

$$wC_{\mathcal{C}\to\mathcal{D}}(X) = w\bar{H}\mathcal{C}(S^1 \wedge X) \times_{w\bar{H}\mathcal{D}(S^1 \wedge X)} w\bar{H}\mathcal{D}(PS^1 \wedge X).$$

Notice also that the construction is natural: if you have a commuting diagram

$$\begin{array}{ccc} \mathcal{C} & \longrightarrow & \mathcal{D} \\ \downarrow & & \downarrow \\ \mathcal{C}' & \longrightarrow & \mathcal{D}' \end{array}$$

you get an induced map $C_{\mathcal{C}\to\mathcal{D}} \to C_{\mathcal{C}'\to\mathcal{D}'}$ by using the universal properties of pullbacks, and the same properties ensure that the construction behaves nicely with respect to composition. Furthermore $C_{*\to\mathcal{D}}(1_+)$ is isomorphic to $\mathcal{D}$, so we get a map $\mathcal{D} \cong C_{*\to\mathcal{D}}(1_+) \to C_{\mathcal{C}\to\mathcal{D}}$, and if we have maps $\mathcal{C} \to \mathcal{D} \to \mathcal{E}$ whose composite is trivial, we get a map $C_{\mathcal{C}\to\mathcal{D}}(1_+) \to \mathcal{E}$.

Recall that, if $\mathcal{C}$ is a category with sum (i.e., with finite coproducts and with a "zero object" which is both final and initial), then an *exact functor* to another category with sums $\mathcal{D}$ is a functor $\mathcal{C} \to \mathcal{D}$ preserving finite coproducts and the zero objects.

In the following lemma we use the fact that the classifying space construction embeds the category of small categories as a full subcategory of the category of spaces; and in this way we apply the language of spaces to categories, cf. Sect. A.2.4.1. For example, that the sequence of functors "is a stable fiber sequence" means that this is true for the sequence of Γ-spaces you get by applying B to the sequence of functors.

Lemma 3.2.1.3 *Let $\mathcal{C} \to \mathcal{D}$ be an exact functor of small categories with sum and weak equivalences. Then there is a stable fiber sequence*

$$w\bar{H}\mathcal{C} \to w\bar{H}\mathcal{D} \to w\bar{H}\big(C_{\mathcal{C}\to\mathcal{D}}(1_+)\big).$$

Proof It is enough to show that

$$w\bar{H}\mathcal{D}(S^1) \to w\bar{H}\big(C_{\mathcal{C}\to\mathcal{D}}(1_+)\big)(S^1) \to w\bar{H}\big(\bar{H}(\mathcal{C})(S^1)\big)(S^1)$$

is a fiber sequence, and by Theorem A.6.0.3 this follows since in each degree n

$$w\bar{H}\mathcal{D}(S^1) \to w\bar{H}\big(C_{\mathcal{C}\to\mathcal{D}}(1_+)_n\big)(S^1) \to w\bar{H}\big(\bar{H}(\mathcal{C})(S^1)_n\big)(S^1)$$

is equivalent to the product fiber sequence

$$w\bar{H}\mathcal{D}(S^1) \to w\bar{H}\big(\mathcal{D}\times\mathcal{C}^{\times n}\big)(S^1) \to w\bar{H}\big(\mathcal{C}^{\times n}\big)(S^1)$$

and all spaces involved are connected. □

We have a canonical map

$$\bar{H}\mathcal{C}(S^1) \to S\mathcal{C}$$

which in dimension q is induced by sending the sum diagram $C \in ob\bar{H}\mathcal{C}(q_+)$ to $c \in obS_q\mathcal{C}$ with $c_{ij} = C_{\{0,i+1,i+2,\dots,j-1,j\}}$ where $c_{i,j-1} \to c_{i,j}$ is induced by the inclusion $\{0,i,\dots,j-1\} \subset \{0,i,\dots,j-1,j\}$ and $c_{i,j} \to c_{i+1,j}$ is the canonical map $C_{\{0,i,\dots,j\}} \to C_{\{0,i+1,\dots,j\}}$ induced by the canonical map $C_{\{0,i\}} \to *$ and the identity on $C_{\{0,i+1,\dots,j\}}$. For instance, the object

$$\begin{array}{ccc} * & \longrightarrow & C_{\{0,1\}} \\ \downarrow & & \downarrow \\ C_{\{0,2\}} & \longrightarrow & C_{\{0,1,2\}} \end{array}$$

in $\bar{H}\mathcal{C}(2_+)$ is sent to

$$\begin{array}{ccc} C_{\{0,1\}} & \longrightarrow & C_{\{0,1,2\}} \\ & & \downarrow \\ & & C_{\{0,2\}} \end{array}$$

in $S_2\mathcal{C}$, where $C_{\{0,1,2\}} \to C_{\{0,2\}}$ is induced by $C_{\{0,1\}} \to *$.

Scholium 3.2.1.4 The additivity theorem for Waldhausen's S-construction says that induced map $iS(S_k\mathcal{C}) \to iS(\mathcal{C}^{\times k})$ is a weak equivalence. We have not used this so far, but in the special case of additive categories it is an immediate consequence of Theorem 3.2.1.5 below. The additivity theorem for Segal's model, saying that $i\bar{H}(S_k\mathcal{C}) \to i\bar{H}(\mathcal{C}^{\times k})$ is a weak equivalence, is at the core of the proof of the theorem, and constitutes the second half of the proof.

Theorem 3.2.1.5 *Let $\mathfrak{C}$ be an additive category. Then the map*

$$i\bar{H}\mathfrak{C}(S^1) \to iS\mathfrak{C}$$

described above is a weak equivalence.

Proof Since both $Bi\bar{H}\mathfrak{C}$ and $BiS\mathfrak{C}$ are connected, the vertical maps in

$$\begin{array}{ccc} Bi\bar{H}\mathfrak{C}(S^1) & \longrightarrow & BiS\mathfrak{C} \\ \simeq\downarrow & & \simeq\downarrow \\ \Omega(Bi\bar{H}(\bar{H}\mathfrak{C}(S^1))(S^1)) & \longrightarrow & \Omega(Bi\bar{H}(S\mathfrak{C})(S^1)) \end{array}$$

are equivalences by Proposition A.6.1.2, and so it is enough to prove that

$$Bi\bar{H}\big(\bar{H}\mathfrak{C}(S^1)\big) \to Bi\bar{H}(S\mathfrak{C})$$

is an equivalence, which again follows if we can show that for every q

$$Bi\bar{H}\big(\bar{H}\mathfrak{C}(q_+)\big) \to Bi\bar{H}(S_q\mathfrak{C})$$

is an equivalence.

Essentially this is the old "triangular matrices vs. diagonal matrices" question, and can presumably be proven directly by showing that $iS_qC \to i\mathfrak{C}^{\times q}$ induces an isomorphism in homology after inverting $\pi_0(iS_qC) \cong \pi_0(i\mathfrak{C}^{\times q})$.

Assume we have proven that the projection $i\bar{H}(S_k\mathfrak{C}) \to i\bar{H}(\mathfrak{C}^{\times k})$ is an equivalence for $k < q$ (this is trivial for $k = 0$ or $k = 1$). We must show that it is also an equivalence for $k = q$. Consider the inclusion by degeneracies $\mathfrak{C} \to S_q\mathfrak{C}$ (sending c to $0 \rightarrowtail 0 \rightarrowtail \cdots \rightarrowtail 0 \rightarrowtail c$), and the last face map $S_q\mathfrak{C} \to S_{q-1}\mathfrak{C}$. We want to show that we have a map of fiber sequences

$$\begin{array}{ccccc} i\bar{H}(\mathfrak{C}) & \longrightarrow & i\bar{H}(S_q\mathfrak{C}) & \longrightarrow & i\bar{H}(S_{q-1}\mathfrak{C}) \\ \| & & \downarrow & & \downarrow \\ i\bar{H}(\mathfrak{C}) & \longrightarrow & i\bar{H}\big(\mathfrak{C}^{\times q}\big) & \longrightarrow & i\bar{H}\big(\mathfrak{C}^{\times q-1}\big) \end{array}.$$

We do have maps of fiber sequences

$$\begin{array}{ccccc} i\bar{H}(\mathfrak{C}) & \longrightarrow & i\bar{H}(S_q\mathfrak{C}) & \longrightarrow & i\bar{H}\big(C_{\mathfrak{C}\to S_q\mathfrak{C}}(1_+)\big) \\ \| & & \downarrow & & \downarrow \\ i\bar{H}(\mathfrak{C}) & \longrightarrow & i\bar{H}\big(\mathfrak{C}^{\times q}\big) & \longrightarrow & i\bar{H}\big(C_{\mathfrak{C}\to \mathfrak{C}^{\times q}}(1_+)\big) \end{array}$$

and the only trouble lies in identifying the base spaces of the fibrations. We have a commuting square

$$\begin{array}{ccc} i\bar{H}\big(C_{\mathfrak{C}\to S_q\mathfrak{C}}(1_+)\big) & \longrightarrow & i\bar{H}(S_{q-1}\mathfrak{C}) \\ \downarrow & & \simeq\downarrow \\ i\bar{H}\big(C_{\mathfrak{C}\to \mathfrak{C}^{\times q}}(1_+)\big) & \xrightarrow{\sim} & i\bar{H}\big(\mathfrak{C}^{\times q-1}\big) \end{array}$$

the bottom map obviously an equivalence (and the right vertical map an equivalence by the induction hypothesis). We have to show that the top map is an equivalence, and for this purpose it is enough to show that

$$iC_{\mathfrak{C}\to S_q\mathfrak{C}}(1_+) \xrightarrow{\ p\ } iS_{q-1}\mathfrak{C}$$

is an equivalence. For every $c \in obS_{q-1}\mathfrak{C}$ the over category p/c is a simplicial category. If we can show that p/c is contractible for all c we are done by Quillen's theorem A [232]. In dimension n, the objects of the category p/c consists of certain sum diagrams of dimension $n+1$ of objects in $S_q\mathfrak{C}$ together with some extra data. Call the vertices of cardinality one $c_0, \dots, c_n$. Part of the data is an isomorphism $d_q c_0 \cong c$, and $c_1, \dots, c_n$ only have nonzero elements in the last column (i.e., $(c_k)(i \le j) = 0$ if $0 < k$ and $j < q$). Hence $(p/c)_n$ is equivalent to the category $iC_{\mathfrak{C}\to\mathfrak{C}_x}(n_+)$ where $x = c_{0,q-1}$ and $\mathfrak{C}_x$ is the category of split inclusions $x \rightarrowtail y \in \mathfrak{C}$ (which is a category with sum by taking pushout over the structure maps from x). The equivalence is induced by sending $c_0, \dots, c_n$ to $x \rightarrowtail (c_0)_{0,q}, (c_1)_{0,q}, \dots, (c_n)_{0,q}$ (considered as objects in $\mathfrak{C}_x \times \mathfrak{C}^{\times n}$). The equivalence is natural in n, and so induces an equivalence $p/c \to iC_{\mathfrak{C}\to\mathfrak{C}_x}(S^1)$, and we show that the latter is contractible.

This is the group completion part: it does not matter what x we put in $\mathfrak{C}_x$. First we show that $\pi_0(iC_{\mathfrak{C}\to\mathfrak{C}_x}(S^1)) = 0$ (which implies that $iC_{\mathfrak{C}\to\mathfrak{C}_x}(S^1) \simeq \Omega i\bar{H}C_{\mathfrak{C}\to\mathfrak{C}_x}(S^1)$) and then that $i\bar{H}\mathfrak{C} \to i\bar{H}\mathfrak{C}_x$ is an equivalence.

The vertices of $iC_{\mathfrak{C}\to\mathfrak{C}_x}(S^1)$ are split inclusions $x \rightarrowtail c$; the 1-simplices in the nerve direction are isomorphisms under x, whereas the 1-simplices in the $\bar{H}$-construction are pushout diagrams

$$\begin{array}{ccc} x & \longrightarrow & c \\ {\scriptstyle in_x}\big\downarrow & & \big\downarrow \\ x \vee c'' & \longrightarrow & c \vee c'' \end{array}$$

Hence, in $\pi_0(iC_{\mathfrak{C}\to\mathfrak{C}_x}(S^1))$ the class of $x \rightarrowtail c$ is equal the class of $x \overset{in_x}{\rightarrowtail} x \vee c/x$ (since the inclusion was splittable), which is equal to the class of the basepoint $x = x$.

Finally, consider the map $i\bar{H}\mathfrak{C} \to i\bar{H}\mathfrak{C}_x$. It is induced by $j\colon \mathfrak{C} \to \mathfrak{C}$ sending c to $in_x\colon x \rightarrowtail x \vee c$, and it has a section $q\colon \mathfrak{C}_x \to \mathfrak{C}$ given by sending $x \rightarrowtail c$ to c/x (there is no danger in choosing quotients). We have to show that jq induces a self map on $i\bar{H}\mathfrak{C}_x$ homotopic to the identity. Note that there is a natural isomorphism $c \coprod_x c \to c \vee c/x \cong c \times c/x$ under x given by sending the first summand by the identity to the first factor, and the second summand to the identity on the first factor and the projection on the second factor. Hence, 2 (twice the identity) is naturally isomorphic to $1 + jq$ in $i\bar{H}\mathfrak{C}_x$, and since this is a connected H-space we have homotopy inverses, giving that jq is homotopic to the identity. □

3.2.2 Segal's Machine and the Plus Construction

We give a brief review of Segal's results on group completion, focusing on the examples that are important to our applications. There are many excellent accounts related to this issue (see e.g., [5, 87, 99, 108, 148, 209] and [217]), but we more or less follow the approach of [257].

Let $\mathcal{C}$ be a symmetric monoidal category with weak equivalences, and consider the simplicial Γ-category $H'\mathcal{C}$ defined by the pullback

$$\begin{array}{ccc} H'\mathcal{C}(X)_q & \longrightarrow & \bar{H}\mathcal{C}(PS^1_q\wedge X) \\ \downarrow & & \downarrow \\ \bar{H}\mathcal{C}(PS^1_q\wedge X) & \longrightarrow & \bar{H}\mathcal{C}(S^1_q\wedge X) \end{array} .$$

By the same considerations as in Corollary 3.2.1.3 (i.e., by reversal of priorities with respect to simplicial directions) we get a fiber sequence

$$w\bar{H}\mathcal{C}(S^1) \longrightarrow wH'\mathcal{C}(S^1) \longrightarrow w\bar{H}\mathcal{C}(PS^1\wedge S^1),$$

but the last simplicial category is contractible, and so $w\bar{H}\mathcal{C}(S^1) \to wH'\mathcal{C}(S^1)$ is an equivalence.

Furthermore, the Γ-category $wH'\mathcal{C}$ is not only special, but very special: it has a homotopy inverse obtained by flipping the defining square around the diagonal.

Recall that a *cofinal* submonoid M' in a symmetric monoid M is a submonoid such that for all $a \in M$ there is a $b \in M$ such that $a \cdot b \in M'$. If M is a multiplicatively closed subset of a commutative ring A, then it is immediate that localizing A with respect to M' or M give isomorphic results.

Lemma 3.2.2.1 *Let $\mathcal{C}$ be a symmetric monoidal category with weak equivalences. Then the map*

$$w\bar{H}\mathcal{C} \to wH'\mathcal{C}$$

is a stable equivalence and $wH'\mathcal{C}$ is very special. Furthermore, if $\mu \subseteq w\mathcal{C}$ is a symmetric monoidal subcategory such that the image of $\pi_0\mu$ in $\pi_0 w\mathcal{C}$ is cofinal and $wT_{\mathcal{C},\mu}$ is defined as the pullback

$$\begin{array}{ccc} wT_{\mathcal{C},\mu} & \longrightarrow & \bar{H}\mathcal{C}(PS^1) \\ \downarrow & & \downarrow \\ \bar{H}\mu(PS^1) & \longrightarrow & \bar{H}\mathcal{C}(S^1) \end{array} ,$$

then the natural map $wT_{\mathcal{C},\mu} \to wH'\mathcal{C}(S^1)$ is an acyclic map.

Consequently there is a chain of natural equivalences

$$(BwT_{\mathcal{C},\mu})^+ \xrightarrow{\sim} (BwH'\mathcal{C}(S^1))^+ \xleftarrow{\sim} BwH'\mathcal{C}(S^1) \xleftarrow{\sim} Bw\bar{H}\mathcal{C}(S^1).$$

Proof Only the part about $wT_{\mathcal{C},\mu} \to wH'\mathcal{C}(S^1)$ being an acyclic map needs explanation. Since $wH'\mathcal{C}(S^1)$ is an H-space, this is equivalent to claiming that the map induces an isomorphism in integral homology.

By coherence theory (see e.g., [198, 4.2] or [161]), we may assume that $w\mathcal{C}$ is "permutative" (the associativity and unitality isomorphisms are identities, while the symmetric structure is still free to wiggle). Hence we are reduced to the following proposition: given a simplicial monoid M (the simplicial set given by the nerve of $w\mathcal{C}$) which is commutative up to all higher homotopies and a submonoid $\mu \subseteq M$ whose image in $\pi_0 M$ is cofinal, then the map $Y \to X$ given by the pullback squares

$$\begin{array}{ccccc} Y & \longrightarrow & X & \longrightarrow & EM \\ \downarrow & & \downarrow & & \downarrow \\ E\mu & \longrightarrow & EM & \longrightarrow & BM \end{array}$$

induces an isomorphism in homology. Analyzing the structures, we see that $Y \to X$ is nothing but the map of one-sided bar-constructions (cf. Sect. A.5.2) $B(M \times \mu, \mu, *) \subseteq B(M \times M, M, *)$ (with the diagonal action). Segal gives an argument why this is an isomorphism in homology in [257, pp. 305–306] by an explicit calculation with arbitrary field coefficients.

The argument is briefly as follows: let k be a field and let $H = H_*(M; k)$; which is a graded ring since M is a monoid, and a Hopf algebra due to the diagonal map:

$$H = H_*(M; k) \xrightarrow{H_*(\text{diagonal})} H_*(M \times M; k) \underset{\cong}{\xleftarrow{\text{Künneth}}} H \otimes_k H.$$

The E^1-term of the spectral sequence for computing $H_*(B(M \times M, M, *); k)$ (in dimension q it is $(H \otimes_k H) \otimes H^{\otimes_k q} \otimes_k k)$ is exactly the standard complex for calculating $Tor_*^H(H \otimes_k H, k)$ (where the H-module structure on $H \otimes_k H$ is induced by the coproduct $H \to H \otimes_k H$).

The map $H \to (H \otimes_k H) \otimes_H k$ given by $h \mapsto h \otimes 1 \otimes 1$ inverts elements in $\pi_0 M \subseteq k[\pi_0 M] \cong H_0(M; k)$: If $p \in \pi_0 M$ then $(p \otimes 1 \otimes 1) \cdot (1 \otimes p \otimes 1) = p \otimes p \otimes 1 = \Delta p \otimes 1 = 1 \otimes 1 \otimes 1$. Induction on the degree and the fact that if $n > 0$ and $h \in H_n$ then $\Delta h = 1 \otimes h + h \otimes 1 + \sum_{i=1}^{n-1} h_i' \otimes h_{n-i}''$ for $h_i', h_i'' \in H_i$ gives that the induced map $H[\pi^{-1}] \to (H \otimes_k H) \otimes_H k$ (where $\pi = \pi_0 M$) is an isomorphism. Furthermore, using that localization in the commutative case is flat we get that $Tor_s^H(H \otimes_k H, k) = 0$ for $s > 0$.

In consequence, we get that $H_*(B(M \times M, M, *); k) \cong H[\pi^{-1}]$. A similar calculation gives $H_*(B(M \times \mu, \mu, *); k) \cong H[\pi_0(\mu)^{-1}]$, and the induced map is an isomorphism since the image of $\pi_0(\mu)$ in $\pi_0(M)$ is cofinal. □

3.2.2.1 Application to the K-Theory of Discrete Rings

As an example we may consider the category of finitely generated free modules over a discrete ring A. For this purpose we use the model $\mathcal{F}_A$ of Definition 1.2.1.4

whose objects are natural numbers and the morphisms are matrices of appropriate sizes with entries in A. Assume for simplicity that A has the invariance of basis number property (IBN, see Example 1.1.3.2(4)). Then $Bi\mathcal{F}_A$ is the simplicial monoid $\coprod_{n\in\mathbf{N}} BGL_n(A)$ under Whitney sum (block sum). If $\mu = ob\mathcal{F}_A = \mathbf{N}$ then $BiT_{\mathcal{F}_A,\mathbf{N}} = B(Bi\mathcal{F}_A \times \mathbf{N}, \mathbf{N}, *)$ is a model for the homotopy colimit over the maps $\coprod_{n\in\mathbf{N}} BGL_n(A) \to \coprod_{n\in\mathbf{N}} BGL_n(A)$ given by Whitney sum (with identity matrices of varying sizes). The homotopy colimit is in turn equivalent to the homotopy colimit over the natural numbers over the maps $\coprod_{n\in\mathbf{N}} BGL_n(A) \to \coprod_{n\in\mathbf{N}} BGL_n(A)$ given by Whitney sum with the rank one identity matrix. This homotopy colimit is equivalent to the corresponding categorical colimit, which simply is $\mathbf{Z} \times BGL(A)$. Hence Lemma 3.2.2.1 says that there is a chain of weak equivalences between $\mathbf{Z} \times BGL(A)^+$ and $\Omega Bi\bar{H}\mathcal{F}_A$. Hence, for the category of finitely generated free modules over a ring A with the invariance of basis number property, the approaches through S, $\bar{H}$ and + are all equivalent:

$$\mathbf{Z} \times BGL(A)^+ \simeq \Omega Bi\bar{H}\mathcal{F}_A(S^1) \xrightarrow{\simeq} \Omega BiS\mathcal{F}_A.$$

If we instead consider the category $\mathcal{P}_A$ of finitely generated projective modules over a ring A, and $\mu = ob\mathcal{F}_A \subseteq \mathcal{P}_A$, then $T_{i\mathcal{P}_A,\mu} \simeq K_0(A) \times BGL(A)$ (since $\mathcal{F}_A \subseteq \mathcal{P}_A$ is cofinal), and we get

Theorem 3.2.2.2 *Let A be a discrete ring. Then there is a chain of equivalences*

$$K_0(A) \times BGL(A)^+ \simeq \Omega Bi\bar{H}\mathcal{P}_A(S^1) \xrightarrow{\simeq} \Omega BiS\mathcal{P}_A.$$ ☺

As to naturality, the best we can say is that Lemma 3.2.2.1 gives a natural equivalence between the connective cover of $\Omega BiS\mathcal{P}_A$ and $BGL(A)^+$.

Notice that comparing the results for $\mathcal{F}_A$ and $\mathcal{P}_A$ gives one proof of *cofinality* in the sense used in e.g., [108]: the connected cover of K-theory does not see the difference between free and projective modules.

Note 3.2.2.3 One should notice that the homotopy equivalence $\Omega BiS\mathcal{P}_A \simeq K_0(A) \times BGL(A)^+$ is not functorial in A. As an example, consider the ring $C(X)$ of continuous maps from a compact topological space X to the complex numbers. There is a functorial (in X) chain of maps

$$\Omega^\infty Bi\bar{H}\mathcal{P}_{C(X)} \to \Omega^\infty Bi\bar{H}\mathcal{P}^{top}_{C(X)} \xleftarrow{\simeq} Bi\mathcal{P}^{top}_{C(X)},$$

where the superscript top means that we shall remember the topology and and consider $\mathcal{P}_{C(X)}$ as a topological category. By a theorem of Swan, the homotopy groups of the rightmost space give the connective cover of Atiyah-Hirzebruch (complex) topological K-theory of X (see Example 1.1.3.2(2),(5)) and is represented by the spectrum $ku = Bi\bar{H}\mathcal{P}^{top}_{C(*)}$. The map from the algebraic K-theory of $C(X)$ to the topological K-theory of X is an isomorphism on path component and a surjection on the fundamental group (see [213, p. 61] or [13]). Consider the map $C(B(\mathbf{Z}/2)) \to$

$C(B(\mathbf{Z}))$ induced by the projection $\mathbf{Z} \to \mathbf{Z}/2$ and let C be the mapping cone of $B\mathbf{Z} \to B\mathbf{Z}/2$. Let F be the homotopy fiber of $K(C(B(\mathbf{Z}/2))) \to K(C(B(\mathbf{Z})))$. By naturality this induces a map of long exact sequences

$$\begin{array}{ccccccc}
K_1\big(C\big(B(\mathbf{Z})\big)\big) & \longrightarrow & \pi_0 F & \longrightarrow & K_0\big(C\big(B(\mathbf{Z}/2)\big)\big) & \longrightarrow & K_0\big(C\big(B(\mathbf{Z})\big)\big) \\
\downarrow & & \downarrow & & \cong\downarrow & & \cong\downarrow \\
K^1\big(B(\mathbf{Z})\big) & \longrightarrow & K^0(C) & \longrightarrow & K^0\big(B(\mathbf{Z}/2)\big) & \longrightarrow & K^0\big(B(\mathbf{Z})\big) \\
\cong\downarrow & & \cong\downarrow & & \cong\downarrow & & \cong\downarrow \\
\mathbf{Z} & \xrightarrow{2} & \mathbf{Z} & \longrightarrow & \mathbf{Z}/2 \oplus \mathbf{Z} & \xrightarrow{0+id} & \mathbf{Z}
\end{array}$$

(since the map $K^1(B(\mathbf{Z})) \to K^1(B(\mathbf{Z})) \cong \tilde{K}^0(C)$ is induced by multiplication by 2). This means that $\pi_0 F \to \tilde{K}_0 C(B(\mathbf{Z}/2)) = \mathbf{Z}/2$ is a non-split surjection, in contrast with what you get if you consider the homotopy fiber of

$$K_0\big(C\big(B(\mathbf{Z}/2)\big)\big) \times BGL\big(C\big(B(\mathbf{Z}/2)\big)\big)^+ \to K_0\big(C\big(B(\mathbf{Z})\big)\big) \times BGL\big(C\big(B(\mathbf{Z})\big)\big)^+.$$

We are grateful to Dan Grayson and John Rognes for assistance with this argument.

3.2.3 The Algebraic K-Theory Space of S*-Algebras*

The definition of K-theory space for **S**-algebras follows the idea for simplicial rings Definition 3.1.2.1. We will later give the spectrum level definitions which agree with this simple definition.

3.2.3.1 The General Linear Group-Like Monoid $\widehat{GL}(A)$

What is to play the rôle of the general linear group for an **S**-algebra? We could of course let it be the group of automorphisms of $A^{\times n}$ (mimicking degreewise K-theory), but this will be much too restrictive for our applications. Instead, we must somehow capture all self-equivalences. The readers who have read Sect. 2.3.3 will recognize the $\widehat{GL}_n(A)$ defined below as the outcome of the functor ω applied to the ΓS_*-category of endomorphisms of $A^{\vee n}$.

Note that we are to perform some unfriendly operations on the monoid of self-equivalences, so we had better ensure that our input is fibrant. Note also that if A is an **S**-algebra, then the multiplication in A gives rise to a simplicial monoid structure on $T_0 A(1_+)$ where T_0 is the fibrant replacement functor of Definition 2.2.2.1. This would not be true if we had used the other fibrant replacement QA of Sect. 2.2.1.4.

Consider the simplicial monoid

$$\widehat{M}_n A = T_0 Mat_n A(1_+) = \underset{\overrightarrow{x \in \mathcal{I}}}{\mathrm{holim}}\, \Omega^x\big(Mat_n(A)\big(S^x\big)\big),$$

where $Mat_n A$ is the **S**-algebra $X \mapsto Mat_n A(X) = \underline{\mathcal{S}_*}(n_+, n_+ \wedge A(X)) \cong \prod_n \bigvee_n A(X)$ of "matrices with only one element in each column" defined in Example 2.1.4.3(6). Its monoid of components is $\pi_0(\widehat{M}_n(A)) = M_n(\pi_0 A)$, and we let $\widehat{GL}_n(A)$ be the grouplike simplicial monoid of homotopy units:

$$\begin{array}{ccc} \widehat{GL}_n(A) & \longrightarrow & \widehat{M}_n(A) \\ \downarrow & & \downarrow \\ GL_n(\pi_0 A) & \longrightarrow & M_n(\pi_0 A) \end{array}$$

This is a (homotopy) pullback diagram (the maps may not be fibrations, but even so, the pullback is a homotopy pullback: picking out components is a homotopy functor).

If R is a simplicial ring with associated Eilenberg–Mac Lane **S**-algebra HR Example 2.1.6.2(2), the inclusion of $\vee$ into $\oplus$ induces a stable equivalence $Mat_n(HR) \to HM_nR$ of **S**-algebras, and hence chains of weak equivalences $T_0 Mat_n(HR)(1_+) \xrightarrow{\sim} T_0 HM_nR(1_+) \xleftarrow{\sim} M_nR$ and $\widehat{GL}(HR) \simeq \widehat{GL}(R)$.

This stabilizes correctly, in the sense that

$$S^0 = \mathbf{S}(1_+) \to A(1_+) = Mat_1(A)(1_+) \to \Omega^n Mat_1(A)\big(S^n\big)$$

and

$$Mat_n A \times Mat_1 A \xrightarrow{\vee} Mat_{n+1} A$$

induce maps

$$\begin{array}{ccc} \widehat{M}_n(A) & = & \underset{\overrightarrow{x\in\mathcal{I}}}{\operatorname{holim}} \Omega^x\big(Mat_n(A)\big(S^x\big)\big) \\ & & \downarrow \\ & & \underset{\overrightarrow{x\in\mathcal{I}}}{\operatorname{holim}} \Omega^x\big(Mat_n(A)\big(S^x\big) \times Mat_1(A)\big(S^x\big)\big) \\ & & \downarrow \vee \\ \widehat{M}_{n+1}(A) & = & \underset{\overrightarrow{x\in\mathcal{I}}}{\operatorname{holim}} \Omega^x\big(Mat_{n+1}(A)\big(S^x\big)\big) \end{array}$$

which in turn induce the usual Whitehead sum

$$M_n(\pi_0 A) \xrightarrow{m \mapsto m \oplus 1} M_{n+1}(\pi_0 A).$$

We let $\widehat{GL}(A)$ denote the colimit of the resulting directed system of $\widehat{GL}_n(A)$'s.

We can now form the classifying space in the usual way and define the algebraic K-theory space just as we did for simplicial rings in Definition 3.1.2.1:

Definition 3.2.3.1 Let A be an $\mathbf{S}$-algebra. Then the *algebraic K-theory space of A* is the space

$$K(A) = B\widehat{GL}(A)^{+}.$$

From the construction we get

Lemma 3.2.3.2 *Let R be a simplicial ring. Then the chain of weak equivalences* $M_n HR \simeq \widehat{M}_n HR$ *induces a chain of natural weak equivalence* $K(HR) \simeq K(R)$, *where* $K(R)$ *is the algebraic K-theory space of R as defined in Definition* 3.1.2.1. ☺

3.2.3.2 Comparison with Waldhausen's Algebraic K-Theory of a Connected Space

A particularly important example is the K-theory of spherical group rings, that is, of an $\mathbf{S}$-algebras of the form $\mathbf{S}[G]$ where G is a simplicial group, see Example 2.1.4.3(2). Then Waldhausen essentially shows that $K(\mathbf{S}[G])$ is equivalent to $A(BG)$, the "algebraic K-theory of the connected space BG".

Thus, the homotopy theoretic invariant $K(\mathbf{S}[G])$ carries deep geometric information. For instance, Waldhausen proves that $\mathbf{Z} \times K(\mathbf{S}[G])$ is equivalent to

$$\Omega^{\infty}(\mathbf{S} \wedge BG_{+}) \times Wh^{\mathrm{Diff}}(BG).$$

It is the last factor, the (smooth) *Whitehead space* that is of geometric significance; its loop space is equivalent to the so-called *stable smooth h-cobordism space*, which among other things carries information about diffeomorphisms of high-dimensional manifolds. See Jahren, Rognes and Waldhausen's [296].

There is a slight difference between the end product in [301, Theorem 2.2.1] and the present definition and we must cover this gap (see also the discussion at the bottom of p. 385 in [301]). For our purposes, we may consider Waldhausen's definition of (the connected cover of) *algebraic K-theory of the connected space* BG, $A(BG)$ to be

$$\varinjlim_{k,m} B\mathcal{H}^k_m(G)^{+}$$

where $\mathcal{H}^k_m(G)$ is the simplicial monoid of pointed $|G|$-equivariant weak self-equivalences of $|m_+ \wedge S^k \wedge G_+|$. More precisely, consider the space of $|G|$-self maps of $|m_+ \wedge S^k \wedge G_+|$

$$M^k_m = \sin Map_{|G|}\big(|m_+ \wedge S^k \wedge G_+|, |m_+ \wedge S^k \wedge G_+|\big).$$

This is a simplicial monoid under composition of maps $(f, g) \mapsto f \circ g$, and $\mathcal{H}^k_m(G)$ is the grouplike submonoid of invertible components. As a simplicial set M^k_m is isomorphic to $\Omega^k Mat_m \mathbf{S}[G](S^k)$. By Bökstedt's approximation Lemma 2.2.2.2 we have a chain of weak equivalences

$$\lim_{\overrightarrow{k}} B\mathcal{H}_m^k(G) \xleftarrow{\simeq} \operatorname{holim}_{\overrightarrow{m}} B\mathcal{H}_m^k(G) \xrightarrow{\simeq} \operatorname{holim}_{\overrightarrow{x\in\mathcal{I}}} B\mathcal{H}_m^x(G)$$

and we want to compare this with $B\widehat{GL}_m(\mathbf{S}[G])$.

We define a map (for convenience, we use the non-pointed homotopy colimit to get an isomorphism below; these are homotopy equivalent to their pointed counterparts since $\mathcal{I}$ is contractible, see Lemma A.7.4.1)

$$\begin{aligned}&\Big(\operatorname{holim}_{\overrightarrow{x\in\mathcal{I}}} \Omega^x Mat_m\big(\mathbf{S}[G](S^x)\big)\Big)^{\times q}\\ &\quad\cong \operatorname{holim}_{\overrightarrow{\mathbf{x}\in\mathcal{I}^q}} \prod_{i=1}^q M_m^{x_i} \to \operatorname{holim}_{\overrightarrow{\mathbf{x}\in\mathcal{I}^q}} \big(M_m^{\vee\mathbf{x}}\big)^{\times q} \to \operatorname{holim}_{\overrightarrow{x\in\mathcal{I}}} \big(M_m^x\big)^{\times q}.\end{aligned}$$

The first map is induced by the ith inclusion $x_i \subseteq \vee\mathbf{x} = x_1 \vee \cdots \vee x_q$ in the ith factor, and the last map is induced by the wedge $\mathcal{I}^q \to \mathcal{I}$. When restricted to homotopy units, this gives by Bökstedt's approximation Lemma 2.2.2.2 the desired equivalence

$$B_q\widehat{GL}_m\big(\mathbf{S}[G]\big) \to \operatorname{holim}_{\overrightarrow{x\in\mathcal{I}}} B_q\mathcal{H}_m^x(G).$$

We must just show that it is a simplicial map.

Note that the diagram

$$\begin{array}{ccc} M_m^x \wedge M_m^y & \xrightarrow{\mathbf{S}\text{-algebra multiplication}} & M_m^{x\vee y} \\ \downarrow & & \| \\ M_m^{x\vee y} \wedge M_m^{x\vee y} & \xrightarrow{\text{composition}} & M_m^{x\vee y} \end{array}$$

is commutative, where the left vertical map is induced by the first and second inclusion $x \subseteq x \vee y$ and $y \subseteq x \vee y$. Thus we see that if $0 < i < q$, then the ith face map in $\operatorname{holim}_{\overrightarrow{\mathbf{x}\in\mathcal{I}^q}} \prod_{i=1}^q M_m^{x_i}$ using the $\mathbf{S}$-algebra multiplication, corresponds to the ith face map in $\operatorname{holim}_{\overrightarrow{x\in\mathcal{I}}} (M_m^x)^{\times q}$, since we have used the ith inclusion in the ith factor, and the $(i+1)$st inclusion in the $(i+1)$st factor. The face maps d_0, d_q just drop the first respectively the last factor in both cases, and the degeneracies include the common unit in the appropriate factor.

3.2.4 *Agreement of the K-Theory of* S*-Algebras Through Segal's Machine and the Definition Through the Plus Construction*

Let A be an $\mathbf{S}$-algebra. Recall from Example 1.1.3.2(2) the Grothendieck group $K_0^f(\pi_0 A)$ of the category of finitely generated free $\pi_0 A$-modules. If $\pi_0 A$ has the invariance of basis number property (i.e., $(\pi_0 A)^{\times k}$ is isomorphic to $(\pi_0 A)^{\times l}$ if and only if $l = k$, which is true for most reasonable rings, and always true for commutative rings), then $K_0^f(\pi_0 A) \cong \mathbf{Z}$, and otherwise it is finite cyclic.

The following is the immediate generalization of the category of finitely generated free modules as in Example 1.2.1.4 to $\mathbf{S}$-algebras.

Definition 3.2.4.1 Let A be an $\mathbf{S}$-algebra. Then the category $\mathcal{F}_A$ of *finitely generated free A-modules* is the $\Gamma\mathcal{S}_*$-category whose objects are the natural numbers, and where $\mathcal{F}_A(m,n) = Mat_{n,m}A \cong \prod_m \bigvee_n A$.

Note that Segal's definition of the algebraic K-theory spectrum of A (with the uniform choice of weak equivalences Sect. 2.3.3) is then

$$K(A) = B\omega\bar{H}\mathcal{F}_A.$$

Theorem 3.2.4.2 *There is a chain of weak equivalences*

$$\Omega^\infty K(A) \simeq K_0^f(\pi_0 A) \times B\widehat{GL}(A)^+.$$

Proof First, note that since $K(A) = B\omega\bar{H}\mathcal{F}_A$ is special, i.e., for each $n_+ \in \Gamma^o$ we have that the natural map $B\omega\bar{H}\mathcal{F}_A(n_+) \to (B\omega\mathcal{F}_A)^{\times n}$ induced by the projections is a weak equivalence, we have a weak equivalence

$$\Omega^\infty K(A) \simeq \Omega B\omega\bar{H}\mathcal{F}_A(S^1).$$

For each $k \geq 0$, let $w\mathcal{F}^k$ be the full subcategory of $\omega\mathcal{F}_A$ whose only object is $k_+ \wedge A$. Note that by definition, this is nothing but $\widehat{GL}_k(A)$ considered as a simplicial category with only one object. Hence we are done, for by Lemma 3.2.2.1 and the discussion prior to Theorem 3.2.2.2 there is a chain of weak equivalences

$$\Omega B\omega\bar{H}\mathcal{F}_A(S^1) \simeq K_0^f(\pi_0 A) \times \left(\lim_{\overrightarrow{k}} Bw\mathcal{F}^k\right)^+ = K_0^f(\pi_0 A) \times B\widehat{GL}(A)^+. \qquad \square$$

If A is a discrete ring, we have a chain of weak equivalences

$$Bi\bar{H}\mathcal{F}_A \xrightarrow{\sim} B\omega\bar{H}\widetilde{\mathcal{F}_A} \xleftarrow{\sim} B\omega\bar{H}\mathcal{F}_{HA}$$

where $\widetilde{\mathcal{F}_A}$ is the construction of Example 2.1.6.2(2) making an Ab-category into a $\Gamma\mathcal{S}_*$-category through the Eilenberg–Mac Lane construction. The first weak equivalence follows by Lemma 2.3.3.2, whereas the second follows from the fact that the natural map $Mat_n HA \to H(M_n A)$ is a stable equivalence (finite wedges are products are stably equivalent).

3.3 Simplicial Rings Are Dense in S-Algebras

The unit $\mathbf{S} \to H\mathbf{Z}$ from the sphere spectrum to the integral Eilenberg–Mac Lane spectrum may either be thought of as the projection onto π_0 or as the Hurewicz map. Either way, we get that it is 1-connected. This implies that there is a very controlled difference between their module categories. The argument which we are going to give for this could equally well be considered in any setting where you have a 1-connected map $A \to B$ of $\mathbf{S}$-algebras. In fact, it is perhaps easiest to see that the result is true in this setting. Assume everything is cofibrant and that $f\colon A \to B$ is a cofibration of $\mathbf{S}$-algebras too (with cofiber B/A), so as to avoid technicalities. Consider the (Quillen) adjoint pair

$$\mathcal{M}_A \underset{f^*}{\overset{B\wedge_A -}{\rightleftarrows}} \mathcal{M}_B$$

where f^* is restriction of scalars, which we will drop from the notation. Let M be any A-module, and consider the unit of adjunction

$$\eta_M\colon M \cong A\wedge_A M \xrightarrow{f\wedge 1} B\wedge_A M.$$

This map has cofiber $(B/A)\wedge_A M$, and since $A \to B$ is 1-connected this gives that $M \to B\wedge_A M$ is 1-connected, and so $B\wedge_A M$ is a B-module giving a rather coarse approximation to M.

We can continue doing this: applying $B\wedge_A -$ to $M \to B\wedge_A M$ gives a square

$$\begin{array}{ccc} M & \xrightarrow{\eta_M} & B\wedge_A M \\ {\scriptstyle \eta_M}\big\downarrow & & \big\downarrow{\scriptstyle \eta_{B\wedge_A M}} \\ B\wedge_A M & \xrightarrow{1_B\wedge_A \eta_M} & B\wedge_A B\wedge_A M \end{array}$$

and a quick analysis gives that this has iterated cofiber $(B/A)\wedge_A(B/A)\wedge_A M$, and so is "2-cartesian", meaning that M is approximated by the pullback of the rest of square, at least up to dimension two. This continues, and gives that any A-module may be approximated to any degree of accuracy by means of B-modules. However, not all the maps connecting the B-modules in these cubes are B-module maps. This is often not dangerous. Because of the rapid convergence, functors satisfying rather weak "continuity" properties and that vanish on B-modules must also vanish on all A-modules.

We will be pursuing this idea, but we will be working non-stably, and our resolutions will in fact be resolutions of $\mathbf{S}$-algebras (in the setup as sketched above, that would require commutativity conditions).

3.3.1 A Resolution of S-Algebras by Means of Simplicial Rings

Recall the adjoint functor pairs of Sect. 2.1.3.1 (briefly, $\bar{H}$ is the Eilenberg–Mac Lane construction, R is evaluation at 1_+, $\tilde{\mathbf{Z}}$ is the free-abelian functor and U the forgetful functor)

$$sAb = \mathcal{A} \underset{R}{\overset{\bar{H}}{\rightleftarrows}} \Gamma\mathcal{A} \underset{U}{\overset{\tilde{\mathbf{Z}}}{\leftrightarrows}} \Gamma\mathcal{S}_*$$

(the left adjoints are on the top). All are symmetric monoidal (all but U are even strong symmetric monoidal), and so all take monoids to monoids. Furthermore, the construction T_0 of Definition 2.2.2.1 could equally well be performed in $\Gamma\mathcal{A}$, where it is called R_0 to remind us that the coproducts involved are now sums and not wedges. In particular, the approximation Lemma 2.2.2.2 works equally well in this setting. If A is an $\bar{H}\mathbf{Z}$-algebra, then R_0A is a special $\bar{H}\mathbf{Z}$-algebra (i.e., its underlying Γ-space is special, Definition 2.1.2.2), and so by Lemma 2.1.3.1 the rightmost map in

$$A \xrightarrow{\sim} R_0A \xleftarrow{\sim} \bar{H}R(R_0A)$$

is a pointwise equivalence. Hence: any $\bar{H}\mathbf{Z}$-algebra is canonically stably equivalent to $\bar{H}$ of a simplicial ring (this has already been noted in Corollary 2.2.2.6). This also works for (bi)modules: if P is an A-bimodule, then R_0P is an R_0A-bimodule, stably equivalent to P (as an A-bimodule); $\bar{H}(RR_0P)$ is an $\bar{H}(RR_0A)$-bimodule and pointwise equivalent to R_0P (as an $\bar{H}(RR_0A)$-bimodule).

In particular, remembering that $H = U\bar{H}$:

Lemma 3.3.1.1 *If A is any* **S**-*algebra and P an A-bimodule, then $(U\tilde{\mathbf{Z}}A, U\tilde{\mathbf{Z}}P)$ is canonically stably equivalent to a pair (HR, HQ) where R is a simplicial ring and Q an R-bimodule*:

$$(U\tilde{\mathbf{Z}}A, U\tilde{\mathbf{Z}}P) \xrightarrow{\sim} (UR_0\tilde{\mathbf{Z}}A, UR_0\tilde{\mathbf{Z}}P) \xleftarrow{\sim} \big(H(RR_0\tilde{\mathbf{Z}}A), H(RR_0\tilde{\mathbf{Z}}P)\big).$$

☺

The adjoint pair connecting $\Gamma\mathcal{A}$ and $\Gamma\mathcal{S}_*$ defines an adjoint pair

$$\bar{H}\mathbf{Z}\text{-algebras} \underset{U}{\overset{\tilde{\mathbf{Z}}}{\leftrightarrows}} \mathbf{S}\text{-algebras}$$

(that is, $U\tilde{\mathbf{Z}}$ is a "triple" in **S**-algebras) and so we have the canonical resolution of Sect. A.1.3 (to be precise and concise, it is the augmented cobar resolution of the monoid $U\tilde{\mathbf{Z}}$ in the category of endofunctors of **S**-algebras).

Lemma 3.3.1.2 *If A is an* **S**-*algebra, then the adjoint pair above gives an augmented cosimplicial object $A \to \{[q] \mapsto (U\tilde{\mathbf{Z}})^{q+1}A\}$, which is equivalent to H of a simplicial ring in each non-negative degree.* ☺

It is fairly straightforward to see that

$$A \to \underset{\overleftarrow{[q]\in\Delta}}{\operatorname{holim}} (U\tilde{\mathbf{Z}})^{q+1} A$$

is an equivalence, but we will not show that now, since we eventually will use the somewhat stronger Hurewicz Theorem A.8.3.4 which tells us that this limit converges fast enough, so that the homotopy limit passes through constructions like K-theory. This has the consequence that these constructions only depend on their value on simplicial rings, and on **S**-algebra maps between simplicial rings. Generally this is bothersome: we would have liked the diagram we are taking the limit of to be contained wholly in the category of simplicial rings. This is of course not possible, since it would imply that all **S**-algebras were stably equivalent to simplicial rings. For instance, **S** itself is not stably equivalent to a simplicial ring, but it **is** the homotopy limit of a diagram

$$H\mathbf{Z} \rightrightarrows U\tilde{\mathbf{Z}}H\mathbf{Z} \Rrightarrow U\tilde{\mathbf{Z}}U\tilde{\mathbf{Z}}H\mathbf{Z}\dots.$$

Remark 3.3.1.3 The categories $sAb = \mathcal{A}$, $\Gamma\mathcal{A}$ and $H\mathbf{Z}$-mod, are all naturally model categories, and the functors

$$\mathcal{A} \xrightarrow{\bar{H}} \Gamma\mathcal{A} \xrightarrow{U} H\mathbf{Z}\text{-mod}$$

induce equivalences between their homotopy categories. This uses the functor $L\colon \Gamma\mathcal{A} \to \mathcal{A}$ of Sect. 2.1.3.3 to construct an adjoint functor pair (see [253]).

3.3.1.1 Review on Cubical Diagrams

We need some language in order to calculate the resolution of Lemma 3.3.1.2 effectively. For a more thorough discussion we refer the reader to Appendix A.8.

Let $\mathcal{P}$ be the category of finite subsets of the natural numbers $\{1, 2, \dots\}$, and inclusions. We let $\mathcal{P}n$ be the subcategory allowing only subsets of $\{1, \dots, n\}$.

Definition 3.3.1.4 An n-cube is a functor $\mathcal{X}$ from the category $\mathcal{P}n$. A cubical diagram is a functor from $\mathcal{P}$.

If we adjoin the empty set $[-1] = \emptyset$ as an initial object to Δ, we get (a skeleton of) Ord, the category of finite ordered sets. A functor from Ord is what is usually called an augmented cosimplicial object. There is a functor $\mathcal{P} \to Ord$ sending a set S of cardinality n to $[n-1]$. Hence any augmented cosimplicial object gives rise to a cubical diagram. In most cases there is no loss of information in considering augmented cosimplicial objects as cubical diagrams (see Lemma A.8.1.1 for further details).

Definition 3.3.1.5 Let $\mathcal{X}$ be an n-cube with values in any of the categories where homotopy (co)limits and connectivity are defined (e.g., the categories considered in Sect. A.7.4: pointed/unpointed spaces, simplicial abelian groups, Γ-spaces or spectra). We say that $\mathcal{X}$ is *k-cartesian* if

$$\mathcal{X}_\emptyset \to \underset{\overleftarrow{S\neq\emptyset}}{\operatorname{holim}} \mathcal{X}_S$$

is k-connected, and *k-cocartesian* if

$$\underset{\overrightarrow{S\neq\{1,\dots,n\}}}{\operatorname{holim}} \mathcal{X}_S \to \mathcal{X}_{\{1,\dots,n\}}$$

is k-connected. It is *homotopy cartesian* if it is k-cartesian for all k, and *homotopy cocartesian* if it is k-cocartesian for all k.

When there is no possibility of confusing with the categorical notions, we write just cartesian and cocartesian. Homotopy (co)cartesian cubes are also called *homotopy pullback cubes* (resp. *homotopy pushout cubes*), and the initial (resp. final) vertex is then called the *homotopy pullback* (resp. *homotopy pushout*).

As a convention we shall say that a 0-cube is k-cartesian (resp. k-cocartesian) if $\mathcal{X}_\emptyset$ is $(k-1)$-connected (resp. k-connected).

So, a 0-cube is an object $\mathcal{X}_\emptyset$, a 1-cube is a map $\mathcal{X}_\emptyset \to \mathcal{X}_{\{1\}}$, and a 1-cube is k-(co)cartesian if it is k-connected as a map. A 2-cube is a square

$$\begin{array}{ccc} \mathcal{X}_\emptyset & \longrightarrow & \mathcal{X}_{\{1\}} \\ \downarrow & & \downarrow \\ \mathcal{X}_{\{2\}} & \longrightarrow & \mathcal{X}_{\{1,2\}} \end{array}$$

and so on. We will regard a natural transformation of n-cubes $\mathcal{X} \to \mathcal{Y}$ as an $(n+1)$-cube. In particular, if $F \to G$ is some natural transformation of functors of simplicial sets, and $\mathcal{X}$ is an n cube of simplicial sets, then we get an $(n+1)$-cube $F\mathcal{X} \to G\mathcal{X}$.

The following definition is useful for book-keeping, and is discussed further in Appendix A.8.

Definition 3.3.1.6 If $f\colon \mathbf{N} \to \mathbf{N}$ is some function, we say that an S-cube $\mathcal{X}$ is f-cartesian if each d-dimensional subcube of $\mathcal{X}$ is $f(d)$-cartesian. Likewise for f-cocartesian.

For instance, if $f(d) = d+1$, the cube is $(id+1)$-cartesian. Some prefer to call this "$(\dim + 1)$-cartesian".

We will need the generalized Hurewicz theorem which we cite from Appendix A.8.3.4:

Theorem 3.3.1.7 *Let $k > 1$. If $\mathcal{X}$ is an $(id+k)$-cartesian cube of simplicial sets, then so is $\mathcal{X} \to \tilde{\mathbf{Z}}\mathcal{X}$.* ☺

Definition 3.3.1.8 Let A be an **S**-algebra and $n > 0$. Define the n-cube of **S**-algebras

$$(A)^n = \{S \mapsto (A)^n_S\}$$

by applying the unit of adjunction $h\colon id \to U\tilde{\mathbf{Z}}$ n times to A (so that $(A)^n_S = (U\tilde{\mathbf{Z}})^{|S|}A$). Carrying this on indefinitely, we get a functor

$$\mathcal{P} \xrightarrow{S\mapsto (A)_S} \mathbf{S}\text{-algebras}$$

such that the restriction of $\{S \mapsto (A)_S\}$ to $\mathcal{P}\mathbf{n} \subseteq \mathcal{P}$ is $(A)^n = \{S \mapsto (A)^n_S\}$.

More concretely $(A)^2$ is the 2-cube

$$\begin{array}{ccc} A & \xrightarrow{h_A} & U\tilde{\mathbf{Z}}A \\ {\scriptstyle h_A}\downarrow & & \downarrow{\scriptstyle h_{U\tilde{\mathbf{Z}}A}} \\ U\tilde{\mathbf{Z}}A & \xrightarrow{U\tilde{\mathbf{Z}}h_A} & (U\tilde{\mathbf{Z}})^2A \end{array} .$$

Corollary 3.3.1.9 *Let $n \geq 0$. The n-cube of spectra $(A)^n$ is id-cartesian.*

Proof For each $k > 1$, the space $A(S^k)$ is $(k-1)$-connected by Lemma 2.2.1.3(2) (and so $(id+k)$-cartesian as a 0-cube). Hence the Hurewicz Theorem 3.3.1.7 says that $S \mapsto (A)^n_S(S^k)$ is $(id+k)$-cartesian, which is stronger than $S \mapsto (A)^n_S$ being id-cartesian as a spectrum. □

The very reason for the interest in this construction stems from the following observation which follows immediately from Lemma 3.3.1.1.

Proposition 3.3.1.10 *Let A be an **S**-algebra. Then $(A)_S$ is canonically equivalent to H of a simplicial ring for all $S \neq \emptyset$.* ☺

3.3.2 K-Theory is Determined by Its Values on Simplicial Rings

First note that K-theory behaves nicely with respect to id-cartesian squares (note that a square being merely highly cartesian is not treated nicely by K-theory, you need good behavior on all subcubes).

Theorem 3.3.2.1 *Let $\mathcal{A}$ be an id-cartesian n-cube of **S**-algebras, $n > 0$. Then $K(\mathcal{A})$ is $(n+1)$-cartesian.*

Proof Let $\mathcal{M} = Mat_m\mathcal{A}$ be the cube given by the $m \times m$ matrices in $\mathcal{A}$. This is id-cartesian, and so $T_0\mathcal{M} = \operatorname{holim}_{\overrightarrow{x\in I}} \Omega^x(Mat_m\mathcal{A})(S^x)$ is an id-cartesian cube of grouplike simplicial monoids. As all maps in $\mathcal{A}$ are 1-connected, they induce isomorphisms on π_0. Hence we get $\mathcal{G} = \widehat{GL}_m(\mathcal{A})$ as a pullback

$$\begin{array}{ccc} \mathcal{G}_T & \longrightarrow & T_0\mathcal{M}_T \\ \downarrow & & \downarrow \\ GL_m(\pi_0\mathcal{A}_{\{1,\dots,n\}}) & \longrightarrow & M_m(\pi_0\mathcal{A}_{\{1,\dots,n\}}) \end{array}$$

for all $T \subset \mathbf{n}$. Consequently, $\widehat{GL}_m(\mathcal{A})$ is id-cartesian, and so $B\widehat{GL}_m(\mathcal{A})$ is $(id+1)$-cartesian. Using Lemma A.8.3.6 we get that also

$$K(\mathcal{A}) = B\widehat{GL}(\mathcal{A})^+ \cong \Big(\varinjlim_{m} B\widehat{GL}_m(\mathcal{A})\Big)^+$$

is $(id+1)$-cartesian. □

Note that with non-connected definitions of algebraic K-theory we still get that the algebraic K-theory of $\mathcal{A}$ is $(n+1)$-cartesian (it is not $(id+1)$-cartesian, but only id-cartesian, because the spaces are not 0-connected). This is so since all the maps of $\mathbf{S}$-algebras involved are 1-connected, and so $K_0(\pi_0\mathcal{A})$ is the constant cube $K_0(\pi_0\mathcal{A}_\emptyset)$.

Theorem 3.3.2.2 *Let A be an* $\mathbf{S}$*-algebra. Then*

$$K(A) \to \underset{\overleftarrow{S\in\mathcal{P}-\emptyset}}{\operatorname{holim}} K\big((A)_S\big)$$

is an equivalence.

Proof We know there is high connectivity to any of the finite cubes: Theorem 3.3.2.1 tells us that $K(A) \to \operatorname{holim}_{\overleftarrow{S\in\mathcal{P}\mathbf{n}-\emptyset}} K((A)^n_S)$ is $(n+1)$-connected, so we just have to know that this assembles correctly. Now, by Lemma A.7.2.4 the map

$$\underset{\overleftarrow{S\in\mathcal{P}\mathbf{n+1}-\emptyset}}{\operatorname{holim}} K\big((A)^{n+1}_S\big) \to \underset{\overleftarrow{S\in\mathcal{P}\mathbf{n}-\emptyset}}{\operatorname{holim}} K\big((A)^n_S\big)$$

induced by restriction along $\mathcal{P}\mathbf{n} \subseteq \mathcal{P}\mathbf{n+1}$ is a fibration. By writing out explicitly the cosimplicial replacement formula of Sect. A.7.3 for the homotopy limit, you get that

$$\underset{\overleftarrow{J}}{\operatorname{holim}} F \cong \underset{\overleftarrow{n\in\mathbf{N}}}{\lim} \underset{\overleftarrow{J_n}}{\operatorname{holim}} F|_{J_n}.$$

Hence, by Lemma A.7.3.2 and Theorem A.7.4.5, you get that $\operatorname{holim}_{\overleftarrow{S\in\mathcal{P}\mathbf{n}-\emptyset}} K((A)^n_S)$ approximates $\operatorname{holim}_{\overleftarrow{S\in\mathcal{P}-\emptyset}} K((A)_S)$. □

Chapter 4
Topological Hochschild Homology

As K-theory is hard to calculate, it is important to know theories that are related to K-theory, but that are easier to calculate. Thus, if somebody comes up with a non-trivial map between K-theory and something one thinks one can compute, it is considered a good thing. For instance, in 1965 Hattori [121] and Stallings [272] defined for a ring A a map $tr_A\colon K_0(A) \to A/[A, A]$—the *Hattori-Stallings trace map*—by sending the class of the projective module defined by an idempotent $p \in M_n(A)$ to the trace of p. The recent preprint of Berrick and Hesselholt [15] where they use that the Hattori-Stallings trace map factors through topological cyclic homology to discover classes of groups where the so-called Bass trace conjecture is true is an interesting application relevant for our setting.

4.1 Introduction

In 1976, R. Keith Dennis observed that there exists a map from the higher K-groups of a ring A to the so-called Hochschild homology $HH(A)$. This map has since been called the *Dennis trace*, and is intimately connected with Hattori–Stalling's trace map, see e.g., Loday [181, 8]. When applied e.g., to number rings, the Dennis trace map is generally a poor invariant rationally, but retains some information when working with finite coefficients [156].

Waldhausen noticed in [299] that there is a connection between the sphere spectrum, stable K-theory (previously discussed in Sect. 1.3) and Hochschild homology. Although the proof appeared only much later ([302]), he also knew before 1980 that stable A-theory coincided with stable homotopy. Motivated by his machine "calculus of functors" and his study of stable pseudo isotopy theory, T. Goodwillie conjectured that there existed a theory sitting between K-theory and Hochschild homology, agreeing integrally with stable K-theory for all "rings up to homotopy", but with a Hochschild-style definition. He called the theory topological Hochschild homology (*THH*), and the only difference between *THH* and *HH* should be that whereas the ground ring in *HH* is the ring of integers, the ground ring of *THH* should be the

B.I. Dundas et al., *The Local Structure of Algebraic K-Theory*,
Algebra and Applications 18,
DOI 10.1007/978-1-4471-4393-2_4, © Springer-Verlag London 2013

sphere spectrum $\mathbf{S}$, considered as a "ring up to homotopy". This would also be in agreement with his proof that stable K-theory and Hochschild homology agreed rationally, as the higher homotopy groups of the sphere spectrum $\mathbf{S}$ are all torsion. He also made some conjectural calculations of $THH(\mathbf{Z})$ and $THH(\mathbf{Z}/p\mathbf{Z})$.

The next step was taken in the mid eighties by M. Bökstedt, who was able to give a definition of THH, satisfying all of Goodwillie's conjectural properties, except possibly the equivalence with stable K-theory. To model rings up to homotopy, he defined functors with smash products which are closely related to the $\mathbf{S}$-algebras defined in Chap. 2.

Theorem 4.1.0.1 (Bökstedt)

$$\pi_k THH(\mathbf{Z}) \cong \begin{cases} \mathbf{Z} & \text{if } k=0, \\ \mathbf{Z}/i\mathbf{Z} & \text{if } k=2i-1, \\ 0 & \text{if } k=2i>0, \end{cases}$$

$$\pi_k THH(\mathbf{Z}/p\mathbf{Z}) \cong \begin{cases} \mathbf{Z}/p\mathbf{Z} & \text{if } k \text{ is even}, \\ 0 & \text{if } k \text{ is odd}. \end{cases}$$

Properly interpreted, work of Breen [41] actually calculated $THH(\mathbf{Z}/p\mathbf{Z})$ already in 1976. The outcome of the two papers of Jibladze, Pirashvili and Waldhausen [150, 225] was that $THH(A)$ could be thought of as the homology of the category $\mathcal{P}_A$ of finitely generated A-modules in the sense of Sect. 1.3, or alternatively as "Mac Lane homology", [190]. This was subsequently used by Franjou, Lannes and Schwartz and Pirashvili to give purely algebraic proofs of Bökstedt's calculations, [84] and [85].

For (flat) rings A, there is a (3-connected) map $THH(A) \to HH(A)$ which should be thought of as being induced by the change of base ring $\mathbf{S} \to H\mathbf{Z}$.

After it became clear that the connection between K-theory and THH is as good as could be hoped, many other calculations of THH have appeared—topological Hochschild homology possesses localization, in the same sense as Hochschild homology does, THH of group rings can be described, and so on. Many calculations have been done in this setting or in the dual Mac Lane cohomology. For instance by Pirashvili in [223, 224] and [222]. For further calculations see Larsen and Lindenstrauss' papers [169, 176] and [168]. For A a ring of integers in a number field, Lindenstrauss and Madsen obtained in [177] the non-canonical isomorphism

$$\pi_i THH(A) \cong \begin{cases} A & \text{if } i=0, \\ A/n\mathcal{D}_A & \text{if } i=2n-1, \\ 0 & \text{otherwise} \end{cases}$$

where $\mathcal{D}_A$ is the different ideal. In [131] Hesselholt and Madsen give a canonical description of the mod p homotopy groups, which we will return to later. A functorial description of the integral homotopy groups is to our knowledge still beyond reach.

For concrete calculations the spectral sequence of Pirashvili and Waldhausen in [225] (see Lemma 4.2.3.7) is very useful. This is especially so since in many cases it degenerates, a phenomenon which is partially explained in [249].

As we have already noted, the first example showing that stable K-theory and *THH* are equivalent is due to Waldhausen, and predates the definition of *THH*. He showed this in the examples arising from his K-theory of spaces; in particular, he showed the so-called "vanishing of the mystery homology": stable K-theory of the sphere spectrum **S** is equivalent to **S**, i.e., $K^S(\mathbf{S}) \simeq THH(\mathbf{S}) \simeq \mathbf{S}$. Based upon this, [250] announced that one could prove $K^S \simeq THH$ in general, but the full proof has not yet appeared.

The second example appeared in [69], and took care of the case of rings, using the interpretation of $THH(A)$ as the homology of $\mathcal{P}_A$. In [63] it was shown how this implies $K^S \simeq THH$ for all **S**-algebras.

When A is a commutative **S**-algebra we get by an appropriate choice of model that $THH(A)$ is also a commutative **S**-algebra (however, you need to change the foundations: Γ-spaces are not good enough, see Sect. 7.3.1.1), and the homotopy groups become a graded commutative ring. For instance, the calculation of $\pi_* THH(\mathbf{Z}/p\mathbf{Z})$ could be summed up more elegantly by saying that it is the graded polynomial ring in $\mathbf{Z}/p\mathbf{Z}$ in one generator in degree 2, see Corollary 7.3.1.2. This is already present in Bökstedt's original calculations, along with the important identification of a preferred polynomial generator for $\pi_* THH(\mathbf{Z}/p\mathbf{Z})$ coming from the cyclic action (which is a central ingredient when we start talking about topological cyclic homology).

Organization In Sect. 4.2 we will give a definition of topological Hochschild homology for **S**-algebras, and prove some basic results with a special view to the ring case. In Sect. 4.3, we will extend our definition to include $\Gamma\mathcal{S}_*$-categories in general as input. This is very similar, and not much more involved; but we have chosen to present the theory for **S**-algebras first so that people not interested in anything but rings can have the definition without getting confused by too much generality. However, this generality is very convenient when one wants to construct the trace map from K-theory, and also when one wants to compare with the homology of additive categories. This is particularly clear when one wants good definitions for the "trace" map from algebraic K-theory, which we present in Chap. 5.

4.1.1 Where to Read

The literature on *THH* is not as well developed as for K-theory; and there is a significant overlap between these notes and most of the other sources. The original paper

[30] is good reading, but has unfortunately not yet appeared. The article [129] develops the ideas in [27] further, and is well worth studying to get an equivariant point of view on the matter. For the *THH* spectrum for exact categories, [70] is slightly more general than these notes.

For a general overview, the survey article of Madsen, [192], is recommended.

4.2 Topological Hochschild Homology of S-Algebras

As topological Hochschild homology is supposed to be a modelled on the idea of Hochschild homology, we recall the standard complex calculating $HH(A)$.

4.2.1 Hochschild Homology of k-Algebras

Recall the definition of Hochschild homology (see Sect. 1.3.2): Let k be a commutative ring, let A be a *flat* k-algebra, and let P be an A-bimodule (i.e., an $A^o \otimes_k A$-module). Then we define the *Hochschild homology* (over k) of A with coefficients in P to be the simplicial k-module

$$HH^k(A, P) = \left\{[q] \mapsto HH^k(A, P)_q = P \otimes_k A^{\otimes_k q}\right\}$$

with face and degeneracies given by

$$d_i(m \otimes a_1 \otimes \cdots \otimes a_q) = \begin{cases} ma_1 \otimes a_2 \cdots \otimes a_q & \text{if } i = 0, \\ m \otimes a_1 \otimes \cdots \otimes a_i a_{i+1} \otimes \cdots \otimes a_q & \text{if } 0 < i < q, \\ a_q m \otimes a_1 \otimes \cdots \otimes a_{q-1} & \text{if } i = q, \end{cases}$$

$$s_i(m \otimes a_1 \otimes \cdots \otimes a_q) = m \otimes a_1 \otimes \cdots \otimes a_i \otimes 1 \otimes a_{i+1} \otimes \cdots \otimes a_q.$$

Just the same definition may be applied to simplicial k-algebras, and this definition of HH^k will preserve weak equivalences. Again we either assume that our ring is flat, or else we substitute it with one that is, and so we are really defining what some call Shukla homology after [263]. To make this functorial in (A, P) we really should choose a functorial flat resolution of rings once and for all, but since our main applications are to rings that are already flat, we choose to suppress this.

4.2.1.1 Cyclic Structure

In the case $P = A$ something interesting happens. Then $HH^k(A) = HH^k(A, A)$ is not only a simplicial object, but also a *cyclic* object (see Sect. 6.2.1 for a more detailed discussion of cyclic objects, and Sect. 4.2.2.4 below for the structure on *THH*). Recall that a *cyclic object* is a functor from Connes' category Λ^o, where Λ

is the category containing Δ, but with an additional endomorphism for each object, satisfying certain relations. In terms of generators, this means that in addition to all maps coming from Δ for each $[q]$ there is a map $t = t_q\colon [q] \to [q]$. In our case t is sent to the map $A^{\otimes_k(q+1)} \to A^{\otimes_k(q+1)}$ sending $a_0 \otimes \cdots \otimes a_q$ to $a_q \otimes a_0 \otimes \cdots \otimes a_{q-1}$.

To be precise:

Definition 4.2.1.1 Connes' category Λ is the category with the same objects as the simplicial category Δ, but with morphism sets

$$\Lambda\big([n],[q]\big) = \Delta\big([n],[q]\big) \times C_{n+1},$$

where C_{n+1} is the cyclic group with generator $t = t_n$ and with $t_n^{n+1} = 1_{[n]}$. Here a pair (σ, t^a) is considered as a composite

$$[n] \xrightarrow{t^a} [n] \xrightarrow{\sigma} [q]$$

(where $t = t_n$ is the generator of C_{n+1}, so that $t_n^{n+1} = 1_{[n]}$). Composition is subject to the extra relations

$$\begin{aligned}
&t_n d^i = d^{i-1} t_{n-1}, \quad 1 \le i \le n,\\
&t_n d^0 = d^n,\\
&t_n s^i = s^{i-1} t_{n+1}, \quad 1 \le i \le n,\\
&t_n s^0 = s^n t_{n+1}^2.
\end{aligned}$$

A *cyclic object* in some category $\mathcal{C}$ is a functor $\Lambda^o \to \mathcal{C}$, and a *cyclic map* is a natural transformation between cyclic objects.

Notice that any map in Λ can be written as a composite ϕt^a where $\phi \in \Delta$. Furthermore, this factorization is unique.

Due to the inclusion $j\colon \Delta \subset \Lambda$, any cyclic object X gives rise to a simplicial object j^*X.

Hochschild homology is just an instance of a general gadget giving cyclic objects: let M be a monoid in a symmetric monoidal category $(\mathcal{C}, \Box, e)$. Then the *cyclic bar construction* is the cyclic object $B^{cy}(M) = \{[q] \mapsto M^{\Box(q+1)}\}$. Hochschild homology is then the example coming from $(k\text{-mod}, \otimes_k, k)$. The most basic example is the *cyclic bar construction* of ordinary monoids: in the symmetric monoidal category of sets with cartesian product, a monoid is just an ordinary monoid, and $B_q^{cy}(M) = M^{\times(q+1)}$. Slightly more fancy are the cases $(\mathcal{C}at, \times, *)$: monoids are strict monoidal categories, or $(\mathcal{S}, \times, *)$: monoids are simplicial monoids. We have already seen an example of the former: the object $\{[q] \mapsto \mathcal{I}^{q+1}\}$ which appeared in Sect. 2.2.2.1 was simply $B^{cy}\mathcal{I}$.

4.2.2 Topological Hochschild Homology of S-Algebras

In analogy with the above definition of HH^k, Bökstedt defined topological Hochschild homology. Of course, $\mathbf{S}$ is initial in the category of $\mathbf{S}$-algebras (as defined in Sect. 2.1.4), just as (for any commutative ring k) k is initial among k-algebras, and the idea is that we should try to replace the symmetric monoidal category $(k\text{-mod}, \otimes_k, k)$ with $(\mathbf{S}\text{-mod}, \wedge, \mathbf{S})$. In other words, instead of taking the tensor product over k, we should take the "tensor product over $\mathbf{S}$", that is, the smash product of Γ-spaces. So we could consider

$$HP \wedge HA \wedge \cdots \wedge HA$$

(or even smashed over some other commutative $\mathbf{S}$-algebra if desirable), and there is nothing wrong with this, except that

1. as it stands it is prone to all the nuisances of the classical case: unless we replace HA with something fairly free in $\Gamma\mathcal{S}_*$ first, this will not preserve equivalences; and
2. without some amendment this will not have enough structure to define the goal of the next chapter: topological cyclic homology.

Inspired by spectra rather than Γ-spaces, Bökstedt defined a compact definition which takes care of both these problems. But before we give Bökstedt's definition, we note that we have already twice encountered one of the obstructions to a too naïve generalization. Let A be a ring. The associated $\mathbf{S}$-algebra HA sending X to $HA(X) = A \otimes \tilde{\mathbf{Z}}[X]$ has a multiplication; but if we want to loop this down we have a problem: the multiplication gives a map from

$$\varinjlim_{k,l \in \mathbf{N}^2} \Omega^{k+l}\big((A \otimes \tilde{\mathbf{Z}}[S^k]) \wedge (A \otimes \tilde{\mathbf{Z}}[S^l])\big)$$

to

$$\varinjlim_{k,l \in \mathbf{N}^2} \Omega^{k+l}\big(A \otimes \tilde{\mathbf{Z}}[S^{k+l}]\big)$$

which sure enough is isomorphic to $\varinjlim_{k \in \mathbf{N}} \Omega^k(A \otimes \tilde{\mathbf{Z}}[S^k])$, but not equal. The problem gets nasty when we consider associativity: we can't get the two maps from the "triple smash" to be **equal**. For Hochschild homology we want a simplicial space which in degree 0 is equivalent to $\varinjlim_{k \in \mathbf{N}} \Omega^k(A \otimes \tilde{\mathbf{Z}}[S^k])$, in degree 1 is equivalent to

$$\varinjlim_{k,l \in \mathbf{N}^2} \Omega^{k+l}\big((A \otimes \tilde{\mathbf{Z}}[S^k]) \wedge (A \otimes \tilde{\mathbf{Z}}[S^l])\big)$$

and so on, and one of the simplicial relations ($d_1^2 = d_1 d_2$) will exactly reflect associativity and it is not clear how to do this, cf. Note 4.2.2.6.

In [30], Bökstedt shows how one can get around this problem by using the category $\mathcal{I}$ (the subcategory of Γ^o with all objects and just injections, see Sect. 2.2.2.1) instead of the natural numbers. To ensure that the resulting colimit has the right homotopy properties, we must use the homotopy colimit, see the approximation Lemma 2.2.2.2.

Recall that, if $x = k_+ = \{0, \dots, k\} \in ob\mathcal{I}$, then an expression like $S^x = S^k$ will mean S^1 smashed with itself k times, and $\Omega^x = \Omega^k$ will mean $Map_*(S^k, -) = \underline{\mathcal{S}_*}(S^k, \sin|-|)$. Furthermore, if $\mathbf{x} = (x_0, \dots, x_q)$ is an object in $\mathcal{I}^{q+1}$, then $\bigvee \mathbf{x} = x_0 \vee \dots \vee x_q$. If $\alpha\colon m_+ \to n_+ \in \mathcal{I}$, let $S^{n_+ - \alpha(m_+)} = \bigwedge_{j=1}^n S(\alpha, j)$ where

$$S(\alpha, j) = \begin{cases} S^0 & \text{if } j \text{ is in the image of } \alpha, \\ S^1 & \text{otherwise,} \end{cases}$$

and let $S^{n_+ - \alpha(m_+)} \wedge S^{m_+} \cong S^{n_+}$ be the shuffle isomorphism: it keeps the order of the S^1's in $S^{n_+ - \alpha(m_+)}$, and inserts the ith smash factor S^1 of S^m in the $\alpha(i)$-factor. Quite concretely, the isomorphism sends a non-basepoint $((a_1 \wedge \dots \wedge a_n) \wedge (b_1 \wedge \dots \wedge b_m))$ to $(c_1 \wedge \dots \wedge c_n)$ where $c_j = a_j$ if j is not in the image of α and $c_j = b_i$ if $j = \alpha(i)$ (remember that α is an injection).

Definition 4.2.2.1 Let A be an **S**-algebra, P an A-bimodule (i.e., an $A^o \wedge A$-module) and X a space, and define for every $q \geq 0$ the assignment $V(A, P)\colon ob\mathcal{I}^{q+1} \to ob\mathcal{S}_*$ by

$$(x_0, \dots, x_q) \mapsto V(A, P)(x_0, \dots, x_q) = P\big(S^{x_0}\big) \wedge \bigwedge_{1 \leq i \leq q} A\big(S^{x_i}\big)$$

This gives rise to a functor $G_q = G(A, P, X)_q \colon \mathcal{I}^{q+1} \to \mathcal{S}_*$ given by

$$\mathbf{x} \mapsto G_q(\mathbf{x}) = \Omega^{\vee \mathbf{x}}\big(X \wedge V(A, P)(\mathbf{x})\big)$$

and if $\phi\colon \mathbf{x} \to \mathbf{y} \in \mathcal{I}^{q+1}$ the map $G_q(\phi)\colon G_q(\mathbf{x}) \to G_q(\mathbf{y})$ is defined as follows:

$$\begin{aligned}
G_q(\mathbf{x}) &= Map_*\left(\bigwedge_{i=0}^{q} S^{x_i}, X \wedge P\big(S^{x_0}\big) \wedge \bigwedge_{i=1}^{q} A\big(S^{x_i}\big)\right) \\
&\longrightarrow Map_*\left(\bigwedge_{i=0}^{q} S^{y_i - \phi(x_i)} \wedge \bigwedge_{i=0}^{q} S^{x_i}, \bigwedge_{i=0}^{q} S^{y_i - \phi(x_i)} \wedge X \wedge P\big(S^{x_0}\big) \wedge \bigwedge_{i=1}^{q} A\big(S^{x_i}\big)\right) \\
&\xrightarrow{\cong} Map_*\left(\bigwedge_{i=0}^{q} \big(S^{y_i - \phi(x_i)} \wedge S^{x_i}\big), X \wedge S^{y_0 - \phi(x_0)} \wedge P\big(S^{x_0}\big) \wedge \bigwedge_{i=1}^{q} \big(S^{y_i - \phi(x_i)} \wedge A\big(S^{x_i}\big)\big)\right) \\
&\longrightarrow Map_*\left(\bigwedge_{i=0}^{q} \big(S^{y_i - \phi(x_i)} \wedge S^{x_i}\big), X \wedge P\big(S^{y_0 - \phi(x_0)} \wedge S^{x_0}\big) \wedge \bigwedge_{i=1}^{q} A\big(S^{y_i - \phi(x_i)} \wedge S^{x_i}\big)\right) \\
&\xrightarrow{\cong} Map_*\left(\bigwedge_{i=0}^{q} S^{y_i}, X \wedge P\big(S^{y_0}\big) \wedge \bigwedge_{i=1}^{q} A\big(S^{y_i}\big)\right) = G_q(\mathbf{y}),
\end{aligned}$$

where the first arrow is induced by smashing with $\bigwedge_{i=0}^{q} S^{y_i - \phi(x_i)}$, the second shuffles smash factors, the third is induced by the structure maps of P and A, and the fourth is the shuffle associated with ϕ described just before the start of the definition. For each q we define

$$THH(A, P, X)_q = \underset{\mathbf{x} \in \mathcal{I}^{q+1}}{\underrightarrow{\text{holim}}}\, G_q(\mathbf{x}).$$

The claim that G_q is actually a functor follows by observing that G_q takes the identity to the identity and considering a $\psi : \mathbf{y} \to \mathbf{z} \in \mathcal{I}^{q+1}$ and filling in the squares and triangles in the diagram shown schematically below

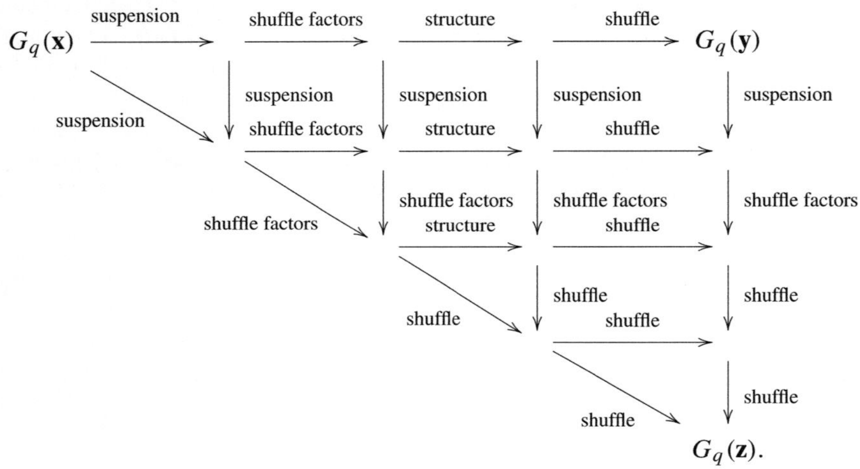

4.2.2.1 The Homotopy Type

We have to know that this has the right homotopy properties, i.e., we want to know that $THH(A, P; X)_q$ is equivalent to

$$\underset{(n_0,\ldots n_q)\in \mathbf{N}^{q+1}}{\underrightarrow{\lim}} \Omega^{\sum n_i}\left(X \wedge P(S^{n_0}) \wedge \bigwedge_{1 \le i \le q} A(S^{n_i}) \right).$$

By the approximation Lemma 2.2.2.2 for $\mathcal{I}$, this will be the case if we can show that a map $\mathbf{x} \subseteq \mathbf{y} \in \mathcal{I}^{q+1}$ will induce a map $G_q(\mathbf{x}) \to G_q(\mathbf{y})$ which gets higher and higher connected with the cardinality of $\mathbf{x}$. Maps in $\mathcal{I}^{q+1}$ can be written as compositions of an isomorphism together with a standard inclusion. The isomorphisms pose no problem, so we are left with considering the standard inclusions which again can be decomposed into successions of standard inclusions involving only one coordinate. Since the argument is rather symmetric, we may assume that we are looking at the standard inclusion

$$\mathbf{x} = (k_+, x_1, \dots, x_q) \subseteq \big((k+1)_+, x_1, \dots, x_q\big).$$

Since P is a Γ-space, Lemma 2.2.1.3(3) says that $S^1 \wedge P(S^k) \to P(S^{k+1})$ is roughly $2k$-connected, and so (by the same Lemma 2.2.1.3(2)) the map

$$S^1 \wedge P(S^k) \wedge \bigwedge A(S^{x_i}) \to P(S^{k+1}) \wedge \bigwedge A(S^{x_i})$$

is roughly $2k + \vee x_i$ connected. The Freudenthal suspension Theorem A.8.2.3 then gives the result.

4.2.2.2 Functoriality

We note that, when varying X in Γ^o, $THH(A, P, X)_q$ becomes a very special Γ-space which we simply call $THH(A, P)_q$ (so that $\sin|THH(A, P)_q)$ is "stably fibrant" in the terminology of Chap. 2, see Corollary 2.2.1.7), and so defines an Ω-spectrum. We also see that it is a functor in the maps of pairs $(A, P) \to (B, Q)$ where $f\colon A \to B$ is a map of **S**-algebras, and $P \to f^*Q$ is a map of A-bimodules—that is, a map of $\Gamma\mathcal{S}_*$*-natural bimodules* in the sense of Appendix A.10.4.1.

4.2.2.3 Simplicial Structure

So far, we have not used the multiplicative structure of our **S**-algebra, but just as for ordinary Hochschild homology this enters when we want to make $[q] \mapsto THH(A, P, X)_q$ into a functor, that is, a simplicial space. The compact way of describing the face and degeneracy maps is to say that they are "just as for ordinary Hochschild homology". This is true and will suffice for all future considerations, and the pragmatic reader can stop here. However, we have seen that it is difficult to make this precise, and the setup of Bökstedt is carefully designed to make this rough definition work.

In detail: Consider the functor $G_q = G(A, P, X)_q\colon \mathcal{I}^{q+1} \to \mathcal{S}_*$ of Definition 4.2.2.1 of $THH(A, P, X)_q$. Homotopy colimits are functors of "$\mathcal{S}_*$-natural modules", in this case restricted to pairs (I, F) where I is a small category and $F\colon I \to \mathcal{S}_*$ is a functor. A map $(I, F) \to (J, G)$ is a functor $f\colon I \to J$ together with a natural transformation $F \to G \circ f$. So to show that $[q] \mapsto THH(A, P, X)_q$ is a functor, we must show that $[q] \mapsto (\mathcal{I}^{q+1}, G_q)$ is a functor from Δ^o to $\mathcal{S}_*$-natural modules. Let $\phi \in \Delta([n], [q])$. The maps $\phi^*\colon \mathcal{I}^{q+1} \to \mathcal{I}^{n+1}$ come from the fact that $\mathcal{I}$ is symmetric monoidal with respect to the pointed sum $m_+ \vee n_+ = (m+n)_+$, and the unitality and associativity isomorphisms are actually identities (see Sect. 2.2.2.1 for details). Hence $\mathcal{I}^{q+1}$ is just a disguise for the q-simplices of the cyclic bar construction $B^{cy}\mathcal{I}$ of Sect. 4.2.1.1, and the ϕ^* are just the structure maps for the cyclic bar construction. The maps $G_\phi(\mathbf{x})\colon G_q(\mathbf{x}) \to G_n(\phi^*\mathbf{x})$ are defined as follows. The loop coordinates are mixed by the obvious isomorphisms $S^{\phi^*\mathbf{x}} \cong S^{\mathbf{x}}$, and the maps $V(A,P)(\mathbf{x}) \to V(A,P)(\phi^*\mathbf{x})$ are given by the following setup:

for	$\phi \in \Lambda([q], ?)$	define $V(A,P)(\mathbf{x}) \to V(A,P)(\phi^*\mathbf{x})$ by means of...
d^0		$P(S^{x_0}) \wedge A(S^{x_1}) \to P(S^{x_0 \vee x_1})$
d^i	for $0 < i < q$	$A(S^{x_i}) \wedge A(S^{x_{i+1}}) \to A(S^{x_i \vee x_{i+1}})$
d^q		$A(S^{x_q}) \wedge P(S^{x_0}) \to P(S^{x_q \vee x_0})$
s^i	for $0 \leq i \leq q$	$S^0 = \mathbf{S}(S^0) \to A(S^0)$ in the $i+1$st slot
t	(only when $A = P$)	cyclic permutation of smash factors

Lemma 4.2.2.2 *Let $\phi\colon [q] \to [n]$ and $\psi\colon [n] \to [m]$ be morphisms in Λ. Then the construction above defines a natural transformation $G_\phi\colon G_n \Rightarrow G_q \circ \phi^*$ of functors $\mathcal{I}^{n+1} \to \mathcal{S}_*$ with $G_{id} = id$ and $G_{\psi\phi} = (G_\phi \psi^*) G_\psi$.*

The last point in the lemma can be depicted as two equal 2-cells

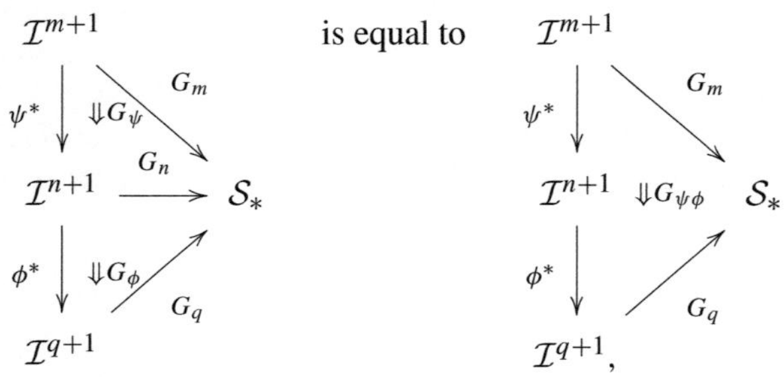

and we will write simply $G_{\psi\phi} = G_\phi G_\psi$.

Now, this is exactly what is needed: if $\phi\colon [q] \to [n] \in \Lambda$, we get a map

$$\underset{\overrightarrow{\mathcal{I}^{n+1}}}{\operatorname{holim}} G_n \xrightarrow{G_\phi} \underset{\overrightarrow{\mathcal{I}^{n+1}}}{\operatorname{holim}} G_q \phi^* \xrightarrow{(\phi^*)_*} \underset{\overrightarrow{\mathcal{I}^{q+1}}}{\operatorname{holim}} G_q$$

(sending $t \in G_n(\mathbf{x}_s)$ in the $\mathbf{x}_0 \leftarrow \cdots \leftarrow \mathbf{x}_s$-summand to $G_\phi(t) \in G_q(\phi^*\mathbf{x}_s)$ in the $\phi^*\mathbf{x}_0 \leftarrow \cdots \leftarrow \phi^*\mathbf{x}_s$-summand), and the equalities $G_{id} = id$ and $G_{\psi\phi} = G_\phi G_\psi$ enter at the crucial moment to prove:

Proposition 4.2.2.3 *Let A be an $\mathbf{S}$-algebra, P an A-bimodule and X a space. With the definitions above, the assignment $[q] \mapsto \operatorname{holim}_{\overrightarrow{\mathcal{I}^{q+1}}} G_q = THH(A,P,X)_q$ defines a functor from Λ^o if $P = A$ and from Δ^o if $P \neq A$.*

Definition 4.2.2.4 Let A be an $\mathbf{S}$-algebra, P an A-bimodule and X a space. Then the topological Hochschild homology is defined as

$$THH(A,P,X) = \operatorname{diag}^*\big\{[q] \mapsto THH(A,P,X)_q\big\}.$$

This gives rise to the very special Γ-space

$$THH(A,P) = \{Y \in ob\Gamma^o \mapsto THH(A,P,Y)\}$$

and the Ω-spectrum

$$\underline{T}(A,P,X) = \{m \mapsto \sin|THH(A,P,S^m \wedge X)|\}$$

The $\sin|-|$ in the definition of $\underline{T}$ will not be of any importance to us now, but will be convenient when discussing the cyclic structure in Chap. 6. We also write $THH(A,P) = THH(A,P,S^0)$ and $THH(A) = THH(A,S^0)$ and so on, where confusion is unlikely.

By just the same formula $THH(A,P)$ should be thought of as a functor from spaces to spaces. Note that by Lemma 4.2.3.1 below,

$$THH(A,P,X) \simeq \operatorname{diag}^*\{[q] \mapsto THH(A,P,X_q)\} = THH(A,P)(X)$$

for all spaces X.

Lemma 4.2.2.5 *$THH(A,P,X)$ is functorial in (A,P) and X, and takes (stable) equivalences to pointwise equivalences. Likewise for THH and $\underline{T}$.*

Proof This follows from the corresponding properties for $THH(A,P,X)_q$ since maps of simplicial spaces inducing weak equivalences in every degree induce weak equivalences on diagonals, Theorem A.6.0.1. □

Note 4.2.2.6 When checking the details of Lemma 4.2.2.2, the reader will discover the importance of using $\mathcal{I}$, and not just the natural numbers. For G_ϕ to be a natural transformation we need the flexibility of "allowing to suspend in more than one coordinate".

For instance, consider any $\alpha\colon (1_+,1_+) \to (x_0,x_1)$ in $\mathcal{I}^2$ (or $\mathbf{N}^2$) with $x_0 \vee x_1 \cong 3_+$, and $f\colon S^1 \wedge S^1 \to \sin|S^1 \wedge S^1|$ (a zero simplex in $G_1(\mathbf{S},\mathbf{S},S^0)(1_+,1_+)$). Then $G_{d^0}(1_+,1_+)(f)$ also equals f. If we allowed only one map $2_+ \to 3_+$ in our index category, then $d_0\alpha$ would have to be independent of α, prohibiting $G_{d^0}(x_0,x_1)G_1(\alpha)(f)$ (which definitely depends on α—the difference is whether the suspension happens after the first or second S^1) to be equal to $G_0(d_0\alpha)G_{d^0}(1_+,1_+)(f)$.

4.2.2.4 The Cyclic Structure

In the case where $P = A$ we have that $THH(A,X) = THH(A,A,X)$ is a cyclic space. Furthermore, $THH(A) = THH(A,A)$ is a cyclic Γ-space and $\underline{T}(A,X) = \underline{T}(A,A,X)$ becomes a spectrum with an $\mathbf{S}^1$-action (where $\mathbf{S}^1 = \sin|S^1|$ and $S^1 = \Delta[1]/\partial\Delta[1]$). This last point needs some explanation, and will become extremely important in the next chapter.

If Z is a cyclic space, then the realization $|Z|$ of the corresponding simplicial space has a natural $|S^1| = \mathbb{T}$-action (see Sect. 6.2.1 for further details), and so $\sin|Z|$ has a natural $\mathbf{S}^1 = \sin|S^1|$-action. Of course, there is no such thing as an "S^1-space", since S^1 is only an innocent space—not a group—before geometric realization (remember that in "space" is a synonym for "(pointed) simplicial set").

In the case where $Z = THH(A, X)$ (considered as a simplicial cyclic set) the actual $\mathbf{S}^1$-fixed points are not very exciting: as we will show in more details in Chap. 6,

$$\sin\big|THH(A,X)\big|^{\mathbf{S}^1} \cong \sin|X|.$$

An important fact in this connection is that, considered as a $\Gamma\mathcal{S}_*$-category, A has only one object. In the next section we will consider more general situations, and get more interesting results.

In Chap. 6 we shall see that, although the $\mathbf{S}^1$-fixed points are not very well behaved, the fixed points of the actions by the finite cyclic subgroups give rise to a very interesting theory.

4.2.2.5 Hochschild Homology over Other Commutative S-Algebras

Bökstedt's definition of topological Hochschild homology is very convenient, and accessible for hands-on manipulations. On the other hand, it is conceptually more rewarding to view topological Hochschild homology as Hochschild homology over $\mathbf{S}$. Let k be a commutative $\mathbf{S}$-algebra. Then $(k\text{-mod}, \wedge_k, k)$ is a symmetric monoidal category, and we may form the cyclic bar construction, see Sect. 4.2.1.1, in this category: if A is a k-algebra which is cofibrant as a k-module and P is an A-bimodule, then $HH^k(A, P)$ is the simplicial k-module

$$HH^k(A, P) = \big\{[q] \to P\wedge_k A\wedge_k \cdots \wedge_k A\big\}$$

By the results of the previous chapter, we see that $HH^{\mathbf{S}}$ and THH have stably equivalent values (the smash product has the right homotopy type when applied to cofibrant Γ-spaces, and so $HH^{\mathbf{S}}(A, P)$ and $THH(A, P)$ are equivalent in every degree). Many of the results we prove in the following section have more natural interpretations in this setting.

If we wish to consider Hochschild homology of k-algebras that are not cofibrant as k-modules, we should apply a functorial cofibrant replacement before using the construction of HH^k above.

Example 4.2.2.7 (*THH* of Spherical Group Rings) Let G be a simplicial group, and consider the spherical group ring $\mathbf{S}[G]$ of Example 2.1.4.3(2) given by sending a finite pointed set X to $\mathbf{S}[G](X) = X\wedge G_+$. Then $THH(\mathbf{S}[G])_q$ has the homotopy type of $\mathbf{S}[G]$ smashed with itself $q+1$ times ($\mathbf{S}[G]$ is a cofibrant Γ-space, so one does not have to worry about cofibrant replacements), with face and degeneracy maps as in Hochschild homology. Hence $THH(\mathbf{S}[G])$ is equivalent to $\mathbf{S}[B^{cy}(G)]$, whose associated infinite loop space calculates the stable homotopy of the cyclic bar construction of G.

A particularly nice interpretation is obtained if we set $X = |BG|$, because there is a natural weak equivalence $|B^{cy}G| \xrightarrow{\sim} \Lambda X$ between the cyclic nerve of the loop group and the free loop space (see e.g., [100, proof of V.1.1]), and so we get a weak equivalence

$$\big|THH\big(\mathbf{S}[G]\big)(1_+)\big| \xrightarrow{\sim} \Omega^\infty\Sigma^\infty\Lambda X_+$$

of pointed topological spaces.

4.2.3 First Properties of Topological Hochschild Homology

An important example is the topological Hochschild homology of an **S**-algebra coming from a (simplicial) ring. We consider *THH* as a functor of rings and bimodules, and when there is no danger of confusion, we write $THH(A, P, X)$, even though we actually mean $THH(HA, HP, X)$ and so on. Whether the ring is discrete or truly simplicial is of less importance in view of the following lemma, which holds for simplicial **S**-algebras in general.

Lemma 4.2.3.1 *Let A be a simplicial* **S***-algebra, P an A-bimodule and X a space. Then there is a chain of natural pointwise equivalences*

$$\operatorname{diag}^*\big\{[q]\mapsto THH(A_q, P_q, X_q)\big\} \simeq THH\big(\operatorname{diag}^*A, \operatorname{diag}^*P, X\big).$$

Proof Let $\mathbf{x}\in\mathcal{I}^{n+1}$. Using that the smash product is formed degreewise, we get that

$$\operatorname{diag}^*\big(X\wedge V(A, P)(\mathbf{x})\big) = X\wedge V\big(\operatorname{diag}^*A, \operatorname{diag}^*P\big)(\mathbf{x}).$$

Since A and P preserve connectivity of their input, the loops in the *THH*-construction may be performed degreewise up to a natural chain of weak equivalences

$$\operatorname{diag}^*\Omega^{\vee\mathbf{x}}\big(X\wedge V(A, P)(\mathbf{x})\big) \simeq \Omega^{\vee\mathbf{x}}\big(X\wedge V\big(\operatorname{diag}^*A, \operatorname{diag}^*P\big)(\mathbf{x})\big)$$

(see Corollary A.6.0.4 and the discussion immediately after, where the chain is described explicitly: the map going "backwards" is simply getting rid of a redundant $\sin|-|$). Since homotopy colimits commute with taking the diagonal, we are done. □

4.2.3.1 Relation to Hochschild Homology (over the Integers)

Since, à priori, Hochschild homology is a simplicial abelian group, whereas topological Hochschild homology is a Γ-space, we could consider *HH* to be a Γ-space by the Eilenberg–Mac Lane construction $H\colon \mathcal{A} = sAb \to \Gamma\mathcal{S}_*$ in order to have maps between them.

We make a slight twist to make the comparison even more straight-forward. Recall the definitions of $\bar{H}\colon \mathcal{A} = sAb \to \Gamma\mathcal{A}$ Example 2.1.2.1(1), and the forgetful functor $U\colon \mathcal{A} \to \mathcal{S}_*$ which is adjoint to the free functor $\tilde{\mathbf{Z}}\colon \mathcal{S}_* \to \mathcal{A}$ of Sect. 2.1.3.1. By definition $H = U\bar{H}$. An $\bar{H}\mathbf{Z}$-algebra A is a monoid in $(\Gamma\mathcal{A}, \otimes, \bar{H}\mathbf{Z})$ (see Sect. 2.1.4.1), and is always equivalent to $\bar{H}$ of a simplicial ring (Corollary 2.2.2.6). As noted in the proof of Corollary 2.2.2.6 the loops and homotopy colimit used to stabilize could be exchanged for their counterpart in simplicial abelian groups if the input has values in simplicial abelian groups. This makes possible the following definition (the loop space of a simplicial abelian group is a simplicial abelian group, and the homotopy colimit is performed in simplicial abelian groups with direct sums instead of wedges, see Sect. A.7.4.1):

Definition 4.2.3.2 Let A be an $\bar{H}\mathbf{Z}$-algebra, P an A-bimodule, and $X \in ob\Gamma^o$. Define the simplicial abelian group

$$HH^{\mathbf{Z}}(A,P,X)_q = \underset{\mathbf{x}\in\mathcal{I}^{q+1}}{\underrightarrow{\operatorname{holim}}}\, \Omega^{\vee\mathbf{x}}\Big(\tilde{\mathbf{Z}}[X] \otimes P\big(S^{x_0}\big) \otimes \bigotimes_{1\le i\le q} A\big(S^{x_i}\big)\Big)$$

with simplicial structure maps as for Hochschild homology. Varying X and q, this defines $\underline{HH}^{\mathbf{Z}}(A,P) \in ob\Gamma\mathcal{A}$.

Remark 4.2.3.3 Again (sigh), should the $\bar{H}\mathbf{Z}$-algebra A not take flat values, we replace it functorially by one that does (for instance by replacing it to $\bar{H}$ of a simplicial ring which may be assumed to be free in every degree). One instance where this is not necessary is when $A = \tilde{\mathbf{Z}}B$ for some $\mathbf{S}$-algebra B. Note that a $\tilde{\mathbf{Z}}B$-module is a special case of a B-module via the forgetful map $U\colon \Gamma\mathcal{A} \to \Gamma\mathcal{S}_*$ (it is a B-module "with values in $\mathcal{A}$").

If A is a simplicial ring and P an A-bimodule, $HH^{\mathbf{Z}}(\bar{H}A, \bar{H}P)$ is clearly (pointwise) equivalent to

$$\bar{H}\big(HH(A,P)\big) = \big\{X \mapsto HH(A,P,X) = HH(A,P) \otimes \mathbf{Z}[X]\big\}.$$

Definition 4.2.3.4 For A an $\bar{H}\mathbf{Z}$-algebra and P an A-bimodule, there is an natural map

$$THH(UA, UP)(X) \to UHH^{\mathbf{Z}}(A,P)(X),$$

called the *linearization map*, given by the Hurewicz map $X \to \tilde{\mathbf{Z}}[X]$ and by sending the smash of simplicial abelian groups to tensor product.

In the particular case of a simplicial ring R and R-bimodule Q, the term *linearization map* refers to the map

$$THH(HR, HR) \to UHH^{\mathbf{Z}}(\bar{H}R, \bar{H}Q) \xleftarrow{\sim} H\big(HH(R,R)\big).$$

Again, if A, P, R or Q should happen to be non-flat, we should take a functorial flat resolution, and in this case the "map" is really the one described preceded by a homotopy equivalence pointing in the wrong direction (i.e., the linearization map is then what is called a *weak* map).

The linearization map is generally far from being an equivalence (it *is* for general reasons always two-connected). If $P = A$ it is a cyclic map.

However, we may factor $THH(UA, UP) \to UHH^{\mathbf{Z}}(A, P)$ through a useful equivalence:

Lemma 4.2.3.5 *Let A be an* **S**-*algebra, P a $\tilde{\mathbf{Z}}A$-bimodule and $X \in ob\Gamma^o$. The inclusion*

$$X \wedge P(S^{x_0}) \wedge \bigwedge_{1 \le i \le q} A(S^{x_i}) \to \tilde{\mathbf{Z}}[X] \otimes P(S^{x_0}) \otimes \bigotimes_{1 \le i \le q} \tilde{\mathbf{Z}}[A(S^{x_i})]$$

induces an equivalence

$$THH(A, UP) \xrightarrow{\sim} UHH^{\mathbf{Z}}(\tilde{\mathbf{Z}}A, P).$$

Proof It is enough to prove it degreewise. If $M \in sAb$ is m-connected, and $Y \in \mathcal{S}_*$ is y-connected, then $M \wedge Y \to M \otimes \tilde{\mathbf{Z}}[Y]$ is $2m + y + 2$ connected (by induction on the cells of Y: assume $Y = S^{y+1}$, and consider $M \to \Omega^{y+1}(M \wedge S^{y+1}) \to \Omega^{y+1} M \otimes \tilde{\mathbf{Z}}[S^{y+1}]$. The composite is an equivalence, and the first map is $2m + 1$ connected by the Freudenthal suspension Theorem A.8.2.3. Setting $M = P(S^{x_0})$ and $Y = X \wedge \bigwedge_{1 \le i \le q} A(S^{x_i})$ we get that the map is $2x_0 - 2 + \sum_{i=1}^{q} (x_i - 1) + \mathrm{conn}(X) + 2$ connected, and so, after looping down the appropriate number of times, $x_0 - q + \mathrm{conn}(X)$ connected, which goes to infinity with x_0. □

In the following we may not always be as pedantic as all this. We will often suppress forgetful functors, and write this as $THH(A, P) \xrightarrow{\sim} HH^{\mathbf{Z}}(\tilde{\mathbf{Z}}A, P)$.

If A is an $\bar{H}\mathbf{Z}$-algebra and P an A-bimodule, this gives a factorization

$$THH(UA, UP) \xrightarrow{\sim} UHH^{\mathbf{Z}}(\tilde{\mathbf{Z}}UA, P) \to UHH^{\mathbf{Z}}(A, P).$$

Remark 4.2.3.6 Some words of caution:

1. Note, that even if $P = A$, $HH^{\mathbf{Z}}(\tilde{\mathbf{Z}}A, P)$ is not a cyclic object.
2. Note that if A is a simplicial ring, then $\tilde{\mathbf{Z}}HA$ is not equal to $H\tilde{\mathbf{Z}}A$. We will discover an interesting twist to this when we apply these lines of thought to additive categories instead of rings (see Sect. 4.3.4).
3. In view of the equivalence $H\mathbf{Z} \wedge A \simeq \tilde{\mathbf{Z}}A$, Lemma 4.2.3.5 should be interpreted as a change of ground ring equivalence

$$HH^{\mathbf{S}}(A, UP) \simeq UHH^{H\mathbf{Z}}(H\mathbf{Z} \wedge A, P).$$

More generally, if $k \to K$ is a map of $\mathbf{S}$-algebras, A a cofibrant k-algebra and P a $K \wedge_k A$-bimodule, then

$$HH^k\bigl(A, f^*P\bigr) \simeq f^* HH^K(K \wedge_k A, P)$$

where $f \colon A \cong k \wedge_k A \to K \wedge_k A$ is the map induced by $k \to K$.

For comparison purposes the following lemmas are important (see [225, 4.2]).

Lemma 4.2.3.7 *If A is a ring and P an A-bimodule, then there is a spectral sequence*

$$E^2_{p,q} = HH_p\bigl(A, \pi_q THH(\mathbf{Z}, P, X), Y\bigr) \Rightarrow \pi_{p+q} THH(A, P, X \wedge Y).$$

Proof For a proof, see Pirashvili and Waldhausen [225]. □

On a higher level, it is just the change of ground ring spectral sequence: let $k \to K$ be a map of commutative $\mathbf{S}$-algebras, A a K-algebra and P a $K \wedge_k A$-bimodule, and assume A and K cofibrant as k-modules, then

$$HH^k(A, P) \simeq HH^K(K \wedge_k A, P) \simeq HH^K\bigl(A, HH^k(K, P)\bigr)$$

where by abuse of notation P is regarded as a bimodule over the various algebras in question through the obvious maps.

In view of Lemma 4.2.3.7 we will need to know the values of $\pi_*(THH(\mathbf{Z}, P))$ for arbitrary abelian groups P. These values follow from Bökstedt's calculations in view of the isomorphism $HH^{\mathbf{Z}}(\tilde{\mathbf{Z}}[\mathbf{Z}], P) \cong HH^{\mathbf{Z}}(\tilde{\mathbf{Z}}[\mathbf{Z}], \mathbf{Z}) \otimes P$ (or the equivalence $THH(H\mathbf{Z}, P) \simeq THH(\mathbf{Z}) \wedge_{H\mathbf{Z}} HP$) and the universal coefficient theorem:

$$\pi_k THH(\mathbf{Z}, P) \cong \begin{cases} P & k = 0, \\ P/iP & k = 2i-1, \\ \operatorname{Tor}_1^{\mathbf{Z}}(\mathbf{Z}/i\mathbf{Z}, P) & k = 2i > 0. \end{cases}$$

Lemma 4.2.3.8 *If A is a ring and P an is A-bimodule, then the linearization map*

$$THH(A, P) \to HH^{\mathbf{Z}}(A, P)$$

(and all the other variants) is a pointwise equivalence after rationalization, and also after profinite completion followed by rationalization.

Proof In the proof of the spectral sequence of Lemma 4.2.3.7, we see that the edge homomorphism is induced by the linearization map $\pi_* THH(A, P) \to \pi_* HH(A, P)$. From the calculation of $\pi_* THH(\mathbf{Z}, P)$ above we get that all terms in the spectral sequence above the base line are torsion groups of bounded order. Thus, $\pi_j THH(A, P)$ and $\pi_j HH^{\mathbf{Z}}(A, P) = \pi_j HH(A, P)$ differ at most by groups of this sort, and so the homotopy groups of the profinite completions $THH(A, P)\widehat{}$

and $HH^{\mathbf{Z}}(A,P)\widehat{}$ will also differ by torsion groups of bounded order, and hence we have an equivalence $THH(A,P)\widehat{}_{(0)} \to HH^{\mathbf{Z}}(A,P)\widehat{}_{(0)}$. □

If the reader prefers not to use the calculation of $THH(\mathbf{Z})$, one can give a direct proof of the fact that the homotopy fiber of $THH(A,P) \to HH(A,P)$ has homotopy groups of bounded order directly from the definition.

Sketch:

1. It is enough to prove the result it in each simplicial dimension.
2. As A and P are flat as abelian groups we may resolve each by free abelian groups (multiplication plays no role), and so it is enough to prove it for free abelian groups.
3. We must show that $\tilde{\mathbf{Z}}[X]\wedge\tilde{\mathbf{Z}}[Y]\wedge Z \to \tilde{\mathbf{Z}}[X\wedge Y]\wedge Z$ has homotopy fiber whose homotopy is torsion of bounded order in a range depending on the connectivity of X, Y and Z. This follows as the homology groups of the integral Eilenberg–Mac Lane spaces are finite in a range.

4.2.4 THH is Determined by Its Values on Simplicial Rings

In Theorem 3.3.2.2, we showed that algebraic K-theory is determined by its values on simplicial rings. In this section we prove the analogous statement of theorem for topological Hochschild homology.

Let A be an $\mathbf{S}$-algebra. Recall the definition of the functorial cube $\mathcal{A} = \{S \mapsto (A)_S\}$ of $\mathbf{S}$-algebras from Definition 3.3.1.8 whose nodes $(A)_S$ were all equivalent to simplicial rings by Proposition 3.3.1.10. In particular, the S'th node was obtained by applying the free-forgetful pair $(\tilde{\mathbf{Z}}, U)$ as many times as there are elements in S. The functor $S \mapsto (-)^n_S$ can clearly be applied to A-bimodules as well, and $S \mapsto (P)^n_S$ will be a cube of $S \mapsto (A)^n_S$-bimodules.

We will need the following result about the smashing of cubes. For the definition of f-cartesian cubes, see Definition 3.3.1.6.

Lemma 4.2.4.1 *Let $\mathcal{X}^i$ be $(id + x_i)$-cartesian cubes of pointed spaces or spectra for $i = 1, \dots, n$. Then*

$$\mathcal{X} = \left\{ S \mapsto \bigwedge_{1\le i\le n} \mathcal{X}^i_S \right\}$$

is $id + \sum_i x_i$ cartesian.

Proof Note that each d-subcube of $\mathcal{X}$ can be subdivided into d-cubes, each of whose maps are the identity on all the smash factors but one. Each of these d-cubes are by induction $2 \cdot id + \sum_i x_i - 1$-cocartesian, and so the d-subcube we started with was $2 \cdot id + \sum_i x_i - 1$-cocartesian. □

Proposition 4.2.4.2 *Let $\mathcal{A}$ be an id-cartesian cube of* **S**-*algebras, and $\mathcal{P}$ an id-cartesian cube of $\mathcal{A}$-bimodules (i.e., each $S \to T$ induces a map of natural bimodules $(\mathcal{A}_S, \mathcal{P}_S) \to (\mathcal{A}_T, \mathcal{P}_T)$) and X a k-connected space. Then $THH(\mathcal{A}, \mathcal{P}, X)$ is $id + k + 1$ cartesian.*

Proof By applying the monoidal fibrant replacement functor T_0 of Definition 2.2.2.1, we may assume that for each S, $\mathcal{A}_S$ and $\mathcal{P}_S$ are stably fibrant, so that the id-cartesian conditions actually hold pointwise: for each finite pointed set Y, the cubes of spaces $\mathcal{A}(Y)$ and $\mathcal{P}(Y)$ are id-cartesian.

Since realization commutes with homotopy colimits, the claim will follow if we can prove that for each $q \geq 0$, $S \mapsto THH(\mathcal{A}_S, \mathcal{P}_S, X)_q$ is $2 \cdot id + k$ cocartesian.

For any $q \geq 0$ the lemma above tells us that

$$S \mapsto X \wedge \mathcal{P}_S\bigl(S^{x_0}\bigr) \wedge \bigwedge_{1 \leq i \leq q} \mathcal{A}_S\bigl(S^{x_i}\bigr)$$

is $id + k + 1 + \sum_{i=0}^{q} x_i$ cartesian. Looping down the appropriate number of times, this is $id + k + 1$ cartesian, and so

$$S \mapsto THH(\mathcal{A}_S, \mathcal{P}_S, X)_q$$

is $id + \mathrm{conn}(X) + 1$ cartesian. □

Theorem 4.2.4.3 (*THH*) *Let A be an* **S**-*algebra and P an A-bimodule. Then the natural map*

$$THH(A, P) \to \underset{S \in \mathcal{P} - \emptyset}{\underleftarrow{\mathrm{holim}}}\, THH\bigl((A)_S, (P)_S\bigr)$$

is an equivalence.

Proof This is a direct consequence of the above proposition applied to

$$\mathcal{A} = \bigl\{S \mapsto (A)_S\bigr\} \quad \text{and} \quad \mathcal{P} = \bigl\{S \mapsto (P)_S\bigr\}$$

since the hypotheses are satisfied by Theorem 3.3.1.9, using the same method as we used in Theorem 3.3.2.2 to pass from finite to infinite cubes. □

This means that we can reduce many questions about *THH* of **S**-algebras to questions about *THH* of (simplicial) rings, which again may often be reduced to questions about integral Hochschild homology by means of the spectral sequence of Lemma 4.2.3.7.

As an example of this technique consider the following proposition.

Proposition 4.2.4.4 *Let A be an* **S**-*algebra and P an A-bimodule, then Morita invariance holds for THH, i.e., there is a natural chain of equivalences*

$$THH(A, P) \simeq THH(Mat_n A, Mat_n P).$$

If B is another **S**-*algebra and Q a B-bimodule, then THH preserves products, i.e., the natural map*

$$THH(A \times B, P \times Q) \xrightarrow{\sim} THH(A, P) \times THH(B, Q)$$

is a pointwise equivalence.

Proof Since all the unmarked arrows in the following composites

$$\tilde{\mathbf{Z}}[Mat_n A(X)] \cong \tilde{\mathbf{Z}}\Big[\prod_n \bigvee_n A(X)\Big] \longleftarrow \tilde{\mathbf{Z}}\Big[\bigvee_n \bigvee_n A(X)\Big] \cong \oplus_n \oplus_n \tilde{\mathbf{Z}}[A(X)]$$
$$\longleftarrow Mat_n \tilde{\mathbf{Z}} A(X)$$

and

$$\tilde{\mathbf{Z}}[A(X) \times B(Y)] \longleftarrow \tilde{\mathbf{Z}}[A(X) \vee B(Y)] \cong \tilde{\mathbf{Z}}[A(X)] \oplus \tilde{\mathbf{Z}}[B(Y)]$$
$$\longleftarrow \tilde{\mathbf{Z}}[A(X)] \vee \tilde{\mathbf{Z}}[B(X)]$$

are induced by injecting finite wedges into products (and so stable equivalences), it is, in view of Theorem 4.2.4.3 and Lemma 4.2.3.1, enough to prove the corresponding statements for rings (this is not strictly true: for Morita invariance one needs to know that the chain of equivalences in question actually is natural in (A, P). With our presentation, this is not really explained before Sect. 4.3.5.4 below. See the note following immediately after the proof). Appealing to the spectral sequence of Lemma 4.2.3.7 together with the easy facts that

$$\pi_q THH(\mathbf{Z}, M_n P) \cong M_n\big(\pi_q THH(\mathbf{Z}, P)\big)$$

and

$$\pi_q THH(\mathbf{Z}, P \oplus Q) \cong \pi_q THH(\mathbf{Z}, P) \oplus \pi_q THH(\mathbf{Z}, Q)$$

it follows from the corresponding statements in Hochschild homology, see e.g. [181, p. 17] (use that matrices (resp. products) of flat resolutions are flat resolutions of matrices (resp. products)). □

Note 4.2.4.5 There are of course direct proofs of these statements, and they are essentially the same as in [181, p. 17], except that one has to remember that finite sums are just equivalent to finite products (not isomorphic), see e.g. [70]. The presence of the "chain of natural equivalences" in the statement of Morita invariance is annoying and stems from the fact that the obvious maps connecting $THH(A, P)$ and $THH(Mat_n A, Mat_n P)$ as presented e.g., in [70] do not respect the degeneracy maps in the simplicial direction of topological Hochschild homology. In [70] this was handled by adding degeneracies artificially as in Segal's fat realization [257]. Another solution using the Barratt-Eccles construction is given in [246]. In Sect. 4.3.5.4 below we see that if we replace the $n \times n$-matrices by the ΓS_*-category

$\mathcal{F}^n_A$ of free A-modules of rank less than or equal to n we get a natural representation $THH(A,P)\xrightarrow{\sim} THH(\mathcal{F}^n_A,P)\xleftarrow{\sim} THH(Mat_n A, Mat_n P)$, where the leftward pointing equivalence is of a simpler sort relating to cofinality.

4.2.5 An Aside: A Definition of the Trace from the K-Theory Space to Topological Hochschild Homology for S*-Algebras*

In Chap. 5 we will give a natural construction of the (Bökstedt–Dennis) trace on the categorical level. However, for those not interested in this construction we give an outline of the trace map construction as it appeared in the unpublished MSRI notes [105], and later in [26]. Some of the elements showing up in the general definitions make an early appearance in the one we are going to give below.

This is only a *weak transformation*, in the sense that we will encounter weak equivalences going the wrong way, but this will cause no trouble in our context. Indeed, such arrows pointing the wrong way can always be rectified by changing our models slightly. Furthermore, as we present it here, this only gives rise to a map of spaces, and not of spectra. We give a quick outline at the end, of how this can be extended to a map of spectra.

For any **S**-algebra A we will construct a weak map (i.e., a chain of maps where the arrows pointing the wrong way are weak equivalences) from $BA^* = B\widehat{GL}_1(A)$, the classifying space of the monoid of homotopy units of A, to $THH(A)(S^0)$. Applying this to the **S**-algebras $Mat_n A$, we get weak maps from $B\widehat{GL}_n(A)$ to $THH(Mat_n A)(S^0)\xleftarrow{\sim} THH(A)(S^0)$. The map produced will respect stabilization, in the sense that

$$\begin{array}{ccc}
B\widehat{GL}_n(A) & \longrightarrow & THH(Mat_n A)(S^0) \\
\downarrow & & \downarrow \\
B\widehat{GL}_n(A)\times B\widehat{GL}_1(A) & \longrightarrow & THH(Mat_n A)(S^0)\times THH(Mat_1 A)(S^0) \\
\simeq\uparrow & & \simeq\uparrow \\
B\big((Mat_n A\times Mat_1 A)^*\big) & \longrightarrow & THH(Mat_n A\times Mat_1 A)(S^0) \\
\downarrow & & \downarrow \\
B\widehat{GL}_{n+1}(A) & \longrightarrow & THH(Mat_{n+1} A)(S^0)
\end{array}$$

commutes, where the upper vertical maps are induced by the identity on the first factor, and the inclusion of $1\in\widehat{M}_1(A)=THH_0(Mat_1 A)$ into the second factor. (Note that the horizontal maps are just weak maps, and that some of the intermediate stages may not have the property that the upwards pointing map is an equivalence, but this does not affect the argument.) Stabilizing this with respect to n and taking the plus construction on both sides to get a weak transformation from $B\widehat{GL}(A)^+$ to $\lim_{n\to\infty} THH(M_n A)^+\simeq THH(A)$.

4.2.5.1 Construction

If M is a monoid, we may use the free forgetful adjoint pair to form a functorial free simplicial resolution $F(M) \overset{\sim}{\twoheadrightarrow} M$. This extends to a functorial free resolution of any simplicial monoid, and in particular of $A^* = \widehat{GL_1}(A)$. The forgetful functor from groups to monoids has a left adjoint $M \mapsto M^{-1}M = \lim_{\leftarrow} G$ where the limit is over the category of groups under M. In the case where M is free, this is obtained by just adjoining formal inverses to all generators, and the adjunction $M \to M^{-1}M$ induces a weak equivalence $BM \to B(M^{-1}M)$ ($|BM|$ is just a wedge of circles, and the "inverses" are already included as going the opposite way around any circle. Alternatively, consider the "fiber" of $M \subset M^{-1}M$, that is, the category C with objects elements in $M^{-1}M$, and a single morphism $m\colon g \cdot m \to g$ for every $m \in M$ and $g \in M^{-1}M$. Now, C is obviously connected, and between any two objects there is at most one morphism, and so C is contractible).

In the case of the simplicial monoids $F(M)$ we get a transformation $F(M) \to G(M) = F(M)^{-1}F(M)$. If M is a group-like, then Corollary A.6.1.3 tells us that the natural map $M \to \Omega BM$ is a weak equivalence. Furthermore, if M is group-like, then so is $F(M)$, and the diagram

$$\begin{array}{ccc} F(M) & \longrightarrow & G(M) \\ \simeq\downarrow & & \simeq\downarrow \\ \Omega BF(M) & \xrightarrow{\simeq} & \Omega BG(M) \end{array}$$

tells us that $F(M) \to G(M)$ is an equivalence.

Now, for any category $\mathcal{C}$, the nerve $\mathbf{N}\mathcal{C}$ may be considered as a simplicial category whose objects in $\mathbf{N}_q\mathcal{C}$ are the q-simples in the classifying space $ob\mathbf{N}_q\mathcal{C} = B_q\mathcal{C} = \{c_0 \leftarrow c_1 \leftarrow \cdots \leftarrow c_q\}$ (see Sect. A.2.4), and morphisms simply diagrams (in $\mathcal{C}$) like

$$\begin{array}{ccccccc} c_0 & \longleftarrow & c_1 & \longleftarrow & \cdots & \longleftarrow & c_q \\ \downarrow & & \downarrow & & & & \downarrow \\ d_0 & \longleftarrow & d_1 & \longleftarrow & \cdots & \longleftarrow & d_q \end{array}.$$

If all morphisms in $\mathcal{C}$ are isomorphisms (i.e., $\mathcal{C}$ is a groupoid), then the face and degeneracies are all equivalences of categories. Hence, for any functor X from categories to simplicial sets sending equivalences to weak equivalences, the natural map

$X(\mathcal{C}) = X(\mathbf{N}_0\mathcal{C}) \xrightarrow{\sim} X(\mathbf{N}\mathcal{C})$ is an equivalence for groupoids $\mathcal{C}$.

Also, just as we extended Hochschild homology from rings to (small) *Ab*-categories in Sect. 1.3.2, the cyclic bar construction can be extended from monoids to categories: If $\mathcal{C}$ is a category and P is a $\mathcal{C}$-bimodule we define the *cyclic nerve* $B^{cy}(\mathcal{C}, P)$ to be the space whose q-simplices are given as

$$B^{cy}_q(\mathcal{C}, P) = \coprod_{c_0, c_1, \ldots, c_q \in ob\mathcal{C}} P(c_0, c_q) \times \prod_{i=1}^{q} \mathcal{C}(c_i, c_{i-1}).$$

In particular, if G is a (simplicial) group regarded as a one point category in the ordinary sense, then we have a chain $BG = ob\mathbf{N}G \longrightarrow B^{cy}\mathbf{N}G \xleftarrow{\sim} B^{cy}G$ where the first map sends $x \in BG$ to $x = x = \cdots = x \in B^{cy}_q\mathbf{N}G$ and the last map is the weak equivalence induced by the equivalences $G \to \mathbf{N}_q G$.

Assembling this information, we have a diagram

$$\begin{array}{ccccccc} BM & \xleftarrow{\sim} & BFM & & B^{cy}FM & \xrightarrow{\sim} & B^{cy}M \\ & & \simeq\downarrow & & \simeq\downarrow & & \\ & & BGM & \longrightarrow\; B^{cy}\mathbf{N}GM \;\xleftarrow{\sim} & B^{cy}GM & & \end{array},$$

where the marked arrows are weak equivalences if M is group-like, giving a weak map $BM \to B^{cy}M$.

Recall the constructions T_0 and R from Chap. 2 (T_0 is like THH_0 used as a "fibrant replacement" for $\mathbf{S}$-algebras, and R takes a Γ-space and evaluates at $1_+ = S^0$). For any $\mathbf{S}$-algebra A, we have a map $B^{cy}RT_0A \to THH(A)(S^0)$ given by

$$B^{cy}_q RT_0(A) = \prod_{0\le i\le q} \underset{\overrightarrow{x_i\in I}}{\operatorname{holim}}\, \Omega^{x_i} A\big(S^{x_i}\big) \to \underset{\overrightarrow{\mathbf{x}\in I^{q+1}}}{\operatorname{holim}}\, \Omega^{\vee\mathbf{x}} \bigwedge_{0\le i\le q} A\big(S^{x_i}\big)$$

where the map simply smashes functions together.

Composing the weak map $BA^* \to B^{cy}A^*$ from the diagram above with the cyclic nerve of the monoid map $A^* = \widehat{GL}_1(A) \to \widehat{M}_1A(S^0) = RT_0(A)$ and $B^{cy}RT_0(A) \to THH(A)(S^0)$ we have the desired "trace map" $BA^* \to THH(A)(S^0)$.

If we insist upon having a transformation on the spectrum level, we may choose a Γ space approach as in [27]. The action on the morphisms is far from obvious, and we refer the reader to [27] for the details.

4.3 Topological Hochschild Homology of $\Gamma\mathcal{S}_*$-Categories

Recall the definition of $\Gamma\mathcal{S}_*$-categories. They were just like categories except that instead of morphism *sets* $\mathcal{C}(c,d)$ we have morphism Γ-spaces $\underline{\mathcal{C}}(c,d)$, the unit is a map $\mathbf{S} \to \underline{\mathcal{C}}(c,c)$ and the composition is a map

$$\underline{\mathcal{C}}(c,d)\wedge\underline{\mathcal{C}}(b,c) \to \underline{\mathcal{C}}(b,d)$$

of Γ-spaces subject to the usual unitality and associativity conditions. See Appendix A.10.2 for details, and Appendix A.10.4 for the natural extension of bimodules to this setting.

Rings are *Ab*-categories with one object, and $\mathbf{S}$-algebras are $\Gamma\mathcal{S}_*$-categories with one object, so just like the extension in Sect. 1.3.2 of Hochschild homology to cover the case of *Ab*-categories, we define topological Hochschild homology of general $\Gamma\mathcal{S}_*$-categories.

Definition 4.3.0.1 Let $\mathcal{C}$ be a $\Gamma\mathcal{S}_*$-category, and P a $\mathcal{C}$-bimodule. For each tuple $\mathbf{x} = (x_0, \dots, x_q) \in ob\Gamma^{q+1}$ let

$$V(\mathcal{C}, P)(\mathbf{x}) = \bigvee_{c_0,\dots,c_q \in ob\mathcal{C}} P(c_0, c_q)\big(S^{x_0}\big) \wedge \bigwedge_{1 \le i \le q} \underline{\mathcal{C}}(c_i, c_{i-1})\big(S^{x_i}\big).$$

For each $X \in ob\Gamma$ and $q \ge 0$, this gives rise to a functor $G_q = G(\mathcal{C}, P, X)_q : \mathcal{I}^{q+1} \to \mathcal{S}_*$ with $G_q(\mathbf{x}) = \Omega^{\vee\mathbf{x}}(X \wedge V(\mathcal{C}, P)(\mathbf{x}))$. Let

$$THH(\mathcal{C}, P, X)_q = \underset{\mathbf{x} \in \mathcal{I}^{q+1}}{\underrightarrow{\text{holim}}}\, \Omega^{\vee\mathbf{x}}\big(X \wedge V(\mathcal{C}, P)(\mathbf{x})\big).$$

This is a simplicial space (checking that G_q is a functor and that $THH(\mathcal{C}, P, X)_q$ is functorial in q proceeds exactly as before). It is functorial in X, and we write $THH(\mathcal{C}, P)$ for the corresponding Γ-object, and $\underline{T}(\mathcal{C}, P, X)$ for the corresponding Ω-spectrum.

4.3.1 *Functoriality*

If $F: \mathcal{C} \to \mathcal{D}$ is a map of $\Gamma\mathcal{S}_*$-categories, P a $\mathcal{C}$-bimodule, Q a $\mathcal{D}$-bimodule and $G: P \to F^*Q$ a $\Gamma\mathcal{S}_*$-natural transformation, we get a map $THH(\mathcal{C}, P) \to THH(\mathcal{D}, Q)$ of Γ-spaces. As a matter of fact, $THH(-,-)$ (as well any of the other versions) is a functor of $\Gamma\mathcal{S}_*$-natural bimodules $(\mathcal{C}, P)$ Sect. A.10.4.1.

Example 4.3.1.1 The example $(\mathcal{C}^\vee, P^\vee)$ of Sect. 2.1.6.1 in the case where $\mathcal{C}$ an additive category is a slight generalization of the case considered in [70, Part 2]. Here $\underline{\mathcal{C}^\vee}(c, d) = H(\mathcal{C}(c, d))$, but $P^\vee(c, d) = H(P(c, d))$ only if P is "bilinear". The restriction that P has to be additive (i.e., send sums in the first variable to products) is sometimes annoying.

Note 4.3.1.2 Since $\Gamma\mathcal{S}_*$-categories are examples of what was called ring functors in [70], it is worth noting that our current definition of *THH* agrees with the old one. In fact, a $\Gamma\mathcal{S}_*$-category is simply a ring functor restricted to Γ^o considered as the category of discrete finite pointed simplicial sets. The distinction between $\Gamma\mathcal{S}_*$-categories and ring functors is inessential in that topological Hochschild homology does not see the difference, and so all the general statements in [70, Part 1] carry over to the new setting.

4.3.1.1 Cyclic Structure and Fixed Points Under the Circle Action

Let $\mathcal{C}$ be a $\Gamma\mathcal{S}_*$-category and X a space. Then, as before, $THH(\mathcal{C}, X) = THH(\mathcal{C}, \mathcal{C}, X)$ is a cyclic space.

We promised in Sect. 4.2.2.4 that we would take a closer look at the $\mathbf{S}^1$-fixed points. Recall that $S^1 = \Delta[1]/\partial\Delta[1]$ and $\mathbf{S}^1 = \sin|S^1|$. We consider $THH(\mathcal{C},X)$ as a simplicial cyclic set, and so if we apply $\sin|-|$ in the cyclic direction we get a simplicial $\mathbf{S}^1$-space which we write $\sin|THH(\mathcal{C},X)|$. As explained in Lemma 6.2.1.4, if Z is a cyclic set, then the space of $\mathbf{S}^1$-fixed points of $\sin|Z|$ is nothing but $\lim_{\overleftarrow{\Lambda^o}} Z$, or more concretely, the set of zero-simplices $z \in Z_0$ such that $ts_0 z = s_0 z \in Z_1$. So, we consider the simplices in the space

$$THH(\mathcal{C},X)_0 = \operatorname*{holim}_{\overrightarrow{x\in\mathcal{I}}} \Omega^x\Big(X \wedge \bigvee_{c\in ob\mathcal{C}} \underline{\mathcal{C}}(c,c)(S^x)\Big)$$

whose degeneracy is invariant under the cyclic action. Note that there is no change if we let the wedge sum run over the subset $\mathcal{O} = \{c \in ob\mathcal{C} \mid id_c \neq *\} \subseteq ob\mathcal{C}$. In dimension q,

$$\begin{aligned}&\operatorname*{holim}_{\overrightarrow{x\in\mathcal{I}}} \Omega^x\Big(X \wedge \bigvee_{c\in\mathcal{O}} \underline{\mathcal{C}}(c,c)(S^x)\Big)_q\\ &\quad= \bigvee_{x_0\leftarrow\cdots\leftarrow x_q\in\mathcal{I}} \mathcal{S}_*\Big(S^{x_q}\wedge\Delta[q]_+, \sin\Big|\Big(X\wedge\bigvee_{c\in\mathcal{O}}\underline{\mathcal{C}}(c,c)(S^{x_q})\Big)\Big|\Big)_0.\end{aligned}$$

The degeneracy sends

$$\Big(x_0 \leftarrow \cdots \leftarrow x_q,\ f\colon S^{x_q}\wedge\Delta[q]_+ \to \sin\Big|X\wedge\bigvee_{c\in\mathcal{O}}\underline{\mathcal{C}}(c,c)(S^{x_q})\Big|\Big)$$

to

$$\left(\begin{matrix}x_0 \leftarrow \cdots \leftarrow x_q\\ 0_+ = \cdots = 0_+\end{matrix}, S^{x_q}\wedge S^0\wedge\Delta[q]_+ \to \sin\Big|\Big(X\wedge\bigvee_{c_0,c_1\in\mathcal{O}}\underline{\mathcal{C}}(c_0,c_1)(S^{x_q})\wedge\underline{\mathcal{C}}(c_1,c_0)(S^0)\Big)\Big|\right)$$

where the map is determined by f and the unit map $S^0 = \mathbf{S}(S^0) \to \underline{\mathcal{C}}(c,c)(S^0)$. For this to be invariant under the cyclic action, we first see that we must have $x_0 = \cdots = x_q = 0_+$. Assume f is a q-simplex in $\sin|X\wedge\bigvee_{c\in\mathcal{O}}\underline{\mathcal{C}}(c,c)(S^0)| \cong \sin(|X|\wedge\bigvee_{c\in\mathcal{O}}|\underline{\mathcal{C}}(c,c)(S^0)|)$ such that

$$\begin{aligned}\big|\Delta[q]\big|_+ &\xrightarrow{f} |X|\wedge\bigvee_{c\in\mathcal{O}}\big|\underline{\mathcal{C}}(c,c)(S^0)\big|\\ &\longrightarrow |X|\wedge\bigvee_{c\in\mathcal{O}}\big|\underline{\mathcal{C}}(c,c)(S^0)\big|\wedge\big|\underline{\mathcal{C}}(c,c)(S^0)\big|\end{aligned}$$

is invariant under permutation, where the last map is induced by the unit map $\underline{\mathcal{C}}(c,c)(S^0) \cong \underline{\mathcal{C}}(c,c)(S^0)\wedge S^0 \to \underline{\mathcal{C}}(c,c)(S^0)\wedge\underline{\mathcal{C}}(c,c)(S^0)$. Hence f only takes the value of the unit and factors through $|\Delta[q]| \to \bigvee_{c\in\mathcal{O}}|X| \cong |X|\wedge(ob\mathcal{C})_+$, i.e.,

$$\lim_{\overleftarrow{\Lambda^o}} THH(\mathcal{C},X) \cong \sin\big|THH(\mathcal{C},X)\big|^{\mathbf{S}^1} \cong \bigvee_{c\in\mathcal{O}} \sin|X|.$$

We may be tempted to say that $\bigvee_{c\in\mathcal{O}} X$ is the "S^1-fixed point space" of $THH(\mathcal{C},X)$ because this is so after applying $\sin|-|$ to everything.

If G is a topological group and X a G-space, then $\sin(X^G) \cong (\sin X)^{\sin G}$. Explicitly, a q-simplex in the latter space is a continuous function $f\colon \Delta^q \to X$ with the property that for any continuous $g\colon \Delta^q \to G$ and any $\sigma \in \Delta^q$ we have that $g(\sigma)f(\sigma) = f(\sigma)$. Since g may be constant, this means that f factors over $X^G \subseteq X$, and so can be considered as a q-simplex in $\sin(X^G)$. This gives a bijection which is clearly simplicial. Likewise for homotopy fixed points (up to homotopy).

4.3.2 The Trace

There is a map, the "Dennis trace map"

$$ob\mathcal{C} \longrightarrow THH(\mathcal{C})(S^0)_0 \xrightarrow{\text{degeneracies}} THH(\mathcal{C})(S^0)$$

sending $d \in ob\mathcal{C}$ to the image of the identity $id_d \in \mathcal{C}(d,d)(S^0)$ via the obvious map

$$\mathcal{C}(d,d)(S^0) \subseteq \bigvee_{c\in ob\mathcal{C}} \mathcal{C}(c,c)(S^0) \to \underset{\overrightarrow{x\in ob\mathcal{I}}}{\operatorname{holim}}\, \Omega^x \bigvee_{c\in ob\mathcal{C}} \mathcal{C}(c,c)(S^x) = THH(\mathcal{C})(S^0)_0.$$

In other words, in view of the discussion in Sect. 4.3.1.1 the trace is the inclusion of the $\mathbf{S}^1$-fixed points exactly when there is a unique object c with $id_c = *$.

This type of definition of the Dennis trace map first appeared in the context of (ordinary) Hochschild homology in [204]. Its main advantage over the other definitions is that it is far easier to transport structure across the categorical definition, and it is also much easier to prove compatibility with the "epicyclic structure" on the fixed points of topological Hochschild homology (see Definition 6.2.3.1).

4.3.3 Comparisons with the Ab-Cases

The statements which were made for $\bar{H}\mathbf{Z}$-algebras in Sect. 4.2.3.1 have their analogues for $\Gamma\mathcal{A}$-categories:

Definition 4.3.3.1 Let $\mathcal{C}$ be a $\Gamma\mathcal{A}$-category, P a $\mathcal{C}$-bimodule and X a finite pointed set. Consider the simplicial abelian group

$$HH^{\mathbf{Z}}(\mathcal{C},P,X)_q$$

$$= \underset{\overrightarrow{\mathbf{x}\in\mathcal{I}^{q+1}}}{\operatorname{holim}}\, \Omega^{\vee\mathbf{x}} \bigoplus_{c_0,\dots,c_q\in ob\mathcal{C}} \left(\tilde{\mathbf{Z}}[X] \otimes P(c_0,c_q)(S^{x_0}) \otimes \bigotimes_{1\le i\le q} \mathcal{C}(c_i,c_{i-1})(S^{x_i})\right),$$

where loop and homotopy colimit are performed in simplicial abelian groups and with face and degeneracies as in Hochschild homology. Varying q and X, this defines $HH^{\mathbf{Z}}(\mathcal{C}, P) \in ob\Gamma\mathcal{A}$.

This is natural in $\Gamma\mathcal{A}$-natural pairs $(\mathcal{C}, P)$ (and is prone to all the irritating nonsense about non-flat values).

Example 4.3.3.2 The prime examples come from ordinary Ab-categories: by using the Eilenberg–Mac Lane construction on every morphism group, an Ab-category $\mathcal{E}$ can be promoted to a $\Gamma\mathcal{A}$-category $\tilde{\mathcal{E}}$ (see Example 2.1.6.2(2), the morphisms Γ-spaces are of the form $X \mapsto \mathcal{E}(c, d) \otimes \tilde{\mathbf{Z}}[X]$). Similarly, we promote an $\mathcal{E}$-bimodule P to an $\tilde{\mathcal{E}}$-bimodule $\tilde{P}$.

Since this construction is so frequent (and often in typographically challenging situations) we commit the small sin of writing $THH(\mathcal{E}, P)$ when we really ought to have written $THH(\tilde{\mathcal{E}}, \tilde{P})$. This conforms with writing $THH(\mathbf{Z})$ instead of $THH(H\mathbf{Z})$.

Also, as in Remark 4.2.3.3, it is clear that if $\mathcal{C}$ is an Ab-category and P a $\mathcal{C}$-bimodule, then $HH^{\mathbf{Z}}(\mathcal{C}, P)$ is pointwise equivalent to $\bar{H}\,(HH(\mathcal{C}, P))$.

The proofs of the following statements are the same as the proofs for Lemma 4.2.3.5 and Lemma 4.2.3.7

Lemma 4.3.3.3 *Let $\mathcal{C}$ be a $\Gamma\mathcal{S}_*$-category, P a $\tilde{\mathbf{Z}}\mathcal{C}$-bimodule and $X \in ob\Gamma^o$. The map $THH(\mathcal{C}, UP) \to UHH^{\mathbf{Z}}(\tilde{\mathbf{Z}}\mathcal{C}, P)$ is an equivalence.* ☺

Lemma 4.3.3.4 *Let $\mathfrak{C}$ be an Ab-category and P a $\mathfrak{C}$-bimodule. Then there is a first quadrant spectral sequence*

$$E^2_{p,q} = HH^{\mathbf{Z}}_p\big(\mathfrak{C}, \pi_q THH(H\mathbf{Z}, HP, X), Y\big) \Rightarrow \pi_{p+q} THH(H\mathfrak{C}, HP, X \wedge Y). \quad ☺$$

4.3.4 Topological Hochschild Homology Calculates the Homology of Additive Categories

There is another fact where the $HH^{\mathbf{Z}}(\tilde{\mathbf{Z}}-, -)$-construction is handy, but which has no analogy for $\mathbf{S}$-algebras.

Let $\mathcal{C}$ be an Ab-category, and let P be a $\mathcal{C}$-bimodule (i.e., an Ab-functor $\mathcal{C}^o \otimes \mathcal{C} \to Ab$). Then, by the results of Sect. 4.3.3 you have that

$$THH(H\mathcal{C}, HP) \simeq U\,HH^{\mathbf{Z}}(\tilde{\mathbf{Z}}\bar{H}\mathcal{C}, \bar{H}P), \quad \text{and}$$

$$HH^{\mathbf{Z}}(\bar{H}\tilde{\mathbf{Z}}\mathcal{C}, \bar{H}P) \simeq \bar{H}\,HH(\tilde{\mathbf{Z}}\mathcal{C}, P),$$

but $HH^{\mathbf{Z}}(\tilde{\mathbf{Z}}\bar{H}\mathcal{C}, \bar{H}P)$ is vastly different from $HH^{\mathbf{Z}}(\bar{H}\tilde{\mathbf{Z}}\mathcal{C}, \bar{H}P)$. As an example, one may note that $THH(\mathbf{Z}, \mathbf{Z})$ is not equivalent to $HH(\tilde{\mathbf{Z}}\mathbf{Z}, \mathbf{Z}) = HH(\mathbf{Z}[t, t^{-1}], \mathbf{Z})$.

However, for additive categories (Ab-categories with sum) something interesting happens. Let $\mathfrak{C}$ be an additive category, and consider it as a $\Gamma\mathcal{A}$-category through the construction Sect. 2.1.6.1: $\mathfrak{C}^{\oplus}(c,d)(k_+) = \mathfrak{C}(c, \overset{k}{\oplus} d)$. Since $\mathfrak{C}$ is additive we see that there is a canonical isomorphism $\tilde{\mathfrak{C}} \cong \mathfrak{C}^{\oplus}$, but this may not be so with the bimodules: if M is a $\tilde{\mathbf{Z}}\mathfrak{C}$-bimodule (which by adjointness is the same as a $U\mathfrak{C}$-bimodule), we define the $\mathfrak{C}^{\oplus}$-bimodule $M^{\oplus}$ by the formula $M^{\oplus}(c,d)(k_+) = M(c, \overset{k}{\oplus} d)$. If M is "linear" in either factor (i.e., M is actually a $\mathfrak{C}^o \otimes \tilde{\mathbf{Z}}\mathfrak{C}$- or $\tilde{\mathbf{Z}}\mathfrak{C}^o \otimes \mathfrak{C}$-module) the canonical map $\tilde{M} \to M^{\oplus}$ is an isomorphism, but for the more general cases it will not even be a weak equivalence.

Theorem 4.3.4.1 *Let $\mathfrak{C}$ be an additive category and let M be a $\mathfrak{C}^o \otimes \tilde{\mathbf{Z}}\mathfrak{C}$-module. Then there is a canonical equivalence*

$$THH\big(U\mathfrak{C}^{\oplus}, UM^{\oplus}\big) \simeq H\left(HH(\tilde{\mathbf{Z}}\mathfrak{C}, M)\right).$$

Proof In this proof we will use the model $HH^{\mathbf{Z}}(\tilde{\mathbf{Z}}(\mathcal{C}^{\oplus}), M^{\oplus})$ instead of $THH(U\mathfrak{C}^{\oplus}, UM^{\oplus})$ (see Lemma 4.3.3.3), and since both expressions are very special it is enough to prove that the stabilization map $HH((\tilde{\mathbf{Z}}\mathfrak{C}, M)) \to HH^{\mathbf{Z}}(\tilde{\mathbf{Z}}(\mathcal{C}^{\oplus}), M^{\oplus})(1_+)$ is an equivalence. Since the functors in the statement are homotopy functors in M, it is enough to prove the theorem for projective M. But all projectives are retracts of sums of projectives of the standard type

$$P_{x,y}(-,-) = \mathfrak{C}(-,y) \otimes \tilde{\mathbf{Z}}\mathfrak{C}(x,-),$$

and hence it is enough to show that the higher homotopy groups vanish, and the map induces an isomorphism on π_0 for these projectives. For $HH^{\mathbf{Z}}(\tilde{\mathbf{Z}}(\mathcal{C}^{\oplus}), M^{\oplus})$ and $HH(\tilde{\mathbf{Z}}\mathfrak{C}, M)$ this vanishing comes from the "extra degeneracy" defined by means of

$$\begin{array}{ccc} P_{x,y}(c,d)(k_+) & = & \mathfrak{C}(c,x) \otimes \tilde{\mathbf{Z}}\mathfrak{C}\big(y, \overset{k}{\oplus} d\big) \\ & & \Big\downarrow {\scriptstyle f\otimes|\sum g_i| \mapsto f\otimes|1_y|\otimes|\sum g_i|} \\ P_{x,y}(c,y)(1_+) \otimes \tilde{\mathbf{Z}}\mathfrak{C}^{\oplus}(y, \oplus d)(k_+) & = & \mathfrak{C}(c,x) \otimes \tilde{\mathbf{Z}}\mathfrak{C}(y,y) \otimes \tilde{\mathbf{Z}}\mathfrak{C}\big(y, \overset{k}{\oplus} d\big) \end{array}$$

(the vertical lines are supposed to remind the reader that whatever is inside these are considered as generators in a free abelian group). This defines a contracting homotopy

$$s_{q+1} \colon HH(\tilde{\mathbf{Z}}\mathfrak{C}, P_{x,y})_q \to HH(\tilde{\mathbf{Z}}\mathfrak{C}, P_{x,y})_{q+1},$$

and likewise for $HH^{\mathbf{Z}}(\tilde{\mathbf{Z}}\mathfrak{C}^{\oplus}, P^{\oplus}_{x,y})$.

On π_0 we proceed as follows. Notice that

$$\pi_0\big(HH^{\mathbf{Z}}\big(\tilde{\mathbf{Z}}\mathfrak{C}^{\oplus}, P^{\oplus}_{x,y}\big)_0\big) \cong \bigoplus_{c \in ob\mathfrak{C}} \mathfrak{C}(c,y) \otimes \mathfrak{C}(x,c)$$

(essentially the Hurewicz theorem: if M is an abelian group $\pi_0 \lim_{\overrightarrow{k}} \Omega^k \tilde{\mathbf{Z}}(M \otimes \tilde{\mathbf{Z}}[S^k]) \cong M$) and $\pi_0(HH^{\mathbf{Z}}(\tilde{\mathbf{Z}}\mathfrak{C}^{\oplus}, P^{\oplus}_{x,y})_1) \cong \bigoplus_{c,d \in ob\mathfrak{C}} \mathfrak{C}(c, y) \otimes \mathfrak{C}(x, d) \otimes \mathfrak{C}(d, c)$. Hence the map $\pi_0 HH(\tilde{\mathbf{Z}}\mathfrak{C}, P_{x,y}) \to \pi_0 HH^{\mathbf{Z}}(\tilde{\mathbf{Z}}\mathfrak{C}^{\oplus}, P_{x,y})$ is the map induced by the map of coequalizers

$$\begin{array}{ccc} \bigoplus_c \mathfrak{C}(c, y) \otimes \tilde{\mathbf{Z}}\mathfrak{C}(x, c) & \leftleftarrows & \bigoplus_{c,d} \mathfrak{C}(c, y) \otimes \tilde{\mathbf{Z}}\mathfrak{C}(x, d) \otimes \tilde{\mathbf{Z}}\mathfrak{C}(d, c) \\ \downarrow & & \downarrow \\ \bigoplus_c \mathfrak{C}(c, y) \otimes \mathfrak{C}(x, c) & \leftleftarrows & \bigoplus_{c,d} \mathfrak{C}(c, y) \otimes \mathfrak{C}(x, d) \otimes \mathfrak{C}(d, c). \end{array}$$

However, both these coequalizers are isomorphic to $\mathfrak{C}(x, y)$, as can be seen by the unit map $\mathfrak{C}(x, y) \to \mathfrak{C}(x, y) \otimes \tilde{\mathbf{Z}}\mathfrak{C}(y, y)$ and the composition $\mathfrak{C}(c, y) \otimes \mathfrak{C}(x, c) \to \mathfrak{C}(x, y)$ (here the linearity in the first factor is crucial: the class of $f \otimes |g| \in \mathfrak{C}(c, y) \otimes \tilde{\mathbf{Z}}\mathfrak{C}(x, c)$ equals the class of $fg \otimes |1_x| \in \mathfrak{C}(x, y) \otimes \tilde{\mathbf{Z}}\mathfrak{C}(x, x)$), and the map comparing the coequalizers is an isomorphism. □

Remark 4.3.4.2 The proof of this theorem is somewhat delicate in that it steers a middle course between variants. We used the non-linearity in the second factor of M to reduce to the projectives $P_{x,y}$ where this non-linearity gave us the contracting homotopy. We then used the linearity in the first factor to identify the π_0 parts. A more general statement is that $THH(U\mathfrak{C}^{\oplus}, UM^{\oplus})$ is $HH(\tilde{\mathbf{Z}}\mathfrak{C}, LM)$ where L is linearization in the second factor. This first/second factor asymmetry is quite unnecessary and due to the fact that we stabilize in the second factor only. We could dualize and stabilize in the first factor only (the opposite of an additive category is an additive category), or we could do both at once. We leave the details to the interested reader.

Corollary 4.3.4.3 (Pirashvili-Waldhausen [225]) *Let A be a discrete ring and M a bimodule. Then there is a natural chain of weak equivalences connecting $THH(HA, HM)$ and (the Eilenberg–Mac Lane spectrum associated to) $HH(\tilde{\mathbf{Z}}\mathcal{P}_A, M)$, where $\mathcal{P}_A$ is the category of finitely generated projective modules, and M is considered as a $\mathcal{P}_A$-bimodule by setting $M(c, d) = \mathcal{P}_A(c, d) \otimes M$.*

Proof Theorem 4.3.5.16 below gives that the inclusion of the rank 1 bimodule gives an equivalence between the topological Hochschild homology of A and of $\mathcal{P}_A$, and Theorem 4.3.4.1 gives the weak equivalence with the homology of the category. □

4.3.5 General Results

Many results are most easily proven directly for $\Gamma\mathcal{S}_*$-categories, and not by referring to a reduction to special cases. We collect a few which will be of importance.

4.3.5.1 *THH* Respects Equivalences

This is the first thing that we should check, so that we need not worry too much about choosing this or that model for our categories.

Lemma 4.3.5.1 *Let $F_0, F_1\colon (\mathcal{C}, P) \to (\mathcal{D}, Q)$ be maps of $\Gamma\mathcal{S}_*$-natural bimodules, and X a space. If there is a natural isomorphism $\eta\colon F_0 \to F_1$, then the two maps*

$$F_0, F_1\colon THH(\mathcal{C}, P)(X) \to THH(\mathcal{D}, Q)(X)$$

are homotopic.

Proof We construct a homotopy $H\colon THH(\mathcal{C}, P)(X)\wedge\Delta[1]_+ \to THH(\mathcal{D}, Q)(X)$ as follows. If $\phi \in \Delta([q], [1])$ and $\mathbf{x} \in \mathcal{I}^{q+1}$ we define the map $H_{\phi,\mathbf{x}}\colon V(\mathcal{C}, P)(\mathbf{x}) \to V(\mathcal{D}, Q)(\mathbf{x})$ by sending the $c_0, \dots, c_q \in \mathcal{C}^{q+1}$ summand into the $F_{\phi(0)}(c_0), \dots, F_{\phi(q)}(c_q) \in ob\mathcal{D}$ summand via the maps

$$\mathcal{C}(c, d) \xrightarrow{F_0} \mathcal{D}(F_0(c), F_0(d)) \xrightarrow{\mathcal{D}\left(\eta_c^{-i}, \eta_d^{j}\right)} \mathcal{D}\big(F_i(c), F_j(d)\big)$$

for $i, j \in \{0, 1\}$ (and $P(c, d) \longrightarrow Q(F_0(c), F_0(d)) \xrightarrow{Q(\eta_c^{-i}, \eta_d^{j})} Q(F_i(c), F_j(d))$).

Clearly, the induced map $\Omega^{\vee\mathbf{x}}(X\wedge H_{\phi,\mathbf{x}})\colon \Omega^{\vee\mathbf{x}}(X\wedge V(\mathcal{C}, P)(\mathbf{x})) \to \Omega^{\vee\mathbf{x}}(X\wedge V(\mathcal{D}, Q)(\mathbf{x}))$ is functorial in $\mathbf{x} \in \mathcal{I}^{q+1}$, and so defines a map $H_\phi\colon THH(\mathcal{C}, P)(X)_q \to THH(\mathcal{D}, Q)(X)_q$. From the construction, we see that if $\psi\colon [p] \to [q] \in \Delta$ then $\psi^* H_\phi = H_{\phi\psi}\psi^*$ (do it separately for ψ's representing face and degeneracies. For the interior face maps (i.e., for $0 < i < q$), use that the diagram

$$\begin{array}{ccc}
\mathcal{C}(c_i, c_{i-1})\wedge\mathcal{C}(c_{i+1}, c_i) & \longrightarrow & \mathcal{C}(c_{i+1}, c_{i-1}) \\
\downarrow{\scriptstyle F_0\wedge F_0} & & \downarrow{\scriptstyle F_0} \\
\mathcal{D}(F_0(c_i), F_0(c_{i-1}))\wedge\mathcal{D}(F_0(c_{i+1}), F_0(c_i)) & \longrightarrow & \mathcal{D}(F_0(c_{i+1}), F_0(c_{i-1})) \\
\downarrow{\scriptstyle \mathcal{D}(\eta_{c_i}^{-\phi(i)}, \eta_{c_{i-1}}^{\phi(i-1)})\wedge\mathcal{D}(\eta_{c_{i+1}}^{-\phi(i+1)}, \eta_{c_i}^{\phi(i)})} & & \downarrow{\scriptstyle \mathcal{D}(\eta_{c_{i+1}}^{-\phi(i+1)}, \eta_{c_{i-1}}^{\phi(i-1)})} \\
\mathcal{D}(F_{\phi(i)}(c_i), F_{\phi(i-1)}(c_{i-1}))\wedge\mathcal{D}(F_{\phi(i+1)}(c_{i+1}), F_{\phi(i)}(c_i)) & \longrightarrow & \mathcal{D}(F_{\phi(i+1)}(c_{i+1}), F_{\phi(i-1)}(c_{i-1}))
\end{array}$$

commutes, where the horizontal maps are composition. The extreme face maps are similar, using the bimodules P and Q). □

Corollary 4.3.5.2 (*THH* Respects $\Gamma\mathcal{S}_*$-Equivalences) *Let $\mathcal{C} \xrightarrow{F} \mathcal{D}$ be $\Gamma\mathcal{S}_*$-equivalence of $\Gamma\mathcal{S}_*$-categories, P a $\mathcal{D}$-bimodule and X a space. Then*

$$THH\big(\mathcal{C}, F^*P\big)(X) \xrightarrow{\simeq} THH(\mathcal{D}, P)(X).$$

Proof Let G be an inverse, and $\eta\colon 1_{\mathcal{C}} \xrightarrow{\cong} GF$ and $\epsilon\colon 1_{\mathcal{D}} \xrightarrow{\cong} FG$ the natural isomorphisms. Consider the (non commutative) diagram

$$\begin{array}{ccc} THH(\mathcal{C}, F^*P)(X) & \xrightarrow{\eta} & THH(\mathcal{C}, (FGF)^*P)(X) \\ \downarrow F & \nwarrow G & \downarrow F \\ THH(\mathcal{D}, P)(X) & \xrightarrow{\epsilon} & THH(\mathcal{D}, (FG)^*P)(X) \end{array}$$

Lemma 4.3.5.1 then states that we get a map homotopic to the identity if we start with one of the horizontal isomorphism and go around a triangle. □

Recall the notion of stable equivalences of $\Gamma\mathcal{S}_*$-categories Definition 2.2.4.1.

Lemma 4.3.5.3 (*THH* Respects Stable Equivalences of $\Gamma\mathcal{S}_*$-Categories) *Consider a map $F\colon (\mathcal{C}, P) \to (\mathcal{D}, Q)$ of $\Gamma\mathcal{S}_*$-natural bimodules, and assume F is a stable equivalence of $\Gamma\mathcal{S}_*$-categories inducing stable equivalences*

$$P(c, c') \to Q(F(c), F(c'))$$

for every $c, c' \in ob\mathcal{C}$. Then F induces a pointwise equivalence

$$THH(\mathcal{C}, P) \to THH(\mathcal{D}, Q).$$

Proof According to Lemma 2.2.4.2 we may assume that F is either a $\Gamma\mathcal{S}_*$-equivalence, or a stable equivalence inducing an identity on the objects. If F is a $\Gamma\mathcal{S}_*$-equivalence we are done by Corollary 4.3.5.2 once we notice that the conditions on P and Q imply that $THH(\mathcal{C}, P) \to THH(\mathcal{C}, F^*Q)$ is a pointwise equivalence.

If F is a stable equivalence inducing the identity on objects, then clearly F induces a pointwise equivalence

$$THH(\mathcal{C}, P)_q \to THH(\mathcal{C}, F^*Q)_q \to THH(\mathcal{D}, Q)_q$$

in every simplicial degree q. □

4.3.5.2 A Collection of Other Results

The approximation in Sect. 4.2.4 of $THH(A)$ for an arbitrary **S**-algebra by means of the topological Hochschild homology of simplicial rings also works, mutatis mutandis, for $\Gamma\mathcal{S}_*$-categories to give an approximation of any $\Gamma\mathcal{S}_*$-category in terms of *sAb*-categories.

The proof of the following lemma is just as for **S**-algebras (Lemma 4.2.3.1)

Lemma 4.3.5.4 *Let $\mathcal{C}$ be a simplicial $\Gamma\mathcal{S}_*$-category and M a $\mathcal{C}$-bimodule (or in other words, $\{[q]\mapsto(\mathcal{C}_q, M_q)\}$ is a natural bimodule). Then there is a natural pointwise equivalence*

$$THH\big(\mathrm{diag}^*\mathcal{C}, \mathrm{diag}^* M\big) \simeq \mathrm{diag}^*\big\{[q]\mapsto THH(\mathcal{C}_q, M_q)\big\}. \qquad ☺$$

Definition 4.3.5.5 Let $\mathcal{A}$ and $\mathcal{B}$ be $\Gamma\mathcal{S}_*$-categories and M an $\mathcal{A}^o-\mathcal{B}$-bimodule. Then the *upper triangular matrix* $\Gamma\mathcal{S}_*$-category

$$\begin{bmatrix}\mathcal{A} & M\\ & \mathcal{B}\end{bmatrix}$$

is the $\Gamma\mathcal{S}_*$-category with objects $ob\mathcal{A}\times ob\mathcal{B}$ and with morphism object from (a,b) to (a',b') given by the matrix

$$\begin{bmatrix}\mathcal{A}(a,a') & M(a,b')\\ & \mathcal{B}(b,b')\end{bmatrix} = \mathcal{A}(a,a')\times\big[M(a,b')\vee\mathcal{B}(b,b')\big]$$

and with obvious matrix composition as in Example 2.1.4.3(6).

The projections from $\left[\begin{smallmatrix}\mathcal{A}(a,a') & M(a,b')\\ & \mathcal{B}(b,b')\end{smallmatrix}\right]$ to $\mathcal{A}(a,a')$ and $\mathcal{B}(b,b')$ induce $\Gamma\mathcal{S}_*$-functors from $\left[\begin{smallmatrix}\mathcal{A} & M\\ & \mathcal{B}\end{smallmatrix}\right]$ to $\mathcal{A}$ and $\mathcal{B}$.

Lemma 4.3.5.6 *With the notation as in the definition, the natural projection*

$$THH\left(\begin{bmatrix}\mathcal{A} & M\\ & \mathcal{B}\end{bmatrix}\right)\to THH(\mathcal{A})\times THH(\mathcal{B})$$

is a pointwise equivalence.

Proof Exchange some products with wedges and do an explicit homotopy as in [70, 1.6.20].

For concreteness and simplicity, let's do the analogous statement for Hochschild homology of k-algebras instead, where k is a commutative ring: let A_{11} and A_{22} be k-algebras, and let A_{12} be an $A_{11}^o\otimes_k A_{22}$-module. The group of q-simplices in $HH(\left[\begin{smallmatrix}A_{11} & A_{12}\\ & A_{22}\end{smallmatrix}\right])$ can be written as

$$\bigoplus\bigotimes_{i=0}^{q} A_{r_i,s_i}$$

where the sum is over the set of all functions $(r,s)\colon\{0,1\dots,q\}\to\{(11),(12),(22)\}$. The projection to $HH(A_{11})\oplus HH(A_{22})$ is split by the inclusion onto the summands where $r_0=\dots r_q=s_0=\dots=s_q$. We make a simplicial homotopy showing that the non-identity composite is indeed homotopic to the identity. Let

$\phi \in \Delta([q],[1])$ and y in the (r,s) summand of the Hochschild homology of the upper triangular matrices. With the convention that $s_{q+1} = r_0$ we set

$$H(\phi, y) = y, \quad \text{if } r_k = s_{k+1} \text{ for all } k \in \phi^{-1}(0)$$

and zero otherwise. We check that for $j \in [q]$ we have equality $d_j H(\phi, y) = H(\phi d^j, d_j y)$, and so we have a simplicial homotopy. Note that $H(1,-)$ is the identity and $H(0,-)$ is the projection ($r_0 = s_1, \dots, r_{q-1} = s_q, r_q = s_0$ implies that all indices are the same due to the upper triangularity).

The general result is proven by just the same method, exchanging products with wedges to use the distributivity of smash over wedge, and keeping track of the objects (this has the awkward effect that you have to talk about non-unital issues. If you want to avoid this you can obtain the general case from the *Ab*-case by approximating as in Sect. 4.2.4). Alternatively you can steal the result from Sect. 1.3.6 via the equivalences

$$THH(\mathfrak{C}) \simeq H\,HH(\tilde{\mathbf{Z}}\mathfrak{C}, \mathfrak{C}) \simeq H\,HH(\mathbf{Z}\mathfrak{C}, \mathfrak{C}) = F(\mathfrak{C}, \mathfrak{C})$$

to get an only slightly weaker result. □

Setting M in Lemma 4.3.5.6 to be the trivial module you get that *THH* preserves products (or again, you may construct an explicit homotopy as in [70, 1.6.15] (replacing products with wedges). There are no added difficulties with the bimodule statement.

Corollary 4.3.5.7 *Let $\mathcal{C}$ and $\mathcal{D}$ be $\Gamma\mathcal{S}_*$-categories, P a $\mathcal{C}$-bimodule, Q a $\mathcal{D}$-bimodule. Then the canonical map is a pointwise equivalence*

$$THH(\mathcal{C} \times \mathcal{D}, P \times Q) \to THH(\mathcal{C}, P, X) \times THH(\mathcal{D}, Q, X). \qquad ☺$$

Recall from Sect. 3.2.1.1 the canonical map $\bar{H}\mathfrak{C}(S^1) \to S\mathfrak{C}$, which in dimension q is induced by sending the sum diagram $C \in ob\bar{H}(\mathfrak{C})(q_+)$ to $c \in obS_q\mathfrak{C}$ with $c_{ij} = C_{\{0,i+1,i+2,\dots,j-1,j\}}$ and obvious maps. This map factors through the (degreewise) equivalence of categories $T\mathfrak{C} \to S\mathfrak{C}$ discussed in Lemma 1.2.2.4, where $T\mathfrak{C}$ is the simplicial category of upper triangular matrices. Since $\bar{H}(\mathfrak{C})$ is equivalent to $\mathfrak{C}^{\times q}$, we get by induction (setting $\mathcal{A} = M = \mathfrak{C}$ and $\mathcal{B} = T_{q-1}C$ in Lemma 4.3.5.6) that, for each q and X, the map $THH(\bar{H}(\mathfrak{C})(q_+), X) \to THH(S_q\mathfrak{C}, X)$ is a weak equivalence. Letting q vary and using that *THH* can be calculated degreewise (just as in Lemma 4.2.3.1), we get the following corollary:

Corollary 4.3.5.8 *Let $\mathfrak{C}$ be an additive category and X a space. Then the map $\bar{H}\mathfrak{C}(S^1) \to S\mathfrak{C}$ induces a weak equivalence $THH(\bar{H}\mathfrak{C}(S^1), X) \to THH(S\mathfrak{C}, X)$.* ☺

4.3.5.3 Cofinality

Another feature which is important is the fact that topological Hochschild homology is insensitive to *cofinal inclusions* (see below). Note that this is very different from the K-theory case where there is a significant difference between the K-theories of the finitely generated free and projective modules: $K_0^f(A) \to K_0(A)$ is not always an equivalence.

Definition 4.3.5.9 Let $\mathcal{C} \subseteq \mathcal{D}$ be a $\Gamma\mathcal{S}_*$-full inclusion of $\Gamma\mathcal{S}_*$-categories. We say that $\mathcal{C}$ is *cofinal* in $\mathcal{D}$ if for every $d \in ob\mathcal{D}$ there exist maps

$$d \xrightarrow{\eta_d} c(d) \xrightarrow{\pi_d} d$$

such that $c(d) \in ob\mathcal{C}$ and $\pi_d \eta_d = 1_d$.

Lemma 4.3.5.10 *Let $j\colon \mathcal{C} \subset \mathcal{D}$ be an inclusion of a cofinal $\Gamma\mathcal{S}_*$-subcategory. Let P be a $\mathcal{D}$-bimodule. Then the induced map*

$$THH(\mathcal{C}, P) \to THH(\mathcal{D}, P)$$

is a pointwise equivalence.

Proof For simplicity we prove it for $P = \mathcal{D}$. For each $d \in ob\mathcal{D}$ choose

$$d \xrightarrow{\eta_d} c(d) \xrightarrow{\pi_d} d,$$

such that η_c is the identity for all $c \in ob\mathcal{C}$. Then for every $\mathbf{x} \in \mathcal{I}^{q+1}$ we have a map $V(\mathcal{D})(\mathbf{x}) \to V(\mathcal{C})(\mathbf{x})$ sending the $d_0, \dots, d_q \in U\mathcal{D}^{q+1}$ summand to the $c(d_0), \dots, c(d_q) \in U\mathcal{C}^{q+1}$ summand via

$$\mathcal{D}(\pi_{d_0}, \eta_{d_q})(S^{x_0}) \wedge \cdots \wedge \mathcal{D}(\pi_{d_q}, \eta_{d_{q-1}})(S^{x_q}).$$

This map is compatible with the cyclic operations and hence defines a map

$$D(\pi, \eta)\colon THH(\mathcal{D}) \to THH(\mathcal{C}).$$

Obviously $D(\pi, \eta) \circ THH(j)$ is the identity on $THH(\mathcal{C})$ and we will show that the other composite is homotopic to the identity. The desired homotopy can be expressed as follows. Let $\phi \in \Delta([q], [1])$ and let

$$d \xrightarrow{\eta_d^i} c^i(d) \xrightarrow{\pi_d^i} d \quad \text{be} \quad \begin{cases} d \xrightarrow{\eta_d} c(d) \xrightarrow{\pi_d} d & \text{if } i = 1, \\ d = d = d & \text{if } i = 0. \end{cases}$$

The homotopy $THH(\mathcal{D}) \wedge \Delta[1]_+ \to THH(\mathcal{D})$ is given by $H_{\phi,\mathbf{x}}\colon V(\mathcal{D})(\mathbf{x}) \to V(\mathcal{D})(\mathbf{x})$ sending the $d_0, \dots, d_q \in obU\mathcal{D}^{q+1}$ summand to the $c^{\phi(0)}(d_0), \dots, c^{\phi(q)}(d_q) \in obU\mathcal{D}^{q+1}$ summand via

$$\mathcal{D}\big(\pi_{d_0}^{\phi(0)}, \eta_{d_q}^{\phi(q)}\big)(S^{x_0}) \wedge \cdots \wedge \mathcal{D}\big(\pi_{d_q}^{\phi(q)}, \eta_{d_{q-1}}^{\phi(q-1)}\big)(S^{x_q}). \qquad \square$$

4.3.5.4 Morita Invariance

If A is an **S**-algebra, let $\mathcal{F}_A$ be the $\Gamma\mathcal{S}_*$ category whose objects are the natural numbers, with n thought of as the free A-module of rank n, and $\mathcal{F}_A(m,n) = \underline{\mathcal{S}_*}(m_+, n_+ \wedge A)$, the $n \times m$ matrices with coefficients in A as in Example 2.1.4.3(6). Let $\mathcal{F}_A^k$ be the full subcategory of objects of rank less than or equal to the natural number k.

This should be compared with the situation when R is a discrete ring. Then $\mathcal{F}_R^k$ is the Ab-category with objects the natural numbers less than or equal to k and a morphism from m to n is an $n \times m$-matrix (in the usual sense) with entries in R. By sending wedges to products, we see that the $\Gamma\mathcal{S}_*$-category $\widetilde{\mathcal{F}_R^k}$ associated with $\mathcal{F}_R^k$ (by taking the Eilenberg–Mac Lane construction on all morphism spaces, cf. Example 2.1.6.2(2)) is stably equivalent to $\mathcal{F}_{HR}^k$, and so $THH(\mathcal{F}_{HR}^k) \xrightarrow{\sim} THH(\widetilde{\mathcal{F}_R^k})$.

Thinking of the **S**-algebra $Mat_k A$ as the full subcategory of $\mathcal{F}_A$ whose only object is k, we get a cofinal inclusion $Mat_k A \subseteq \mathcal{F}_A^k$: for $n \in ob\mathcal{F}_A^k$, the maps η_n and π_n of Definition 4.3.5.9 ensuring cofinality are given by the matrices representing inclusion into and projection onto the first n coordinates. Hence we get that

Lemma 4.3.5.11 *Let A be an* **S**-*algebra and P an A-bimodule. Then the inclusion $Mat_k A \to \mathcal{F}_A^k$ induces a pointwise equivalence $THH(Mat_k A, Mat_k P) \xrightarrow{\sim} THH(\mathcal{F}_A^k, P)$.* ☺

Here we have written P also for the $\mathcal{F}_A^k$-bimodule given by $P(m,n) = \underline{\mathcal{S}_*}(m_+, n_+ \wedge P)$ with matrix multiplication from left and right by matrices with entries in A.

The inclusion of the rank one A-module, $A \subseteq \mathcal{F}_A^k$, is not cofinal, unless $k = 1$, but still induces an equivalence:

Lemma 4.3.5.12 *Let A be an* **S**-*algebra and P an A-bimodule. Then the inclusion $A \subseteq \mathcal{F}_A^k$ induces a pointwise equivalence*

$$THH(A, P) \xrightarrow{\sim} THH\big(\mathcal{F}_A^k, P\big).$$

Proof By Sect. 4.2.4, Lemma 4.2.3.1 and Lemma 4.2.3.7 it is enough to prove the lemma for Hochschild homology of discrete rings (alternatively, you must work with homotopies not respecting degeneracies as in [70]).

For the rest of the proof A will be a discrete ring and P an A-bimodule. It is helpful to write out $HH(\mathcal{F}_A^k, P)_q$ by means of distributivity as

$$\bigoplus_{(\mathbf{n},\mathbf{r},\mathbf{s})} P \otimes A^{\otimes q}$$

where the sum is over the tuples $\mathbf{n} = (n_0, \dots, n_q)$, $\mathbf{r} = (r_0, \dots, r_q)$ and $\mathbf{s} = (s_0, \dots, s_q)$ of natural numbers where $r_i, s_i \le n_i \le k$ for $i = 0, \dots, q$. The isomorphism to $HH(\mathcal{F}^k_A, P)_q$ is given by sending $p \otimes a_1 \otimes \cdots \otimes a_q$ in the $(\mathbf{n}, \mathbf{r}, \mathbf{s})$-summand to $(p\, pr_{s_q} in_{r_0}) \otimes \bigotimes_{i=1}^q (a_i\, pr_{s_{i-1}} in_{r_i})$ in the $\mathbf{n} \in (ob\mathcal{F}^k_A)^{\times q+1}$ summand. Here in_i and pr_i represent the ith injection and projection matrices.

There is a map $tr\colon HH(\mathcal{F}^k, P) \to HH(A, P)$ sending $\mathbf{a} = p \otimes a_1 \otimes \cdots \otimes a_q$ in the $(\mathbf{n}, \mathbf{r}, \mathbf{s})$-summand to

$$tr(\mathbf{a}_{(\mathbf{n},\mathbf{r},\mathbf{s})}) = \begin{cases} \mathbf{a} & \text{if } r_k = s_k \text{ for all } k \in [q], \\ 0 & \text{otherwise.} \end{cases}$$

Clearly, the composite $HH(A, P) \to HH(\mathcal{F}^k_A, P) \to HH(A, P)$ is the identity, and we are done if we can construct a concrete simplicial homotopy $H\colon HH(\mathcal{F}^k_A, P) \otimes \tilde{\mathbf{Z}}[\Delta[1]] \to HH(\mathcal{F}^k_A, P)$ between the other composite and the identity.

If $\phi\colon [q] \to [1] \in \Delta$, and $(\mathbf{n}, \mathbf{r}, \mathbf{s})$ is a tuple as above, let $(\mathbf{n}^\phi, \mathbf{r}^\phi, \mathbf{s}^\phi)$ be the tuple where the ith factor in each of the three entries is unchanged if $\phi(i) = 1$ and set to 1 if $\phi(i) = 0$. Then we define

$$H(\mathbf{a}_{(\mathbf{n},\mathbf{r},\mathbf{s})} \otimes \phi) = \begin{cases} \mathbf{a}_{(\mathbf{n}^\phi,\mathbf{r}^\phi,\mathbf{s}^\phi)} & \text{if } r_k = s_k \text{ for all } k \in \phi^{-1}(0), \\ 0 & \text{otherwise.} \end{cases}$$

A direct check reveals that this defines the desired simplicial homotopy. □

Remark 4.3.5.13 We noted earlier that in our presentation there was an unfortunate lack of a map $THH(A) \to THH(Mat_k(A))$ realizing Morita invariance. The natural substitute is $THH(A) \xrightarrow{\sim} THH(\mathcal{F}^k_A) \xleftarrow{\sim} THH(Mat_k(A))$.

Since topological Hochschild homology commutes with filtered colimits (loops respect filtered colimits (Lemma A.2.5.3) and $V(\mathcal{F}_A, P)(\mathbf{x}) = \lim_{k\to\infty} V(\mathcal{F}^k_A, P)(\mathbf{x})$ for all $\mathbf{x} \in \mathcal{I}^{q+1}$) we get the following corollary:

Corollary 4.3.5.14 *Let A be an* **S***-algebra and P an A-bimodule. Then the inclusion of A as the rank one module in $\mathcal{F}_A$ induces a pointwise equivalence $THH(A, P) \to THH(\mathcal{F}_A, P)$.* ☺

4.3.5.5 Application to the Case of Discrete Rings

As an easy application, we will show how these theorems can be used to analyze the topological Hochschild homology of a discrete ring.

For a discrete ring A recall the category $\mathcal{P}_A$ of finitely generated projective modules (Definition 1.2.1.3) and the category $\mathcal{F}_A$ of finitely generated free modules (Definition 1.2.1.4). Again, if P is an A-bimodule, we also write P for the $\mathcal{P}_A$-bimodule $\mathrm{Hom}_A(-, - \otimes_A P) \cong \mathcal{P}_A(-, -) \otimes_A P\colon \mathcal{P}_A \times \mathcal{P}^o_A \to Ab$.

By sending wedges to products, we see that the $\Gamma\mathcal{S}_*$-category associated with $\mathcal{F}_A$ (by taking the Eilenberg-Mac Lane construction on all morphism spaces to achieve what would be recorded as $\widetilde{\mathcal{F}}_A$) is stably equivalent to the $\Gamma\mathcal{S}_*$-category $\mathcal{F}_{HA}$ of finitely generated free HA-modules, and so the results for the latter found in the previous section give exactly the same results for the former. In particular, $THH(A, P) \to THH(\mathcal{F}_A, P)$ is a pointwise equivalence. Here we have again used the shorthand of writing $THH(A, P)$ when we really mean $THH(HA, HP)$, and likewise for $THH(\mathcal{F}_A, P)$.

Since $\mathcal{F}_A \subseteq \mathcal{P}_A$ is a cofinal inclusion we get by Lemma 4.3.5.10 that

Lemma 4.3.5.15 *Let A be a discrete ring, and let P be an A-bimodule. Then the inclusion $\mathcal{F}_A \subseteq \mathcal{P}_A$ induces a pointwise equivalence*

$$THH(\mathcal{F}_A, P) \xrightarrow{\sim} THH(\mathcal{P}_A, P).$$ ☺

Collecting the results, we get

Theorem 4.3.5.16 *The (full and faithful) inclusion of A in $\mathcal{P}_A$ as the rank 1 free module induces a pointwise equivalence*

$$THH(A, P) \xrightarrow{\sim} THH(\mathcal{P}_A, P).$$ ☺

Chapter 5
The Trace $K \to THH$

In this chapter we explain how the Dennis trace map Sect. 4.3.2 can be lifted to a trace map from algebraic K-theory to topological Hochschild homology. We first concentrate on the *Ab*-case since this is somewhat easier. This case is, however, sufficient to define the trace for discrete rings, and carries all the information we need in order to complete our proofs. The general construction is more complex, but this needs not really concern us: the only thing we actually use it for is that it exists and is as functorial as anybody can wish.

The general construction occupies the second section, and tries to reconcile this construction with the others we have seen. In the third section we have another look at stable K-theory and verify that it agrees with topological Hochschild homology for **S**-algebras in general. In the last section we give an outline of another construction of the trace which has several advantages. For instance, with this formulation it is easier to prove that the trace preserves operad actions, and in particular that it is multiplicative when evaluated at commutative ring spectra.

Common to all these approaches is that they contain nothing resembling a "trace" in the usual sense. The only vestige of a trace can be found in the very last step: for a given ring, the inclusion of the rank 1 module into the category of free finitely generated modules induces an equivalence in topological Hochschild homology. For Hochschild homology this equivalence has a concrete homotopy inverse given by a trace construction, and this feature was much more prominent in the early developments of Hattori, Stallings and Dennis.

5.1 *THH* and K-Theory: The Linear Case

In this section we define the trace map from algebraic K-theory to the topological Hochschild homology of an additive or exact category, much as was done in [70].

Before we do so, we have to prepare the ground a bit, and since these results will be used later we work in a wider generality for a short while.

B.I. Dundas et al., *The Local Structure of Algebraic K-Theory*,
Algebra and Applications 18,
DOI 10.1007/978-1-4471-4393-2_5, © Springer-Verlag London 2013

Algebraic K-theory is preoccupied with the weak equivalences, topological Hochschild homology with the enrichment. The Dennis trace map Sect. 4.3.2 should seek to unite these points of view.

Let $\mathcal{C}$ be a symmetric monoidal $\Gamma\mathcal{S}_*$-category, and recall from Sect. 2.3.1.1 the adaption $\bar{H}\mathcal{C}$ of Segal's construction. This is a functor from Γ^o to symmetric monoidal $\Gamma\mathcal{S}_*$-categories such that for each $k_+ \in ob\Gamma^o$ the canonical map

$$\bar{H}\mathcal{C}(k_+) \to \mathcal{C}^{\times k}$$

is a $\Gamma\mathcal{S}_*$-equivalence. Hence

$$THH(\bar{H}\mathcal{C})$$

is a functor from Γ^o to $\Gamma\mathcal{S}_*$ or more symmetrically: a functor $\Gamma^o \times \Gamma^o \to \mathcal{S}_*$. For such functors we have again a notion of stable equivalences: if X and Y are functors $\Gamma^o \times \Gamma^o \to \mathcal{S}_*$, a map $X \to Y$ is a *stable equivalence* if

$$\varinjlim_{k,l} \Omega^{k+l} X\big(S^k, S^l\big) \to \varinjlim_{k,l} \Omega^{k+l} Y\big(S^k, S^l\big)$$

is a weak equivalence.

If X is a Γ-space, we will write $\Sigma^\infty X$ for the functor $\Gamma^o \times \Gamma^o \to \mathcal{S}_*$ sending (k_+, l_+) to $k_+ \wedge X(l_+)$. Notice that, by Lemma 2.2.1.3(3), the maps $k_+ \wedge X(l_+) \to X(k_+ \wedge l_+)$ give rise to a stable equivalence $\Sigma^\infty X \xrightarrow{\sim} X \circ \wedge$, and $\Sigma^\infty X$ should be thought of as a bispectrum representing the same spectrum as X.

For each $k_+ \in ob\Gamma^o$ there is a map $k_+ \wedge THH(\mathcal{C}) \to THH(\bar{H}\mathcal{C}(k_+))$ (induced by the k functors $\mathcal{C} \to \bar{H}\mathcal{C}(k_+)$ given by the injections $1_+ \to k_+$). Varying k_+, these maps assemble to a natural map $\Sigma^\infty THH(\mathcal{C}) \to THH(\bar{H}\mathcal{C})$ of functors $\Gamma^o \to \Gamma\mathcal{S}_*$.

Proposition 5.1.0.1 *Let $\mathcal{C}$ be a symmetric monoidal $\Gamma\mathcal{S}_*$-category. Then for each $l_+ \in \Gamma^o$ the Γ-space*

$$k_+ \mapsto THH\big(\bar{H}\mathcal{C}(k_+)\big)(l_+)$$

is special, and the natural map

$$\Sigma^\infty THH(\mathcal{C}) \to THH(\bar{H}\mathcal{C})$$

is a stable equivalence.

Proof For each $k_+, l_+ \in ob\Gamma^o$ the map

$$THH\big(\bar{H}\mathcal{C}(k_+)\big)(l_+) \to THH\big(\mathcal{C}^{\times k}\big)(l_+)$$

is a weak equivalence (since $\bar{H}\mathcal{C}$ is special and THH sends $\Gamma\mathcal{S}_*$-equivalences to pointwise equivalences Lemma 4.3.5.3), and so is

$$THH\big(\mathcal{C}^{\times k}\big)(l_+) \to THH(\mathcal{C})(l_+)^{\times k}$$

(since *THH* respects products Corollary 4.3.5.7), and so the first part of the proposition is shown: $THH(\bar{H}\mathcal{C})(l_+)$ is special. For each k_+, the composite

$$k_+ \wedge THH(\mathcal{C}) \longrightarrow THH\big(\bar{H}\mathcal{C}(k_+)\big) \longrightarrow THH(\mathcal{C})^{\times k}$$

is a stable equivalence, and the last map is a pointwise equivalence, hence the first map is a stable equivalence, assembling to the stated result. □

This is a special case of a more general statement below which is proved similarly. A functor $(\mathcal{C}, P)$ from Γ^o to $\Gamma\mathcal{S}_*$-natural bimodules is nothing but a functor $\mathcal{C}\colon \Gamma^o \to \Gamma\mathcal{S}_*$-categories and for each $X \in ob\Gamma^o$ a $\mathcal{C}(X)$-bimodule $P(X)$, such that for every $f\colon X \to Y \in \Gamma^o$ there is a map of $\mathcal{C}(X)$-bimodules $\bar{f}\colon P(X) \to f^*P(Y)$ such that $\overline{gf} = f^*(\bar{g}) \circ \bar{f}$ (i.e., if in addition $g\colon Y \to Z$, then the diagram

$$\begin{array}{ccc} P(X) & \xrightarrow{\bar{f}} & f^*P(Y) \\ & \overline{gf}\searrow & \downarrow f^*(\bar{g}) \\ & & (gf)^*P(Z) = f^*\big(g^*P(Z)\big) \end{array}$$

commutes). In particular, $(\mathcal{C}, \mathcal{C})$ will serve as an easy example.

Proposition 5.1.0.2 *Let* $(\mathcal{C}, P)$ *be a functor from* Γ^o *to* $\Gamma\mathcal{S}_*$*-natural bimodules. Assume that* $\mathcal{C}$ *is quite special* (*see Definition* 2.3.2.1) *and that for all* $X, Y \in ob\Gamma^o$ *the map*

$$P(X \vee Y) \xrightarrow{(\overline{p_X}, \overline{p_Y})} p_X^* P(X) \times p_Y^* P(Y)$$

(*induced by the projections* $p_X\colon X \vee Y \to X$ *and* $p_Y\colon X \vee Y \to Y$) *is a stable equivalence of* $\mathcal{C}(X \vee Y)$*-bimodules. Then*

$$THH(\mathcal{C}, P) \xleftarrow{\sim} \Sigma^\infty THH\big(\mathcal{C}(1_+), P(1_+)\big)$$

is a stable equivalence. ☺

Preparing the way for the trace from the algebraic K-theory of exact categories, we make the following preliminary nerve construction (a more worked-out version will be needed later, see Sects. 5.2.1.2 or 5.4 below, but this will do for now). Note the connections to the nerve construction used in the proof of Corollary 1.2.3.2.

Definition 5.1.0.3 Let $\mathcal{C}$ be a category. The *nerve of* $\mathcal{C}$ *with respect to the isomorphisms* is the simplicial category $\mathbf{N}(\mathcal{C}, i)$ whose simplicial set of object is the classifying space $Bi\mathcal{C}$ of the subcategory of isomorphisms, and whose set of morphisms between $c_0 \leftarrow c_1 \leftarrow \cdots \leftarrow c_q$ and $c_0' \leftarrow c_1' \leftarrow \cdots \leftarrow c_q'$ is the set of all commuting diagrams

$$\begin{array}{ccccccc} c_0 & \xleftarrow{\cong} & c_1 & \xleftarrow{\cong} & \cdots & \xleftarrow{\cong} & c_q \\ \downarrow & & \downarrow & & & & \downarrow \\ c'_0 & \xleftarrow{\cong} & c'_1 & \xleftarrow{\cong} & \cdots & \xleftarrow{\cong} & c'_q \end{array}$$

in $\mathcal{C}$.

Note that the vertical maps need not be isomorphisms. Furthermore we have that

Lemma 5.1.0.4 *For all q the map $s\colon \mathcal{C} = \mathbf{N}_0(\mathcal{C}, i) \to \mathbf{N}_q(\mathcal{C}, i)$ induced by the unique $[q] \to [0]$ in Δ (sending c to $c = c = \cdots = c$) is an equivalence of categories.*

Proof Let $d\colon \mathbf{N}_q(\mathcal{C}, i) \to \mathcal{C}$ be the functor induced by the function $[0] \to [q]$ sending 0 to 0 (so that $c_0 \leftarrow \cdots \leftarrow c_q$ is sent to c_0). We see that the composite ds is the identity functor $\mathcal{C} = \mathcal{C}$, whereas we have a natural transformation from the identity on $\mathbf{N}_q\mathcal{C}$ to sd by

$$\begin{array}{ccccccc} c_0 & \xleftarrow{\alpha_0} & c_1 & \xleftarrow{\alpha_2} & \cdots & \xleftarrow{\alpha_q} & c_q \\ \| & & {\scriptstyle\alpha_1}\downarrow & & & & {\scriptstyle\alpha_1\circ\cdots\circ\alpha_q}\downarrow \\ c_0 & = & c_0 & = & \cdots & = & c_0 \end{array},$$

which is a natural *isomorphism* since all the αs are assumed to be isomorphisms. □

Lastly, if $\mathcal{C}$ is an *Ab*-category, $\mathbf{N}(\mathcal{C}, i)$ will be a simplicial *Ab*-category.

If $\mathcal{C}$ is an *Ab*-category we will abuse notation by writing $THH(\mathcal{C})$ when we really should have written $THH(\tilde{\mathcal{C}})$ (where the functor $\mathcal{C} \mapsto \tilde{\mathcal{C}}$ from *Ab*-categories to $\Gamma\mathcal{S}_*$-categories of Example 2.1.6.2(2) allows us to consider all *Ab*-categories as $\Gamma\mathcal{S}_*$-categories).

A consequence of Lemma 5.1.0.4 is that if $\mathcal{C}$ is an *Ab*-category the map

$$THH(\mathcal{C}) \to THH\big(\mathbf{N}(\mathcal{C}, i)\big)$$

induced by the degeneracies becomes a pointwise equivalence (since the functor $\mathcal{C} \mapsto \tilde{\mathcal{C}}$ sends *Ab*-equivalences to $\Gamma\mathcal{S}_*$-equivalences and *THH* sends $\Gamma\mathcal{S}_*$-equivalences to pointwise equivalences).

This paves the way for our first definition of the trace from algebraic K-theory to topological Hochschild homology:

Definition 5.1.0.5 (The Trace for Additive Categories) Let $\mathcal{E}$ be an additive category. The trace map for $\mathcal{E}$ in the Segal formalism is the following chain of natural transformations where the leftward pointing arrows are all stable equivalences

$$\Sigma^\infty Bi\bar{H}\mathcal{E} = \Sigma^\infty ob\mathbf{N}(\bar{H}\mathcal{E}, i) \longrightarrow THH\big(\mathbf{N}(\bar{H}\mathcal{E}, i)\big) \overset{\sim}{\longleftarrow} THH(\bar{H}\mathcal{E})$$
$$\overset{\sim}{\longleftarrow} \quad \Sigma^\infty THH(\mathcal{E})$$

where the first map is the Dennis trace of Sect. 4.3.2, the second is the equivalence coming from the equivalences of categories $\mathcal{E} \to \mathbf{N}_q(\mathcal{E}, i)$ and the third from from Lemma 5.1.0.1.

5.1.1 The Dennis Trace with the S-Construction

We may also use the S-construction of Waldhausen (see Definition 1.2.2.1). This has simplicial exact categories as output, and we may apply *THH* degreewise to these categories.

Let $\mathfrak{C}$ be an exact category and X a space. We saw in Sect. 1.2.4 that if Y is a simplicial set with $Y_0 = *$, there is a canonical map $S^1 \wedge Y_1 \to Y$, and so the fact that $S_0\mathfrak{C} = *$ gives rise to a map $S^1 \wedge ob\mathfrak{C} \to S\mathfrak{C}$. Ultimately this gives that $\{m \mapsto obS^{(m)}\mathfrak{C}\}$ forms a spectrum $ob\underline{S}\mathfrak{C}$. For exactly the same reason ($THH(S_0\mathcal{C}, X)$ is trivial) we get a map $S^1 \wedge THH(S^{(k)}\mathfrak{C}, X) \to THH(S^{(k+1)}\mathfrak{C}, X)$, and so $THH(\underline{S}\mathfrak{C}, X) = \{k \mapsto THH(S^{(k)}\mathfrak{C}, X)\}$ defines a spectrum. It is proven in [70] that the adjoint

$$THH\big(S^{(k)}\mathfrak{C}, X\big) \to \Omega THH\big(S^{(k+1)}\mathfrak{C}, X\big)$$

of the structure map is an equivalence for $k > 0$. Furthermore if $\mathfrak{C}$ is split exact, that is, all short exact sequences split, then it is an equivalence also for $k = 0$. Note that any additive category can be viewed as a split exact category by choosing exactly the split exact sequences as the admissible exact sequences. In fact, if we apply the S-construction to an additive category with no mention of exact sequences, this is what we mean.

5.1.1.1 Split Exact Categories

Let $\mathfrak{C}$ be an additive category. We defined the $n \times n$ *upper triangular matrices*, $T_n\mathfrak{C}$, in Definition 1.2.2.3, to be the category with objects $ob\mathfrak{C}^{\times n}$, and morphisms

$$T_n\mathfrak{C}\big((c_1, \dots, c_n), (d_1, \dots, d_n)\big) = \bigoplus_{1 \le j \le i \le n} \mathfrak{C}(c_i, d_j)$$

with composition given by matrix multiplication. Since $\mathfrak{C}$ is additive, so is $T_n\mathfrak{C}$. Consider the two functors

$$\mathfrak{C}^{\times n} \to T_n\mathfrak{C} \to \mathfrak{C}^{\times n}.$$

The first is the inclusion of $\mathfrak{C}^{\times n}$ as the diagonal subcategory of $T_n\mathfrak{C}$, the second forgets about off-diagonal entries, and the composite is the identity.

Proposition 5.1.1.1 *Let $\mathfrak{C}$ be an additive category. Then the inclusion of the diagonal $\mathfrak{C}^{\times n} \to T_n\mathfrak{C}$ induces a pointwise equivalence*

$$THH\bigl(\mathfrak{C}^{\times n}\bigr) \to THH(T_n\mathfrak{C}).$$

Proof Using the stable equivalence of products and wedges, we see that the map of $\Gamma\mathcal{S}_*$-categories

$$\begin{bmatrix} \mathfrak{C}^{\oplus} & (\mathfrak{C}^{\times n-1})^{\oplus} \\ & (T_{n-1}\mathfrak{C})^{\oplus} \end{bmatrix} \to (T_n\mathfrak{C})^{\oplus},$$

(where the left hand category is defined in Definition 4.3.5.5), is a stable equivalence. Hence the statement follows by induction on n from Lemma 4.3.5.3 and Lemma 4.3.5.6.

Alternatively you can steal the result from Sect. 1.3.6 via the equivalences of Sect. 4.3.3

$$THH(\mathfrak{C}) \simeq U\,HH^{\mathbf{Z}}(\tilde{\mathbf{Z}}\mathcal{C}, \mathcal{C}) \simeq H\bigl(HH(\tilde{\mathbf{Z}}\mathfrak{C}, \mathfrak{C})\bigr) \simeq H\bigl(HH(\mathbf{Z}\mathfrak{C}, \mathfrak{C})\bigr) = F(\mathfrak{C}, \mathfrak{C}). \quad \square$$

Considering the additive category $\mathfrak{C}$ as a split exact category, the forgetful map $T_n\mathfrak{C} \to \mathfrak{C}^{\times n}$ factors through $S_n\mathfrak{C}$

$$T_n\mathfrak{C} \to S_n\mathfrak{C} \to \mathfrak{C}^{\times n}.$$

The first map is given by sending $(c_1, \dots, c_n)$ to $i \le j \mapsto c_{i+1} \oplus \dots \oplus c_j$, and the second projects $i \le j \mapsto c_{ij}$ onto $i \mapsto c_{i-1,i}$.

Corollary 5.1.1.2 *Let $\mathfrak{C}$ be a additive category. Then*

$$THH\bigl(\mathfrak{C}^{\times n}\bigr) \to THH(S_n\mathfrak{C})$$

is a pointwise equivalence, and so for every $X \in \Gamma^o$ the natural map

$$THH(\mathfrak{C}, X) \to \Omega THH(S\mathfrak{C}, X)$$

is a weak equivalence.

Proof This follows by Proposition 5.1.1.1 since by Lemma 1.2.2.4 $T_n\mathfrak{C}$ is equivalent to $S_n\mathfrak{C}$, and THH sends equivalences to pointwise equivalences. $\square$

Corollary 5.1.1.3 *Let $\mathfrak{C}$ be an additive category. Then for every $k \ge 0$ the natural map $\bar{H}\mathfrak{C}(S^k) \to S^{(k)}\mathfrak{C}$ induces a pointwise equivalence*

$$THH\bigl(\bar{H}\mathfrak{C}(S^k)\bigr) \xrightarrow{\sim} THH\bigl(S^{(k)}\mathfrak{C}\bigr).$$

Substituting $\mathfrak{C}$ with $S^{(k)}\mathfrak{C}$ in Corollary 5.1.1.2 we get

Corollary 5.1.1.4 *Let $\mathfrak{C}$ be an additive category. Then the natural map*

$$THH\big(S^{(k)}\mathfrak{C}\big) \to \Omega THH\big(S^{(k+1)}\mathfrak{C}\big)$$

is a pointwise equivalence for all $k \geq 0$. ☺

Exactly the same proof gives the

Corollary 5.1.1.5 *Let $\mathfrak{C}$ be an additive category, and M a bilinear $\mathfrak{C}$-bimodule. Then the natural map*

$$THH\big(S^{(k)}\mathfrak{C}, S^{(k)}M\big) \to \Omega THH\big(S^{(k+1)}\mathfrak{C}, S^{(k+1)}M\big)$$

is a pointwise equivalence for all $k \geq 0$. ☺

These results allow us to define the trace used in [70], competing with the one we gave in Definition 5.1.0.5. Just as we converted Γ-spaces M to spectra $\underline{M} = \{n \mapsto M(S^n)\}$ in Definition 2.2.1.8, we can view a functor $X\colon \Gamma^o \times \Gamma^o \to \mathcal{S}_*$ as a bispectrum with (n, m)-space $X(S^n, S^m)$. If E is a spectrum, we have a bispectrum $\Sigma^\infty E$ with (n, m)-space $S^n \wedge E_m$. If M is a Γ-space, the bispectrum corresponding to $\Sigma^\infty M$ will be exactly $\Sigma^\infty \underline{M}$.

Definition 5.1.1.6 (The Nerveless Trace for Split Exact Categories) Let $\mathcal{E}$ be an additive category. The trace map for $\mathcal{E}$ in the Waldhausen formalism is the following chain of natural transformations (of functors from additive categories to Γ-spectra)

$$\Sigma^\infty ob\underline{\mathbf{S}}\mathcal{E} \longrightarrow THH(\underline{\mathbf{S}}\mathcal{E}) \xleftarrow{\;\sim\;} \Sigma^\infty THH(\mathcal{E})$$

where the first map is the Dennis trace of Sect. 4.3.2 and the leftwards pointing map is the stable equivalence coming from Corollary 5.1.1.2.

As before, we have used the notation $\underline{\mathbf{S}}\mathcal{E}$ for $\{m \mapsto S^{(m)}\mathcal{E}\}$.

5.1.1.2 Comparison of Traces for the Waldhausen and Segal Approaches

As a last step, we want to know that the two definitions of the trace for additive categories agree. This information is collected in the following commutative diagram of bispectra (the Γ-spaces are tacitly evaluated on spheres)

$$\begin{array}{ccccccc}
\Sigma^\infty ob\underline{\mathbf{S}}\mathcal{E} & & \longrightarrow & \longrightarrow & \longrightarrow & & \\
{\scriptstyle 1.2.3.2}\downarrow{\scriptstyle\sim} & & & & \searrow & & \\
\Sigma^\infty Bi\underline{\mathbf{S}}\mathcal{E} & \longrightarrow & \underline{T}\big(\mathbf{N}(\underline{\mathbf{S}}\mathcal{E},i)\big) & \xleftarrow[4.3.5.2]{\sim} & \underline{T}(\underline{\mathbf{S}}\mathcal{E}) & & \\
{\scriptstyle 3.2.1.5}\uparrow{\scriptstyle\sim} & & {\scriptstyle\sim}\uparrow & & {\scriptstyle\sim}\uparrow & \nwarrow{\scriptstyle\sim}\,{\scriptstyle 5.1.1.2} & \\
\Sigma^\infty Bi\bar{H}\mathcal{E} & \longrightarrow & \underline{T}\big(\mathbf{N}(\bar{H}\mathcal{E},i)\big) & \xleftarrow[4.3.5.2]{\sim} & \underline{T}(\bar{H}\mathcal{E}) & \xleftarrow[5.1.0.1]{\sim} & \Sigma^\infty\underline{T}(\mathcal{E}),
\end{array}$$

where each number refer to the result showing that the corresponding arrow is a weak equivalence, and the arrows leaving the left hand column are instances of the Dennis trace Sect. 4.3.2.

5.1.2 Comparison with the Homology of an Additive Category and the S-Construction

One thing that needs clarification is the relationship with the homology $F(\mathfrak{C},M)$ of a category which we used in Sect. 1.3, and which we showed is equivalent to stable K-theory when applied to an additive category. We used the S-construction there, and we use it here, and in both places the outcomes are Ω-spectra, and these coincide. As a comparison tool we use the model for topological Hochschild homology by means of the simplicial abelian groups $HH^{\mathbf{Z}}(\tilde{\mathbf{Z}}\mathcal{C},M)$ of Lemma 4.3.3.3.

Remark 5.1.2.1 If $\mathfrak{C}$ is an additive category, and M an additive bimodule, then we have levelwise equivalences of spectra (indexed by m)

$$F\big(S^{(m)}\mathfrak{C},S^{(m)}M\big)\xrightarrow{\sim}HH^{\mathbf{Z}}\big(\tilde{\mathbf{Z}}S^{(m)}\mathfrak{C},S^{(m)}M\big)\xleftarrow{\sim}THH\big(S^{(m)}\mathfrak{C},S^{(m)}M\big).$$

We have two independent proofs that these spectra are Ω-spectra (the first was given in Proposition 1.3.6.5). Furthermore, the maps

$$F_0\big(S^{(m)}\mathfrak{C},S^{(m)}M\big)\xrightarrow{\sim}HH^{\mathbf{Z}}\big(\tilde{\mathbf{Z}}S^{(m)}\mathfrak{C},S^{(m)}M\big)_0\xleftarrow{\sim}THH\big(S^{(m)}\mathfrak{C},S^{(m)}M\big)_0$$

are also levelwise equivalences, and so all maps in

$$\begin{array}{ccccc}
F_0\big(S^{(m)}\mathfrak{C},S^{(m)}M\big) & \xrightarrow{\sim} & HH^{\mathbf{Z}}\big(\tilde{\mathbf{Z}}S^{(m)}\mathfrak{C},S^{(m)}M\big)_0 & \xleftarrow{\sim} & THH\big(S^{(m)}\mathfrak{C},S^{(m)}M\big)_0 \\
\simeq\downarrow & & \simeq\downarrow & & \simeq\downarrow \\
F\big(S^{(m)}\mathfrak{C},S^{(m)}M\big) & \xrightarrow{\sim} & HH^{\mathbf{Z}}\big(\tilde{\mathbf{Z}}S^{(m)}\mathfrak{C},S^{(m)}M\big) & \xleftarrow{\sim} & THH\big(S^{(m)}\mathfrak{C},S^{(m)}M\big)
\end{array}$$

are (stable) equivalences of spectra (the leftmost vertical stable equivalence is that of Corollary 1.3.3.2).

In general we have the following definition.

Definition 5.1.2.2 If $\mathcal{C}$ is an additive category, M a $\mathcal{C}$-bimodule and X a space, let $\mathbf{T}(\mathcal{C}, M, X)$ be the Ω-spectrum

$$\{k \mapsto THH(S^{(k)}\mathcal{C}, S^{(k)}M, X)\}.$$

5.1.3 More on the Trace Map K → THH for Rings

For comparison with earlier constructions, it is often fruitful to give a slightly different view of the trace map, where the cyclic nerve plays a more prominent rôle. Furthermore, the comparison with the map defining the equivalence between stable K-theory and topological Hochschild homology has not yet been seen to relate to the trace. This will be discussed further in the next section.

If A is a ring and P an A-bimodule, we let $\mathbf{T}(A, P, X)$ be the Ω-spectrum

$$\mathbf{T}(\mathcal{P}_A, P, X) = \{k \mapsto THH(S^{(k)}\mathcal{P}_A, S^{(k)}\mathcal{M}_A(-, - \otimes_A P), X)\}.$$

We use the obvious abbreviations $\mathbf{T}(A, X) = \mathbf{T}(A, A, X)$ and $\mathbf{T}(A) = \mathbf{T}(A, S^0)$.

Consider

$$\begin{array}{ccc} ob\underline{\mathbf{S}}\mathcal{P}_A & \xrightarrow{c \mapsto c = c} & \operatorname{holim}_{\overrightarrow{x \in I}} \Omega^x \bigvee_{c \in ob\underline{\mathbf{S}}\mathcal{P}_A} \underline{\mathbf{S}}\mathcal{P}_A(c, c) \otimes_{\mathbf{Z}} \tilde{\mathbf{Z}}[S^x] \\ & & \| \\ \mathbf{T}(A) & \xleftarrow{\text{inclusion of the zero skeleton}} & \mathbf{T}_0(A) \end{array}.$$

This map agrees with the trace map given in the previous section, and displays the map as the composite $\mathbf{K}(A) \cong \mathbf{T}(A)^{\mathbf{S}^1} \subset \mathbf{T}(A)$ by Sect. 4.3.1.1, and so tells you that the circle action on *THH* is important. You do not expect to be able to calculate fixed point sets in general (it is not even a homotopy invariant notion), and so any approximation to the fixed points which are calculable should be explored.

If one want maps from $Bi\underline{\mathbf{S}}\mathcal{P}_A$ instead of from $ob\underline{\mathbf{S}}\mathcal{P}_A$, one can either do as we did in Sect. 5.1.1.2, or one may rewrite this slightly. As for groups, there is a map $Bi\mathfrak{C} \to B^{cy}i\mathfrak{C}$ for any category $\mathfrak{C}$, where B^{cy} is the cyclic nerve construction introduced in Sect. 4.2.5.1, given by sending $c_0 \xleftarrow{\alpha_1} c_1 \xleftarrow{\alpha_2} \cdots \xleftarrow{\alpha_q} c_q \in B_q i\mathfrak{C}$ to

$$c_q \xleftarrow{(\prod \alpha_i)^{-1}} c_0 \xleftarrow{\alpha_1} c_1 \xleftarrow{\alpha_2} \cdots \xleftarrow{\alpha_q} c_q \in B_q^{cy} i\mathfrak{C}.$$

This splits the natural map $B^{cy}i\mathfrak{C} \to Bi\mathfrak{C}$ given by forgetting (which is there regardless of maps being isomorphisms). If $\mathfrak{C}$ is a linear category we have a map $B^{cy}i\mathfrak{C} \to B^{cy}\mathfrak{C} \to THH(\mathfrak{C})$, where the first one is given by the inclusion of the isomorphism into all of $\mathfrak{C}$, and the second is stabilization. The diagram

$$\begin{array}{ccc} Bi\mathfrak{C} & \longleftarrow & B_0 i\mathfrak{C} = ob\mathfrak{C} \\ \downarrow & & \downarrow \\ B^{cy}i\mathfrak{C} & \longrightarrow & THH(\mathfrak{C}, S^0) \end{array}$$

commutes, where the rightmost map is defined as above. Setting $\mathfrak{C} = S^{(m)}\mathcal{P}_A$ and letting m vary, we obtain the commutative diagram

$$\begin{array}{ccc} Bi\underline{S}\mathcal{P}_A & \longleftarrow & ob\underline{S}\mathcal{P}_A \\ \downarrow & & {\scriptstyle tr}\downarrow \\ B^{cy}i\underline{S}\mathcal{P}_A & \longrightarrow & THH(A, S^0) \end{array}.$$

Note that the fact that

$$\begin{array}{ccc} Bi\underline{S}\mathcal{P}_A & \longrightarrow & THH(\mathbf{N}(\underline{S}\mathcal{P}_A, i), S^0) \\ \downarrow & & \uparrow\wr \\ B^{cy}i\underline{S}\mathcal{P}_A & \longrightarrow & THH(A, S^0) \end{array}$$

does not commute does not give rise to a contradiction.

5.1.4 The Trace and the K-Theory of Endomorphisms

Let $\mathfrak{C}$ be an exact category and let $\mathrm{End}(\mathfrak{C})$ be the category of endomorphisms in $\mathfrak{C}$. That is, it is the exact category with objects pairs (c, f), with $f: c \to c$ a morphism in $\mathfrak{C}$, and where a morphism $(c, f) \to (d, g)$ is a commuting diagram

$$\begin{array}{ccc} c & \longrightarrow & d \\ {\scriptstyle f}\downarrow & & {\scriptstyle g}\downarrow \\ c & \longrightarrow & d \end{array}.$$

A sequence $(c', f') \to (c, f) \to (c'', f'')$ in $\mathrm{End}(\mathfrak{C})$ is exact if the underlying sequence $c' \to c \to c''$ in $\mathfrak{C}$ is exact. We note that

$$obS\,\mathrm{End}(\mathfrak{C}) \cong \coprod_{c \in obS\mathfrak{C}} \mathrm{End}(c)$$

(by which we mean the simplicial object with q-simplices $\coprod_{c\in obS_q\mathfrak{C}}\mathrm{End}(c)$). There are two functors $\mathfrak{C}\to\mathrm{End}(\mathfrak{C})$ given by $c\mapsto\{c\overset{0}{\to}c\}$ and $c\mapsto\{c=c\}$ splitting the forgetful projection $\mathrm{End}(\mathfrak{C})\to\mathfrak{C}$ given by $(c,f)\mapsto c$. We let

$$\mathbf{End}(\mathfrak{C})=\bigvee_{c\in ob\underline{\mathbf{S}}\mathfrak{C}}\mathrm{End}(c)\simeq\mathrm{hofib}\bigl\{ob\underline{\mathbf{S}}\,\mathrm{End}(\mathfrak{C})\to ob\underline{\mathbf{S}}\mathfrak{C}\bigr\}$$

(again, there is a simplicial direction hidden in the summation over $c\in ob\underline{\mathbf{S}}\mathfrak{C}$). The $\mathrm{End}(c)$s are pointed at the zero maps.

Note that the first step in the trace, $ob\underline{\mathbf{S}}\mathfrak{C}\to\mathbf{T}(\mathfrak{C})_0$ factors through $ob\underline{\mathbf{S}}\mathfrak{C}\to\mathbf{End}(\mathfrak{C})$ via the map $c\mapsto c=c$.

If $\mathfrak{C}\subseteq\mathfrak{D}$ is cofinal then $\mathrm{End}(\mathfrak{C})\subseteq\mathrm{End}(\mathfrak{D})$ is also cofinal, and a quick calculation tells us that $K_0(\mathrm{End}(\mathfrak{D}))/K_0(\mathrm{End}(\mathfrak{C}))\cong K_0(\mathfrak{D})/K_0(\mathfrak{C})$, and hence by [271] we get that $\mathbf{End}(\mathfrak{C})\to\mathbf{End}(\mathfrak{D})$ is an equivalence. This tells us that the "strong" cofinality of *THH* Lemma 4.3.5.10 appears at a very early stage in the trace; indeed before we have started to stabilize.

5.2 The General Construction of the Trace

In order to state the trace in the full generality we need, it is necessary to remove the dependence on the enrichment in abelian groups we have used so far. This is replaced by an enrichment in Γ-spaces, which is always present for categories with sum by Sect. 2.1.6.1. The second thing we have to relax is our previous preoccupation with isomorphisms. In general this involves a choice of weak equivalences, but in order to retain full functoriality of our trace construction we choose to restrict to the case where the weak equivalences come as a natural consequence of the $\Gamma\mathcal{S}_*$-category structure. This is sufficient for all current applications of the trace, and the modifications one would want in other (typically geometric) applications are readily custom built from this.

5.2.1 Localizing at the Weak Equivalences

The weak equivalences are typically independent of the enrichment of our categories, in the sense that they only form an $\mathcal{S}$-category; much like the units in a ring only form a group, disregarding the additive structure. When one wants to localize a ring A, one considers "multiplicatively closed subsets", or in other words, submonoids of A considered as a monoid under multiplications. In order to compare $\Gamma\mathcal{S}_*$-categories and $\mathcal{S}$-categories we use the functor R from $\Gamma\mathcal{S}_*$-categories to $\mathcal{S}$-categories, sending a $\Gamma\mathcal{S}_*$-category $\mathcal{C}$ to the $\mathcal{S}$-category $R\mathcal{C}$ with the same objects, but with morphism spaces $R\mathcal{C}(c,c')=\mathcal{C}(c,c')(1_+)$. In the analogy with rings, R is like the forgetful functor from rings to monoids (under multiplication).

We will need to talk about various categories of pairs $(\mathcal{C}, w)$, where $\mathcal{C}$ is a $\Gamma\mathcal{S}_*$-category and $w\colon \mathcal{W} \to R\mathcal{C}$ is a $\mathcal{S}$-functor specifying the "weak equivalences" (hence the choice of the letter w) we want to invert. Note that it is *not* required that $w\colon \mathcal{W} \to R\mathcal{C}$ is an inclusion of a subcategory; w is free to vary over $\mathcal{S}$-functors with target $R\mathcal{C}$.

Definition 5.2.1.1 The category of *free pairs* $\mathfrak{P}^{\mathrm{free}}$ is the category whose objects are pairs $(\mathcal{C}, w)$ where $\mathcal{C}$ is a small $\Gamma\mathcal{S}_*$-category and $w\colon \mathcal{W} \to R\mathcal{C}$ an $\mathcal{S}$-functor of small $\mathcal{S}$-categories.

A morphism $(\mathcal{C}, w) \to (\mathcal{C}', w')$ in $\mathfrak{P}^{\mathrm{free}}$ is a pair $(F\colon \mathcal{C} \to \mathcal{C}',\ G\colon \mathcal{W} \to \mathcal{W}')$, where F is a $\Gamma\mathcal{S}_*$-functor and G is an $\mathcal{S}$-functor such that the diagram

$$\begin{array}{ccc} \mathcal{W} & \xrightarrow{w} & R\mathcal{C} \\ {\scriptstyle G}\downarrow & & \downarrow{\scriptstyle RF} \\ \mathcal{W}' & \xrightarrow{w'} & R\mathcal{C}' \end{array}$$

commutes. The morphism (F, G) is a *weak equivalence* of free pairs if F is a stable equivalence of $\Gamma\mathcal{S}_*$-categories and G is a weak equivalence of $\mathcal{S}$-categories.

Definition 5.2.1.2 The *category of pairs* (without qualifications) is in our context the full subcategory $\mathfrak{P} \subseteq \mathfrak{P}^{\mathrm{free}}$ whose objects are the pairs $(\mathcal{C}, w)$ that have the property that $w\colon \mathcal{W} \to R\mathcal{C}$ is the identity on objects.

The subcategory of *fixed pairs* $\mathfrak{P}^{\mathrm{fix}} \subseteq \mathfrak{P}$ contains all objects, but a morphism of fixed pairs is a morphism of pairs $(F, G)\colon (\mathcal{C}, w) \to (\mathcal{C}', w')$ where F (and hence also G) is the identity on objects.

A morphism of pairs or of fixed pairs is a *weak equivalence* if it is so when considered as a morphism of free pairs.

If $(\mathcal{C}, w\colon \mathcal{W} \to R\mathcal{C})$ is a free pair, we let the set of objects, $ob(\mathcal{C}, w)$, be the set of objects of the small $\mathcal{S}$-category $\mathcal{W}$.

5.2.1.1 Making a Free Pair a Pair

The inclusion $\mathfrak{P} \subseteq \mathfrak{P}^{\mathrm{free}}$ has a right adjoint (it is "coreflective") $\phi\colon \mathfrak{P}^{\mathrm{free}} \to \mathfrak{P}$, "forcing the set of objects of the $\mathcal{S}$-categories on the $\Gamma\mathcal{S}_*$-category". That is to say, if $(\mathcal{C}, w\colon \mathcal{W} \to R\mathcal{C})$ is a free pair, then $\phi(\mathcal{C}, w)$ is the pair $(\phi_w\mathcal{C}, \phi w)$ defined as follows: The set of objects in $\phi_w\mathcal{C}$ is $ob\mathcal{W}$ and given two objects c and d the Γ-space of morphism is defined by $\phi_w\mathcal{C}(c, d) = \mathcal{C}(wc, wd)$, and finally ϕw is given by

$$\mathcal{W}(c, d) \xrightarrow{w} R\mathcal{C}(wc, wd) = R\phi_w\mathcal{C}(c, d)\cdot$$

Note that the composite

$$\mathfrak{P} \subseteq \mathfrak{P}^{\mathrm{free}} \xrightarrow{\phi} \mathfrak{P}$$

is the identity. Considered as an endofunctor of free pairs, ϕ is idempotent ($\phi^2 = \phi$) and there is a natural transformation $\phi \to id_{\mathfrak{P}^{\text{free}}}$ given by the $\Gamma\mathcal{S}_*$-functor $\phi_w\mathcal{C} \to \mathcal{C}$ which is w on objects and the identity on morphisms.

5.2.1.2 The Nerve of a Pair

For each non-negative integer q, let $[q] = \{0 < 1 < \cdots < q\}$, and consider it as the category $\{0 \leftarrow 1 \leftarrow \cdots \leftarrow q\}$.

If $\mathcal{W}$ is an $\mathcal{S}$-category, we get a bisimplicial category $\mathbf{N}\mathcal{W}$ which in bidegree p, q is the category $(\mathbf{N}_q\mathcal{W})_p = [[q], \mathcal{W}_p]$ of functors $[q] \to \mathcal{W}_p$. Here $\mathcal{W}_p$ is the category with the same objects as $\mathcal{W}$, and with morphisms the set of p-simplices of morphisms in $\mathcal{W}$, or in other words, $\mathcal{W}_p$ is the category of p-simplices of $\mathcal{W}$ considered as a simplicial category.

Note that $\mathbf{N}\mathcal{W}$ is not a *simplicial $\mathcal{S}$-category* since the set of objects may vary in both the p and q direction. However, it is convenient to consider $\mathbf{N}\mathcal{W}$ as a bisimplicial $\mathcal{S}$-category with discrete morphism spaces. Likewise, if $\mathcal{C}$ is a $\Gamma\mathcal{S}_*$-category, $\mathbf{N}\mathcal{C}$ is the bisimplicial $\Gamma\mathcal{S}_*$-category $[p], [q] \mapsto [[q], \mathcal{C}_p]$.

Definition 5.2.1.3 The *free nerve* $\mathbf{N}^{\text{free}} \colon \mathfrak{P}^{\text{free}} \to [\Delta^o \times \Delta^o, \mathfrak{P}^{\text{free}}]$ is the functor which sends $(\mathcal{C}, w \colon \mathcal{W} \to R\mathcal{C})$ to the bisimplicial pair which in bidegree q, p is given by

$$\mathbf{N}_q^{\text{free}}(\mathcal{C}, w)_p = \big(\mathbf{N}_q\mathcal{C}_p, \mathbf{N}_q\mathcal{W}_p \xrightarrow{\mathbf{N}_q w} \mathbf{N}_q R\mathcal{C}_p = R\mathbf{N}_q\mathcal{C}_p\big).$$

The *nerve* $\mathbf{N} \colon \mathfrak{P} \to [\Delta^o \times \Delta^o, \mathfrak{P}]$ is the composite

$$\mathfrak{P} \subseteq \mathfrak{P}^{\text{free}} \xrightarrow{\mathbf{N}^{\text{free}}} [\Delta^o \times \Delta^o, \mathfrak{P}^{\text{free}}] \xrightarrow{\phi} [\Delta^o \times \Delta^o, \mathfrak{P}].$$

When we need to have names for the individual components of the bisimplicial $\Gamma\mathcal{S}_*$-category $\mathbf{N}(\mathcal{C}, w)$ we will write $(\mathbf{N}^w\mathcal{C}, \mathbf{N}w)$ instead.

Note that $ob\mathbf{N}(\mathcal{C}, w)$ is (the simplicial set of objects of) the degreewise nerve of $\mathcal{W}$, that is $ob\mathbf{N}(\mathcal{C}, w) = B\mathcal{W}$.

If $\mathcal{B}$ is an $\mathcal{S}$-category we let $\pi_0\mathcal{B}$ be the category with the same objects as $\mathcal{B}$, but with morphism sets from a to b the path components $\pi_0\mathcal{B}(a, b)$. Recall that a category is a *groupoid* is all its morphisms are isomorphisms.

Definition 5.2.1.4 We say that an $\mathcal{S}$-category $\mathcal{B}$ is a *groupoid* if for every q the category $\mathcal{B}_q$ is a groupoid. We say that $\mathcal{B}$ is *groupoid-like* if $\pi_0\mathcal{B}$ is a groupoid. A pair $(\mathcal{C}, w \colon \mathcal{W} \to R\mathcal{C}) \in \mathfrak{P}$ is called a *groupoid pair* (resp. a *groupoid-like pair*) if $\mathcal{W}$ is a groupoid (resp. groupoid-like). We say that a functor

$$(\mathcal{C}, w \colon \mathcal{W} \to R\mathcal{C}) \colon \Gamma^o \longrightarrow \mathfrak{P}$$

is *groupoid-like* if $\mathcal{W}(X)$ is groupoid-like for all $X \in \Gamma^o$.

5.2.1.3 Localization of Pairs

Given a pair $(\mathcal{C}, w)$ we think of the map $w\colon \mathcal{W} \to R\mathcal{C}$ as an inclusion of a subcategory of "weak equivalences". The purpose of the localization functor is to "invert the weak equivalences". We list the properties we will need, see [66] for further details:

Theorem 5.2.1.5 *There are two functors $L, \Phi\colon \mathfrak{P} \to \mathfrak{P}$ connected by natural transformations*

$$(\mathcal{C}, w) \longleftarrow \Phi(\mathcal{C}, w) \longrightarrow L(\mathcal{C}, w)$$

consisting of maps of fixed pairs, satisfying the following properties

lo1 *Given a pair $(\mathcal{C}, w)$ the map $\Phi(\mathcal{C}, w) \to (\mathcal{C}, w)$ is a weak equivalence.*
lo2 *Given a pair $(\mathcal{C}, w)$ the localization $L(\mathcal{C}, w)$ is a groupoid pair.*
lo3 *If $(\mathcal{C}, w) \to (\mathcal{C}', w')$ is a weak equivalence of fixed pairs, then $L(\mathcal{C}, w) \to L(\mathcal{C}', w')$ is a weak equivalence.*
lo4 *If $(\mathcal{C}, w)$ is a groupoid-like pair, then $\Phi(\mathcal{C}, w) \to L(\mathcal{C}, w)$ is a weak equivalence.*
lo5 *On the subcategory of $\mathfrak{P}^{fix}$ of groupoid pairs $(\mathcal{C}, w)$ there is a natural weak equivalence $L(\mathcal{C}, w) \xrightarrow{\sim} (\mathcal{C}, w)$ such that*

$$\begin{array}{ccc} L(\mathcal{C}, w) & \longleftarrow & \Phi(\mathcal{C}, w) \\ & \searrow & \downarrow \\ & & (\mathcal{C}, w) \end{array}$$

commutes. ☺

We call $L(\mathcal{C}, w)$ the *localization* of $(\mathcal{C}, w)$.

Note that in the triangular diagram of the last property, all the arrows are equivalences.

5.2.1.4 A Definition Giving the K-Theory of a Symmetric Monoidal $\Gamma\mathcal{S}_*$-Category Using the Uniform Choice of Weak Equivalences

We are ready for yet another definition of algebraic K-theory to be used in this book. This formulation uses the uniform choice of weak equivalences we made in Sect. 2.3.3, which to a $\Gamma\mathcal{S}_*$-category $\mathcal{C}$ associates a map of $\mathcal{S}$-categories $w_{\mathcal{C}}\colon \omega\mathcal{C} \to RT_0\mathcal{C}$ by pulling back along the inclusion of the isomorphisms in $\pi_0\mathcal{C}$. Here T_0 is the monoidal fibrant replacement discussed in Definition 2.2.2.1.

Definition 5.2.1.6 Let $W\colon \Gamma\mathcal{S}_* - \text{categories} \to \mathfrak{P}$ be the functor with $W(\mathcal{C}) = (T_0\mathcal{C}, w_\mathcal{C})$. Let

$$\mathfrak{k}\colon \text{symmetric monoidal } \Gamma\mathcal{S}_*\text{-categories} \longrightarrow [\Gamma^o, \mathfrak{P}]_*$$

(the $_*$ subscript means that the functors are pointed) be the composite

$$\text{symmetric monoidal } \Gamma\mathcal{S}_*\text{-categories} \xrightarrow{\bar{H}} [\Gamma^o, \Gamma\mathcal{S}_*\text{-categories}]_* \xrightarrow{W} [\Gamma^o, \mathfrak{P}]_*.$$

This is the first part of our model $\mathcal{K}$ of algebraic K-theory: If $\mathcal{D}$ is a symmetric monoidal $\Gamma\mathcal{S}_*$-category, then we call $\mathcal{K}(\mathcal{D}) = \mathbf{N}L\mathfrak{k}\mathcal{D}$ the *algebraic K-theory category* of $\mathcal{D}$, where the functors $\mathbf{N}$ and L were defined in Sect. 5.2.1.2 and Theorem 5.2.1.5.

The set of objects $ob\mathcal{K}(\mathcal{D})$ is referred to as the *algebraic K-theory spectrum* of $\mathcal{D}$.

5.2.2 Comparison with Other Definitions of Algebraic K-Theory

The morphism objects in $\mathcal{K}(\mathcal{D})$ are there for the purpose of the trace, but if one is only interested in the objects, i.e., in the algebraic K-theory spectrum, much of the structure is superfluous.

Lemma 5.2.2.1 *Let $(\mathcal{C}, w)$ be a groupoid-like pair. Then the natural maps*

$$ob\mathbf{N}(\mathcal{C}, w) \longleftarrow ob\mathbf{N}\Phi(\mathcal{C}, w) \longrightarrow ob\mathbf{N}L(\mathcal{C}, w)$$

are weak equivalences.

Proof This follows from the properties Theorem 5.2.1.5 of the localization since, by [75, 9.5], the nerve preserves weak equivalences of $\mathcal{S}$-categories (with fixed sets of objects). □

Together with Lemma 2.3.3.2 this gives the following theorem, justifying our claim that $\mathcal{K}$ measures algebraic K-theory. Recall that $Bi\bar{H}\mathcal{C}$ (the nerve of the isomorphisms of the Segal construction which we call $\bar{H}$) is one of the formulae for the algebraic K-theory, which we in Sect. 3.2 compared with Waldhausen's construction and in Sect. 3.3 compared with the plus construction.

Theorem 5.2.2.2 *Let $\mathcal{D}$ be a symmetric monoidal $\Gamma\mathcal{S}_*$-category. Then $ob\mathcal{K}(\mathcal{D})$ is connected to $ob\mathbf{N}\mathfrak{k}(\mathcal{D}) = B\omega(\bar{H}\mathcal{D})$ by a chain of natural weak equivalences, where ω is the uniform choice of weak equivalences of Sect. 2.3.3. If $\mathcal{D}$ has discrete morphism spaces it is naturally isomorphic to $Bi\bar{H}\mathcal{D}$.* ☺

5.2.3 The Trace

We define topological Hochschild homology on the level of pairs $(\mathcal{C}, w) \in \mathfrak{P}$ by

$$THH(\mathcal{C}, w) = \{[q] \mapsto THH(\mathcal{C}_q)\}.$$

Notice that $THH(\mathcal{C}, w)$ is independent of w. By an argument just like the proof of Lemma 4.2.3.1 regarding *THH* of simplicial **S**-algebras, we get that there is a chain of natural stable equivalences between (the diagonal of) $THH(\mathcal{C}, w)$ and $THH(\mathcal{C})$.

Consider the transformation

$$THH(\mathbf{N}(\mathcal{C}, w)) \longleftarrow THH(\mathbf{N}_0(\mathcal{C}, w)) = THH(\mathcal{C}, w)$$

induced by the degeneracies.

Lemma 5.2.3.1 *If* $(\mathcal{C}, w\colon \mathcal{W} \to R\mathcal{C}) \in \mathfrak{P}$ *is a groupoid pair, then the natural map*

$$THH(\mathcal{C}, w) \longrightarrow THH(\mathbf{N}(\mathcal{C}, w))$$

is an equivalence.

Proof Fix a simplicial dimension p. Note that since all maps in $\mathcal{W}_p$ are isomorphisms, the map induced by the degeneracy maps

$$(\mathcal{C}, w)_p = \mathbf{N}_0(\mathcal{C}, w)_p \to \mathbf{N}_q(\mathcal{C}, w)_p$$

gives an equivalence $\mathcal{C}_p = \mathbf{N}_0^w \mathcal{C}_p \to \mathbf{N}_q^w \mathcal{C}_p$ of (Γ-set)-categories for each $q \geq 0$. The statement follows immediately. □

Because of naturality, we immediately get the result also if $(\mathcal{C}, w)\colon \Gamma^o \to \mathfrak{P}$ is a Γ-groupoid pair.

5.2.3.1 The Dennis Trace for $\Gamma\mathcal{S}_*$-Categories

Recall the Dennis trace map of Sect. 4.3.2. For a pair $(\mathcal{C}, w) \in \mathcal{P}$ it takes the form

$$ob\mathcal{C} \longrightarrow THH(\mathcal{C}_0, w)(S^0)_0 \xrightarrow{\text{degeneracies}} THH(\mathcal{C}, w)(S^0).$$

Here the first map is defined by sending $d \in ob\mathcal{C}$ to the image of the non-base point under the unit map

$$S^0 = 1_+ = \mathbf{S}(1_+) \longrightarrow \mathcal{C}_0(d, d)(1_+)$$

composed with the obvious map

$$\mathcal{C}_0(d,d)\big(S^0\big) \subseteq \bigvee_{c\in ob\mathcal{C}} \mathcal{C}_0(c,c)\big(S^0\big) \longrightarrow \Big(\underset{\overrightarrow{x\in ob\mathcal{I}}}{\text{holim}}\Omega^x \bigvee_{c\in ob\mathcal{C}} \mathcal{C}_0(c,c)\big(S^x\big)\Big)_0$$
$$= THH(\mathcal{C}_0, w)\big(S^0\big)_0.$$

If $\mathcal{C}$ has an initial object, then the Dennis trace is a pointed map, and we get a map of Γ-spaces $\Sigma^\infty ob\mathcal{C} \to THH(\mathcal{C}, w)$ given by the "assembly"

$$X \wedge ob\mathcal{C} \longrightarrow X \wedge THH(\mathcal{C}, w)\big(S^0\big) \longrightarrow THH(\mathcal{C}, w)(X).$$

Note that since the morphism spaces in $\mathcal{C}$ are pointed by a map we may call 0, being an initial object in $\mathcal{C}$ is the same as being final. It simply means that the identity morphism of the object equals the 0-map.

Definition 5.2.3.2 The *Dennis trace of symmetric monoidal* $\Gamma\mathcal{S}_*$*-categories* is the natural transformation of $\Gamma^o \times \Gamma^o$-spaces which to a symmetric monoidal $\Gamma\mathcal{S}_*$-category $\mathcal{D}$ gives the map

$$\Sigma^\infty ob\mathcal{K}(\mathcal{D}) \longrightarrow THH\big(\mathcal{K}(\mathcal{D})\big)$$

induced by the Dennis trace.

We think of, and may occasionally refer to, these $\Gamma^o \times \Gamma^o$-spaces as "bispectra". This definition is relevant in view of a natural equivalence between $THH(\mathcal{K}(\mathcal{D}))$ and $\Sigma^\infty THH(\mathcal{D})$ which we will establish below as Theorem 5.2.3.5. The proof of Theorem 5.2.3.5 contains ingredients that are important in their own right and we will need to refer to later. Let

$$(\mathcal{C}, w)\colon \Gamma^o \longrightarrow \mathfrak{P}$$

be a Γ-pair and consider the commutative diagram of $\Gamma^o \times \Gamma^o$-spaces

$$\begin{array}{ccccc}
\Sigma^\infty ob\mathbf{N}(\mathcal{C}, w) & \longrightarrow & THH\big(\mathbf{N}(\mathcal{C}, w)\big) & \longleftarrow & THH(\mathcal{C}, w) \\
\simeq\uparrow & & \uparrow & & \simeq\uparrow \\
\Sigma^\infty ob\mathbf{N}\big(\Phi(\mathcal{C}, w)\big) & \longrightarrow & THH\big(\mathbf{N}\big(\Phi(\mathcal{C}, w)\big)\big) & \longleftarrow & THH\big(\Phi(\mathcal{C}, w)\big), \\
i_K\downarrow & & \downarrow & & i_{THH}\downarrow \\
\Sigma^\infty ob\mathbf{N}\big(L(\mathcal{C}, w)\big) & \longrightarrow & THH\big(\mathbf{N}\big(L(\mathcal{C}, w)\big)\big) & \overset{\simeq}{\longleftarrow} & THH\big(L(\mathcal{C}, w)\big)
\end{array} \quad (5.1)$$

where the vertical arrows are induced by the natural transformations $id \Leftarrow \Phi \Rightarrow L$, the left horizontal maps are instances of the Dennis trace and the right horizontal arrows are induced by inclusion of zero simplices in the nerve.

Lemma 5.2.3.3 *The arrows marked with* $\simeq$ *in diagram* (5.1) *are stable equivalences.*

If $(\mathcal{C}, w)$ *is groupoid-like (Definition* 5.2.1.4), *then also the arrows marked* i_K *and* i_{THH} *in diagram* (5.1) *are stable equivalences.*

Proof First consider the two maps induced by $\Phi(\mathcal{C}, w) \to (\mathcal{C}, w) \in \mathfrak{P}^{\text{fix}}$. By Theorem 5.2.1.5 $\Phi(\mathcal{C}, w) \to (\mathcal{C}, w)$ is a weak equivalence, giving the result since both $\Sigma^\infty ob\mathbf{N}$ and THH send weak equivalences to stable equivalences.

The lower right hand map

$$THH\bigl(\mathbf{N}(L(\mathcal{C}, w))\bigr) \xleftarrow{\sim} THH\bigl(L(\mathcal{C}, w)\bigr)$$

is a stable equivalence by Lemma 5.2.3.1 since $L(\mathcal{C}, w)$ is a groupoid pair.

Assume that $(\mathcal{C}, w)$ is groupoid-like. That i_K is a stable equivalence follows from Lemma 5.2.2.1. That i_{THH} is a stable equivalence follows from Theorem 5.2.1.5 since THH preserves stable equivalences by Lemma 4.3.5.3. □

Lemma 5.2.3.4 *Consider a* Γ*-pair* $(\mathcal{C}, w)\colon \Gamma^o \to \mathfrak{P}$. *If* $\mathcal{C}$ *is quite special, then there is a chain of natural stable equivalences between* $THH(\mathcal{C}, w)$ *and* $\Sigma^\infty THH(\mathcal{C}(1_+))$.

Proof If $\mathcal{C}$ is quite special we obtain a chain of stable equivalences since THH preserves products and stable equivalences, following the idea of the proof of Proposition 5.1.0.1 and lastly using the chain of equivalences between $THH(\mathcal{C}, w)$ and $THH(\mathcal{C})$ (cf. the beginning of Sect. 5.2.3). □

If $\mathcal{D}$ is a symmetric monoidal $\Gamma\mathcal{S}_*$-category and $(\mathcal{C}, w) = \mathfrak{k}(\mathcal{D})$ as in Sect. 5.2.1.4, diagram (5.1) takes the form

$$\begin{array}{ccccc}
\Sigma^\infty ob\mathbf{N}\mathfrak{k}\mathcal{D} & \longrightarrow & THH(\mathbf{N}\mathfrak{k}\mathcal{D}) & \longleftarrow & THH(\mathfrak{k}\mathcal{D}) \\
\simeq\uparrow & & \uparrow & & \simeq\uparrow \\
\Sigma^\infty ob\mathbf{N}\bigl(\Phi(\mathfrak{k}\mathcal{D})\bigr) & \longrightarrow & THH\bigl(\mathbf{N}\bigl(\Phi(\mathfrak{k}\mathcal{D})\bigr)\bigr) & \longleftarrow & THH\bigl(\Phi(\mathfrak{k}\mathcal{D})\bigr), \\
i_K\downarrow & & \downarrow & & i_{THH}\downarrow \\
\Sigma^\infty ob\mathcal{K}(\mathcal{D}) & \longrightarrow & THH\bigl(\mathcal{K}(\mathcal{D})\bigr) & \xleftarrow{\simeq} & THH\bigl(L(\mathfrak{k}\mathcal{D})\bigr)
\end{array}$$

and the lower left hand horizontal map in the diagram is exactly the trace $\Sigma^\infty ob\mathcal{K}(\mathcal{D}) \to THH(\mathcal{K}\mathcal{D})$ of Definition 5.2.3.2.

Theorem 5.2.3.5 *Let* $\mathcal{D}$ *be a symmetric monoidal* $\Gamma\mathcal{S}_*$*-category. Then* $THH(\mathcal{K}(\mathcal{D}))$ *is naturally equivalent to* $\Sigma^\infty THH(\mathcal{D})$.

Proof This is essentially Lemmas 5.2.3.3 and 5.2.3.4 applied to the group-like and quite special case $(\mathcal{C}, w) = \mathfrak{k}(\mathcal{D})$, since $\mathcal{C}(1_+) = T_0\mathcal{D}$. In essence, if $\mathcal{D}$ is a symmetric monoidal $\Gamma\mathcal{S}_*$-category, then diagram (5.1) gives a natural chain of stable equivalences

$$\begin{array}{ccc}
 & THH(\mathfrak{t}\mathcal{D}) \simeq \Sigma^\infty THH(T_0\mathcal{D}) & \xleftarrow{\sim} \Sigma^\infty THH(\mathcal{D}) \\
 & \simeq\uparrow & \\
 & THH\big(\Phi(\mathfrak{t}\mathcal{D})\big) & \\
 & \simeq\downarrow i_{THH} & \\
THH\big(\mathcal{K}(\mathcal{D})\big) \xleftarrow{\sim} & THH\big(L(\mathfrak{t}\mathcal{D})\big) &
\end{array}$$

□

5.2.4 The Weak Trace

The price we have to pay for having a single map representing our trace is that the models of either side are more involved than their classical counterparts. At the cost of having to talk about *weak transformations* (a zig-zag of natural transformations where the transformations pointing in the "wrong" direction are weak equivalences) this can be remedied by just exchanging the complicated models with their simpler, but equivalent, cousins.

Definition 5.2.4.1 If

$$\begin{array}{ccccccccccccc}
A_0 & \longrightarrow & B_0 & \xleftarrow{\sim} & C_0 & \longrightarrow & \cdots & \xleftarrow{\sim} & Y_0 & \longrightarrow & Z_0 \\
\| & & \simeq\downarrow & & \simeq\downarrow & & & & \simeq\downarrow & & \| \\
A_1 & \longrightarrow & B_1 & \xleftarrow{\sim} & C_1 & \longrightarrow & \cdots & \xleftarrow{\sim} & Y_1 & \longrightarrow & Z_1
\end{array}$$

is a diagram of natural transformations of functors to a category with a notion of weak equivalences, we say that the weak transformations $A_0 \to Z_0$ given at the top and at the bottom *agree up to homotopy*. More generally, we use the term *agree up to homotopy* for the equivalence relation this generates.

This is useful when we want to compare our definition to previous definitions of traces which were all examples of quite special groupoid-like pairs (cf. Definitions 2.3.2.1 and 5.2.1.4).

Definition 5.2.4.2

1. Let

$$(\mathcal{C}, w)\colon \Gamma^o \to \mathfrak{P}$$

be a quite special and groupoid-like Γ-pair. Then the *weak trace* is the functorial weak composite

$$\Sigma^\infty ob\mathbf{N}(\mathcal{C}, w) \to THH(\mathcal{C}, w) \simeq \Sigma^\infty THH\big(\mathcal{C}(1_+)\big).$$

Along the lower outer edge of diagram (5.1).

2. If $\mathcal{D}$ is a symmetric monoidal $\Gamma\mathcal{S}_*$-category, then the *weak trace* of $\mathcal{D}$ is the composite weak map

$$\Sigma^\infty ob\mathbf{N}\mathfrak{k}(\mathcal{D}) \longrightarrow \Sigma^\infty THH\big(\mathfrak{k}(\mathcal{D})(1_+)\big) = \Sigma^\infty THH(T_0\mathcal{D}) \xleftarrow{\sim} \Sigma^\infty THH(\mathcal{D})$$

where the leftmost weak map is the weak trace of $\mathfrak{k}(\mathcal{D})$ (which is quite special and groupoid-like) and the rightmost map is induced by the stable equivalence $T_0\mathcal{D} \leftarrow \mathcal{D}$.

Note that the only map in the weak trace of $\mathcal{D}$ that is not a weak equivalence is the trace $\Sigma^\infty ob\mathcal{K}(\mathcal{D}) \to THH(\mathcal{K}(D))$ of Definition 5.2.3.2 (recall that by definition $\mathcal{K}(\mathcal{D}) = \mathbf{N}L\mathfrak{k}(\mathcal{D})$).

5.2.4.1 The Quite Special Groupoid Case

If $(\mathcal{C}, w) \in \mathfrak{P}$ is a groupoid pair, then Lemma 5.2.3.1 says that

$$THH\big(\mathbf{N}(\mathcal{C}, w)\big) \longleftarrow THH(\mathcal{C}, w)$$

is an equivalence, and we are free to consider the weak map

$$\Sigma^\infty ob\mathbf{N}(\mathcal{C}, w) \longrightarrow THH\big(\mathbf{N}(\mathcal{C}, w)\big) \xleftarrow{\sim} THH(\mathcal{C}, w)$$

from the upper line of the main diagram (5.1). This gives rise to the simpler definition of the trace which was used in the *Ab*-case in Sect. 5.1 using either Segal's or Waldhausen's constructions. In our context we have to keep the nerves in place, and in view of the commutativity of

$$\begin{array}{ccc} obS\mathcal{E} & \longrightarrow & THH(S\tilde{\mathcal{E}})(S^0) \\ \simeq\downarrow & & \simeq\downarrow \\ BiS\mathcal{E} & \longrightarrow & THH\big(\mathbf{N}(S\tilde{\mathcal{E}}, iS\mathcal{E} \subseteq S\mathcal{E})\big)(S^0) \\ \simeq\uparrow & & \simeq\uparrow \\ obBi\bar{H}\mathcal{E} & \longrightarrow & THH\big(\mathbf{N}(\bar{H}\tilde{\mathcal{E}}, i\bar{H}\mathcal{E} \subseteq \bar{H}\mathcal{E})\big)(S^0) \end{array}$$

the relevant translation is the following:

Definition 5.2.4.3 Let $\mathcal{E}$ be a symmetric monoidal *Ab*-category. The weak trace of $\mathcal{E}$ is the weak map

$$\Sigma^\infty Bi\bar{H}\mathcal{E} \longrightarrow THH\big(\mathbf{N}(\bar{H}\tilde{\mathcal{E}}, i\bar{H}\mathcal{E} \subseteq \bar{H}\mathcal{E})\big) \xleftarrow{\sim} THH(\bar{H}\tilde{\mathcal{E}}) \xleftarrow{\sim} \Sigma^\infty THH(\tilde{\mathcal{E}})$$

obtained from the top row of the main diagram (5.1) with $(\mathcal{C}, w) = (\bar{H}\tilde{\mathcal{E}}, i\bar{H}\mathcal{E} \subseteq \bar{H}\mathcal{E})$.

Proposition 5.2.4.4 *Let $\mathcal{E}$ be a symmetric monoidal Ab-category. Then the weak trace of $\tilde{\mathcal{E}}$ precomposed with the map $\Sigma^\infty Bi\bar{H}\mathcal{E} \xrightarrow{\sim} \Sigma^\infty ob\mathbf{N}\mathfrak{k}\tilde{\mathcal{E}}$ agrees up to homotopy with the weak trace of $\mathcal{E}$.*

Proof If we let $(\mathcal{C}, w) = (\bar{H}\tilde{\mathcal{E}}, i\bar{H}\mathcal{E} \subseteq \bar{H}\mathcal{E})$, in the main diagram (5.1), we have by property *lo5* of Theorem 5.2.1.5 that there are natural vertical equivalences from the bottom to the top rows making everything commute

$$\begin{array}{ccccccc} \Sigma^\infty ob\mathbf{N}(\mathcal{C}, w) & \longrightarrow & THH\big(\mathbf{N}(\mathcal{C}, w)\big) & \xleftarrow{\sim} & THH(\mathcal{C}, w) & \simeq & \Sigma^\infty THH\big(\mathcal{C}(1_+)\big) \\ \simeq\uparrow & & \uparrow & & \simeq\uparrow & & \\ \Sigma^\infty ob\mathbf{N}\big(L(\mathcal{C}, w)\big) & \longrightarrow & THH\big(\mathbf{N}\big(L(\mathcal{C}, w)\big)\big) & \xleftarrow{\simeq} & THH\big(L(\mathcal{C}, w)\big). & & \end{array}$$

The top row is the weak trace of $\mathcal{E}$ whereas going around the lower edge agrees up to homotopy with the weak trace of $(\mathcal{C}, w)$. But all nodes of the weak trace are homotopy invariants, and so the weak equivalence $(\mathcal{C}, w) \to \mathfrak{k}\tilde{\mathcal{E}}$ shows that the weak trace of $(\mathcal{C}, w)$ agrees up to homotopy with the weak trace of $\tilde{\mathcal{E}}$ precomposed with the map $\Sigma^\infty Bi\bar{H}\mathcal{E} \xrightarrow{\sim} \Sigma^\infty ob\mathbf{N}\mathfrak{k}\tilde{\mathcal{E}}$. □

5.2.5 The Category of Finitely Generated Free A-Modules

Recall the definition of the category of "finitely generated free" A-modules Definition 3.2.4.1 for an $\mathbf{S}$-algebra A. Consider the $\Gamma\mathcal{S}_*$-full subcategory of the category of A-modules with objects $k_+ \wedge A$ for $k \geq 0$. More precisely, we could equally well describe it as the $\Gamma\mathcal{S}_*$-category whose objects are the natural numbers (including zero), and where the morphisms are given by

$$\mathcal{F}_A(k, l) = \mathcal{M}_A(k_+ \wedge A, l_+ \wedge A) \cong \prod_k \bigvee_l A.$$

This forms a symmetric monoidal $\Gamma\mathcal{S}_*$-category via the sum. Let

$$\mathfrak{k}(\mathcal{F}_A) = (\mathcal{C}_A, w_A) \colon \Gamma^o \to \mathfrak{P}$$

be the functor produced by the $\bar{H}$-construction followed by the uniform choice of weak equivalences of Sect. 2.3.3: $\mathcal{C}_A = T_0\bar{H}\mathcal{F}_A$ and $w_A \colon \mathcal{W}_A \to R\mathcal{C}_A$ the pullback of $i\pi_0\mathcal{C}_A \to \pi_0\mathcal{C}_A \leftarrow R\mathcal{C}_A$.

Considering A as a $\Gamma\mathcal{S}_*$-category with a single object 1, we get an inclusion of $\Gamma\mathcal{S}_*$-categories, naturally referred to as the inclusion of the rank one module. By Morita invariance Proposition 4.2.4.4, the induced map $THH(\mathcal{F}_A) \leftarrow THH(A)$ is a stable equivalence.

Definition 5.2.5.1 The algebraic K-theory of an $\mathbf{S}$-algebra A is the Γ-space

$$K(A) = ob\mathbf{N}\mathfrak{k}(\mathcal{F}_A)$$

and the *trace* for **S**-algebras is the weak natural transformation

$$\Sigma^\infty K(A) \longrightarrow \Sigma^\infty THH(A)$$

given by the weak trace for $\mathcal{F}_A$ followed by the equivalence induced by the inclusion of the rank one module

$$THH(\mathcal{F}_A) \xleftarrow{\sim} THH(A).$$

This definition of K-theory agrees with the one given by the plus construction, and this definition of the trace agrees with the one already defined for discrete rings:

Theorem 5.2.5.2 *Let A be an* **S**-*algebra There is a natural chain of weak equivalences*

$$\Omega^\infty ob\mathcal{K}(\mathcal{F}_A) \simeq \Omega^\infty K(A) \simeq K_0^f(\pi_0 A) \times B\widehat{GL}(A)^+,$$

where $\widehat{GL}_k(A)$ was defined in Sect. 3.2.3.1.

Proof The first weak equivalence follows from Lemma 5.2.2.1 (and is itself a chain of weak equivalences, and not a direct map). To simplify notation, let $W = \omega\bar{H}\mathcal{F}_A$. Recall from Theorem 5.2.2.2 the chain of natural equivalences $K(A) \simeq BW$. Since the associated spectrum is special

$$\Omega^\infty K(A) \simeq \Omega BW(S^1),$$

and for each $n_+ \in \Gamma^o$ we have that $BW(n_+) \simeq (BW(1_+))^{\times n}$. For each $k \geq 0$, let W^k be the full subcategory of $W(1_+)$ whose only object is $k_+ \wedge A$. Note that by definition, this is nothing but $\widehat{GL}_k(A)$ considered as a simplicial category with only one object. Hence we are done, for by Segal [257] there is a chain of weak equivalences

$$\Omega BW(S^1) \simeq K_0^f(\pi_0 A) \times \varinjlim_k (BW^k)^+ = K_0^f(\pi_0 A) \times B\widehat{GL}(A)^+. \qquad \square$$

Theorem 5.2.5.3 *Let A be a discrete ring. The definition of the weak trace Definition* 5.2.4.2 *of the $\Gamma\mathcal{S}_*$-category $\mathcal{F}_{HA}$ agrees up to homotopy with the one given in Sect.* 5.1.1.2 *with $\mathcal{E}$ the category of free finitely generated A-modules.*

Proof There are two steps to this. The first is to note that if A is a discrete ring, then the definition we have given of the category $\mathcal{F}_{HA}$ of finitely generated free HA-modules, agrees with the more down-to-earth definition of the category $\mathcal{F}_A$ of finitely generated free A-modules. We choose the usual skeleton for $\mathcal{F}_A$: the objects are the natural numbers (including zero), and a morphism from m to n is an $n \times m$-matrix with entries in A. Let $\mathcal{F}_A^\oplus$ be the $\Gamma\mathcal{S}_*$-category obtained through the procedure described in Sect. 2.1.6.1. We see that there is a $\Gamma\mathcal{S}_*$-weak equivalence $\mathcal{F}_A \to \mathcal{F}_A^\oplus$ given by sending $\vee$ to $\oplus$, and so also an equivalence

$$RT_0\mathcal{F}_A \xrightarrow{\sim} RT_0\mathcal{F}_A^\oplus \xleftarrow{\sim} \mathcal{F}_A.$$

Hence the K-theory and THH as given here are naturally equivalent to the usual ones in the discrete case when we choose the weak equivalences to be the isomorphisms, since $\widehat{GL_k}(HA) \simeq GL_k(A)$.

The second thing we have to see is that the two definitions of the trace agree, but this follows from Proposition 5.2.4.4. □

5.3 Stable K-Theory and Topological Hochschild Homology

In this section we are going to give a proof of Waldhausen and Goodwillie's conjecture $K^S \simeq THH$ for $\mathbf{S}$-algebras. For rings, this is almost clear from the results of Sect. 1.3 and Sect. 5.2, but for $\mathbf{S}$-algebras we need to know that some of the maps used in the ring case have their analog in the $\mathbf{S}$-algebra world. These considerations run parallel with a need which will be apparent in Chap. 7, namely: we need to know what consequences the equivalence $K^S \simeq THH$ has for the trace map.

5.3.1 Stable K-Theory

Let A be a simplicial ring and P an A-bimodule. Recall the discussion of stable K-theory in Sect. 1.3.5, and in particular the equivalence between $\mathbf{K}^S(A, P)$ and the first differential of the functor C_A defined in Definition 1.3.4.3, and the homology $\mathbf{F}(A, P)$ of (1.3).

As before $\mathbf{T}(A, P)$ is the Ω-spectrum $\{k \mapsto THH(S^{(k)}\mathcal{P}_A, S^{(k)}\mathcal{M}_A(-, - \otimes_A P))\}$. Notice that there is a map

$$\begin{aligned} D_1\mathsf{C}_A(P) &= \left\{n \mapsto \varinjlim_{k} \Omega^k \bigvee_{c \in obS^{(n)}\mathcal{P}_A} S^{(n)}\mathrm{Hom}_A\big(c, c \otimes_A B^k P\big)\right\} \\ &\to \left\{n \mapsto \underset{\overrightarrow{x \in I}}{\mathrm{holim}}\, \Omega^x \bigvee_{c \in obS^{(n)}\mathcal{P}_A} S^{(n)}\mathrm{Hom}_A\big(c, c \otimes_A P \otimes \tilde{\mathbf{Z}}\big[S^x\big]\big)\right\} \\ &= \mathbf{T}(A, P)_0 \end{aligned}$$

which is an equivalence by Bökstedt's approximation Lemma 2.2.2.2.

Theorem 5.3.1.1 *Let A be a simplicial ring and P a simplicial A bimodule. Then we have an equivalence*

$$\mathbf{K}^S(A, P) \simeq \mathbf{T}(A, P)$$

induced by

$$\mathbf{K}^S(A, P) \simeq D_1\mathsf{C}_A(P) \xrightarrow{\simeq} \mathbf{T}(A, P)_0 \xrightarrow{\simeq} \mathbf{T}(A, P).$$

This equivalence is compatible with the equivalence to the $\mathbf{F}$*-construction of Theorem* 1.3.5.2.

Proof As both K-theory (of radical extensions) and *THH* may be computed degree-wise we may assume that A and P are discrete. Then the only thing which needs verification is the compatibility. Recall that the equivalence $\mathbf{K}^S(A, P) \simeq \mathbf{F}(A, P)$ of Theorem 1.3.5.2 was given by a chain

$$D_1\mathrm{C}_A(-) \xrightarrow{\simeq} D_1\mathbf{F}_0(A, -) \xleftarrow{\simeq} \mathbf{F}_0(A, -) \xrightarrow{\simeq} \mathbf{F}(A, -)$$

of equivalences. Consider the diagram

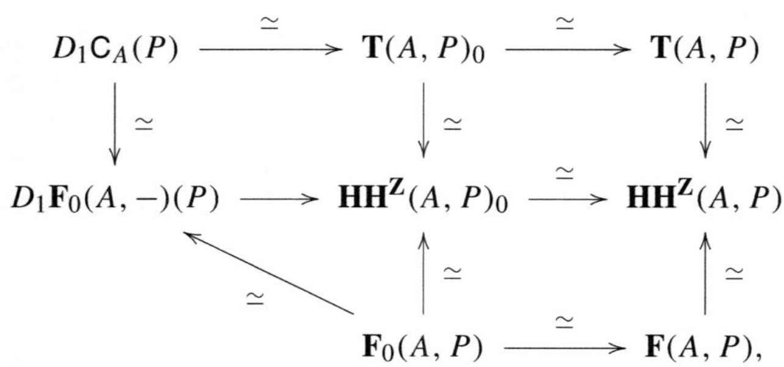

where $\mathbf{HH}^{\mathbf{Z}}(A, P)$ represents the spectrum $HH^{\mathbf{Z}}(\mathbf{Z}\underline{\mathbf{S}}\mathcal{P}_A, \underline{\mathbf{S}}\mathcal{M}_A(-, - \otimes_A P))$ and $HH^{\mathbf{Z}}$ is the abelian group version of *THH* as in Definition 4.3.3.1. The right side of the diagram is simply the diagram of Remark 5.1.2.1 (rotated), and the map from F_0 to $HH_0^{\mathbf{Z}}$ is stabilization and so factors through the map to the differential. □

5.3.2 THH of Split Square Zero Extensions

Let A be an $\mathbf{S}$-algebra and P an A-bimodule. Let $A \vee P$ be given the $\mathbf{S}$-algebra structure we get by declaring that the multiplication $P \wedge P \to P$ is trivial. We want to study $THH(A \vee P)$ closer. If R is a simplicial ring and Q an R-bimodule, we get that the inclusion of wedge into product, $HR \vee HQ \to H(R \ltimes Q)$, is a stable equivalence of $\mathbf{S}$-algebras, and so $A \vee P$ will cover all the considerations for split square zero extensions of rings.

The first thing one notices, is that the natural distributivity of smash and wedge give us a decomposition of $THH(A \vee P, X)$, or more precisely a decomposition of $V(A \vee P)(\mathbf{x})$ for every $\mathbf{x} \in \mathcal{I}^{q+1}$, as follows. Let

$$V^{(j)}(A, P)(\mathbf{x}) = \bigvee_{\phi \in \Delta_m([j-1],[q])} \bigwedge_{0 \le i \le q} F_{i,\phi}(x_i),$$

where $\Delta_m([j-1], [q])$ is the set of order preserving injections $[j-1] \to [q]$ and

$$F_{i,\phi}(x) = \begin{cases} A(S^x) & \text{if } i \text{ is not in the image of } \phi, \\ P(S^x) & \text{if } i \text{ is in the image of } \phi. \end{cases}$$

Then distributivity induces an isomorphism

$$V(A\vee P)(\mathbf{x})\cong\bigvee_{j\geq 0}V^{(j)}(A,P)(\mathbf{x})$$

(note that $V^{(j)}(A,P)(\mathbf{x})=*$ for $j>q+1$). Set

$$THH^{(j)}(A,P,X)_q=\underset{\mathbf{x}\in\mathcal{I}^{q+1}}{\underrightarrow{\operatorname{holim}}}\,\Omega^{\vee\mathbf{x}}\big(X\wedge V^{(j)}(A,P)(\mathbf{x})\big)$$

and

$$\underline{T}^{(j)}(A,P,X)=\big\{k\mapsto THH^{(j)}\big(A,P,S^k\wedge X\big)\big\}.$$

We see that this defines cyclic objects (the transformations used to define THH respect the number of occurrences of the bimodule), when varying q. The inclusions and projections

$$V^{(j)}(A,P)(\mathbf{x})\subseteq V(A\vee P)(\mathbf{x})\to V^{(i)}(A,P)(\mathbf{x})$$

define cyclic maps

$$\bigvee_{j\geq 0}THH^{(j)}(A,P,X)\to THH(A\vee P,X)\to\prod_{j\geq 0}THH^{(j)}(A,P,X).$$

The approximation Lemma 2.2.2.2 assures us that

$$\underset{\mathbf{x}\in\mathcal{I}^{q+1}}{\underrightarrow{\operatorname{holim}}}\prod_{j\geq 0}\Omega^{\vee\mathbf{x}}\big(X\wedge V^{(j)}(A,P)(\mathbf{x})\big)\to\prod_{j\geq 0}THH^{(j)}(A,P,X)_q$$

is an equivalence. In effect, we have shown the first statement in the proposition below, and the second statement follows since $THH^{(j)}(A,P,X)$ is $(j-1)$-reduced.

Proposition 5.3.2.1 *Let A be a connected* **S***-algebra and P an A-bimodule. Then the cyclic map*

$$THH(A\vee P,X)\xrightarrow{\sim}\textstyle\prod_{0\leq j}THH^{(j)}(A,P,X)$$

is a weak equivalence.

If P is $(k-1)$-connected and X is $(m-1)$-connected, then $THH^{(j)}(A,P,X)$ is $(jk+m-1)$-connected, and so

$$THH(A\vee P,X)\to THH(A,X)\times THH^{(1)}(A,P,X)$$

is $(2k+m)$-connected. ☺

This means that the space $THH^{(1)}(A,P,X)$ merits special attention as a first approximation to the difference between $THH(A\vee P,X)$ and $THH(A,X)$.

Also, since $THH^{(j)}(A,P,X)$ is $(j-1)$-reduced, the product is equivalent to the weak product, and we obtain

Corollary 5.3.2.2 *Both maps in*

$$\bigvee_{j\geq 0} \underline{T}^{(j)}(A,P,X) \to \underline{T}(A\vee P,X) \to \prod_{j\geq 0} \underline{T}^{(j)}(A,P,X)$$

are equivalences. ☺

See also Sect. 7.1.2 for the effect on fixed points.

5.3.3 Free Cyclic Objects

In this section we review the little we need at this stage about free cyclic objects. See Sect. 6.2.1 for a more thorough treatment. Recall that Λ is Connes' cyclic category. Let $\mathcal{C}$ be a category with finite coproducts. The forgetful functor from cyclic $\mathcal{C}$ objects to simplicial $\mathcal{C}$ objects has a left adjoint, the free cyclic functor j_* defined as follows.

If $\phi \in \Lambda$ we can write $\tau^{-s}\phi\tau^s = \psi\tau^r$ in a unique fashion with $\psi \in \Delta$. If X is a simplicial object, j_*X is given in dimension q by $\coprod_{C_{q+1}} X_q$, and with ϕ^* sending x in the $s \in C_{q+1}$ summand to ψ^*x in the $(r+s)$th summand.

Example 5.3.3.1 If X is a pointed set, then $j_*(X) \cong S^1_+\wedge X$. If A is a commutative ring, then $j_*(A) \cong HH(A)$.

Lemma 5.3.3.2 *The map*

$$j_*\underline{T}(A,P,X) \to \underline{T}^{(1)}(A,P,X)$$

adjoint to the inclusion $\underline{T}(A,P,X) \subseteq \underline{T}^{(1)}(A,P,X)$ is an equivalence. More precisely, if P is $(k-1)$-connected and X is $(m-1)$-connected, then

$$j_*THH(A,P,X) \to THH^{(1)}(A,P,X)$$

is a $2(k+m)$-connected cyclic map.

Proof Note that $V(A,P)(\mathbf{x}) \subseteq V^{(1)}(A,P)(\mathbf{x})$ defines the summand in which the P appears in the zeroth place. There are q other possibilities for placing P, and we may encode this by defining the map

$$C_{q+1_+}\wedge THH(A,P,X)_q \to THH^{(1)}(A,P,X)_q$$

taking $t^i \in C_{q+1}$, $\mathbf{x} \in \mathcal{I}^{q+1}$ and $f\colon S^{\vee\mathbf{x}} \to X\wedge V(A,P)(\mathbf{x})$ and sending it to

$$t^i\mathbf{x},\quad \begin{array}{ccccc} S^{\vee t^i\mathbf{x}} & & & & X\wedge V^{(1)}(A,P)(t^i\mathbf{x}) \\ \cong\uparrow & & & & \uparrow \\ S^{\vee\mathbf{x}} & \xrightarrow{f} & X\wedge V(A,P)(\mathbf{x}) & \xrightarrow{\subseteq} & X\wedge V^{(1)}(A,P)(\mathbf{x}). \end{array}$$

Varying q, this is the cyclic map

$$j_* THH(A, P, X) \to THH^{(1)}(A, P, X).$$

Let $V^{(1,i)}(A, P)(\mathbf{x}) \subset V^{(1)}(A, P)(\mathbf{x})$ be the summand with the P at the ith place. The map may be factored as

$$\begin{array}{ccc} \bigvee_{t^i \in C_{q+1}} \operatorname*{holim}_{\overrightarrow{\mathbf{x}\in\mathcal{I}^{q+1}}} \Omega^{\vee\mathbf{x}}\big(X \wedge V(A, P)(\mathbf{x})\big) & \xrightarrow{\cong} & \operatorname*{holim}_{\overrightarrow{\mathbf{x}\in\mathcal{I}^{q+1}}} \bigvee_{t^i \in C_{q+1}} \Omega^{\vee\mathbf{x}}\big(X \wedge V^{(1,i)}(A, P)(\mathbf{x})\big) \\ & & \downarrow \\ & & \operatorname*{holim}_{\overrightarrow{\mathbf{x}\in\mathcal{I}^{q+1}}} \Omega^{\vee\mathbf{x}}\big(X \wedge V^{(1)}(A, P)(\mathbf{x})\big) \end{array}$$

where the first map is given by the same formula with $V^{(1,i)}$ instead of $V^{(1)}$, and where the latter is induced by the inclusions

$$V^{(1,j)}(A, P)(\mathbf{x}) \subseteq \bigvee_{t^i \in C_{q+1}} V^{(1,i)}(A, P)(\mathbf{x}) \cong V^{(1)}(A, P)(\mathbf{x}).$$

We may exchange the wedges by products

$$\begin{array}{ccc} \operatorname*{holim}_{\overrightarrow{\mathbf{x}\in\mathcal{I}^{q+1}}} \bigvee_{t^i \in C_{q+1}} \Omega^{\vee\mathbf{x}}\big(X \wedge V^{(1,i)}(A, P)(\mathbf{x})\big) & \longrightarrow & \operatorname*{holim}_{\overrightarrow{\mathbf{x}\in\mathcal{I}^{q+1}}} \Omega^{\vee\mathbf{x}}\big(X \wedge V^{(1)}(A, P)(\mathbf{x})\big) \\ \downarrow & & \simeq\downarrow \\ \operatorname*{holim}_{\overrightarrow{\mathbf{x}\in\mathcal{I}^{q+1}}} \prod_{t^i \in C_{q+1}} \Omega^{\vee\mathbf{x}}\big(X \wedge V^{(1,i)}(A, P)(\mathbf{x})\big) & \xrightarrow{\cong} & \operatorname*{holim}_{\overrightarrow{\mathbf{x}\in\mathcal{I}^{q+1}}} \Omega^{\vee\mathbf{x}}\big(X \wedge \prod_{t^i \in C_{q+1}} V^{(1,i)}(A, P)(\mathbf{x})\big) \end{array}$$

and the left vertical arrow is $2(k+m)$-connected and the right vertical arrow is an equivalence by the Blakers–Massey Theorem A.8.2.1. □

When A is a discrete ring and P an A-bimodule (not necessarily discrete), these considerations carry over to the $\mathbf{T}(A \ltimes P)$ spectra. Recall the notation from Sect. 1.2.5 where we defined a category $\mathcal{D}_A P$ with objects $ob\underline{\mathbf{S}}\mathcal{P}_A$, and where $\mathcal{D}_A P(c, d) \cong \underline{\mathbf{S}}\mathcal{P}_A(c, d) \oplus \underline{\mathbf{S}}\mathcal{M}_A(c, d \otimes_A P)$ (where we have suppressed the index n running in the spectrum direction, and identified the morphism objects via Lemmas 1.2.5.1 and 1.2.5.2).

We saw in Theorem 1.2.5.4 that $\mathcal{D}_A^{(m)} P \subseteq S^{(m)}\mathcal{P}_{A\ltimes P}$ is a degreewise equivalence of categories, so $THH(\mathcal{D}_A P) \xrightarrow{\sim} \mathbf{T}(A \ltimes P)$. Substituting $X \mapsto \mathcal{D}_A P(c, d) \otimes_{\mathbf{Z}} \tilde{\mathbf{Z}}[X]$ with the stably equivalent $X \mapsto \underline{\mathbf{S}}\mathcal{P}_A(c, d) \otimes_{\mathbf{Z}} \tilde{\mathbf{Z}}[X] \vee \underline{\mathbf{S}}\mathcal{M}_A(c, d \otimes_A P) \otimes_{\mathbf{Z}} \tilde{\mathbf{Z}}[X]$ we may define $\mathbf{T}^{(j)}(A, P)$ as we did in Sect. 5.3.2, and we get that the cyclic map

$$\bigvee_{j \geq 0} \mathbf{T}^{(j)}(A, P, X) \to THH(\mathcal{D}_A P) \to \mathbf{T}(A \ltimes P)$$

is an equivalence. If P is $(k-1)$-connected then

$$\mathbf{T}(A,X) \vee \mathbf{T}^{(1)}(A,P,X) \to \mathbf{T}(A \ltimes P,X)$$

is $(2k-1)$-connected. Furthermore, as j_* preserves equivalences (see Lemma 6.2.1.3), we have that the composite

$$S^1_+ \wedge \mathbf{T}_0(A,P,X) \cong j_*\big(\mathbf{T}_0(A,P,X)\big) \to j_*\mathbf{T}(A,P,X) \to \mathbf{T}^{(1)}(A \ltimes P)$$

is an equivalence, and so the weak natural transformation $j_*\mathbf{T}(A,P,X) \to S^1_+ \wedge \mathbf{T}(A,P,X)$ is an equivalence.

5.3.4 Relations to the Trace $\tilde{\mathbf{K}}(A \ltimes P) \to \tilde{\mathbf{T}}(A \ltimes P)$

Let A be a discrete ring and let P be a simplicial A-bimodule. Our definition of the ("nerveless") trace $\tilde{K}(A \ltimes P) \to \widetilde{THH}(A \ltimes P)$ in Definition 5.1.1.6 is the map

$$\tilde{\mathbf{K}}(A \ltimes P) = \tilde{ob}\underline{\mathbf{S}}\mathcal{P}_{A \ltimes P} \xrightarrow{tr} \widetilde{THH}(\underline{\mathbf{S}}\mathcal{P}_{A \ltimes P}) = \tilde{\mathbf{T}}(A \ltimes P).$$

Recall that, by Sect. 1.3.4.1 $\tilde{\mathbf{K}}(A \ltimes P) \simeq \mathsf{C}_A(BP)$, so another definition of this map could be via

$$\mathsf{C}_A(BP) \longrightarrow \mathsf{C}_A\big(B^{cy}P\big) \cong \widetilde{B^{cy}}t\mathcal{D}_A P \longrightarrow \widetilde{THH}(\mathcal{D}_A P) \xrightarrow{\simeq} \widetilde{THH}(\underline{\mathbf{S}}\mathcal{P}_{A \ltimes P}).$$

The two are related by the diagram

$$\begin{array}{ccccc}
\mathsf{C}_A(BP) & \xrightarrow{\sim} & \tilde{N}t\underline{\mathbf{S}}\mathcal{P}_{A \ltimes P} & \xleftarrow{\sim} & \tilde{ob}\underline{\mathbf{S}}\mathcal{P}_{A \ltimes P} \\
\downarrow & & \downarrow & & \downarrow \\
\mathsf{C}_A\big(B^{cy}P\big) & \xrightarrow{\sim} & \widetilde{B^{cy}}t\underline{\mathbf{S}}\mathcal{P}_{A \ltimes P} & \longrightarrow & \widetilde{THH}(\underline{\mathbf{S}}\mathcal{P}_{A \ltimes P}).
\end{array} \tag{5.2}$$

Lemma 5.3.4.1 *If P is $(k-1)$-connected, and X a finite pointed simplicial set, then*

$$X \wedge \mathsf{C}_A(P) \to \mathsf{C}_A\big(P \otimes_{\mathbf{Z}} \tilde{\mathbf{Z}}[X]\big)$$

is $2k$-connected.

Proof It is enough to prove it for a finite set X. The smash moves past the wedges in the definition of C_A, and the map is simply $\bigvee_{c \in ob S^{(m)}_q \mathcal{P}_A}$ of the inclusion

$$\begin{array}{ccc}
X \wedge S^{(m)}_q \mathcal{M}(c, c \otimes_A P) & \xrightarrow{\cong} & \bigvee_{X-*} S^{(m)}_q \mathcal{M}(c, c \otimes_A P) \\
 & & \subseteq \big\downarrow \\
\tilde{\mathbf{Z}}[X] \otimes_{\mathbf{Z}} S^{(m)}_q \mathcal{M}(c, c \otimes_A P) & \xleftarrow{\cong} & \prod_{X-*} S^{(m)}_q \mathcal{M}(c, c \otimes_A P)
\end{array}$$

which is $2k$-connected by Blakers–Massey. The usual considerations about m-reducedness in the q direction(s), give the lemma. □

Lemma 5.3.4.2 *If P is $(k-1)$-connected, then the middle map in*

$$\mathsf{C}_A(BP) \longrightarrow \mathsf{C}_A\big(B^{cy}P\big) \longleftarrow S^1_+\wedge \mathsf{C}_A(P) \longrightarrow S^1\wedge \mathsf{C}_A(P)$$

is $2k$-connected, and in degrees less than $2k$ the induced composite on homotopy groups is an isomorphism.

Proof This follows from Lemma 5.3.4.1, and the commuting diagram

$$\begin{array}{ccccc}
\mathsf{C}_A(BP) & \longrightarrow & \mathsf{C}_A(B^{cy}P) & \longleftarrow & S^1_+\wedge \mathsf{C}_A(P) \\
 & \diagdown\!\!\diagdown & \downarrow & & \downarrow \\
 & & \mathsf{C}_A(BP) & \longleftarrow & S^1\wedge \mathsf{C}_A(P).
\end{array}$$

□

Consider the diagram (of bispectra)

$$\begin{array}{ccccc}
\tilde{\mathbf{T}}(A\ltimes P) & \xrightarrow{\sim} & \underline{\tilde{\mathbf{T}}}(A\ltimes P) & \xleftarrow{\sim} & \underline{\tilde{T}}(A\ltimes P) \\
\uparrow & & \uparrow & & \uparrow \\
j_*\mathbf{T}(A,P) & \xrightarrow{\sim} & j_*\underline{\mathbf{T}}(A,P) & \xleftarrow{\sim} & j_*\underline{T}(A,P) \\
\simeq\uparrow & & \uparrow & & \uparrow \\
S^1_+\wedge\mathbf{T}(A,P)_0 & \xrightarrow{\sim} & S^1_+\wedge\underline{\mathbf{T}}(A,P)_0 & \longleftarrow & S^1_+\wedge\underline{T}(A,P)_0 \\
\simeq\downarrow & & \downarrow & & \downarrow \\
S^1_+\wedge\mathbf{T}(A,P) & \xrightarrow{\sim} & S^1_+\wedge\underline{\mathbf{T}}(A,P) & \xleftarrow{\sim} & S^1_+\wedge\underline{T}(A,P) \\
\downarrow & & \downarrow & & \downarrow \\
S^1\wedge\mathbf{T}(A,P) & \xrightarrow{\sim} & S^1\wedge\underline{\mathbf{T}}(A,P) & \xleftarrow{\sim} & S^1\wedge\underline{T}(A,P).
\end{array}$$

The upwards pointing arrows are induced by the inclusion $V(A,P)(\mathbf{x})\subseteq V(A\ltimes P)(\mathbf{x})$ (likewise with $V(\underline{\mathbf{S}}\mathcal{P}_A,P)$ instead of $V(A,P)$). The rightmost upper vertical map is $2k$-connected by the considerations in Sect. 5.3.2, and so all up-going arrows are $2k$-connected. Note that the middle layer of 0-simplices could have been skipped if we performed geometric realization all over the place, using the equivalence $|j_*X|\simeq|S^1_+\wedge X|$ of Lemma 6.2.1.3.

Proposition 5.3.4.3 *If A is a discrete ring and P an $(k-1)$-connected A-bimodule, then the undecorated leftward pointing arrows in*

$$\tilde{\mathbf{K}}(A\ltimes P) \longrightarrow \tilde{\mathbf{T}}(A\ltimes P) \longleftarrow S^1_+\wedge\mathbf{T}(A,P)_0 \longrightarrow S^1\wedge\mathbf{T}(A,P)$$

and

$$\tilde{\mathbf{K}}(A \ltimes P) \longrightarrow \underline{\tilde{\mathbf{T}}}(A \ltimes P) \longleftarrow S^1_+ \wedge \underline{\mathbf{T}}(A, P)_0 \longrightarrow S^1 \wedge \underline{\mathbf{T}}(A, P) \overset{\sim}{\longleftarrow} S^1 \wedge \underline{T}(A, P)$$

are $2k$-connected, and in degrees less than $2k$, the induced composites on homotopy groups are isomorphisms.

Proof The second statement follows from the first. As $\mathsf{C}_A(P) \to (D_1\mathsf{C}_A)(P) \simeq \mathbf{T}(A, P)_0$ is $2k$-connected (Theorem 1.3.5.2), Lemma 5.3.4.2 gives that all the induced composites on homotopy groups in degree less than $2k$ from top left to bottom right in

$$\begin{array}{ccccccc}
\mathsf{C}_A(BP) & \longrightarrow & \mathsf{C}_A(B^{cy}P) & \longleftarrow & S^1_+ \wedge \mathsf{C}_A(P) & \longrightarrow & S^1 \wedge \mathsf{C}_A(P) \\
 & & \downarrow & & \downarrow & & \downarrow \\
 & & \tilde{\mathbf{T}}(A \ltimes P) & \longleftarrow & S^1_+ \wedge \mathbf{T}(A, P)_0 & \longrightarrow & S^1 \wedge \mathbf{T}(A, P)_0 \\
 & & & & \simeq\downarrow & & \simeq\downarrow \\
 & & & & S^1_+ \wedge \mathbf{T}(A, P) & \longrightarrow & S^1 \wedge \mathbf{T}(A, P)
\end{array}$$

are isomorphisms. □

5.3.5 Stable K-Theory and THH for S-Algebras

The functor $S \mapsto A^n_S$ from Sect. 3.3.1.8, can clearly be applied to A-bimodules as well, and $S \mapsto P^n_S$ will be a cube of $S \mapsto A^n_S$-bimodules, which ultimately gives us a cube $S \mapsto A^n_S \vee P^n_S$ of $\mathbf{S}$-algebras. If P is an A-bimodule, so is $X \mapsto \Sigma^m P(X) = P(S^m \wedge X)$. We defined

$$K^S(A, P) = \operatorname*{holim}_{\overrightarrow{k}} \Omega^k \operatorname{hofib}\{K(A \vee \Sigma^{k-1} P) \to K(A)\}.$$

The trace map induces a map to

$$\operatorname*{holim}_{\overrightarrow{k}} \Omega^k \operatorname{hofib}\{THH(A \vee \Sigma^{k-1} P) \to THH(A)\}$$

and we may compose with the weak map to

$$\operatorname*{holim}_{\overrightarrow{k}} \Omega^k (S^1 \wedge THH(A, \Sigma^{k-1} P))$$

given by the discussion of the previous section. From the previous section we know that this is an equivalence when A is (the Eilenberg–Mac Lane spectrum associated with) a discrete ring and P a simplicial A-bimodule.

Theorem 5.3.5.1 *Let A be an* $\mathbf{S}$*-algebra and P an A-bimodule. Then the trace induces an equivalence $K^S(A,P)\simeq THH(A,P)$.*

Proof If A is discrete and P a simplicial A-bimodule this has already been covered. If A is a simplicial ring P a simplicial A-bimodule this follows by considering each degree at a time, using that K-theory of simplicial radical extensions may be calculated degreewise, Proposition 3.1.4.2. In the general case we reduce to the simplicial case as follows. There is a stable equivalence $A^n_S\vee P^n_S\to(A\vee P)^n_S$, consisting of repeated applications of the $2k$-connected map $\tilde{\mathbf{Z}}[A(S^k)]\vee\tilde{\mathbf{Z}}[P(S^k)]\to\tilde{\mathbf{Z}}[A(S^k)]\oplus\tilde{\mathbf{Z}}[P(S^k)]\cong\tilde{\mathbf{Z}}[A(S^k)\vee P(S^k)]$. The non-initial nodes in these cubes are all equivalent to a simplicial ring case, and is hence taken care of by Theorem 5.3.1.1 (or rather Proposition 5.3.4.3 since the identification of the equivalence in Theorem 5.3.1.1 with the trace map is crucial in order to have functoriality for $\mathbf{S}$-algebras), and all we need to know is that

$$K(A\vee P)\to\underset{\overleftarrow{S\neq\emptyset}}{\operatorname{holim}}\,K\bigl(A^n_S\vee P^n_S\bigr)$$

is $(n+1)$-connected, and that

$$THH(A\vee P)\to\underset{\overleftarrow{S\neq\emptyset}}{\operatorname{holim}}\,THH\bigl(A^n_S\vee P^n_S\bigr)$$

and

$$THH(A,P)\to\underset{\overleftarrow{S\neq\emptyset}}{\operatorname{holim}}\,THH\bigl(A^n_S,P^n_S\bigr)$$

are n-connected. These follow from Theorems 3.3.2.2 and 4.2.4.3. □

5.4 The Normal Trace

In this section we give a more direct construction of the trace than the one given in Definition 5.2.3.2 making the roundabout with localizations in the weak trace of Definition 5.2.4.2 redundant. It depends on a refined definition of the nerve, called the *homotopy nerve* Definition 5.4.2.9 $\mathrm{h\tilde{o}N}(\mathcal{C},w)$ of a pair $(\mathcal{C},w\colon\mathcal{W}\to R\mathcal{C})$.

There are several advantages to this construction apart from it being simpler. Most importantly, it puts weak equivalences on the same footing as isomorphisms in the following sense. In Lemma 5.1.0.4 we saw that if $\mathcal{C}$ is a category then there is an equivalence of categories between $\mathcal{C}$ and the category $N_n(\mathcal{C},i)$ of n-simplexes along the isomorphisms. In Lemma 5.4.2.10 below we will see that under certain conditions on a $\Gamma\mathcal{S}_*$-category $\mathcal{C}$ with a choice of weak equivalences there is a stable equivalence of $\Gamma\mathcal{S}_*$-categories (Definition 2.2.4.1) between $\mathcal{C}$ and $\mathrm{h\tilde{o}N}_n(\mathcal{C},w)$. Hence the schism between the discrete and the $\Gamma\mathcal{S}_*$-enriched case is all but eradicated. Another advantage is that it is straightforward to see that the improved trace is multiplicative (it preserves operad actions in general, but in any case you have to replace Γ-spaces with a more suitable monoidal model for spectra such as symmetric spectra). The reason this construction is not our official trace map is that it only occurred to us very late, and only after Andrew Blumberg dropped the hint that he and Mike Mandell were considering a "Moore type nerve", and we set out to see whether Moore paths could simplify our life as well (actually, it was used to simplify an argument that the cyclotomic trace preserves operad actions). After the appearance of Blumberg and Mandell's very recommendable preprint [24] it is clear that they have precedence for the construction (which they trace back to McClure and Smith) we offer below and we are indebted to them. Indeed, several of the constructions in this book could have benefited from improvements offered in [24]. A further reason for not rewriting the presentation is that with results as the one suggested in the preprint [21] of Blumberg, Gepner and Tabuada, concrete constructions may be needed only for technical considerations.

We should point out that the homotopy nerve construction below is different, but have some similarities with the simplicial nerve construction of [186, 1.1.5].

The Normal Trace We postpone the construction of the homotopy nerve a bit to give the consequences. Recall the notion of a *pair* $(\mathcal{C}, w\colon \mathcal{W} \to R\mathcal{C})$ Definition 5.2.1.2; $\mathcal{C}$ is a $\Gamma\mathcal{S}_*$-category, $\mathcal{W}$ a $\mathcal{S}$-category and $w\colon \mathcal{W} \to R\mathcal{C}$ a $\mathcal{S}$-functor which is the identity on objects.

As always, the functor w should be viewed as an inclusion of the weak equivalences, which should make the following situation seem not too unnatural.

Definition 5.4.0.1 A *normal* pair is a pair $(\mathcal{C}, w\colon \mathcal{W} \to R\mathcal{C})$ such that

1. every isomorphism (in any degree) in $R\mathcal{C}$ is the image under w of an isomorphism in $\mathcal{W}$ and
2. if $x \to y$ is an isomorphism in the underlying category of $\mathcal{W}$ and z an object in $\mathcal{W}$, then the induced maps $\mathcal{C}(y, z) \to \mathcal{C}(x, z)$ and $\mathcal{C}(z, x) \to \mathcal{C}(z, y)$ are stable equivalences.

We will eventually use that w is surjective on objects to ensure that we may identify the zero simplices of the nerve $\mathrm{h\tilde{o}N}(\mathcal{C}, w)$ Definition 5.4.2.9 of a normal pair $(\mathcal{C}, w)$ with $\mathcal{C}$. The reason this is extended to a surjectivity condition on isomorphisms in general is to render the notion invariant under the K-theory construction in Lemma 5.4.0.3 below.

We note that the category of finitely generated free modules over an $\mathbf{S}$-algebra gives rise to an example:

Example 5.4.0.2 Let A be an $\mathbf{S}$-algebra, $\mathcal{C} = \mathcal{F}_A$ the $\Gamma\mathcal{S}_*$-category of finitely generated free A modules defined in Definition 3.2.4.1 with $w_A\colon \mathcal{W} = w\mathcal{F}_A \to R\mathcal{F}_A$ the uniform choice of weak equivalences of Sect. 2.3.3 (pullback along $i\pi_0\mathcal{F}_A \to \pi_0\mathcal{F}_A$), giving exactly the A-module maps $\alpha\colon k_+\wedge A \overset{\sim}{\to} k_+\wedge A$ that are stable equivalence. Then $(\mathcal{F}_A, w_A)$ is a normal pair (remember that $\mathcal{F}_A(k_+ \wedge A, k_+ \wedge A) \cong \prod_k \bigvee_k A \simeq \prod_k \prod_k A \cong \mathcal{F}_A(k_+ \wedge A, \prod_k A)$).

Recall the K-theory construction $\bar{H}$ of Definition 2.3.1.1 for symmetric monoidal $\Gamma\mathcal{S}_*$-categories.

Lemma 5.4.0.3 *If $\mathcal{C}$ is a symmetric monoidal $\Gamma\mathcal{S}_*$-category and $w\colon \mathcal{W} \to R\mathcal{C}$ is a symmetric monoidal $\mathcal{S}$-functor such that $(\mathcal{C}, w)$ is a normal pair, then for each $k_+ \in ob\Gamma^o$ the pair $(\bar{H}\mathcal{C}(k_+), \bar{H}w(k_+))$ is normal.*

Proof Checking the Definition 2.3.1.1 it is immediate that if w is surjective when restricting to the isomorphisms, so is $\bar{H}w$. Since the property of being normal is preserved under products and the $\bar{H}$-construction has special values, the condition on the morphism objects follows. □

In Definition 5.4.2.9 below we define the homotopy nerve as a bisimplicial $\Gamma\mathcal{S}_*$-category $\text{h}\tilde{\text{o}}\text{N}(\mathcal{C}, w)$. The topological Hochschild homology of a (bi)simplicial $\Gamma\mathcal{S}_*$-category is the diagonal of the topological Hochschild homology applied in every degree.

The following important result tells us that the homotopy nerve does its job for normal pairs.

Proposition 5.4.0.4 *Let $(\mathcal{C}, w)$ be a normal pair. Then the inclusion of zero simplices $\mathcal{C} = \text{h}\tilde{\text{o}}\text{N}_0(\mathcal{C}, w) \to \text{h}\tilde{\text{o}}\text{N}(\mathcal{C}, w)$ defines a pointwise equivalence*

$$THH\big(\text{h}\tilde{\text{o}}\text{N}(\mathcal{C}, w)\big) \leftarrow THH(\mathcal{C}).$$

Proof By Lemma 5.4.2.10 below, for each n all face maps $\text{h}\tilde{\text{o}}\text{N}_n(\mathcal{C}, w) \to \text{h}\tilde{\text{o}}\text{N}_{0,0}(\mathcal{C}, w)$ are stable equivalences of $\Gamma\mathcal{S}_*$-categories. Hence, by Lemma 4.3.5.3 they induce pointwise equivalences $THH(\text{h}\tilde{\text{o}}\text{N}_n(\mathcal{C}, w)) \xrightarrow{\sim} THH(\text{h}\tilde{\text{o}}\text{N}_{0,0}(\mathcal{C}, w))$, inverse to the inclusion by degeneracies. However, by Lemma 5.4.2.10 again, we have a canonical stable equivalence of $\Gamma\mathcal{S}_*$-categories $\mathcal{C} \to \text{h}\tilde{\text{o}}\text{N}_{0,0}(\mathcal{C}, w)$, and we are done. □

Definition 5.4.0.5 If $\mathcal{C}$ is a symmetric monoidal $\Gamma\mathcal{S}_*$-category and $w\colon \mathcal{W} \to R\mathcal{C}$ is a symmetric monoidal $\mathcal{S}$-functor such that $(\mathcal{C}, w)$ is a normal pair, then the *normal* trace is the weak map

$$\begin{aligned}\Sigma^\infty B\bar{H}\mathcal{W} = \Sigma^\infty ob\, \text{h}\tilde{\text{o}}\text{N}(\bar{H}\mathcal{C}, \bar{H}w) &\to THH\big(\text{h}\tilde{\text{o}}\text{N}(\bar{H}\mathcal{C}, \bar{H}w)\big)\\ &\xleftarrow{\sim} THH(\bar{H}\mathcal{C}) \xleftarrow{\sim} \Sigma^\infty THH(\mathcal{C}),\end{aligned}$$

where the first map is the Dennis trace map Sect. 4.3.2 the second map is induced by the inclusion of the zero skeleton in the homotopy nerve (and is a pointwise equivalence by Lemma 5.4.0.3 and Lemma 5.4.0.4) the and the third is the pointwise equivalence of Proposition 5.1.0.1.

If A is an **S**-algebra we compose with Morita invariance Corollary 4.3.5.14 to get the normal trace

$$\Sigma^\infty B\bar{H}w_A\mathcal{F}_A \to THH\big(\text{hõN}(\bar{H}\mathcal{F}_A, \bar{H}w_A)\big) \overset{\sim}{\leftarrow} \Sigma^\infty THH(\mathcal{F}_A) \overset{\sim}{\leftarrow} \Sigma^\infty THH(A)$$

from a model for the (free) algebraic K-theory of A to topological Hochschild homology of A.

Remark 5.4.0.6 When we get as far as defining topological cyclic homology, TC, in Chap. 6, it will become apparent that Proposition 5.4.0.4 holds equally well for TC to give a *normal cyclotomic trace*

$$\Sigma^\infty B\bar{H}\mathcal{W} \to TC\big(\text{hõN}(\bar{H}\mathcal{C}, \bar{H}w)\big) \overset{\sim}{\leftarrow} TC(\bar{H}\mathcal{C}) \overset{\sim}{\leftarrow} \Sigma^\infty TC(\mathcal{C})$$

from a model of algebraic K-theory to TC, and for **S**-algebras

$$\Sigma^\infty B\bar{H}w_A\mathcal{F}_A \to TC\big(\text{hõN}(\bar{H}\mathcal{F}_A, \bar{H}w_A)\big) \overset{\sim}{\leftarrow} TC(\bar{H}\mathcal{F}_A) \overset{\sim}{\leftarrow} \Sigma^\infty TC(\mathcal{F}_A)$$
$$\overset{\sim}{\leftarrow} \Sigma^\infty TC(A).$$

5.4.1 Moore Singular Simplices

5.4.1.1 Arrow Categories

Recall that the *twisted arrow category* $\mathcal{A}_I$ of a category I is the category whose objects are the arrows (morphisms) of I and where a morphism from $\phi\colon c_1 \to c_0 \in ob\mathcal{A}_I$ to $\psi\colon d_1 \to d_0 \in ob\mathcal{A}_I$ is a commutative square

$$\begin{array}{ccc} c_0 & \xleftarrow{\ \phi\ } & c_1 \\ \uparrow{\scriptstyle\delta} & & \downarrow{\scriptstyle\gamma} \\ d_0 & \xleftarrow[\ \psi\ ]{} & d_0. \end{array}$$

We are going to use the opposite so frequently that we allow ourselves to introduce special notation: let $\mathcal{A}^I = (\mathcal{A}_I)^o$. The obvious definition of $\mathcal{A}^F : \mathcal{A}^I \to \mathcal{A}^J$ for a functor $F\colon I \to J$ (apply F to the diagrams in $\mathcal{A}^I$, so that for instance the object $f\colon i \to i'$ is sent to $F(f)\colon F(i) \to F(i')$) displays $I \mapsto \mathcal{A}^I$ as an endofunctor on the category of categories. Notice that there is a functor

$$\text{forget}^I : \mathcal{A}^I \to I \times I^o$$

(natural in I) sending $c \leftarrow d$ to (c, d), and a canonical isomorphism of categories

$$D^I : \mathcal{A}^{(I^o)} \cong \mathcal{A}^I$$

given by turning all arrows around.

We need to iterate this construction, and define $\mathcal{A}^I_k = \mathcal{A}^{\mathcal{A}^I_{k-1}}$ for $k > 0$, with $\mathcal{A}^I_0 = I$.

If $0 \le k < l \le 3$ with k even and l odd, consider the functors $p_{kl}\colon \mathcal{A}^I_2 \to \mathcal{A}^I$ sending the object $(i_0 \leftarrow i_1 \leftarrow i_2 \leftarrow i_3)$ in $\mathcal{A}^I_2$ to $p_{kl}(i_0 \leftarrow i_1 \leftarrow i_2 \leftarrow i_3) = (i_k \leftarrow i_l)$. Similarly, with $k = 1$ and $l = 2$ the same formula gives a functor $p_{12}\colon \mathcal{A}^I_2 \to (\mathcal{A}^I)^o = \mathcal{A}_I$. Also, observe that the functor $(p_{01}, p_{23})\colon \mathcal{A}^I_2 \to \mathcal{A}^I \times \mathcal{A}^I$ equals the composite functor

$$\mathcal{A}^I_2 = \mathcal{A}^{\mathcal{A}^I} \xrightarrow{\mathcal{A}^{\mathrm{forget}^I}} \mathcal{A}^{I\times I^o} \cong \mathcal{A}^I \times \mathcal{A}^{I^o} \cong \mathcal{A}^I \times \mathcal{A}^I,$$

where the isomorphisms are the canonical isomorphisms $\mathcal{A}^{I\times J} \cong \mathcal{A}^I \times \mathcal{A}^J$ sending $(i_0, j_0) \leftarrow (i_1, j_1)$ to $((i_0 \leftarrow i_1), (j_0 \leftarrow j_1))$ and $D^I : \mathcal{A}^{(I^o)} \cong \mathcal{A}^I$.

We use the shorthand $\mathcal{A}^n_k = \mathcal{A}^{[n]}_k$. Explicitly, an object in $\mathcal{A}^n_k$ is a sequence $i = \{0 \le i_0 \le i_1 \le \cdots i_{2^k-1} \le n\}$ and there is a unique map from i to j provided $i_k \ge j_k$ when k is even and $i_k \le j_k$ when k is odd.

Example 5.4.1.1 For the examples that are to follow, the following construction of a functor $\mathcal{O}^n\colon \mathcal{A}^n \to \mathit{Cat}$ is illustrating. For integers $0 \le i \le j \le n$ let $\mathcal{O}^n(i \le j)$ be the ordered set of all integers t with $i \le t \le j$ (so that $\mathcal{O}^n(0 \le n) = [n]$), considered as a category. That you have a morphism $(i \le j) \to (i' \le j')$ in $\mathcal{A}^n$ means exactly that $i' \le i \le j \le j'$, and we let $\mathcal{O}^n((i \le j) \to (i' \le j'))$ be the inclusion $\mathcal{O}^n(i \le j) \subseteq \mathcal{O}^n(i' \le j')$. If $\phi\colon [m] \to [n]$, then the function $\mathcal{O}^m(i \le j) \to \mathcal{O}^n(\phi(i) \le \phi(j)) = \mathcal{O}^n\mathcal{A}^\phi(i \le j)$ with $t \mapsto \phi(t)$ gives rise to a natural transformation $\mathcal{O}^\phi\colon \mathcal{O}^m \to \mathcal{O}^n\mathcal{A}^\phi$. Note that $\mathcal{O}^{id} = id$, and if $\phi'\colon [m'] \to [m] \in \Delta$ then $\mathcal{O}^{\phi'\phi} = \mathcal{O}^{\phi'}\mathcal{O}^\phi$.

5.4.1.2 Monoidal Structure in $[\mathcal{A}^I, T]$

Let $T = (T, \otimes, e)$ be a monoidal category containing all finite colimits, and let I be a small category. Then the category $[\mathcal{A}^I, T]$ of functors $\mathcal{A}^I \to T$ is monoidal: the unit is the constant functor with value e, and if $X, Y \in [\mathcal{A}^I, T]$ we define $X \otimes Y \in [\mathcal{A}^I, T]$ by declaring that

$$(X \otimes Y)(i \leftarrow l) = \varinjlim_{i\leftarrow j\leftarrow k\leftarrow l} X(k \leftarrow l) \otimes Y(i \leftarrow j)$$

where the colimit runs over the subcategory $p_{03}^{-1}(i \leftarrow l)$ of $\mathcal{A}^I_2$. The unitality and associativity follows from the universal property of the colimit and the monoidality of T.

Notice that since $[\mathcal{A}_I, T] \cong [\mathcal{A}^I, T^o]^o$ this is equivalent to the statement that if $T = (T, \otimes, e)$ is a monoidal category containing all finite limits, and I is a small category, then the functor category $[\mathcal{A}_I, T]$ is monoidal with product

$$(X \otimes Y)(i \leftarrow l) = \varprojlim_{i \leftarrow j \leftarrow k \leftarrow l} X(k \leftarrow l) \otimes Y(i \leftarrow j).$$

We need both variants, but in the following we spell out the details for just one of the cases.

If $f\colon I \to J$ is a functor of small categories, the canonical map $(X\mathcal{A}^f) \otimes (Y\mathcal{A}^f) \to (X \otimes Y)\mathcal{A}^f$ (induced by the subfunctor $p_{03}^{-1}(i \leftarrow l) \to p_{03}^{-1}(f(i) \leftarrow f(l))$ of $\mathcal{A}_2^f : \mathcal{A}_2^I \to \mathcal{A}_2^J$) displays $(\mathcal{A}^f)^* : [\mathcal{A}^J, T] \to [\mathcal{A}^I, T]$ as lax monoidal.

If P is a monoidal category, we consider the strong monoidal functors $B^n : P \to [\mathcal{A}^n, T]$ that come equipped with monoidal natural transformations $B^\phi\colon B^m \to (\mathcal{A}^\phi)^* B^n$ for every $\phi\colon [m] \to [n] \in \Delta$, such that if $\phi'\colon [m'] \to [m]$ then $B^{\phi'\phi} = B^{\phi'} B^\phi$ and $B^{id} = id$

$$\begin{array}{ccc} P & \xrightarrow{B^n} & [\mathcal{A}^n, T] \\ & \overset{\Rightarrow}{\scriptstyle B^\phi} & \\ & {\scriptstyle B^m}\searrow & \downarrow {\scriptstyle (\mathcal{A}^\phi)^*} \\ & & [\mathcal{A}^m, T] \end{array}$$

(this gives that B is a right lax natural transformation from the functor $[n] \mapsto P \times \mathcal{A}^n$ to the constant functor T). We write B_r^n for $B^n(r)$, and the structure map $B_r^n \otimes B_s^n \to B_{r+s}^n$ is denoted $B_{r,s}^n$.

For r, s in $\mathbf{P}$, the natural transformation $B_{r,s}^n : B_r^n \otimes B_s^n \to B_{r+s}^n$ consists of natural transformations $B_{r,s}^n : B_r^n p_{23} \times B_s^n p_{01} \to B_{r+s}^n p_{03}$ of functors $\mathcal{A}_2^n \to T$. Spelling out, this implies for instance that we have morphisms $B_{r,s}^n(i_0 \leq i_1 \leq i_2 \leq i_3) : B_r^n(i_2 \leq i_3) \times B_s^n(i_0 \leq i_1) \to B_{r+s}^n(i_0 \leq i_3)$, natural in $(i_0 \leq i_1 \leq i_2 \leq i_3) \in \mathcal{A}_2^n$. The monoidality of B^n implies that if $r, s, t \in \mathbf{P}$, then the diagram

$$\begin{array}{ccc} (B_r^n(i_4 \leq i_5) \times B_s^n(i_2 \leq i_3)) \times B_t^n(i_0 \leq i_1) & \xrightarrow{B_{r,s}^n(i_2 \leq i_3 \leq i_4 \leq i_5) \times 1} & B_{r+s}^n(i_2 \leq i_5) \times B_t^n(i_0 \leq i_1) \\ & & \downarrow {\scriptstyle B_{r+s,t}^n(i_0 \leq i_1 \leq i_2 \leq i_5)} \\ \downarrow {\scriptstyle \text{assoc}} & & B_{r+s+t}^n(i_0 \leq i_5) \\ & & \uparrow {\scriptstyle B_{r,s+t}^n(i_0 \leq i_3 \leq i_4 \leq i_5)} \\ B_r^n(i_4 \leq i_5) \times (B_s^n(i_2 \leq i_3) \times B_t^n(i_0 \leq i_1)) & \xrightarrow{1 \times B_{s,t}^n(i_0 \leq i_1 \leq i_2 \leq i_3)} & B_r^n(i_4 \leq i_5) \times B_{s+t}^n(i_0 \leq i_3) \end{array}$$

commutes. That B^n is strong monoidal implies that for each pair of natural numbers $i_0 \le i_3$ the induced morphism

$$\begin{aligned} B^n_{r,s}(i_0 \le i_3)\colon & \big(B^n_r \otimes B^n_s\big)(i_0 \le i_3) \\ &= \varinjlim_{i_0 \le i_1 \le i_2 \le i_3} B^n_r(i_2 \le i_3) \times B^n_s(i_0 \le i_1) \to B^n_{r+s}(i_0 \le i_3) \end{aligned}$$

to be an isomorphism, where the colimit runs over the subcategory $p_{03}^{-1}(i_0 \le i_3)$ of $\mathcal{A}^n_2$.

Example 5.4.1.2 Our prime example is directly related to McClure and Smith's prismatic subdivision which they used in their proof of Deligne's Hochschild cohomology conjecture [208], but which has appeared in diverse situations. Let $\mathbf{P}$ be the monoid of non-negative real numbers under addition, and consider the standard topological n-simplex $\Delta^n = \{(x_0, \dots, x_n) \in \mathbf{R}^{n+1} \mid \sum_{k=0}^n x_k = 1,\ \text{all } x_k \ge 0\} \subseteq \mathbf{R}^{n+1}$. For r a non-negative real number, consider the scaled n-simplex $\Delta^n_r = \{ru \mid u \in \Delta^n\}$. If $r \in \mathbf{P}$, consider the subspace of Δ^n_r given by

$$\Delta^n_r(i \le j) = \left\{ (x_0, \dots x_n) \in \Delta^n_r \;\middle|\; \sum_{k=i}^{j} x_k = r \right\}.$$

That is, claiming that $(x_0, \dots, x_n) \in \Delta^n_r$ is in $\Delta^n_r(i \le j)$ means that $x_k \ne 0$ only if $i \le k \le j$. This is clearly functorial in $i \le j \in \mathcal{A}^n$: if $i' \le i \le j \le j'$, then $\Delta^n_r(i \le j) \subseteq \Delta^n_r(i' \le j')$.

Note that $\Delta^n_r(i \le j)$ does not depend on n (upon identifying $\mathbf{R}^n$ with its image in $\mathbf{R}^\infty$), and we may occasionally drop the n from the notation.

Consider the embeddings $\Delta^n_{r,s}(i_0 \le i_1 \le i_2 \le i_3)\colon \Delta^n_r(i_2 \le i_3) \times \Delta^n_s(i_0 \le i_1) \to \Delta^n(i_0 \le i_3)$ induced by addition in $\mathbf{R}^{n+1}$, sending

$$\big((0, \dots, 0, y_{i_2}, \dots, y_{i_3}, 0, \dots, 0), (0, \dots, 0, x_{i_0}, \dots, x_{i_1}, 0, \dots, 0)\big)$$

to

$$(0, \dots, 0, x_{i_0}, \dots, x_{i_1}, 0, \dots, 0) + (0, \dots, 0, y_{i_2}, \dots, y_{i_3}, 0, \dots, 0).$$

In particular, the extreme case $\Delta^n_{r,s}(0 \le i = i \le n)$ sends the point

$$\big((0, \dots, 0, y_i, \dots, y_n), (x_0, \dots, x_i, 0, \dots, 0)\big)$$

to $(x_0, \dots, x_{i-1}, x_i + y_i, y_{i+1}, \dots, y_n)$. The associativity criterion

$$\Delta^n_{r,s+t}\big(1 \times \Delta^n_{s,t}\big) = \Delta^n_{r+s,t}\big(\Delta^n_{r,s} \times 1\big)$$

follows directly from the formula.

Also, we recognize this as part of the (weighted) second edgewise subdivision and get that these embeddings glue together to give us coherent homeomorphisms

$$\begin{aligned}\Delta^n_{r,s}(i_0 \le i_3)\colon & \big(\Delta^n_r \times \Delta^n_s\big)(i_0 \le i_3) \\ &= \varinjlim_{i_0 \le i_1 \le i_2 \le i_3} \Delta^n_r(i_2 \le i_3) \times \Delta^n_s(i_0 \le i_1) \cong \Delta^n_{r+s}(i_0 \le i_3),\end{aligned}$$

endowing Δ^n with the structure of a strong monoidal functor.

5.4.1.3 Moore Singular Simplices

In the following we are working in the category CG of compactly generated spaces (see e.g., [139, 2.4], following the convention that compactly generated spaces are weak Hausdorff Kelley spaces), and all limits and function spaces are taken in this category. Recall that geometric realization takes values in compactly generated spaces, and it is with respect to this category that geometric realization preserves finite limits [139, 3.2.4].

We will now introduce a generalization of the Moore path space. For $r \in \mathbf{P}$, let $H_r\colon \mathbf{R}^n \times I \to \mathbf{R}^n$ be the function given by sending (x, t) to

$$\begin{aligned}(1-t)x + t\,\big(&\max\big(0, \min(r, x_1, \dots, x_n)\big), \max\big(0, \min(r, x_2, \dots, x_n)\big), \dots, \\ &\max\big(0, \min(r, x_n)\big)\big).\end{aligned}$$

This defines a retraction p^n_r of $\mathbf{R}^n$ onto the subspace Simp^n_r of points $(x_1, \dots, x_n)$ with $0 \le x_1 \le \dots \le x_n \le r$. Varying n we get a cosimplicial space $[n] \mapsto \mathrm{Simp}^n_r$ with structure maps given by repeating or deleting coordinates. The map $\Delta^n_r \to \mathrm{Simp}^n_r$ given by sending $(x_0, \dots, x_n)$ to $(x_0, x_0 + x_1, \dots, x_0 + \dots + x_{n-1})$ gives a natural homeomorphism of cosimplicial spaces.

For $r \in \mathbf{P}$ and $0 \le i \le j \le n$, let $\mathrm{Simp}^n_r(i \le j)$ be the subspace of $\mathbf{R}^n$ of points $(x_1, \dots, x_n)$ such that $0 \le x_1 \le \dots \le x_n \le r$ and such that $x_k = 0$ for $k \le i$ and $x_k = r$ for $k > j$. Recall the scaled simplices Δ^n_r of Example 5.4.1.2. The homeomorphism $\Delta^n_r(0 \le n) = \Delta^n_r \to \mathrm{Simp}^n_r$ restricts to a natural homeomorphism $h^{i\le j}_r\colon \Delta^n_i(i \le j) \cong \mathrm{Simp}^n_r(i \le j)$.

Consider the embedding $e^{i\le j}_r\colon \mathrm{Simp}^n_r(i \le j) \to \mathbf{R}^{j-i}$ sending $(x_1 \le \dots \le x_n) \mapsto (x_{i+1}, \dots, x_j)$ with section $p^{i\le j}_r(x_{i+1}, \dots, x_j) = (0 = \dots = 0 \le x_{i+1} \le \dots \le x_j \le r = \dots = r)$.

Definition 5.4.1.3 If Y is a compactly generated space, and $0 \le i \le j \le n$ we let the space of *Moore singular $(i \le j)$-simplices*, $P^{i\le j}_n Y$, be the set $\coprod_{r\in\mathbf{P}} Y^{\Delta^n_r(i\le j)}$ topologized so that the injective function $P^{i\le j}_n Y \to P \times Y^{\mathbf{R}^{j-i}}$ sending $f\colon \Delta^n_r(i \le j) \to Y$ in the rth summand to $(r, f(h^{i\le j}_r)^{-1} p^{i\le j}_r)$ is an embedding where $Y^{\mathbf{R}^{j-i}}$ is given the compact-open topology.

If in addition $0 \le k \le l \le n$ we let the space $P_n^{\substack{i\le j\\ k\le l}} Y$ as the set $\coprod_{(r,s)\in\mathbf{R}^r} Y^{\Delta^n_r(i\le j)\times\Delta^n_s(k\le l)}$ endowed with the topology given by declaring that the injection into $\mathbf{R}^2 \times Y^{\mathbf{R}^{j-i}\times\mathbf{R}^{l-k}}$ given by the above retractions is an embedding.

To take care of the extreme cases, notice that the maps $P_n^{i=i} Y \to \mathbf{P} \times Y$ sending (r, f) to $(r, f(0))$ and $P_n^{i\le j}(*) \to \mathbf{P}$ sending (r, f) to r are homeomorphisms and that $P_n^{i\le j}\emptyset = \emptyset$.

If $0 \le i \le j \le k \le l \le n$ then the inclusions $\Delta^n_r(j \le k) \subseteq \Delta^n_r(i \le l)$ for varying r induce a projection $P_n^{i\le l} Y \to P_n^{j\le k} Y$ which is continuous since it extends via the embeddings defining the topology to the continuous function $\mathbf{R} \times Y^{\mathbf{R}^{i\le l}} \to \mathbf{R} \times Y^{\mathbf{R}^{j\le k}}$ sending (r, f) to $(r, f\, inc_r)$, where $inc_r\colon \mathbf{R}^{k-j} \to \mathbf{R}^{l-i}$ is the embedding sending $(x_{j+1}, \dots, x_k)$ to $(0, \dots, 0, x_{j+1}, \dots, x_k, r, \dots, r)$.

In this way, $i \le j \mapsto P_n^{i\le j} Y$ becomes a functor $P_n Y$ from the category $\mathcal{A}_{[n]} = (\mathcal{A}^n)^o$. Likewise, $P_n^{\substack{i\le j\\ k\le l}} Y$ defines a functor from $\mathcal{A}_{[n]} \times \mathcal{A}_{[n]}$.

For $\phi\colon [m] \to [n] \in \Delta$, the natural transformation $\Delta^\phi\colon \Delta^n \mathcal{A}^\phi \Rightarrow \Delta^m$ induces a natural transformation

$$P_\phi\colon P_n Y \mathcal{A}_\phi \Rightarrow P_m Y$$

of functors from $\mathcal{A}_{[m]}$ such that if $\psi\colon [l] \to [m] \in \Delta$, then we have an equality $P_\psi P_\phi = P_{\phi\psi}$ of natural transformations of functors from $\mathcal{A}_{[l]}$.

Lemma 5.4.1.4 *Let $Y_0 \subseteq Y_1 \subseteq Y_2 \subseteq \cdots$ be injections of compactly generated spaces and $0 \le i \le j \le n$. Then the canonical map $\varinjlim_k P_n^{i\le j} Y_k \to P_n^{i\le j}(\varinjlim_k Y_k)$ is a homeomorphism.*

Proof The claim follows since the $(i \le j)$-simplices are compact and unions of inclusion of compactly generated spaces are compactly generated. □

By the usual arguments for the Moore path space (see e.g., Whitehead [313, III.2.15]) we get

Lemma 5.4.1.5 *Let Y be a compactly generated space. Then the projections*

$$P_n^{i\le j} Y \to \mathbf{P}$$

sending (r, f) to r and

$$P_n^{i\le j} Y \to P_n^{i\le a} Y \times_{\mathbf{P}} P_n^{a+1\le j} Y$$

induced by the functoriality are Hurewicz fibrations.

5.4.1.4 The hoN-Construction

Now, let Y be a **functor** $\mathcal{A}^n \to CG$. Consider the functor $P_nY\colon \mathcal{A}_2^n \to CG$ given by

$$P_nY = P_n^{p_{12}}(Yp_{03}) = \{i \le j \le k \le l \mapsto P_n^{j\le k}Y(i \le l)\}$$

and its limit $\mathrm{hoN}_nY = \lim_{\overleftarrow{\mathcal{A}_2^n}} P_nY$. Given $\phi\colon [m] \to [n] \in \Delta$ we get a map

$$\begin{aligned}\mathrm{hoN}_\phi Y\colon \mathrm{hoN}_nY = \lim_{\overleftarrow{\mathcal{A}_2^n}} P_nY &\xrightarrow{\phi^*} \lim_{\overleftarrow{\mathcal{A}_2^m}}(P_nY)\mathcal{A}_2^\phi \xrightarrow{P_\phi} \lim_{\overleftarrow{\mathcal{A}_2^m}} P_m\big(Y\mathcal{A}^\phi\big)\\ &= \mathrm{hoN}_m(Y\mathcal{A}^\phi),\end{aligned}$$

where the first map is given by the functoriality of limits and the second is given by the naturality of the Moore singular simplex functor. Notice that if $\psi\colon [l] \to [m] \in \Delta$, then $\mathrm{hoN}_{\phi\psi}Y = \mathrm{hoN}_\psi Y \circ \mathrm{hoN}_\phi Y$ as maps from hoN_nY to $\mathrm{hoN}_l(Y\mathcal{A}^{\phi\psi})$.

Example 5.4.1.6 A model for the homotopy pullback. The category $\mathcal{A}^1$ can be depicted as $(0=0) \to (0<1) \leftarrow (1=1)$, and so a functor $X\colon \mathcal{A}^1 \to CG$ is the same as a diagram $X_{00} \to X_{01} \leftarrow X_{11}$. We see that

$$\mathrm{hoN}_1X = \lim_{\leftarrow}\{P_1^{0=0}X_{00} \to P_1^{0=0}X_{01} \leftarrow P_1^{0\le 1}X_{01} \to P_1^{1=1}X_{01} \leftarrow P_1^{1=1}X_{11}\}$$

is homeomorphic to the pullback of

$$P_1^{0\le 1}X_{01} \twoheadrightarrow P_1^{0=0}X_{01} \times_{\mathbf{P}} P_1^{1=1}X_{01} \leftarrow P_1^{0=0}X_{00} \times_{\mathbf{P}} P_1^{1=1}X_{11}$$

(which is a homotopy pullback since the leftmost arrow is a fibration by Lemma 5.4.1.5), which is the space of tuples $(x_{00}, x_{11}, r, \gamma)$ where $(x_{00}, x_{11}) \in X_{00} \times X_{11}$, $r \in \mathbf{P}$ and $\gamma\colon [0,r] \to X_{01}$ is a path with pair of endpoints $(\gamma(0), \gamma(r))$ the image of (x_{00}, x_{11}).

Example 5.4.1.7 The category $\mathcal{A}^0$ is the trivial one-point category, so a functor $X\colon \mathcal{A}^0 \to CG$ is nothing but a compactly generated space X and $\mathrm{hoN}_0X = \mathbf{P} \times X$.

Lemma 5.4.1.8 *Let $Y \to Y'$ be a natural transformation of functors $\mathcal{A}^n \to CG$ consisting of weak equivalences. Then the induced map $\mathrm{hoN}_nY \to \mathrm{hoN}_nY'$ is a weak equivalence. In particular, if Y is a functor $\mathcal{A}^n \to CG$ sending morphisms in $\mathcal{A}^n$ to weak equivalences, then the canonical map*

$$\mathrm{hoN}_nY \to P_n^{0\le n}Y(0 \le n) \to Y(0 \le n)$$

is a weak equivalence.

Proof Rewrite the limit so that it appears as an iterated cube and use Lemma 5.4.1.5. □

As usual, we extend to simplicial sets X by declaring that $P^{i\le j}X = \sin P^{i\le j}|X|$ and likewise for functors $\mathcal{A}^n \to \mathcal{S}$.

5.4.1.5 Monoidality

Let Y be a compactly generated space. Recall the spaces $P_n{}^{\substack{i\le j\\k\le l}}Y$ from Definition 5.4.1.3. Given $i_0 \le i_3 \in \mathcal{A}_{[n]}$, the monoidality of Δ^n followed by addition in $\mathbf{P}$ yields a map

$$G_{i_0\le i_3}\colon \lim_{\overleftarrow{p_{03}^{-1}(i_0\le i_3)}} P_n{}^{\substack{p_{23}\\p_{01}}} Y \to P^{i_0\le i_3}Y$$

which is natural in Y and $i_0 \le i_3$. Explicitly, an element in $\lim_{\overleftarrow{p_{03}^{-1}(i_0\le i_3)}} P_n{}^{\substack{p_{23}\\p_{01}}} Y$ is represented by a pair $(r,s) \in \mathbf{P}^2$ and a compatible collection of maps

$$\left(f_{i_0\le i_1\le i_2\le i_3}\colon \Delta^n_r(i_2\le i_3)\times\Delta^n_s(i_0\le i_1)\to Y\right)_{i_0\le i_1\le i_2\le i_3}.$$

The image of this element in $P^{i_0\le i_3}Y$ is given by the sum $r+s$ and the amalgamation

$$\Delta^n_{r+s}(i_0\le i_3)\cong \varinjlim \Delta^n_r(i_2\le i_3)\times\Delta^n_s(i_0\le i_1)\to Y$$

of the $f_{i_0\le i_1\le i_2\le i_3}$ (the only part of this assignment that is not bijective is $(r,s)\mapsto r+s$).

Let Y_1 and Y_2 be compactly generated spaces. The product map $P^{i_2\le i_3}Y_1 \times P^{i_0\le i_1}Y_2 \to P^{\substack{i_2\le i_3\\i_0\le i_1}}(Y_1\times Y_2)$ followed by $G_{i_0\le i_3}$ defines a pairing

$$\mu_{Y_1,Y_2}\colon (PY_1\times PY_2)(i_0\le i_3) = \lim_{\overleftarrow{p_{03}^{-1}(i_0\le i_3)}} \left(P_n^{p_{23}}Y_1\right)\times\left(P_n^{p_{01}}Y_2\right)\to P_n^{i_0\le i_3}(Y_1\times Y_2)$$

which is natural in Y_1, Y_2 and $i_0\le i_3\in\mathcal{A}_{[n]}$.

The map $Y\to P_n^{i\le j}Y$ sending $y\in Y$ to $(0,c_y)$ where $c_y\colon \Delta^n_0(i\le j)=*\to Y$ has the single value y defines a natural transformation from the identity to P_n, and in particular, when Y is the one point space $*$ defines a unit map $*\to P_n(*)$.

Lemma 5.4.1.9 *The functor $P_n\colon CG \to [\mathcal{A}_{[n]}, CG]$ together with the pairing and the unit is a monoidal functor. In particular, given $Y_1, Y_2, Y_3 \in CG$, the diagram*

$$\begin{array}{ccccc}
(P_nY_1 \times P_nY_2) \times P_nY_3 & \xrightarrow{\mu_{Y_1,Y_2}\times 1} & P_n(Y_1 \times Y_2) \times P_nY_3 & \xrightarrow{\mu_{Y_1\times Y_2,Y_3}} & P_n((Y_1 \times Y_2) \times Y_3) \\
{\scriptstyle assoc.}\downarrow\cong & & & & {\scriptstyle assoc.}\downarrow\cong \\
P_nY_1 \times (P_nY_2 \times P_nY_3) & \xrightarrow{1\times\mu_{Y_2,Y_3}} & P_nY_1 \times P_n(Y_2 \times Y_3) & \xrightarrow{\mu_{Y_1,Y_2\times Y_3}} & P_n(Y_1 \times (Y_2 \times Y_3))
\end{array}$$

commutes.

Now, assume Y_1, Y_2, Y_3 are functors $\mathcal{A}^n \to CG$ and $m\colon Y_1p_{01} \times Y_2p_{23} \to Y_3p_{03}$ is a natural transformation of functors $\mathcal{A}_2^n \to CG$. The considerations above give a pairing

$$m\colon \mathrm{hoN}_nY_1 \times \mathrm{hoN}_nY_2 \to \mathrm{hoN}_nY_3.$$

Explicitly, $(p_{0123}, p_{4567})\colon \mathcal{A}_3^n \to \mathcal{A}_2^n \times \mathcal{A}_2^n$ and the product map defines a map

$$\begin{aligned}
\mathrm{hoN}_nY_1 \times \mathrm{hoN}_nY_2 &= \varprojlim_{\mathcal{A}_2^n} PY_1 \times \varprojlim_{\mathcal{A}_2^n} PY_2 \\
&\cong \varprojlim_{\mathcal{A}_2^n\times\mathcal{A}_2^n} (PY_1 \times PY_2) \to \varprojlim_{\mathcal{A}_3^n} P^{\,p_{12}}_{\,p_{56}}(Y_1p_{03} \times Y_2p_{47}),
\end{aligned}$$

which is followed by the pairing m and the above defined natural transformation G to give a map to $\lim_{\leftarrow} PY_3 = \mathrm{hoN}_nY_3$. To check commutative diagrams we write this map out on individual elements. Notice that an element of $\varprojlim_{\mathcal{A}_2^n} PY_k$ may be given by giving an $r \in \mathbf{P}$ and a compatible collection of maps $f_{i_0\le i_1\le i_2\le i_3}\colon \Delta_r^n(i_1 \le i_2) \to Y_k(i_0 \le i_3)$. Given an element $((r, \{f_{i_0\le i_1\le i_2\le i_3}\}), (s, \{g_{i_0\le i_1\le i_2\le i_3}\}))$ in $\varprojlim_{\mathcal{A}_2^n} PY_1 \times \varprojlim_{\mathcal{A}_2^n} PY_2$, restricting our attention to the case $j_3 \le i_0$ we get maps

$$\Delta_r^n(j_1 \le j_2) \times \Delta_s^n(i_1 \le i_2) \xrightarrow{f\times g} Y_1(j_0 \le j_3) \times Y_2(i_0 \le i_3) \xrightarrow{m} Y_3(j_0 \le i_3)$$

and the compatibilities guarantee that these glue together to give a map $\Delta_{r+s}^n(j_1 \le i_2) \to Y_3(j_0 \le i_3)$.

Using this point-set description we see that the monoidality of Δ^n imply the following associativity property for hoN_n.

Lemma 5.4.1.10 *For $\emptyset \neq S \subseteq \{1, 2, 3\}$ let Y_S be functors $\mathcal{A}^n \to CG$, together with pairings such that the diagram*

$$\begin{array}{ccc}
Y_1 \times (Y_2 \times Y_3) & \xrightarrow{m_{12}\times 1} & Y_{1,2} \times Y_3 \\
 & & \downarrow{\scriptstyle m_{12,3}} \\
\downarrow{\scriptstyle assoc.} & & Y_{1,2,3} \\
 & & \uparrow{\scriptstyle m_{1,23}} \\
(Y_1 \times Y_2) \times Y_3 & \xrightarrow{1\times m_{23}} & Y_1 \times Y_{2,3}
\end{array}$$

commutes. Then the induced diagram

$$\begin{array}{ccc}
\mathrm{hoN}_n Y_1 \times (\mathrm{hoN}_n Y_2 \times \mathrm{hoN}_n Y_3) & \xrightarrow{m_{12}\times 1} & \mathrm{hoN}_n Y_{1,2} \times \mathrm{hoN}_n Y_3 \\
 & & \downarrow{\scriptstyle m_{12,3}} \\
\downarrow{\scriptstyle assoc.} & & \mathrm{hoN}_n Y_{1,2,3} \\
 & & \uparrow{\scriptstyle m_{1,23}} \\
(\mathrm{hoN}_n Y_1 \times \mathrm{hoN}_n Y_2) \times \mathrm{hoN}_n Y_3 & \xrightarrow{1\times m_{23}} & \mathrm{hoN}_n Y_1 \times \mathrm{hoN}_n Y_{2,3}
\end{array}$$

commutes too.

5.4.1.6 The Simplicial Case

In view of our intended applications (and to avoid technicalities like degenerate base points) we now shift solely attention to the simplicial cases and fetch the constructions above to this setting via the geometric realization/singular complex adjoint pair.

In particular, if X is a space (simplicial set), the space of singular $(i \leq j)$-simplices $P_n^{i\leq j} X$ is defined as $\sin(P_n^{i\leq j}|X|)$. If $X \colon \mathcal{A}^n \to \mathcal{S}$ is a functor to spaces, then we define $\mathrm{hoN}_n X = \lim_{\overleftarrow{\mathcal{A}_s^n}} PX$ and note that there is a natural isomorphism to $\sin \mathrm{hoN}_n |X|$. The conclusions of Lemma 5.4.1.8 and Lemma 5.4.1.9 still hold in the simplicial case (add a $\sin|-|$ to the last space in Lemma 5.4.1.8).

5.4.1.7 The Pointed Case

Let $X: \mathcal{A}^n \to \mathcal{S}_*$ be a functor to pointed spaces. We define the pointed version of hoN by setting

$$\mathrm{h\tilde{o}N}_n X = \mathrm{hoN}_n X / \mathrm{hoN}_n *,$$

where the inclusion $\mathrm{hoN}_n * \to \mathrm{hoN}_n X$ is induced by the inclusion of the base point in X. Note that $\mathrm{hoN}_n * \cong \sin \mathbf{P}$ is contractible. The conclusions of Lemma 5.4.1.8 and Lemma 5.4.1.9 still hold in the pointed case for $\mathrm{h\tilde{o}N}_n$ with the obvious modifications.

Note that if X is a pointed simplicial set, then $\mathrm{h\tilde{o}N}_0 X \cong \sin \mathbf{P}_+ \wedge \sin |X|$ and we have a natural weak equivalence $X \xrightarrow{\sim} \mathrm{h\tilde{o}N}_0 X$ induced by the $\sin/|-|$-adjointness and $0 \in \mathbf{P}$.

5.4.2 The Homotopy Nerve

5.4.2.1 The Homotopy Nerve of $\mathcal{S}_*$-Categories

We first deal with the unpointed situation.

Definition 5.4.2.1 Let $\mathcal{W}$ be a small $\mathcal{S}$-category with underlying category $U\mathcal{W}$. The *homotopy nerve* $\mathrm{hoN}\mathcal{W} = \{[n] \mapsto \mathrm{hoN}_n \mathcal{W}\}$ is the simplicial $\mathcal{S}$-category with space of objects the classifying space of $\mathcal{W}$ (i.e., an n-simplex in the space of objects is a functor $[n] \to U\mathcal{W}$), if $a, b: [n] \to U\mathcal{W}$ are two objects, then we define the space of morphisms as $\mathrm{hoN}_n \mathcal{W}(a, b)$, where hoN_n is the construction introduced in Sect. 5.4.1.4 and $\mathcal{W}(a, b) = \{i \leq j \mapsto \mathcal{W}(a_j, b_i)\}$ considered as a functor $\mathcal{A}^n \to \mathcal{S}$. The unit and composition in $\mathrm{hoN}_n \mathcal{W}$ are defined by applying the pairings constructed in Sect. 5.4.1.5 to the composition in $\mathcal{W}$.

The simplicial structure is given by sending $\phi: [m] \to [n] \in \Delta$ to the $\mathcal{S}$-functor

$$\phi^* = \mathrm{hoN}_\phi \mathcal{W}: \mathrm{hoN}_n \mathcal{W} \to \mathrm{hoN}_m \mathcal{W}$$

defined as follows. If $a: [n] \to U\mathcal{W}$ is an object, $\phi^* a$ is the composition $a\phi: [m] \to [n] \to U\mathcal{W}$, and the map of morphism spaces $\mathrm{hoN}_n \mathcal{W}(a, b) \to \mathrm{hoN}_m \mathcal{W}(\phi^* a, \phi^* b)$ is the $\mathrm{hoN}_\phi \mathcal{W}(a, b)$ of Sect. 5.4.1.4.

From Lemma 5.4.1.8 we get that the homotopy nerve only depends on the homotopy type of the morphism spaces in the following sense.

Lemma 5.4.2.2 *Let $F: \mathcal{W}_1 \to \mathcal{W}_2$ be an $\mathcal{S}$-functor inducing the identity on the underlying category and weak equivalences on morphism spaces, then so is* $\mathrm{hoN}_n F: \mathrm{hoN}_n \mathcal{W}_1 \to \mathrm{hoN}_n \mathcal{W}_2$.

The considerations above also works for $\mathcal{S}_*$-categories $\mathcal{C}$ to give a simplicial $\mathcal{S}_*$-category $\mathrm{h\tilde{o}N}\mathcal{C}$ by declaring that the morphisms spaces in degree n are of the form $\mathrm{h\tilde{o}N}_n \mathcal{C}(a, b) = \mathrm{hoN}_n \mathcal{C}(a, b) / \mathrm{hoN}_n *$.

5.4.2.2 The Homotopy Nerve of $\Gamma\mathcal{S}_*$-Categories

We now define the homotopy nerve of a $\Gamma\mathcal{S}_*$-category. The reason we do it for $\Gamma\mathcal{S}_*$-categories is that this is what is needed for our application, but if you for instance wanted to prove that the cyclotomic trace is multiplicative you should exchange this for a more suitable model of spectra, e.g., symmetric spectra.

Definition 5.4.2.3 Let $\mathcal{C}$ be a small $\Gamma\mathcal{S}_*$-category with underlying category $U\mathcal{C}$. The (pointwise) *homotopy nerve* $\mathrm{h\tilde{o}N}\mathcal{C}$ is the simplicial $\Gamma\mathcal{S}_*$-category with space of objects the classifying space of $U\mathcal{C}$, and with morphism objects defined as follows. If $a, b\colon [n] \to U\mathcal{C}$ are two objects in simplicial dimension n, then $\mathrm{h\tilde{o}N}_n\mathcal{C}(a,b)$ is the Γ-space which evaluated on the finite pointed set T is given by

$$\big(\mathrm{h\tilde{o}N}_n\mathcal{C}(a,b)\big)(T) = \mathrm{h\tilde{o}N}_n\big(\mathcal{C}(a,b)(T)\big),$$

where $\mathcal{C}(a,b)(T)$ is the functor

$$\mathcal{A}^n \xrightarrow{\mathrm{forget}^{[n]}} [n]^o \times [n] \xrightarrow{a\times b} U^o\mathcal{C} \times U\mathcal{C} \xrightarrow{\mathcal{C}(-,-)(T)} \mathcal{S}_*.$$

Composition, unitality and simplicial structure is defined as for the simplicial case Definition 5.4.2.1 using Lemma 5.4.1.4 and the pairings of Sect. 5.4.1.5.

Lemma 5.4.2.4 *Given a $\Gamma\mathcal{S}_*$-category $\mathcal{C}$ and two n-simplices $a, b\colon [n] \to U\mathcal{C}$ so that for all $0 \le i \le j \le n$ the induced maps $\mathcal{C}(a_j, b_j) \to \mathcal{C}(a_j, b_i) \leftarrow \mathcal{C}(a_i, b_i)$ are stable equivalences for all $0 \le i \le j \le n$. Then the canonical map $\mathrm{h\tilde{o}N}_n\mathcal{C}(a,b) \to \sin|\mathcal{C}(a_n, b_0)|$ is a stable equivalence.*

Proof In the special case where the induced maps $\mathcal{C}(a_j, b_j) \to \mathcal{C}(a_j, b_i) \leftarrow \mathcal{C}(a_i, b_i)$ are pointwise equivalences this follows from Lemma 5.4.1.8, and even gives that for every finite pointed set T the map $\mathrm{h\tilde{o}N}_n\mathcal{C}(a,b)(T) \to \sin|\mathcal{C}(a_n, b_0)(T)|$ is a pointwise equivalence.

For any pointed compactly generated space Y we point $P_n^{i\le j}Y$ in $(0,*)$ and note that the canonical inclusion $P_n^{i\le j}\Omega^k Y \subseteq \Omega^k P_n^{i\le j} Y$ (the difference is that in $\Omega^k P_n^{i\le j} Y$ the scaling factor of the singular simplex is allowed to vary over S^k) is a weak equivalence, and the maps that are fibrations in view of Lemma 5.4.1.5 remain fibrations if we replace $P_n^{i\le j}Y$ by either $P_n^{i\le j}\Omega^k Y$ or $\Omega^k P_n^{i\le j} Y$.

For simplicity of notation, let $X^k = \mathcal{C}(a,b)(S^k \wedge T)\colon \mathcal{A}^n \to \mathcal{S}_*$ for some fixed finite pointed set T and $k \ge 0$ and consider the diagram

$$\begin{array}{ccccccc}
\varinjlim_k \Omega^k \varprojlim_{\mathcal{A}_2^n} P X^k & \xrightarrow{\sim} & \varinjlim_k \varprojlim_{\mathcal{A}_2^n} \sin\Omega^k P|X^k| & \xleftarrow{\sim} & \varinjlim_k \varprojlim_{\mathcal{A}_2^n} \sin P\Omega^k|X^k| & \xleftarrow{\sim} & \varinjlim_k \varprojlim_{\mathcal{A}_2^n} P\Omega^k X^k \\
\downarrow & & \downarrow & & \downarrow & & \downarrow \\
\varinjlim_k \Omega^k \sin|X^k| & \xrightarrow{\sim} & \varinjlim_k \Omega^k X^k & = & \varinjlim_k \Omega^k X^k & \xleftarrow{\sim} & \varinjlim_k \sin|\Omega^k X^k|,
\end{array}$$

where the left hand horizontal arrows unravel the use of singular complex and geometric realizations in our definitions of Ω^k and P in the simplicial setting and that loops commute with finite limits, the middle top horizontal map is the comparison of $\Omega^k P$ with $P\Omega^k$, and the right hand horizontal maps sort out the definitions of Ω^k and P for spaces and topological spaces.

Notice up to canonical isomorphisms, the right hand vertical map is the map

$$\lim_{\overleftarrow{\mathcal{A}_2^n}} P \lim_{\overrightarrow{k}} \Omega^k X^k \to \sin\left|\lim_{\overrightarrow{k}} \Omega^k X^k\right|,$$

which by Lemma 5.4.1.8 is a weak equivalence, and so the left vertical map is a weak equivalence too. However, since the left vertical map factors as the canonical projection $Q\mathrm{hoN}_n X^k \xrightarrow{\sim} Q\mathrm{h\tilde{o}N}_n X^k = Q(\mathrm{h\tilde{o}N}_n \mathcal{C}(a,b))(T)$ followed by the map $Q(\mathrm{h\tilde{o}N}_n \mathcal{C}(a,b))(T) \to Q \sin|\mathcal{C}(a,b)|(T)$ which we sought to show was a weak equivalence, we are done. □

5.4.2.3 The Relative Case

We are really interested in the homotopy nerve mainly in the relative case.

If $(\mathcal{C}, w\colon \mathcal{W} \to U\mathcal{C})$ is a "discrete" pair (so that $\mathcal{W}$ is an ordinary category, w an ordinary functor and $\mathcal{C}$ a $\Gamma\mathcal{S}_*$-category) we define the homotopy nerve as follows.

Definition 5.4.2.5 The *homotopy nerve* $\mathrm{h\tilde{o}N}(\mathcal{C}, w)$ of a discrete pair $(\mathcal{C}, w\colon \mathcal{W} \to U\mathcal{C})$ is the simplicial $\Gamma\mathcal{S}_*$-category whose space of objects is the classifying space $B\mathcal{W}$ and whose morphism objects are given by

$$\mathrm{h\tilde{o}N}_n(\mathcal{C}, w)(a,b) = \mathrm{h\tilde{o}N}_n \mathcal{C}(wa, wb)$$

for $a, b\colon [n] \to \mathcal{W}$ two objects in simplicial dimension n.

Notice that (apart from the contractible noise introduced to the morphism objects by the application of the $(0=0)$-singular simplex functor) $\mathrm{h\tilde{o}N}_0(\mathcal{C}, w)$ is nothing but the $\Gamma\mathcal{S}_*$-full subcategory of $\mathcal{C}$ generated by the image of w.

Recall the monoidal stabilization functor $T_0\colon \Gamma\mathcal{S}_* \to \Gamma\mathcal{S}_*$ from Definition 2.2.2.1. If $\mathcal{C}$ is a $\Gamma\mathcal{S}_*$-category $T_0\mathcal{C}$ is the $\Gamma\mathcal{S}_*$-category obtained by letting $obT_0\mathcal{C} = ob\mathcal{C}$, but employing T_0 on all the morphism objects. Essentially the same reasoning as employed in the proof of Lemma 5.4.2.4 gives the following lemmas

Lemma 5.4.2.6 *Given a discrete pair $(\mathcal{C}, w\colon \mathcal{W} \to U\mathcal{C})$. Then the natural transformation $\eta\colon 1 \to T_0$ induces a simplicial $\Gamma\mathcal{S}_*$-functor* $\mathrm{h\tilde{o}N}(\mathcal{C}, w) \to \mathrm{h\tilde{o}N}(T_0\mathcal{C}, (R\eta)w)$ *which is the identity on objects and stable equivalences on all morphism objects.*

So, for all practical purposes, we can always assume that our $\Gamma\mathcal{S}_*$-categories have stably fibrant morphism objects.

Lemma 5.4.2.7 *Let $(\mathcal{C}, w\colon \mathcal{W} \to U\mathcal{C})$ be a discrete pair such that for all morphisms $\alpha\colon x \to y$ and objects z in $\mathcal{W}$ the induced maps $\mathcal{C}(wy, wz) \to \mathcal{C}(wx, wz)$ and $\mathcal{C}(wz, wx) \to \mathcal{C}(wz, wy)$ are stable equivalences. Then the face maps*
$\mathrm{h\tilde{o}N}_n(\mathcal{C}, w) \to \mathrm{h\tilde{o}N}_0(\mathcal{C}, w)$ *are all stable equivalences of $\Gamma\mathcal{S}_*$-categories.*

Proof Follows from Lemma 5.4.2.4 since the face maps are surjective on objects. □

5.4.2.4 The Simplicial Case

Let $\mathcal{W}$ be an $\mathcal{S}$-category. We define the simplicial category $\{[n] \mapsto \mathcal{W}[n]\}$ by declaring that the objects of $\mathcal{W}[n]$ are the objects of $\mathcal{W}$ and if a and b are objects in $\mathcal{W}$, the set of morphisms in $\mathcal{W}[n]$ from a to b is the set of n-simplices in the function space $\mathcal{W}(a, b)$. Assuming that $\mathcal{W}$ is tensored over $\mathcal{S}$ this implies that $\mathcal{W}[n](a, b) = \mathcal{W}(a \otimes \Delta[n], b)$.

Recall that if X is a Γ-space, we defined the space $RX = X(1_+)$. It is really a pointed space, but we will make no notational distinction and consider R as a functor $\Gamma\mathcal{S}_* \to \mathcal{S}$. The functor is monoidal, so that a $\Gamma\mathcal{S}_*$-category $\mathcal{C}$ gives rise to a $\mathcal{S}$-category $R\mathcal{C}$.

We define a simplicial pair $\{[n] \mapsto (\mathcal{C}[n], w[n]\colon \mathcal{W}[n] \to U\mathcal{C}[n])\}$, where $\mathcal{W}[n]$ was defined above and $\mathcal{C}[n]$ is the $\Gamma\mathcal{S}_*$-category with objects the same as $\mathcal{C}$, but with morphism objects given by $\mathcal{C}[n](a, b) = \mathcal{S}_*(\Delta[n]_+, \mathcal{C}(a, b))$ (this is different from picking out the Γ-set of n-simplices. In particular, the standard contraction of $\Delta[n]$ gives a simplicial homotopy equivalence $\mathcal{C}[n](a, c) \simeq \mathcal{C}(a, b)$). Notice that $U\mathcal{C}[n]$ is nothing but the category of n-simplexes in $R\mathcal{C}$, and so w induces a functor $w[n]\colon \mathcal{W}[n] \to U\mathcal{C}[n]$.

Lemma 5.4.2.8 *If $(\mathcal{C}, w)$ is a normal pair, then so is $(\mathcal{C}[n], w[n])$ for each $n \geq 0$.*

Proof We must show that if $\alpha \in \mathcal{W}[n](x, y)$, then $\alpha_*\colon \mathcal{C}[n](z, x) \to \mathcal{C}[n](z, y)$ and $\alpha^*\colon \mathcal{C}[n](y, z) \to \mathcal{C}[n](x, z)$ are stable equivalences for every object z in $\mathcal{W}$. If $\phi\colon \Delta[0] \to \Delta[n]$ is the inclusion of the initial vertex, consider the induced map of pairs

$$\phi^*\colon \big(\mathcal{C}[n], w[n]\big) \to \big(\mathcal{C}[0], w[0]\big).$$

On morphism objects we get a homotopy equivalence

$$\mathcal{C}[n](z, x) = \mathcal{C}(z, x)^{\Delta[n]} \xrightarrow{(\phi_*)^*} \mathcal{C}(z, x)^{\Delta[0]} = \mathcal{C}(z, x)$$

and the diagram

$$\begin{array}{ccc} \mathcal{C}(z, x)^{\Delta[n]} & \xrightarrow{\alpha_*} & \mathcal{C}(z, y)^{\Delta[n]} \\ {\scriptstyle (\phi_*)^*}\downarrow\simeq & & {\scriptstyle (\phi_*)^*}\downarrow\simeq \\ \mathcal{C}(z, x) & \xrightarrow{(\phi^*\alpha)_*} & \mathcal{C}(z, y) \end{array}$$

commutes. Since, by assumption, $(\phi^*\alpha)_*$ is a stable equivalence we get that α_* is a stable equivalence. Likewise for α^*. □

Definition 5.4.2.9 If $(\mathcal{C}, w\colon \mathcal{W} \to R\mathcal{C})$ is a pair, we define its *homotopy nerve* to be the bisimplicial $\Gamma\mathcal{S}_*$-category $\mathrm{h\tilde{o}N}(\mathcal{C}, w) = \{([m],[n]) \mapsto \mathrm{h\tilde{o}N}_{m,n}(\mathcal{C}, w) = \mathrm{h\tilde{o}N}_m(\mathcal{C}[n], w[n])\}$.

Lemma 5.4.2.10 *Let $(\mathcal{C}, w\colon \mathcal{W} \to R\mathcal{C})$ be a normal pair. Then the face maps* $\mathrm{h\tilde{o}N}_{m,n}(\mathcal{C}, w) \to \mathrm{h\tilde{o}N}_0(\mathcal{C}, w)$ *are all stable equivalences of $\Gamma\mathcal{S}_*$-categories.*

Furthermore $\mathrm{h\tilde{o}N}_0(\mathcal{C}, w) = \mathrm{h\tilde{o}N}_0\mathcal{C}$, *and there is a stable (even pointwise) $\Gamma\mathcal{S}_*$-equivalence $\mathcal{C} \to \mathrm{h\tilde{o}N}_0\mathcal{C}$.*

Proof The conclusion does not use all of normality: the behavior on isomorphisms can be weakened to demanding that w is onto on objects, which is enough for getting the equality $\mathrm{h\tilde{o}N}_0(\mathcal{C}, w) = \mathrm{h\tilde{o}N}_0\mathcal{C}$. The last claim then follows from the pointwise equivalences $\mathcal{C}(a,b) \xrightarrow{\sim} \mathrm{h\tilde{o}N}_0\mathcal{C}(a,b)$ of Sect. 5.4.1.7.

Lemma 5.4.2.7 gives that the face maps $\mathrm{h\tilde{o}N}_m(\mathcal{C}[n], w[n]) \to \mathrm{h\tilde{o}N}_0(\mathcal{C}[n], w[n])$ are all $\Gamma\mathcal{S}_*$-equivalences. Furthermore, since $\Delta[n]$ is contractible, we get that the face maps $\mathrm{h\tilde{o}N}_0(\mathcal{C}[n], w[n]) \to \mathrm{h\tilde{o}N}_0(\mathcal{C}[0], w[0])$ are $\Gamma\mathcal{S}_*$-equivalences (inducing the identity on objects). □

Chapter 6
Topological Cyclic Homology

A motivation for the definitions to come can be found by looking at the example of a $\Gamma\mathcal{S}_*$-category $\mathcal{C}$. Consider the trace map

$$ob\mathcal{C} \to THH(\mathcal{C})(S^0)$$

Topological Hochschild homology is a cyclic space, $ob\mathcal{C}$ is merely a set. However, the trace Sect. 4.3.2 is universal in the sense that $ob\mathcal{C} \cong \lim_{\overleftarrow{\Lambda^o}} THH(\mathcal{C})(S^0)$. A more usual way of putting this, is to say that $ob\mathcal{C} \to |THH(\mathcal{C})(S^0)|$ is **the inclusion of the $\mathbb{T} = |S^1|$-fixed points**, which also makes sense since the realization of a cyclic space is a topological space with a circle action (see Sect. 6.2.1 below).

In particular, the trace from K-theory has this property. The same is true for the other definition of the trace (Sect. 4.2.5), but this follows more by construction than by fate. In fact, any reasonable definition of the trace map should factor through the $\mathbb{T}$-fixed point space, and so, if one wants to approximate K-theory one should try to mimic the $\mathbb{T}$-fixed point space by any reasonable means. The awkward thing is that forming the $\mathbb{T}$-fixed point space as such is really not a reasonable thing to do, in the sense that it does not preserve weak equivalences. Homotopy fixed point spaces are nice approximations which are well behaved, and strangely enough it turns out that so are the actual fixed point spaces with respect to finite subgroups of the circle. The aim is now to assemble as much information from these nice constructions as possible.

6.1 Introduction

6.1.1 Connes' Cyclic Homology

The first time the circle came into action for trace maps, was when Alain Connes defined his cyclic cohomology [56]. We are mostly concerned with homology theories, and in one of its many guises, *cyclic homology* is just the $\mathbb{T}$-homotopy orbits of the

B.I. Dundas et al., *The Local Structure of Algebraic K-Theory*, Algebra and Applications 18,
DOI 10.1007/978-1-4471-4393-2_6,

Hochschild homology spectrum. This is relevant to K-theory for several reasons, and one of the more striking reasons is the fact discovered by Loday and Quillen [182] and Tsygan [290]: just as the K-groups are rationally the primitive part of the group homology of $GL(A)$, cyclic homology is rationally the primitive part of the Lie-algebra homology of $\mathfrak{gl}(A)$.

However, in the result above there is a revealing dimension shift, and, for the purposes of comparison with K-theory via trace maps, it is not the homotopy orbits, but the homotopy fixed points which play the central rôle. The homotopy fixed points of Hochschild homology give rise to Goodwillie and J.D.S. Jones' *negative cyclic homology* $HC^{-}(A)$. In [102] Goodwillie proves that if $A \to B$ is a map of simplicial $\mathbf{Q}$-algebras inducing a surjection $\pi_0(A) \to \pi_0(B)$ with nilpotent kernel, then the relative K-theory $K(A \to B)$ is equivalent to the relative negative cyclic homology $HC^{-}(A \to B)$.

All told, the cyclic theories associated with Hochschild homology seem to be right rationally, but just as for the comparison with stable K-theory, we must replace Hochschild homology by topological Hochschild homology to obtain integral results.

6.1.2 Bökstedt, Hsiang, Madsen and $TC^{\wedge}_p$

Topological cyclic homology, also known as TC, appears for the first time in Bökstedt, Hsiang and Madsen's proof on the algebraic K-theory analog of the Novikov conjecture [27], and is something of a surprise. The obvious generalization of negative cyclic homology would be the homotopy fixed point space of the circle action on topological Hochschild homology, but this turns out not to have all the desired properties. Instead, they consider actual fixed points under the actions of the finite subgroups of $\mathbb{T}$.

After completing at a prime, looking only at the action of the finite subgroups is not an unreasonable thing to do, since you can calculate the homotopy fixed points of the entire circle action by looking at a tower of homotopy fixed points with respect to cyclic groups of prime power order (see Example A.7.6.4). The equivariant nature of Bökstedt's formulation of *THH* is such that the actual fixed point spaces under the finite groups are nicely behaved Corollary 6.2.4.6, and in one respect they are highly superior to the homotopy fixed point spaces: The fixed point spaces with respect to the finite subgroups of $\mathbb{T}$ are connected by more maps than you would think of by considering the homotopy fixed points or the linear analogs (in particular, the "restriction maps" of Sect. 6.2.3), and the interplay between these maps can be summarized in topological cyclic homology to give an amazingly good approximation of K-theory.

Topological cyclic homology, as we define it, is a non-connective spectrum, but its completions $\underline{TC}(-)^{\wedge}_p$ are all -2-connected. As opposed to topological Hochschild homology, the topological cyclic homology of a discrete or simplicial ring is generally not an Eilenberg–Mac Lane spectrum.

In [27] the problem at hand is reduced to studying topological cyclic homology and trace maps of **S**-algebras of the form $\mathbf{S}[G]$, where **S** is the sphere spectrum and G is some simplicial group (see Example 2.1.4.3), i.e., the **S**-algebras associated to Waldhausen's A theory of spaces (see Sect. 3.2.3.2). In this case, TC is particularly easy to describe: for each prime p, there is a cartesian square

$$\begin{array}{ccc} \underline{TC}(\mathbf{S}[G])^{\wedge}_{p} & \longrightarrow & (\Sigma \underline{T}(\mathbf{S}[G])_{h\mathbf{S}^1})^{\wedge}_{p} \\ \downarrow & & \downarrow \\ \underline{T}(\mathbf{S}[G])^{\wedge}_{p} & \longrightarrow & \underline{T}(\mathbf{S}[G])^{\wedge}_{p} \end{array}$$

(in the homotopy category) where the right vertical map is the "circle transfer", and the lower horizontal map is analogous to something like the difference between the identity and a pth power map. The nature of the top horizontal map in the diagram is not well understood.

6.1.3 *TC of the Integers*

Topological cyclic homology is much harder to calculate than topological Hochschild homology, but—and this is the main point of this book—it exhibits the same "local" behavior as algebraic K-theory, and so is well worth the extra effort. The first calculation to appear is in fact one of the hardest ones produced to date, but also the most prestigious: in [28] Bökstedt and Madsen set forth to calculate $TC(\mathbf{Z})^{\wedge}_{p}$ for $p > 2$, and found that they could describe $TC(\mathbf{Z})^{\wedge}_{p}$ in terms of objects known to homotopy theorists:

$$TC(\mathbf{Z})^{\wedge}_{p} \simeq imJ^{\wedge}_{p} \times BimJ^{\wedge}_{p} \times SU^{\wedge}_{p},$$

where imJ is the image of J, cf. [3], and SU is the infinite special unitary group—provided a certain spectral sequence behaved as they suspected it did. In his thesis "The equivariant structure of topological Hochschild homology and the topological cyclic homology of the integers", [Ph.D. Thesis, Brown Univ., Providence, RI, 1994] Stavros Tsalidis proved that the spectral sequence was as Bökstedt and Madsen had supposed, by adapting an argument in G. Carlsson's proof of the Segal conjecture [52] to suit the present situation. Using this Bökstedt and Madsen calculated in [29] $TC(A)^{\wedge}_{p}$ for A the Witt vectors of finite fields of odd characteristic, and in particular got the above formula for $TC(\mathbf{Z})^{\wedge}_{p} \simeq TC(\mathbf{Z}^{\wedge}_{p})^{\wedge}_{p}$. See also Tsalidis' papers [288] and [289]. Soon after Rognes showed in [240] that an analogous formula holds for $p = 2$ (you do not have the splitting, and the image of J should be substituted with the complex image of J).

A bit more on the story behind this calculation, and also the others briefly presented in this introduction, can be found in Sect. 7.3.

6.1.4 Other Calculations of TC

All but the last of the calculations below are due to the impressive effort of Hesselholt and Madsen. As the calculations below were made after the p-complete version of Theorem 7.0.0.2 on the correspondence between K-theory and TC was known for rings, they were stated for K-theory whenever possible, even though they were actually calculations of TC.

For a ring A, let $W(A)$ be the p-typical Witt vectors, see [259] or more briefly Sect. 6.4.2.2 for the commutative case and [124, 125] for the general case. Let $W(A)_F$ be the coinvariants under the Frobenius action, i.e., the cokernel of $1-F\colon W(A)\to W(A)$. Note that $W(\mathbf{F}_p)=W(\mathbf{F}_p)_F=\mathbf{Z}\widehat{_p}$.

1. Hesselholt [124] Let A be a discrete ring. Then there is an isomorphism $\pi_{-1}\underline{TC}(A)\widehat{_p}\cong W(A)_F$.
2. Hesselholt and Madsen (cf. [128] k a perfect field of characteristic $p>0$). Then $\underline{TC}(A)$ is an Eilenberg–Mac Lane spectrum for any k-algebra A. Furthermore, we have isomorphisms

$$\pi_i\underline{TC}(k)\widehat{_p}\cong\begin{cases}W(k)_F & \text{if } i=-1,\\ \mathbf{Z}\widehat{_p} & \text{if } i=0,\\ 0 & \text{otherwise,}\end{cases}$$

and

$$\pi_i\underline{TC}\big(k[t]/(t^n)\big)\widehat{_p}\cong\begin{cases}\pi_i\underline{TC}(k)\widehat{_p} & \text{if } i=-1 \text{ or } i=0,\\ \mathbf{W}_{nm-1}(k)/V_n\mathbf{W}_{m-1}(k) & \text{if } i=2m-1>0,\\ 0 & \text{otherwise,}\end{cases}$$

where $\mathbf{W}_j(k)=(1+tk[[t]])^\times/(1+t^{j+1}k[[t]])^\times$ is the ring of truncated big Witt vectors, and $V_n\colon\mathbf{W}_{m-1}(k)\to\mathbf{W}_{nm-1}(k)$ is the Verschiebung map sending $f(t)=1+t\sum_{i=1}^\infty a_it^i$ to $f(t^n)$.
3. Hesselholt ([124]). Let A be a free associative $\mathbf{F}_p$-algebra. Then

$$\pi_i\underline{TC}(A)\widehat{_p}\cong\begin{cases}W(A)_F & \text{if } i=-1,\\ \mathbf{Z}\widehat{_p} & \text{if } i=0,\\ 0 & \text{otherwise.}\end{cases}$$

On the other hand, the topological cyclic homology groups of the free *commutative* $\mathbf{F}_p$-algebras are generally not concentrated in non-positive degrees:

$$\pi_i\underline{TC}\big(\mathbf{F}_p[t_1,\dots t_n]\big)\widehat{_p}\cong\begin{cases}(\bigoplus_{g\in G_m}\mathbf{Z}\widehat{_p})\widehat{_p} & \text{for } -1\le i\le n-2,\\ 0 & \text{otherwise,}\end{cases}$$

where G_m is some explicit (non-empty) set (see [124, p. 140]).

4. Hesselholt and Madsen [131]. Let K be a complete discrete valuation field of characteristic zero with perfect residue field k of characteristic $p > 2$. Let A be the valuation ring of K. Hesselholt and Madsen analyze $TC(A)^{\widehat{}}_p$, and in particular they give very interesting algebraic interpretations of the relative term of the transfer map $TC(k)^{\widehat{}}_p \to TC(A)^{\widehat{}}_p$ (obtained by inclusion of the category of k-vector spaces into the category of torsion modules of A). See Theorem 7.3.3.1 for some further details.
5. Ausoni and Rognes [10, 11]. In order to calculate the algebraic K-theory $K(ku)$ of connective complex K-theory Ausoni calculated the topological cyclic homology of ku. Previously, Ausoni and Rognes calculated the topological cyclic homology of the Adams summand ℓ_p. See Sect. 7.3.1 where the Adams summand is one of the examples.

6.1.5 Where to Read

The literature on TC is naturally even more limited than on *THH*. Bökstedt, Hsiang and Madsen's original paper [27] is still very readable. The first chapters of Hesselholt and Madsen's [129] can serve as a streamlined introduction for those familiar with equivariant G-spectra. For more naïve readers, the unpublished lecture notes [105] can be of help. Again, the survey article of Madsen [192] is recommendable.

6.2 The Fixed Point Spectra of *THH*

We will define TC by means of a homotopy cartesian square of the type (i.e., it will be the homotopy limit of the rest of the diagram)

$$\begin{array}{ccc} TC(-) & \longrightarrow & THH(-)^{h\mathbf{S}^1} \\ \downarrow & & \downarrow \\ \prod_{p \text{ prime}} TC(-;p)^{\widehat{}}_p & \longrightarrow & \left(\prod_{p \text{ prime}} THH(-)^{\widehat{}}_p\right)^{h\mathbf{S}^1} \end{array}$$

(as it stands, this strictly does not make sense: there are some technical adjustments we shall return to). The $\mathbf{S}^1$-homotopy fixed points are formed with respect to the cyclic structure.

In this section we will mainly be occupied with preparing the ground for the lower left hand corner of this diagram. Let $C_n \subseteq \mathbf{S}^1$ be the subgroup consisting of the nth roots of unity. We do a consistent choice of generators t_{n-1} of the cyclic groups C_n such that $t^m_{mn-1} = t_{n-1}$ under the inclusion $C_n \subseteq C_{mn}$. For each prime number p, the functor $TC(-;p)$ is defined as the homotopy limit of a diagram of fixed point spaces $|THH(-)|^{C_{p^n}}$. The maps in the diagrams are, partially, inclusion of fixed points $|THH(-)|^{C_{p^{n+1}}} \subseteq |THH(-)|^{C_{p^n}}$, and partially, some more exotic maps—the "restriction maps"—which we will describe in Sect. 6.2.3 below. The contents of this section is mostly fetched from the unpublished MSRI notes [105]. If desired, the reader can consult Appendix A.9 for some facts on group actions.

6.2.1 Cyclic Spaces and the Edgewise Subdivision

Recall Connes' category Λ (see e.g., Definition 4.2.1.1). Due to the inclusion $j\colon \Delta \subset \Lambda$, any cyclic object X gives rise to a simplicial object j^*X.

As noted by Connes [55], cyclic objects are intimately related to objects with a circle action (see also [19, 27, 74, 151] and [268]). In analogy with the standard n-simplices $\Delta[n] = \{[q] \mapsto \Delta([q],[n])\}$, we define the cyclic sets

$$\Lambda[n] = \Lambda\big(-,[n]\big)\colon \Lambda^o \to Ens.$$

We identify the circle group $\mathbb{T} = |S^1|$ and $\mathbf{R}/\mathbf{Z}$ under the isomorphism induced by the inclusion $\Delta^1 \cong [0,1] \subseteq \mathbf{R}$ (the first homeomorphism is projection onto the first coordinate from the 1-simplex $\Delta^1 = \{(x,y) \in \mathbf{R}^2 | x+y=1, x \geq 0, y \geq 0\}$).

The starting point for the connection to $\mathbb{T}$-spaces (topological spaces with a circle action) is the following lemma, whose proof may be found, for instance, in [74, 2.7].

Lemma 6.2.1.1 *For all n, $|j^*\Lambda[n]|$ is a $\mathbb{T}$-space naturally (in $[n] \in ob\Lambda^o$) homeomorphic to $\mathbb{T} \times |\Delta[n]|$.* ☺

This gives us the building blocks for a realization/singular functor pair connecting the category $\mathbb{T}$-Top_* of (pointed) $\mathbb{T}$-spaces with (pointed) cyclic sets:

$$\mathbb{T}\text{-}Top_* \underset{\sin_\Lambda}{\overset{|-|_\Lambda}{\leftrightarrows}} Ens_*{}^{\Lambda^o}$$

given by sending a cyclic set X to

$$|X|_\Lambda = \int^{[q]\in\Lambda^o} \big|j^*\Lambda[q]\big|_+ \wedge X_q \cong \bigvee_{[q]\in\Lambda^o} \big|j^*\Lambda[q]\big|_+ \wedge X_q/\sim$$

(where $\sigma \wedge \phi^* x \sim \phi_*\sigma \wedge x$ for $\phi \in \Lambda([p],[q])$, $\sigma \in |j^*\Lambda[p]|$ and $x \in X_q$) considered as a pointed $\mathbb{T}$-space through Lemma 6.2.1.1, and by sending a pointed $\mathbb{T}$-space Z to

$$\sin_\Lambda Z = \big\{[q] \mapsto (\mathbb{T}\text{-}Top_*)\big(\big|j^*\Lambda[q]\big|_+, Z\big)\big\}.$$

An equivalent way of stating this is to say that the realization is the left Kan extension in

$$\begin{array}{ccc} \Lambda & \xrightarrow{q\mapsto|j^*\Lambda[q]|_+} & \mathbb{T}\text{-}Top_* \\ \Big\downarrow{\scriptstyle y} & \nearrow & \\ Ens_*{}^{\Lambda^o} & & \end{array} .$$

Letting U be the forgetful functor from $\mathbb{T}$-spaces to pointed topological spaces (right adjoint to $\mathbb{T}_+\wedge-$) we get

Lemma 6.2.1.2 *There are natural isomorphisms*

$$j^*\sin_\Lambda Z \cong \sin(UZ) \quad \textit{and} \quad U|X|_\Lambda \cong \left|j^*X\right|$$

where X is a pointed cyclic set and Z a pointed $\mathbb{T}$-space.

Proof The first follows by the isomorphism $|\Lambda[q]|_\Lambda \cong \mathbb{T}\times|\Delta[n]|$, and the adjointness of U with $\mathbb{T}_+\wedge-$; and the second is formal and follows by writing out the definitions. □

If one is familiar with coends a formal writeup of the second isomorphism reads quite compactly

$$\begin{aligned} U|X|_\Lambda &= \int^{[q]\in\Lambda^o} \left|j^*\Lambda[q]\right|_+\wedge X_q \\ &\cong \int^{[s]\in\Delta^o} \left|\Delta[s]\right|_+\wedge \int^{[q]\in\Lambda^o} \left|\Lambda\big(j[s],[q]\big)\right|_+\wedge X_q \\ &\cong \int^{[s]\in\Delta^o} \left|\Delta[s]\right|_+\wedge j^*X_s = \left|j^*X\right|, \end{aligned}$$

where we have used that $|j^*\Lambda[q]| = \int^{[s]\in\Delta^o} |\Delta[s]|\times\Lambda(j[s],[q])$ and "Fubini's theorem" (saying that coends commute) in the first isomorphism and the "dual Yoneda lemma" in the second isomorphism.

These isomorphisms will mean that we won't be fanatical about remembering to put the subscript Λ on sin and $|-|$.

The functor j^* from cyclic to simplicial sets given by precomposition with $j\colon \Delta^o\subseteq\Lambda^o$ has a left adjoint j_* (it also has a right adjoint, but that is not important to us right now). We have already encountered j_* in Sect. 5.3.3.

If $\mathcal{C}$ is a category with finite coproducts we get an adjoint pair

$$\mathcal{C}^{\Lambda^o} \underset{j^*}{\overset{j_*}{\leftrightarrows}} \mathcal{C}^{\Delta^o}$$

where j_* is the cyclic bar construction (with respect to the coproduct $\vee$) $j_* = B^{cy}_{\vee}$ given in degree q by $(j_*X)_q = \bigvee_{C_{q+1}} X_q$, but with a twist in the simplicial structure. To be precise, consider the bijection

$$\Lambda\big([m],[n]\big) \xrightarrow{f\mapsto\psi(f)=(\psi_\Delta(f),\psi_C(f))} \Delta\big([m],[n]\big)\times C_{m+1} \cong \big(j_*\Delta[n]\big)_m,$$

where the components are given by the unique factorization of maps in Λ. The inverse of ψ is given by composition: $\psi^{-1}(\sigma, t^a) = \sigma t^a$. Hence we can identify $\Lambda[n]$ with $j_*\Delta[n]$ where the latter has the cyclic structure $\phi^*((\sigma, t^a)) = \psi(\sigma t^a \phi)$. In general, for $y \in (j_*X)_m$ in the t^a-summand this reads $\phi^*(y) = \phi_\Delta(t^a\phi)^* y$ in the $\psi_C(t^a\phi)$-summand.

The adjoint of the first isomorphism in Lemma 6.2.1.2 then reads

Lemma 6.2.1.3 *There is a natural isomorphism*

$$|j_*Y|_\Lambda \cong \mathbb{T}_+ \wedge |Y|$$

where Y is a simplicial set.

Lemma 6.2.1.4 *Let X be a pointed cyclic set. Then*

$$\varprojlim_{\Lambda^o} X \cong \{x \in X_0 | s_0 x = t s_0 x\} \cong |X|^{\mathbb{T}}_\Lambda$$

Proof The first equation is a direct calculation, and the second equation follows from the adjunction isomorphism $|X|^{\mathbb{T}}_\Lambda = (\mathbb{T}\text{-}\mathcal{T}op_*)(S^0, |X|_\Lambda) \cong \mathcal{E}ns_*^{\Lambda^o}(S^0, X) = \varprojlim_{\Lambda^o} X$. □

Note in particular that if we consider a cyclic space as a simplicial cyclic set, then the formula always holds true if applied degreewise. For those who worry about the difference between spaces (simplicial sets) and topological spaces, we note that if G is a finite discrete group and X a simplicial G-set, then the two fixed-point constructions $|X^G|$ and $|X|^G$ are naturally homeomorphic (realization commutes with finite limits, Proposition A.2.1.1), and if K is a topological group and Y is a K-space, then $\sin(Y^K)$ and $(\sin Y)^{\sin K}$ are equal as subspaces of $\sin Y$: a simplex $y\colon |\Delta[n]| \to Y$ factors through Y^K if and only if for all $k\colon |\Delta[n]| \to K$ and $t \in |\Delta[n]|$ we have $k(t)\cdot y(t) = y(t)$.

6.2.2 The Edgewise Subdivision

If S and T are finite ordered sets, then their concatenation $S \sqcup T$ is the disjoint union of S and T with the ordering given by declaring that the canonical inclusions S and T into $S \sqcup T$ are order preserving, and that any element in the image of S is considered to be less than any element in the image of T.

Let a be a natural number. The *edgewise subdivision* functor $sd^a\colon \Delta \to \Delta$ is the composite of the diagonal $\Delta \to \Delta^{\times a}$ with the functor $\Delta^{\times a} \to \Delta$ which sends $(S_1, \dots S_a)$ to the concatenation $S_1 \sqcup \cdots \sqcup S_a$. Note that $sd^a[k-1] = [ka-1]$.

If X is a simplicial object, then $sd_a X$ is the simplicial object obtained by precomposing with sd^a. As an example, let X be the one-simplex $\Delta[1]$. Then $sd_a\Delta[1]$ has $a+1$ vertices (namely the elements of $\Delta([a-1],[1])$), it has a non-degenerate 1-simplices (namely the elements in $\Delta([2a-1],[1])$ that take the value 0 an odd number of times) and no non-degenerate k-simplices for $k>1$. Explicitly, $sd_a\Delta[1]$ is the result of gluing a copies of $\Delta[1]$ end-to-end.

The edgewise subdivision is a subdivision in the following sense. Recall the topological standard n-simplex $\Delta^n = \{(x_0, \dots, x_n) \in \mathbf{R}^{n+1} | x_i \geq 0 \text{ for all } i = 0, \dots, n,\ \sum_{i=0}^n x_i = 1\}$. Consider the map $d_a\colon \Delta^{n-1} \to \Delta^{an-1}$ sending x to the diagonal $\frac{1}{a}(x, \dots, x) \in (\mathbf{R}^n)^a = \mathbf{R}^{an}$.

Lemma 6.2.2.1 *Let X be a simplicial set. The map of geometric realizations $D_a\colon |sd_a X| \to |X|$ induced by $1 \times d_a\colon X_{an-1} \times \Delta^{n-1} \to X_{an-1} \times \Delta^{an-1}$ is a homeomorphism.*

Proof One first checks directly that D_a is well defined, and then that it induces a homeomorphism when X is the one-simplex $\Delta[1]$. Then one uses that $\Delta[n]$ is a retract of the n-fold product $\Delta[1]^n$ to establish the lemma in the case when $X = \Delta[n]$. Now, since any simplicial set is built out of simplices and the map D_a is natural in X, the general result follows. □

This construction extends to the cyclic world as follows. Let Λ_a be the category defined exactly as Λ, except that whereas $\mathrm{Aut}_\Lambda([n-1]) = C_n$ (with generator t_{n-1}), $\mathrm{Aut}_{\Lambda_a}([n-1]) = \mathrm{Aut}_\Lambda([an-1]) = C_{an}$, see [27, Sect. 1]. In more detail, Λ_a and Δ have the same objects and $\Lambda_a([n],[q]) = \Delta([n],[q]) \times C_{a(n+1)}$ where $C_{a(n+1)}$ is the cyclic group with generator $t = t_{(a,n)}$ and with $t_{(a,n)}^{a(n+1)} = 1_{[n]}$. Here a pair (σ, t^k) is considered as a composite

$$[n] \xrightarrow{t^k} [n] \xrightarrow{\sigma} [q]\cdot$$

Composition is subject to the extra relations

$$\begin{aligned} t_{(a,n)} d^i &= d^{i-1} t_{(a,n-1)}, \quad 1 \leq i \leq n, \\ t_{(a,n)} d^0 &= d^n, \\ t_{(a,n)} s^i &= s^{i-1} t_{(a,n+1)}, \quad 1 \leq i \leq n, \\ t_{(a,n)} s^0 &= s^n t_{(a,n+1)}^2. \end{aligned}$$

Notice that Λ_1 is nothing but Connes' category Λ, and that the displayed equations above are common for all a (which justifies that we occasionally will write t_n instead of $t_{(a,n)}$). As before, $j = j_a\colon \Delta \subseteq \Lambda_a$ is the inclusion, and if X is a functor from Λ_a^o to some category, the simplicial object obtained by precomposing with j is denoted j^*X. In particular, if $\Lambda_a[q]$ is the functor $[n] \mapsto \Lambda_a([n],[q])$, the simplicial set $j^*\Lambda_a[q]$ has $a(q+1)$ nondegenerate $(q+1)$-simplices, namely $s^{q-i}t_{(a,q+1)}^{k(q+2)+i+1} \in \Lambda_a([q+1],[q])$ for $i = 0,\dots,q$ and $k = 0,\dots,a-1$. From the relations above, we see that $t_{(a,n)}^{n+1}$ commutes with the face and degeneracy maps, and so if $\phi \in \Delta([m],[n])$, then we get that $\phi t_{(a,m)}^{m+1} = t_{(a,n)}^{n+1}\phi$. Hence Λ_a contains a subcategory which we may identify with $\Delta \times C_a$, and any functor from Λ_a^o comes naturally with a C_a-action when viewed as a simplicial object. For instance, $j^*\Lambda_a[q]$ has the structure of a simplicial C_a-set. To look at examples we can visualize, $j^*\Lambda_2[0]$ has two nondegenerate 1-simplices s^0t_1 and $s^0t_1^3$ joined at the vertices $1 = t_0^2$ and t_0 to form a circle, and we may picture the simplicial C_2-set $j^*\Lambda_2[1]$ as follows:

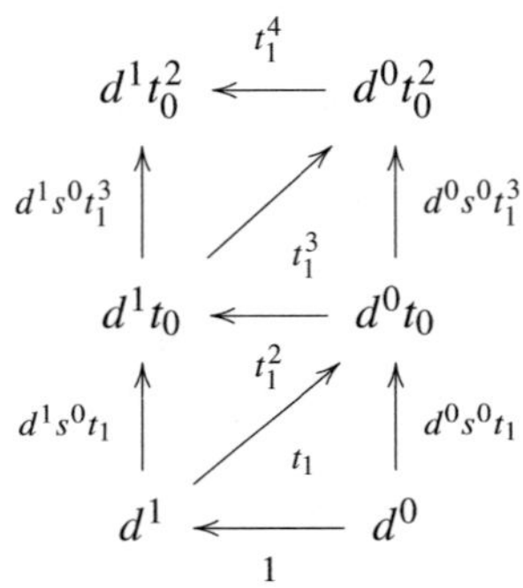

(where the unmarked 2-simplices are $\begin{smallmatrix} s^0t_2^5 & s^1t_2^4 \\ s^0t_2^2 & s^1t_2 \end{smallmatrix}$ and where the top 1-simplex $t_1^4\colon [1] \to [1]$ is identified with the bottom 1-simplex), which we recognize as a model for the cylinder $\mathbf{R}/2\mathbf{Z} \times \Delta^1$. For $a > 2$ the picture is just the same: take the bottom square and precompose its j-simplices with $t_j^{k(j+1)}$ for $k = 0,\dots,a-1$ and stack the squares on top of each other to see that we have a homeomorphism of realizations $|j^*\Lambda_a[1]| \cong \mathbf{R}/a\mathbf{Z} \times \Delta^1$.

The subdivision functor extends as

$$\begin{array}{ccc} \Delta & \xrightarrow{sd_a} & \Delta \\ {\scriptstyle j_{ab}}\downarrow & & \downarrow{\scriptstyle j_b} \\ \Lambda_{ab} & \xrightarrow{sd_a^{\Lambda_{ab}}} & \Lambda_b \end{array}$$

by declaring that $sd_a^{\Lambda_{ab}}$ is the identity on the automorphisms $\mathrm{Aut}_{\Lambda_{ab}}([n-1]) = \mathrm{Aut}_{\Lambda_b}([an-1]) = C_{abn}$.

All together, we get the following lemma (see also [27, Sect. 1]):

Lemma 6.2.2.2 *There is a homeomorphism $|j^*\Lambda_b[q]| \cong \mathbf{R}/b\mathbf{Z} \times \Delta^q$ of C_b-spaces (under the isomorphism $\mathbf{Z}/b\mathbf{Z} \cong C_b$ taking 1 to t_b). If X is a Λ_b-set, this homeo-*

morphism induces an $\mathbf{R}/b\mathbf{Z}$*-action on the realization* $|j^*X|$ *which agrees with the simplicial* C_b*-action. Furthermore, the homeomorphism* $D_a\colon |sd_a j^*X| \to |j^*X|$ *of Lemma* 6.2.2.1 *is equivariant in the sense that* $D_a(s+ab\mathbf{Z})\cdot x = (\frac{s}{a}+b\mathbf{Z})\cdot D_a(x)$.

Exactly the same considerations hold in the pointed case.

6.2.3 The Restriction Map

Let A be an $\mathbf{S}$-algebra and X a space. We will now define an important cyclic map

$$R\colon sd_q THH(A,X)^{C_q} \to THH(A,X),$$

called the *restriction map*. This map is modeled on the fact that if C is a group and $f\colon Z\to Y$ is a C-map, then f sends the C-fixed points to C-fixed points; and hence we get a map

$$Map_*(Z,Y)^C \to Map_*\big(Z^C, Y^C\big)$$

by restricting to fixed points. Notice that the $(j-1)$-simplices of $sd_a THH(A,X)$ are given by

$$THH(A,X)_{aj-1} = \underrightarrow{\operatorname{holim}}_{x_{k,l}\in\mathcal{I},1\le k\le a,1\le l\le j} Map_*\Big(\bigwedge_{k,l} S^{x_{k,l}}, X\wedge V(A)(x_{k,l})\Big),$$

where we, as before, use the notation $V(A)(x_{k,l}) = \bigwedge_{k,l} A(S^{x_{k,l}})$ and where an $a\times j$-tuple $\{x_{k,l}\}$ of elements in $\mathcal{I}$ represents the element in $\mathcal{I}^{aj}$ whose $(jk+l)$-th coordinate is $x_{k,j}$. The C_a-fixed points under the action on $\mathcal{I}^{aj}$ are exactly the image of the diagonal $\mathcal{I}^j \to \mathcal{I}^{aj}$ sending $\mathbf{x}$ to $\mathbf{x}^a = (\mathbf{x},\dots,\mathbf{x})$, and the C_a-fixed points are given by

$$\begin{aligned}&THH(A,X)^{C_a}_{aj-1}\\ &\quad\cong \underrightarrow{\operatorname{holim}}_{(x_1,\dots,x_j)\in\mathcal{I}^j} Map_*\Bigg(\Big(\bigwedge_{1\le i\le j} S^{x_i}\Big)^{\wedge a}, X\wedge V(A)\big((x_1,\dots,x_j)^a\big)\Bigg)^{C_a}.\end{aligned}$$

Note the C_a-equivariant isomorphism

$$V(A)\big((x_1,\dots,x_j)^a\big) \cong V(A)(x_1,\dots,x_j)^{\wedge a} \cong \Big(\bigwedge_{1\le i\le j} A\big(S^{x_i}\big)\Big)^{\wedge a}.$$

In the mapping space, both the domain and target are a-fold smash products with C_a-action given by permutation (except for the C_a-fixed space X which just stays on for the ride) and so we get a restriction map to the mapping space of the fixed points:

$$Map_*\Big(\bigwedge_{1\le i\le j} S^{x_i}, X\wedge V(A)(x_1,\dots,x_j)\Big).$$

Taking the homotopy colimit we get a map $sd_a THH(A,X)_{j-1}^{C_a} \to THH(A,X)_{j-1}$ which assembles to a cyclic map

$$R\colon sd_a THH(A,X)^{C_a} \to THH(A,X)$$

giving the pair $(THH(A,X), R)$ the structure of an *epicyclic space* in the following sense:

Definition 6.2.3.1 An *epicyclic space* (Y,ϕ) is a pointed cyclic space Y equipped with pointed simplicial maps

$$\phi_q\colon (sd_q Y)^{C_q} \to Y$$

for all $q \ge 1$ satisfying

1. $\phi_q t = t\phi_q$ (which implies that $\phi_q(Y_{qaj-1}^{C_{aq}}) \subseteq Y_{aj-1}^{C_a}$),
2. $\phi_a\phi_q = \phi_{aq}\colon Y_{aqj-1}^{C_{aq}} \to Y_{j-1}$ and
3. $\phi_1 = 1$.

Note that ϕ_q can be regarded as a cyclic map $(sd_q Y)^{C_q} \to Y$, and also as a C_a-equivariant simplicial map $(sd_{aq}Y)^{C_q} = sd_a((sd_q Y)^{C_q}) \to sd_a Y$ for any a. For $a \ge 1$, consider the $\mathbb{T}$-space

$$Y\langle a\rangle = \big|(sd_a Y)^{C_a}\big|.$$

In addition to the map $\phi_q\colon Y\langle aq\rangle \to Y\langle a\rangle$ we have a map—the "inclusion of fixed points"—given as $i_q\colon Y\langle aq\rangle \cong |Y|^{C_{qa}} \subseteq |Y|^{C_a} \cong Y\langle a\rangle$. By the definition of an epicyclic space we get that these maps obey the following relations

$$\phi_q\phi_r = \phi_{qr}, \qquad \phi_1 = i_1 = id,$$
$$i_q i_r = i_{qr}, \qquad i_q\phi_r = \phi_r i_q.$$

In other words, $a \mapsto Y\langle a\rangle$ is a functor to topological spaces from the category $\mathcal{RF}$:

Definition 6.2.3.2 Let $\mathcal{RF}$ be the category whose objects are the positive integers, and where there is a morphism $f_{r,s}\colon a \to b$ whenever $a = rbs$ for positive integers r and s, with composition $f_{r,s}\circ f_{p,q} = f_{rp,qs}$. An epicyclic space (Y,ϕ) gives rise to a functor from $\mathcal{RF}$ to spaces by sending a to $Y\langle a\rangle$, $f_{q,1}$ to ϕ_q and $f_{1,q}$ to i_q. Sloppily, we write $R = R_r = f_{r,1}$ and $F = F^r = f_{1,r}$ for any unspecified r (and range), hence the name of the category. For any given prime p, the full subcategory of $\mathcal{RF}$ containing only the powers of p is denoted $\mathcal{RF}_p$.

Example 6.2.3.3 We have seen that topological Hochschild homology defines an epicyclic space, and a map of **S**-algebras gives rise to a map respecting the epicyclic structure.

Another example is the cyclic nerve. Let $\mathcal{C}$ be any (small simplicial) category, and consider the *cyclic nerve* $B^{cy}\mathcal{C}$ discussed in Sect. 4.2.5.1. This is a straight-forward generalization of the cyclic bar construction of a monoid:

$$B_q^{cy}\mathcal{C} = \{c_q \leftarrow c_0 \leftarrow c_1 \leftarrow \cdots \leftarrow c_{q-1} \leftarrow c_q \in \mathcal{C}\}$$

with face and degeneracies given by composition and insertion of identities, and with cyclic structure given by cyclic permutation. This is a cyclic set, and $|B^{cy}\mathcal{C}|^{\mathbb{T}} \cong \lim_{\overleftarrow{\Lambda^o}} B^{cy}\mathcal{C} = ob\mathcal{C}$ where an object is identified with its identity morphism in $B_0^{cy}\mathcal{C}$. The fixed point sets under the finite subgroups of the circle are more interesting as $(sd_r B^{cy}\mathcal{C})^{C_r} \cong B^{cy}\mathcal{C}$. In fact, an element $x \in (sd_r B^{cy}\mathcal{C})_{q-1} = B^{cy}_{rq-1}\mathcal{C}$ which is fixed by the C_r-action must be of the form

$$c_q \xleftarrow{f_1} c_1 \xleftarrow{f_2} \cdots \xleftarrow{f_q} c_q \xleftarrow{f_1} c_1 \xleftarrow{f_2} \cdots \xleftarrow{f_q} c_q \xleftarrow{f_1} c_1 \xleftarrow{f_2} \cdots \xleftarrow{f_q} c_q$$

and we get an isomorphism $\phi_r \colon (sd_r B^{cy}\mathcal{C})^{C_r} \cong B^{cy}\mathcal{C}$ by forgetting the repetitions. This equips the cyclic nerve with an epicyclic structure, and a functor of categories gives rise to a map of cyclic nerves respecting the epicyclic structure.

An interesting example is the case where A is an **S**-algebra and $\mathcal{C}$ is the simplicial monoid $M = THH_0(A) = \operatorname{holim}_{\overrightarrow{x \in \mathcal{I}}} \Omega^x A(S^x)$. We have a map $B^{cy}M \to THH(A)$ given by smashing together functions

$$\prod_{0 \le i \le q} \operatorname*{holim}_{\overrightarrow{x_i \in \mathcal{I}}} \Omega^{x_i} A\big(S^{x_i}\big) \to \operatorname*{holim}_{\overrightarrow{\mathbf{x} \in \mathcal{I}^{q+1}}} \Omega^{\vee \mathbf{x}} \bigwedge_{0 \le i \le q} A\big(S^{x_i}\big).$$

This map preserves the epicyclic structure.

Remark 6.2.3.4 Our notion of an epicyclic space is **not** the same as the one proposed in [103], and which later was used by Burghelea, Fiedorowicz, and Gajda in [48] to compare Adams operators. This older definition generalized the so-called power maps $P_q = \phi_q^{-1} \colon B^{cy}\mathcal{C} \to (sd_q B^{cy}\mathcal{C})^{C_q}$ instead. Cyclic nerves are epicyclic spaces under either definition, but topological Hochschild homology only gives rise to an epicyclic structure under the current definition.

Remark 6.2.3.5 An epicyclic space (Y, ϕ) is more than a functor from $\mathcal{RF}$ to spaces. In fact, as each $(sd_a Y)^{C_a}$ is again a cyclic space, each $Y\langle a\rangle = |(sd_a Y)^{C_a}|$ comes equipped with a $\mathbb{T}$-action. However, $Y\langle a\rangle$ is not a functor to $\mathbb{T}$-spaces: the inclusion of fixed point spaces under the finite subgroups of $\mathbb{T}$ is not $\mathbb{T}$-equivariant, but speeds up the action. We may encode this as a continuous functor sending $\theta \in \mathbf{R}/\mathbf{Z} = \mathbb{T}$ to $\rho_\theta \colon Y\langle a\rangle \to Y\langle a\rangle$ with the additional relations

$$\phi_q \rho_\theta = \rho_\theta \phi_q, \qquad i_q \rho_\theta = \rho_{q\theta} i_q, \qquad \rho_\theta \rho_\tau = \rho_{\theta+\tau}.$$

This can again be encoded in a topological category $\mathcal{SRF}$ with objects the natural numbers and morphisms $\mathcal{SRF}(a,b) = \mathbb{T} \times \mathcal{RF}(a,b)$. Composition is given by

$$(\theta, f_{r,s})(\tau, f_{p,q}) = (\theta + s\tau, f_{rp,qs}).$$

Sending θ to ρ_θ we see that any epicyclic space give rise to a continuous functor $a \mapsto Y\langle a\rangle$ from $\mathcal{SRF}$ to topological spaces. In the MSRI notes [105] Goodwillie defines

$$\underline{TC}(A, X) = \left\{k \mapsto \underset{\overleftarrow{a \in \mathcal{SRF}}}{\text{holim}} \left|sd_a THH\big(A, S^k \wedge X\big)^{C_a}\right|\right\}$$

(the homotopy limit remembers the topology in $\mathbb{T}$), and gives a proof that this elegant definition agrees with the one we are going to give. The only reasons we have chosen to refrain from giving this as our definition is that our definition is custom built for our application (and for computations), and the proof that they agree would lengthen the discussion further.

6.2.4 Properties of the Fixed Point Spaces

We now make a closer study of the C_q-fixed point spaces of THH. The most important result is Proposition 6.2.4.2, often referred to as "the fundamental cofibration sequence" which guarantees that the actual (and not just the homotopy) fixed point spaces will have good homotopical properties.

Definition 6.2.4.1 Let

$$T\langle\ \rangle(A, X) \colon \mathcal{RF} \to \mathcal{S}_*$$

with $T\langle a\rangle(A,X) = \sin|sd_a THH(A,X)^{C_a}|$, be the functor associated with the epicyclic space $(THH(A,X), R)$. We set $R = R_r = T\langle f_{r,1}\rangle$ (for "Restriction", which it is) and $F = F^r = T\langle f_{1,r}\rangle$ (for "Frobenius", see Sect. 6.4.2.2), which here is the inclusion of fixed points

$$T\langle rq\rangle(A,X) \cong \sin\big|THH(A,X)\big|^{C_{rq}} \subseteq \sin\big|THH(A,X)\big|^{C_q} \cong T\langle q\rangle(A,X).$$

Since this is a simplicial functor in X we have an associated spectrum

$$\underline{T}\langle a\rangle(A,X) = \big\{k \mapsto T\langle a\rangle\big(A, S^k \wedge X\big)\big\}.$$

Variants of $\underline{T}\langle a\rangle(A,X)$ are denoted $TR^a(A,X)$ by some authors.

Remember that each $\underline{T}\langle a\rangle$ can be considered as functors to cyclic spaces (but they do not assemble when varying a). We will not distinguish notationally whether we think of $\underline{T}\langle a\rangle(A,X)$ as a simplicial or cyclic space, and we offer the same ambiguity to $\underline{T}(A,X) = \underline{T}\langle 1\rangle(A,X)$.

The spectra $\underline{T}\langle a\rangle(A,X)$ are Ω-spectra for all positive integers a and are homotopy functors in A. This important fact can be seen by analyzing the restriction maps as in Proposition 6.2.4.2 below, establishing the "fundamental cofiber sequence" ("cofiber" since the result is most often used in the spectrum version where cofiber and fiber sequences agree). A variant of this proposition was proven by Madsen in a letter to Hsiang around 1988. It does play a role in [27], but only in the form of the Segal–tom Dieck splitting of the fixed point set of an equivariant suspension spectrum, cf. [256] and [286, 287].

The fundamental cofibration sequence is vital for all calculations of TC, and appears as Theorem 1.10 in Bökstedt and Madsen's first paper [28] on the topological cyclic homology of the integers. In [105] it is used to simplify many of the arguments in [27]. This is how we will use it. For instance, the mentioned properties of the $\underline{T}\langle a\rangle(A,X)$-spectra follows as corollaries, noting that homotopy orbits preserve equivalences.

Proposition 6.2.4.2 *Let p be a prime. Then there is a chain of natural equivalences from the homotopy fiber of*

$$\underline{T}\langle p^n\rangle(A,X) \xrightarrow{R} \underline{T}\langle p^{n-1}\rangle(A,X)$$

to $sd_{p^n}\underline{T}(A,X)_{hC_{p^n}}$. *Indeed, for each* j, *the homotopy fiber of*

$$\left(sd_{p^n}THH(A,X)^{C_{p^n}}\right)_{j-1} \xrightarrow{R} \left(sd_{p^{n-1}}THH(A,X)^{C_{p^{n-1}}}\right)_{j-1}$$

is naturally weakly equivalent to $\operatorname{holim}_{\overrightarrow{k}} \Omega^k((sd_{p^n}THH(A,S^k\wedge X)_{j-1})_{hC_{p^n}})$.

More generally, if q *is a positive integer, consider the category* $D(q)$ *whose objects are positive integers dividing* q, *and where there is a unique map* $r\to s$ *if* s *divides* r. *The homotopy fiber of the map*

$$\underline{T}\langle q\rangle(A) \to \underset{\overleftarrow{1\neq r\in D(q)}}{\operatorname{holim}} \underline{T}\langle q/r\rangle(A)$$

induced by the restriction maps is connected to $\underline{T}(A)_{hC_q}$ *by a natural chain of levelwise equivalences.*

Proof Since maps of simplicial spaces that induce weak equivalences in every degree induce weak equivalences on diagonals (Theorem A.6.0.1) the first statement follows from the second. Let $q=p^n$, $G=C_q$ and $H=C_p$. For $\mathbf{x}\in\mathcal{I}^j$, let

$$Z(\mathbf{x})=\left(\bigwedge_{1\le i\le j} S^{x_i}\right)^{\wedge q}, \quad\text{and}\quad W(\mathbf{x})=X\wedge\left(\bigwedge_{1\le i\le j} A\left(S^{x_i}\right)\right)^{\wedge q}.$$

By the approximation Lemma 2.2.2.2, the (homotopy) fiber of the restriction map R is naturally equivalent to

$$\underset{\overrightarrow{\mathbf{x}\in\mathcal{I}^j}}{\operatorname{holim}} \operatorname{hofib}\left\{Map_*\left(Z(\mathbf{x}),W(\mathbf{x})\right)^G \to Map_*\left(Z(\mathbf{x})^H,W(\mathbf{x})^H\right)^{G/H}\right\}$$

which, by the canonical isomorphism

$$Map_*\big(Z(\mathbf{x})^H, W(\mathbf{x})^H\big)^{G/H} \cong Map_*\big(Z(\mathbf{x})^H, W(\mathbf{x})\big)^G$$

is isomorphic to

$$\underset{\overrightarrow{\mathbf{x}\in\mathcal{I}^j}}{\operatorname{holim}} Map_*\big(U(\mathbf{x}), W(\mathbf{x})\big)^G$$

where $U(\mathbf{x}) = Z(\mathbf{x})/(Z(\mathbf{x})^H)$. As $U(\mathbf{x})$ is a free finite based G-complex, Corollary 6.3.2.7 below tells us that there is a natural chain of maps

$$\begin{aligned} Map_*\big(U(\mathbf{x}), W(\mathbf{x})\big)^G &\longrightarrow Map_*\big(U(\mathbf{x}), \lim_{\overrightarrow{k}} \Omega^k(S^k \wedge W(\mathbf{x}))\big)^G \\ &\overset{\sim}{\longleftarrow} \lim_{\overrightarrow{k}} \big(\Omega^k Map_*\big(U(\mathbf{x}), S^k \wedge W(\mathbf{x})\big)_{hG}\big), \end{aligned}$$

and that the first map is $\sum \mathbf{x} - 1$ connected. Furthermore, the cofiber sequence $Z(\mathbf{x})^H \subseteq Z(\mathbf{x}) \to U(\mathbf{x}) = Z(\mathbf{x})/(Z(\mathbf{x})^H)$ induces a fiber sequence

$$\begin{aligned} Map_*\big(U(\mathbf{x}), S^k \wedge W(\mathbf{x})\big) &\longrightarrow Map_*\big(Z(\mathbf{x}), S^k \wedge W(\mathbf{x})\big) \\ &\longrightarrow Map_*\big(Z(\mathbf{x})^H, S^k \wedge W(\mathbf{x})\big). \end{aligned}$$

Since $Z(\mathbf{x})^H$ is $\mathbf{x}q/p$-dimensional and $S^k \wedge W(\mathbf{x})$ is $\mathbf{x}q + k - 1$-connected, the first map in the fiber sequence is $\mathbf{x}(q - q/p) + k - 1$-connected, and since, by Lemma A.7.3.1, taking homotopy orbits preserves connectivity we get that the map

$$\Omega^k\big(Map_*\big(U(\mathbf{x}), S^k \wedge W(\mathbf{x})\big)_{hG}\big) \to \Omega^k\big(Map_*\big(Z(\mathbf{x}), S^k \wedge W(\mathbf{x})\big)_{hG}\big)$$

is $\mathbf{x}(q - q/p) - 1$-connected. Taking the homotopy colimit over $\mathcal{I}^j$, this gives the statement for fixed j and prime p.

The proof of the statement for composite q is obtained quite similarly, letting $G = C_q$, $Z(\mathbf{x})$ and $W(\mathbf{x})$ be as before, but forgetting that q was a prime power. Assume by induction that the statement has been proven for all groups of cardinality less than G, and so that all for these groups the fixed point spectra of *THH* are homotopy functors and Bökstedt's approximation lemma applies.

This means that the canonical map

$$\begin{aligned} &\underset{\overrightarrow{\mathbf{x}\in\mathcal{I}^j}}{\operatorname{holim}}\ \underset{\overleftarrow{0\neq H\subset G}}{\operatorname{holim}} Map_*\big(Z(\mathbf{x})^H, W(\mathbf{x})^H\big)^{G/H} \\ &\qquad \longrightarrow \underset{\overleftarrow{0\neq H\subset G}}{\operatorname{holim}}\ \underset{\overrightarrow{\mathbf{x}\in\mathcal{I}^j}}{\operatorname{holim}} Map_*\big(Z(\mathbf{x})^H, W(\mathbf{x})^H\big)^{G/H} \end{aligned}$$

is an equivalence. The right hand side is isomorphic to

$$\underset{\overleftarrow{0\neq H\subset G}}{\operatorname{holim}} \left(\underset{\overrightarrow{\mathbf{x}\in\mathcal{I}^{j\cdot|G/H|}}}{\operatorname{holim}} Map_*\big(Z(\mathbf{x})^H, W(\mathbf{x})^H\big) \right)^{G/H} = \underset{\overleftarrow{0\neq H\subset G}}{\operatorname{holim}} sd_{|G/H|} THH(A, X)^{G/H},$$

and the left hand side is isomorphic to

$$\underrightarrow{\operatorname{holim}}_{\mathbf{x}\in\mathcal{I}^j}\ \underleftarrow{\operatorname{holim}}_{0\neq H\subset G} Map_*\big(Z(\mathbf{x})^H, W(\mathbf{x})\big)^G$$

which is equivalent to

$$\underrightarrow{\operatorname{holim}}_{\mathbf{x}\in\mathcal{I}^j} Map_*\left(\bigcup_{0\neq H\subset G} Z(\mathbf{x})^H, W(\mathbf{x})\right)^G$$

(the union can be replaced by the corresponding homotopy colimit). Via this equivalence, the homotopy fiber of

$$sd_{|G|}THH(A,X)^G_{j-1} \longrightarrow \underleftarrow{\operatorname{holim}}_{0\neq H\subset G} sd_{|G/H|}THH(A,X)^{G/H}_{j-1}$$

is equivalent to

$$\underrightarrow{\operatorname{holim}}_{\mathbf{x}\in\mathcal{I}^j} Map_*\big(U(\mathbf{x}), W(\mathbf{x})\big)^G,$$

where $U(\mathbf{x}) = Z(\mathbf{x})/\bigcup_{0\neq H\subset G} Z(\mathbf{x})^H$. Then the same argument leads us to our conclusion, using that $U(\mathbf{x})$ is a free finite based G-space. □

Corollary 6.2.4.3 *Let q be a positive integer, $A \to B$ a map of* $\mathbf{S}$*-algebras inducing an equivalence $THH(A) \to THH(B)$ and X a space (in particular, $A \to B$ may be a stable equivalence). Then*

1. *the induced map $\underline{T}\langle q\rangle(A,X) \to \underline{T}\langle q\rangle(B,X)$ is an equivalence,*
2. *$\underline{T}\langle q\rangle(A,X)$ is a connective Ω-spectrum,*
3. *$\underline{T}\langle q\rangle(-,X)$ is Morita invariant and*
4. *$\underline{T}\langle q\rangle(-,X)$ preserves products up to levelwise equivalence.*

Proof This follows by Proposition 6.2.4.2 and the corresponding properties of *THH*, plus the fact that homotopy orbits preserve loops and products of spectra up to stable equivalence. □

6.2.4.1 $\Gamma\mathcal{S}_*$-Categories

Essentially just the same construction can be applied to the case of $\Gamma\mathcal{S}_*$-categories.

If $\mathcal{C}$ is a $\Gamma\mathcal{S}_*$-category, $THH(\mathcal{C},X)$ also has its restriction map R, and $(THH(\mathcal{C},X),R)$ is an epicyclic space: If $\mathbf{x}\in\mathcal{I}^q$, then we have a restriction map

$$\big(\Omega^{\vee\mathbf{x}^a}\big(X\wedge V(\mathcal{C})(\mathbf{x}^a)\big)\big)^{C_a} \to \Omega^{\vee\mathbf{x}}\big(X\wedge V(\mathcal{C})\big((\mathbf{x}^a)\big)^{C_a}\big)$$

as before, and note the canonical isomorphism $V(\mathcal{C})(\mathbf{x}^a)^{C_a} \cong V(\mathcal{C})(\mathbf{x})$. Proceeding just as for $\mathbf{S}$-algebras we see that

$$a \mapsto T\langle a\rangle(\mathcal{C}, X) = \sin\left|sd_a THH(\mathcal{C}, X)^{C_a}\right|$$

defines a functor from the category $\mathcal{RF}$ of Definition 6.2.3.2 (or better: a continuous functor from the topological category $S\mathcal{RF}$ of Remark 6.2.3.5) to spaces.

Lemma 6.2.4.4 *Let $\mathcal{C}$ be a $\Gamma\mathcal{S}_*$-category and X a space. If p is a prime, the restriction map R fits into a fiber sequence*

$$\underline{T}(\mathcal{C}, X)_{hC_{p^n}} \longrightarrow \underline{T}\langle p^n\rangle(\mathcal{C}, X) \xrightarrow{R} \underline{T}\langle p^{n-1}\rangle(\mathcal{C}, X).$$

More generally, if a is a positive integer we have a fiber sequence

$$\underline{T}(\mathcal{C}, X)_{hC_a} \to \underline{T}\langle a\rangle(\mathcal{C}, X) \to \operatorname*{holim}_{\overleftarrow{1\neq r|q}} \underline{T}\langle q/r\rangle(\mathcal{C}, X)$$

induced by the restriction map and where the homotopy limit is over the positive integers r dividing q.

Proof Exactly the same proof as for the $\mathbf{S}$-algebra case proves that this is indeed true in every simplicial degree. □

As before, this gives a series of corollaries.

Corollary 6.2.4.5 *Let a be a positive integer and X a space. Any $\Gamma\mathcal{S}_*$-functor $F\colon \mathcal{C} \to \mathcal{D}$ inducing an equivalence $THH(\mathcal{C}) \to THH(\mathcal{D})$ induces an equivalence*

$$\underline{T}\langle a\rangle(\mathcal{C}, X) \to \underline{T}\langle a\rangle(\mathcal{D}, X).$$ ☺

Corollary 6.2.4.6 *Let $\mathcal{C}$ be a $\Gamma\mathcal{S}_*$-category, a a positive integer and X a space. Then*

1. *$\underline{T}\langle a\rangle(\mathcal{C}, X)$ is a connective Ω-spectrum,*
2. *the functor $\underline{T}\langle p^n\rangle(-, X)$ takes $\Gamma\mathcal{S}_*$-equivalences of categories to equivalences,*
3. *if A is a ring, then the inclusion $A \subseteq \mathcal{P}_A$ as a rank one module induces an equivalence $\underline{T}\langle a\rangle(A, X) \xrightarrow{\sim} \underline{T}\langle a\rangle(\mathcal{P}_A, X)$ and*
4. *$\underline{T}\langle a\rangle(-, X)$ preserves products up to equivalence.* ☺

Corollary 6.2.4.7 *Let a be a positive integer and X a space. If $\mathcal{C}$ is a symmetric monoidal $\Gamma\mathcal{S}_*$-category, then $T\langle a\rangle(\bar{H}\mathcal{C}, X) = \{k \mapsto T\langle a\rangle(\bar{H}\mathcal{C}(S^k), X)\}$ is an Ω-spectrum, equivalent to $\underline{T}\langle a\rangle(\mathcal{C}, X)$.*

Proof That $T\langle a\rangle(\bar{H}\mathcal{C}, X)$ is an Ω-spectrum follows for instance from Corollary 6.2.4.6(2) and 6.2.4.6(4) since $\bar{H}\mathcal{C}(k_+)$ is $\Gamma\mathcal{S}_*$-equivalent to $\mathcal{C}^{\times k}$. That the two Ω-spectra are equivalent follows by comparing both to the bispectrum $\underline{T}\langle a\rangle(\bar{H}\mathcal{C}, X)$. □

Corollary 6.2.4.8 *If $\mathcal{C}$ is an additive category, then*

$$T\langle a\rangle(\bar{H}\mathcal{C}, X) \to \left\{k \mapsto T\langle a\rangle\left(S^{(k)}\mathcal{C}, X\right)\right\}$$

is an equivalence of Ω-spectra.

Proof Follows by Corollary 6.2.4.5 since $THH(\bar{H}\mathcal{C}(S^k), X) \to THH(S^{(k)}\mathcal{C}, X)$ is an equivalence by Corollary 4.3.5.8. □

6.2.5 Spherical Group Rings

In the special case of spherical group rings Example 2.1.4.3(2) the restriction maps split, making it possible to give explicit models for the C_{p^n}-fixed point spectra of topological Hochschild homology.

Lemma 6.2.5.1 *The restriction maps split canonically for spherical group rings.*

Proof Let G be a simplicial group. We will prove that, for each space X and positive integers a and b, the restriction map $sd_{ab}THH(\mathbf{S}[G], X)^{C_{ab}} \to sd_a THH(\mathbf{S}[G], X)^{C_a}$ splits. We fix an object $\mathbf{x} \in \mathcal{I}^j$, and consider the restriction map

$$\begin{array}{c} Map_*\left(\left(\bigwedge_{i=1}^{j} S^{x_i}\right)^{\wedge ab}, X\wedge\left(\bigwedge_{i=1}^{j}\left(S^{x_i}\wedge G_+\right)\right)^{\wedge ab}\right)^{C_{ab}} \\ \downarrow \\ Map_*\left(\left(\bigwedge_{i=1}^{j} S^{x_i}\right)^{\wedge a}, X\wedge\left(\bigwedge_{i=1}^{j}\left(S^{x_i}\wedge G_+\right)\right)^{\wedge a}\right)^{C_a} \end{array}.$$

Let $S = (\bigwedge_{i=1}^{j} S^{x_i})^{\wedge a}$ and $n = a \cdot |\vee \mathbf{x}|$, and consider the isomorphism

$$\left|S^{\wedge b}\right| \cong |S|\wedge S^{\perp}$$

coming from the one-point compactification of

$$\mathbf{R}^n \otimes \mathbf{R}^b \cong \mathbf{R}^n \otimes \left(\text{diag} \oplus \text{diag}^{\perp}\right) \cong \mathbf{R}^n \oplus \left(\mathbf{R}^n \otimes \text{diag}^{\perp}\right)$$

where $\text{diag} \subseteq \mathbf{R}^b$ is the diagonal line. The desired splitting

$$Map_*\left(S, X\wedge S\wedge G_+^{\times ja}\right)^{C_a} \to Map_*\left(S^{\wedge b}, X\wedge S^{\wedge b}\wedge G_+^{\times jab}\right)^{C_{ab}}$$

is determined by sending $f\colon |S| \to |X \wedge S \wedge G_+^{\times ja}|$ to

$$\begin{aligned}|S^{\wedge b}| &\cong |S| \wedge S^\perp \xrightarrow{f\wedge id} |X \wedge S \wedge G_+^{\times ja}| \wedge S^\perp \cong |X| \wedge |S| \wedge S^\perp \wedge |G_+^{\times ja}| \\ &\cong |X \wedge S^{\wedge b} \wedge G_+^{\times ja}| \xrightarrow{id\wedge \mathrm{diag}} |X \wedge S^{\wedge b} \wedge G_+^{\times jab}|,\end{aligned}$$

where the isomorphisms beside the above chosen splitting of $|S^{\wedge b}|$ are induced by permuting smash factors and using that smash commutes with realization. □

Example 6.2.5.2 To see how the isomorphism $|S|^{\wedge b} \cong |S| \wedge S^\perp$ of the above proof works, consider the following example.

Let $S = S^1$ and identify $|S|$ with the one-point compactification of $\mathbf{R}$, let $b = 2$. Then $|S^{\wedge 2}| \cong |S| \wedge S^\perp$ is obtained by one-point compactification of

$$\mathbf{R}^2 \cong \mathbf{R} \oplus \mathbf{R}, \qquad \begin{bmatrix} a \\ b \end{bmatrix} \mapsto \big((a+b)/2, (a-b)/2\big)$$

and the $\mathbf{Z}/2$-action is trivial in the first factor and mult by -1 in the other. Notice that if $f\colon |S| \to |X \wedge S \wedge G_+|$ sends a to $x_a \wedge s_a \wedge g_a$, then the composite

$$|S^2| \cong |S| \wedge S^\perp \xrightarrow{f\wedge id} |X \wedge S \wedge G_+| \wedge S^\perp \cong |X \wedge S^2 \wedge G_+| \to |X \wedge S^2 \wedge G_+^{\times 2}|$$

sends $[{a \atop b}]$ to

$$x_{(a+b)/2} \wedge s_{(a+b)/2} \wedge g_{(a+b)/2} \wedge (a-b)/2$$

in $|X \wedge S \wedge G_+| \wedge S^\perp$ to

$$x_{(a+b)/2} \wedge \begin{bmatrix} s_{(a+b)/2} + (a-b)/2 \\ s_{(a+b)/2} - (a-b)/2 \end{bmatrix} \wedge g_{(a+b)/2}$$

in $|X \wedge S^2 \wedge G_+|$, and finally to

$$x_{(a+b)/2} \wedge \begin{bmatrix} s_{(a+b)/2} + (a-b)/2 \\ s_{(a+b)/2} - (a-b)/2 \end{bmatrix} \wedge g_{(a+b)/2} \wedge g_{(a+b)/2}$$

in $|X \wedge S^2 \wedge G_+^{\times 2}|$.

Exchanging a and b in this formula transforms it to

$$x_{(a+b)/2} \wedge \begin{bmatrix} s_{(a+b)/2} - (a-b)/2 \\ s_{(a+b)/2} + (a-b)/2 \end{bmatrix} \wedge g_{(a+b)/2} \wedge g_{(a+b)/2}.$$

From the splitting of Lemma 6.2.5.1 and from the fundamental cofibration sequence Proposition 6.2.4.2 we get the following "Segal–tom Dieck" calculation of the fixed points:

Corollary 6.2.5.3 *Let a be a positive integer. The splitting of the restriction map induces a natural equivalence*

$$\bigvee_{r|a} |THH(\mathbf{S}[G])|_{hC_r} \to |THH(\mathbf{S}[G])|^{C_a},$$

where the sum ranges over the positive integers r dividing a. Under this equivalence, the restriction map corresponds to the projection. ☺

Coupled with the equivalence $THH(\mathbf{S}[G]) \simeq \mathbf{S}[B^{cy}(G)]$ of Example 4.2.2.7, this gives an effective calculation of the fixed points of topological Hochschild homology of spherical group rings. The inclusion of fixed points maps F and their relation to this splitting are discussed in Sect. 6.4.2.3 below.

6.3 (Naïve) G-Spectra

Let G be a simplicial monoid. The category of *G-spectra*, $G\mathcal{S}pt$ is the category of simplicial functors from G to the category of spectra. A map of G-spectra is called a levelwise (resp. stable) equivalence if the underlying map of spectra is.

For a compact Lie group G, this notion of G-spectra is much less rigid than what most people call G-spectra (see e.g., [173]), and they would prefer to call these spectra something like "naïve pre-G-spectra". To make it quite clear: in our setup, a G-spectrum X is just a normal spectrum with an action by G: a sequence of G-spaces together with structure maps $S^1 \wedge X^n \to X^{n+1}$ that are G-maps. A map of G-spectra $X \to Y$ is simply a collection of G-maps $X^n \to Y^n$ commuting with the structure maps.

Again, G-spectra form a simplicial category, with function spaces given by

$$\underline{G\mathcal{S}pt}^0(X,Y) = \{[q] \mapsto G\mathcal{S}pt(X \wedge \Delta[q]_+, Y)\}.$$

Even better, it has function spectra

$$\underline{G\mathcal{S}pt}(X,Y) = \{k \mapsto \underline{G\mathcal{S}pt}^0(X, Y^{k+?})\},$$

where $Y^{k+?}$ is the shifted spectrum $\{n \mapsto Y^{k+n}\}$.

If X is a G-spectrum we could define the homotopy orbit and fixed point spectra levelwise, i.e., X_{hG} would be $\{k \mapsto (X^k)_{hG}\}$ and "X^{hG}" would be $\{k \mapsto (X^k)^{hG}\}$. These construction obviously preserve levelwise equivalences, but just as the homotopy limit naïvely defined (without the $\sin|-|$, see Appendix A.7) may not preserve weak equivalences, some care is needed in the stable case.

Levelwise homotopy orbits always preserve stable equivalences (they preserve levelwise equivalences by Lemma A.9.2.2 stable equivalences by the method of the proof of Lemma A.7.4.7). On the other hand, without some adjustments levelwise homotopy fixed points may not. However, if the spectrum X is an Ω-spectrum, stable and levelwise equivalences coincide, and this may always be assured by applying the construction $QX = \{k \mapsto Q^k X = \lim_{\overrightarrow{n}} \Omega^n X^{k+n}\}$ of Appendix A.3.2.1 (if we were operating with "genuine G-spectra" the colimit would not have been over the natural numbers, but rather over G-representations). This is encoded in the real definition.

Definition 6.3.0.1 Let G be a simplicial monoid and X a G-spectrum. Then the *homotopy orbit spectrum* is given by

$$X_{hG} = \{k \mapsto (X^k)_{hG}\}$$

whereas the *homotopy fixed point spectrum* is given by

$$X^{hG} = \{k \mapsto (Q^k X)^{hG}\}.$$

Since Q transforms stable equivalences to levelwise equivalences and homotopy fixed points preserve weak equivalences of G-spaces we get.

Lemma 6.3.0.2 *Let G be a simplicial monoid and $f: X \to Y$ a map of G-spectra. If f is a stable equivalence of spectra, then $f_{hG}: X_{hG} \to Y_{hG}$ and $f^{hG}: X^{hG} \to Y^{hG}$ are stable equivalences.* ☺

If G is a simplicial group, we let EG be the free contractible G-space obtained by the one sided bar construction, as Definition A.9.2.1 (since here G is a group and not a monoid, we can choose either of the two sides of the bar construction), and recall that the homotopy orbits and fixed point spaces are given by $EG_+ \wedge_G X^k$ and $Map_*(EG_+, X^k)^G$. The skeleton filtration of EG gives rise to spectral sequences:

Lemma 6.3.0.3 *Let G be a simplicial group and X a G-spectrum. The skeleton filtration of EG gives rise to spectral sequences*

$$E^2_{s,t} = H_s(BG; \pi_t X) \Rightarrow \pi_{s+t}(X_{hG})$$

and

$$E^2_{s,t} = H^{-s}(BG; \pi_t X) \Rightarrow \pi_{s+t}\left(X^{hG}\right).$$

The homotopy orbit spectral sequence is a first quadrant spectral sequence and is strongly convergent, whereas the homotopy fixed point spectral sequence is a second quadrant spectral sequence and is conditionally convergent. The indexing is so that the differentials are of the form $d^r: E^r_{s,t} \to E^r_{s-r,t+r-1}$. ☺

6.3.1 Circle and Finite Cyclic Actions

If X is an $\mathbf{S}^1$-spectrum, we can also consider the homotopy fixed points under the finite subgroups $C \subset \mathbf{S}^1$. For any simplicial group G, let EG be the free contractible

G-space obtained by applying the one-sided bar construction to G. Notice that $E\mathbf{S}^1$ is also a free contractible C-space, and by Corollary A.9.2.4 $Map_*(E\mathbf{S}^1{}_+, X)^C \to Map_*(EC_+, X)^C$ is an equivalence so we can calculate X^{hC} equally well as $Map_*(E\mathbf{S}^1{}_+, X)^C$. Thus, if $C' \subseteq C$ is a subgroup, we can think of $X^{hC} \to X^{hC'}$ most conveniently as the inclusion $Map_*(E\mathbf{S}^1{}_+, X)^C \subseteq Map_*(E\mathbf{S}^1{}_+, X)^{C'}$.

Lemma 6.3.1.1 *If X is an $\mathbf{S}^1$-spectrum and p some prime, then the natural map*

$$X^{h\mathbf{S}^1} \to \underset{\overleftarrow{r}}{\operatorname{holim}}\, X^{hC_{p^r}}$$

is an equivalence after p-completion.

Proof This is just a reformulation of Corollary A.7.6.4. □

6.3.2 The Norm Map

The theory for finite groups has a nice continuation to a theory for compact Lie groups. We will only need one case beyond finite groups: $G = \mathbf{S}^1 = \sin|S^1|$, and in an effort to be concrete, we cover that case in some detail. In these cases the theory simplifies considerably, so although some of the considerations to follow have more general analogs (see e.g., [173, 195] or [116]) we shall restrict our statements to these cases, and we use ideas close to [311]. We have to define the *norm map* $\Sigma^{Ad(G)} X_{hG} \to X^{hG}$ (where "$Ad(G)$" is a place holder for the "adjoint representation", which is of dimension 0 for finite groups and is trivial and of dimension 1 for the circle group).

Lemma 6.3.2.1 *Let G be a finite discrete group. Then the inclusion*

$$\underline{\mathbf{S}} \wedge G_+ \to \prod_G \underline{\mathbf{S}} \cong \underline{\mathcal{S}_*}(G_+, \underline{\mathbf{S}})$$

of the finite wedge into the finite product is a stable equivalence and a $G \times G$-map where $G \times G$ acts on G to the left via $(a, b) \cdot g = a \cdot g \cdot b^{-1}$, and permutes the factors to the right accordingly.

Proof This is a special case of Corollary A.8.2.4. □

Lemma 6.3.2.2 *Let Y be a spectrum, and let the functorial (in Y) $\mathbf{S}^1 \times \mathbf{S}^1$-map of spectra*

$$f' : \mathbf{S}^1{}_+ \wedge Y \longrightarrow Map_*\big(\mathbf{S}^1{}_+, S^1 \wedge Y\big)$$

be the adjoint of the composite

$$\mathbf{S}^1{}_+ \wedge \mathbf{S}^1{}_+ \wedge Y \xrightarrow{\mu \wedge 1} \mathbf{S}^1{}_+ \wedge Y \xrightarrow{pr \wedge 1} \mathbf{S}^1 \wedge Y \longrightarrow \sin\big|S^1 \wedge Y\big|$$

(where the last map is the adjoint of the map $Y \to \underline{S}_*(\sin|S^1|, \sin|S^1 \wedge Y|)$ *induced by the functor* $\sin|S^1 \wedge - |$*). Then* f' *is an equivalence of spectra.*

Proof The diagram

$$\begin{array}{ccccc} Y^l & \longrightarrow & \mathbf{S}^1{}_+ \wedge Y^l & \xrightarrow{pr} & \mathbf{S}^1 \wedge Y^l \\ \downarrow & & \downarrow & & \downarrow \simeq \\ Map_*(\mathbf{S}^1, S^1 \wedge Y^l) & \xrightarrow{pr^*} & Map_*(\mathbf{S}^1{}_+, S^1 \wedge Y^l) & \longrightarrow & \sin\left|S^1 \wedge Y^l\right| \end{array}$$

commutes, and both horizontal sequences are (stable) fiber sequences of spectra (when varying l). The outer vertical maps are both stable equivalences, and the so the middle map (which is the map in question) must also be a stable equivalence. □

Corollary 6.3.2.3 *If* Y *is a spectrum, then there are natural chains of stable equivalences* $(\mathbf{S}^1{}_+ \wedge Y)^{h\mathbf{S}^1} \simeq S^1 \wedge Y$ *and* $(G_+ \wedge Y)^{hG} \simeq Y$*, for* G *a finite group.*

Proof Lemma 6.3.2.2 gives us that

$$\begin{aligned} (\mathbf{S}^1{}_+ \wedge Y)^{h\mathbf{S}^1} &\xrightarrow{\sim} Map_*(\mathbf{S}^1{}_+, \mathbf{S}^1 \wedge Y)^{h\mathbf{S}^1} \simeq Map_*(E\mathbf{S}^1{}_+ \wedge \mathbf{S}^1{}_+, Q(\mathbf{S}^1 \wedge Y))^{\mathbf{S}^1} \\ &\cong Map_*(E\mathbf{S}^1{}_+, Q(\mathbf{S}^1 \wedge Y)) \simeq Q(\mathbf{S}^1 \wedge Y) \simeq S^1 \wedge Y. \end{aligned}$$

Likewise for the discrete case. □

Let G be a simplicial group and assume given a functorial (in the spectrum Y) $G \times G$-map

$$G_+ \wedge Y \to Map_*(G_+, \Sigma^a Y)$$

which is a stable equivalence when Y is the sphere spectrum. When G is finite we use the map from Lemma 6.3.2.1 and $a = 0$, and when G is the circle group we use the map from Lemma 6.3.2.2 with $a = 1$.

The construction $Q^n X = \lim_{\overrightarrow{k}} \Omega^k X^{k+n}$ of Appendix A.3.2.1 sends this stable equivalence to a weak equivalence $Q^0(G_+ \wedge Y) \xrightarrow{\sim} Map_*(G_+, Q^0(\Sigma^a Y))$.

Since homotopy fixed points preserve equivalences we get an equivalence

$$\underline{S}_*(S^a, Q^0(G_+ \wedge \underline{\mathbf{S}}))^{h(G\times G)} \xrightarrow{\sim} Map_*(G_+ \wedge S^a, Q^0(\Sigma^a \underline{\mathbf{S}}))^{h(G\times G)}.$$

We have a preferred point Δ in the latter space, namely the one defined by

$$E(G \times G)_+ \wedge G_+ \wedge S^a \to S^a \to \Omega^k S^{k+a}$$

where the first map is the (ath suspension of the) projection and the second map is the adjoint to the identity. Note that, when G is finite, the homotopy class of Δ represents the "norm" in the usual sense:

$$[\Delta] \cong \sum_{g\in G} g \in \mathbf{Z}[G] = \pi_0 \Omega^l \prod_G S^l.$$

Now, pick the $G \times G$-map

$$f\colon E(G\times G)_+ \wedge S^a \to Q^0(G_+ \wedge \underline{\mathbf{S}})$$

in $\mathcal{S}_*(E(G\times G)_+ \wedge S^a, Q^0(\underline{\mathbf{S}} \wedge G_+))^{G\times G}$ of your choice in the component represented by Δ (in the preliminary draft there was a lousy joke at this point).

Consider the composite ϕ_X

$$E(G\times G)_+ \wedge S^a \wedge X^k \xrightarrow{f\wedge 1} Q^0(G_+ \wedge \underline{\mathbf{S}}) \wedge X^k \longrightarrow Q^k(G_+ \wedge X) \xrightarrow{\mu} Q^k X,$$

where the middle map is induced by $\Omega^n(S^n) \wedge X^k \to \Omega^n(S^n \wedge X^k) \to \Omega^n X^{n+k}$ and the last by the G-action on X. This is a $G \times G$-map if we let the action on the source be given by $(g_1, g_2)\cdot((e_1, e_2)\wedge s \wedge x) = (g_1 e, g_2 e_2) \wedge s \wedge g_2 x$ and on the target by $(g_1, g_2)\cdot x = g_1 x$.

However, these actions are complementary and we get a factorization through the orbit and fixed point spaces:

$$\begin{array}{ccccccc} & & EG_+ \wedge S^a \wedge X^k & \longrightarrow & \underline{\mathcal{S}_*}(EG_+, Q^k X) & & \\ & & \downarrow & & \uparrow & & \\ (X_{hG})_k & = & (EG_+ \wedge S^a \wedge X^k)/G & \xrightarrow{\nu_X} & \underline{\mathcal{S}_*}(EG_+, Q^k X)^G & = & (X^{hG})^k. \end{array}$$

Proposition 6.3.2.4 *Let G be a simplicial group, and assume given a choice of a map f as above. Then the* norm map

$$\nu_X\colon S^a \wedge X_{hG} \to X^{hG}$$

given above is natural in the G-spectrum X. In the homotopy category, the norm map is independent of the choice of f, up to isomorphism.

Proof The functoriality follows since all choices involved in producing this map were done before we introduced X on the scene. The independence between two choices of f follows by choosing a $G\times G$-homotopy and tracing it through the construction. □

Proposition 6.3.2.5 *If G is a finite and discrete group and X a G-spectrum, then the composite*

$$X \twoheadrightarrow X_{hG} \xrightarrow{\nu_X} X^{hG} \hookrightarrow QX$$

induces the endomorphism of π_*X given by multiplication with the norm element $[\Delta]=\sum_{g\in G} g\in\pi_0(G_+\wedge\underline{\mathbf{S}})$. If $X=G_+\wedge Y$ with trivial G-action on Y, then the norm map is an equivalence.

Proof By the choice of f, we see that if G is finite, the map induces multiplication by $[\Delta]=\sum_{g\in G} g$ on the homotopy groups π_*X, in the sense that

$$\begin{array}{ccccccc}
\pi_{*+k}(E(G\times G)_+\wedge S^0\wedge X^k) & \xrightarrow{f_*\wedge 1} & \pi_{*+k}Q^0(G_+\wedge\underline{\mathbf{S}})\wedge X^k & \longrightarrow & \pi_{*+k}Q^k(G_+\wedge X) & \xrightarrow{\mu} & \pi_{*+k}Q^kX \\
\downarrow & & & & \cong\downarrow & \nearrow_{\text{multiplication}} & \\
\pi_{*+k}Q^kX & & \xrightarrow{\text{diagonal}} & & \mathbf{Z}[G]\otimes\pi_{*+k}Q^kX & &
\end{array}$$

commutes, and the left vertical map stabilizes to an isomorphism. Since EG is a contractible space, the same holds for the adjoint used in the definition of the norm.

The last statement may be proven as follows. If $X=G_+\wedge Y$ then consider the commutative diagram

$$\begin{array}{ccccccc}
\pi_*(G_+\wedge Y) & \longrightarrow & \pi_*(G_+\wedge Y)_{hG} & \longrightarrow & \pi_*(G_+\wedge Y)^{hG} & \longrightarrow & \pi_*(G_+\wedge Y) \\
\cong\downarrow & & \cong\downarrow & & \cong\downarrow & & \cong\downarrow \\
\bigoplus_G\pi_*Y & \xrightarrow{\nabla} & \pi_*Y & & \pi_*Y & \xrightarrow{\Delta} & \bigoplus_G\pi_*Y
\end{array}$$

where $\nabla(g\mapsto y_g)=\sum_g y_g$, and $\Delta(y)=\{g\mapsto y\}$. The "missing" arrow can of course be filled in as the vertical maps are isomorphisms, but there is only one map $\pi_*Y\to\pi_*Y$ making the bottom composite the norm, namely the identity. □

We list two corollaries that we will need in the finite case.

Corollary 6.3.2.6 *Let G be a finite discrete group, let U be a finite free pointed G-space and Y a G-spectrum. Then the norm maps*

$$(U\wedge Y)_{hG}\to(U\wedge Y)^{hG}$$

and

$$\underline{S_*}(U,Y)_{hG}\to\underline{S_*}(U,Y)^{hG}$$

are both equivalences.

Proof Recall that a finite free pointed G-space is the result of adjoining finitely many G-cells, cf. Sect. A.9, so by induction on the number of G-cells it is enough to consider the case $U=S^n_+\wedge G_+$. Use a shear map as in the proof Lemma A.9.2.3 to remove action from $S^n_+\wedge Y$ and $\underline{S_*}(S^n_+,Y)$ in the resulting expressions. Note the stable product to sum shift in the last case. Finally, use Proposition 6.3.2.5. □

We have one very important application of this corollary:

Corollary 6.3.2.7 *Let G be a finite discrete group, let U be a finite free pointed G-space, and X any pointed G-space. Then there is a chain of natural equivalences*

$$\varinjlim_k \Omega^k\big(Map_*\big(U, S^k\wedge X\big)_{hG}\big) \simeq Map_*\big(U, \varinjlim_k \Omega^k\big(S^k\wedge X\big)\big)^{hG}.$$

If U is d-dimensional and X n-connected, then

$$Map_*(U, X)^G \to Map_*\big(U, \varinjlim_k \Omega^k\big(S^k\wedge X\big)\big)^{hG}$$

is $2n - d + 1$ connected.

Proof Recall that $Map_*(-,-) = \underline{\mathcal{S}_*}(-, \sin|-|)$. Corollary 6.3.2.6 tells us that the norm map

$$\varinjlim_k \Omega^k\big(Map_*\big(U, S^k\wedge X\big)_{hG}\big) \xrightarrow{\sim} \varinjlim_k \Omega^k\big(\varinjlim_l \Omega^l Map_*\big(U, S^l\wedge S^k\wedge X\big)\big)^{hG}$$

is an equivalence, and the latter space is equivalent to $Map_*(U, \varinjlim_k \Omega^k S^k\wedge X)^{hG}$ by Lemma A.2.5.1 since U and EG_+ (and G) are finite. The last statement is just a reformulation of Lemma A.9.2.3 since $X \to \varinjlim_k \Omega^k(S^k\wedge X)$ is $2n+1$ connected by the Freudenthal suspension Theorem A.8.2.3. □

6.4 Topological Cyclic Homology

In this section we will finally give a definition of topological cyclic homology. We first will fix a prime p and define the pieces $TC(-;p)$ which are relevant to the p-complete part of TC, and later merge this information with the rational information coming from the homotopy fixed points of the whole circle action.

6.4.1 The Definition and Properties of $TC(-;p)$

As an intermediate stage, we define the functors $TC(-;p)$ which capture the information of topological cyclic homology when we complete at the prime p. We continue to list the case of an **S**-algebra separately, in case the reader feels uncomfortable with $\Gamma\mathcal{S}_*$-categories.

Recall from Definition 6.2.3.2 that $\mathcal{RF}_p \subset \mathcal{RF}$ is the full subcategory of powers of p.

Definition 6.4.1.1 Let p be a prime, A an **S**-algebra and X a space. We define

$$TC(A, X; p) = \underset{\overleftarrow{p^n \in \mathcal{RF}_p}}{\operatorname{holim}}\ T\langle p^n\rangle(A, X).$$

This gives rise to the spectrum

$$\underline{TC}(A, X; p) = \operatorname*{holim}_{\overleftarrow{p^n \in \mathcal{RF}_p}} \underline{T}\langle p^n\rangle(A, X) = \{k \mapsto TC(A, S^k \wedge X; p)\}.$$

If $\mathcal{C}$ is a $\Gamma\mathcal{S}_*$-category we define

$$TC(\mathcal{C}, X; p) = \operatorname*{holim}_{\overleftarrow{p^n \in \mathcal{RF}_p}} T\langle p^n\rangle(\mathcal{C}, X)$$

with associated spectrum

$$\underline{TC}(\mathcal{C}, X; p) = \operatorname*{holim}_{\overleftarrow{p^n \in \mathcal{RF}_p}} \underline{T}\langle p^n\rangle(\mathcal{C}, X) = \{k \mapsto TC(\mathcal{C}; S^k \wedge X; p)\}.$$

If $\mathcal{C}$ is a symmetric monoidal $\Gamma\mathcal{S}_*$-category we have a spectrum

$$TC(\bar{H}\mathcal{C}, X; p) = \{k \mapsto TC(\bar{H}\mathcal{C}(S^k), X; p)\}$$

We get the analogs of the results in the previous chapter directly:

Lemma 6.4.1.2 *Let $\mathcal{C}$ be a $\Gamma\mathcal{S}_*$-category, X a pointed space and p a prime. Then*

1. *the spectrum $\underline{TC}(\mathcal{C}, X; p)$ is an Ω-spectrum,*
2. *the functor $\underline{TC}(-, X; p)$ takes $\Gamma\mathcal{S}_*$-equivalences of categories to equivalences,*
3. *if A is a ring, then the inclusion $A \subseteq \mathcal{P}_A$ as a rank one module induces an equivalence $\underline{TC}(A, X; p) \xrightarrow{\sim} \underline{TC}(\mathcal{P}_A, X; p)$,*
4. *the functor $\underline{TC}(-, X; p)$ preserves products up to levelwise equivalence,*
5. *if $\mathcal{C} \to \mathcal{D}$ is a $\Gamma\mathcal{S}_*$-functor inducing an equivalence $THH(\mathcal{C}) \to THH(\mathcal{D})$, then it induces an equivalence*

$$\underline{TC}(\mathcal{C}, X; p) \to \underline{TC}(\mathcal{D}, X; p),$$

6. *if $\mathcal{C}$ is a symmetric monoidal $\Gamma\mathcal{S}_*$-category, then $TC(\bar{H}\mathcal{C}, X; p)$ is an Ω-spectrum equivalent to $\underline{TC}(\mathcal{C}, X; p)$.*

Proof This follows from the corresponding properties for $\underline{T}\langle p^n\rangle$ from Sect. 6.2.4, cf. in particular Corollary 6.2.4.5, Corollary 6.2.4.6 and Corollary 6.2.4.7, and the properties of homotopy limits. □

Here we see that it made a difference that we considered $T\langle a\rangle(\bar{H}\mathcal{C}, X)$ as the spectrum $\{n \mapsto T\langle a\rangle(\bar{H}\mathcal{C}(S^n), X)\}$, and not as a Γ-space $\{k_+ \mapsto T\langle a\rangle(\bar{H}\mathcal{C}(k_+), X)\}$: the spectrum associated to the (pointwise) homotopy limit of a Γ-space is not the same as the (levelwise) homotopy limit of the spectrum, since the homotopy limits can destroy connectivity. We will shortly see that this is not a real problem, since it turns out that $\underline{TC}(\mathcal{C}, X; p)$ is always -2-connected, and so $TC(\bar{H}\mathcal{C}, X; p)$ and the spectrum associated with $\{k_+ \mapsto TC(\bar{H}\mathcal{C}(k_+), X; p)\}$ will be equivalent once X is connected. In any case, it may be that the correct way of thinking of this is to view TC of symmetric monoidal $\Gamma\mathcal{S}_*$-categories as Γ-spectra:

$$\{k_+ \mapsto \underline{TC}(\bar{H}\mathcal{C}(k_+), X; p)\}.$$

This point will become even more acute when we consider the homotopy fixed point spectra for the entire circle actions since these are not even bounded below.

If $\mathcal{C}$ is exact we have an equivalent Ω-spectrum

$$\mathbf{TC}(\mathcal{C}, X; p) = \underset{p^n \in \mathcal{RF}_p}{\underleftarrow{\operatorname{holim}}}\, T\langle p^n\rangle(S\mathcal{C}, X) = \{k \mapsto TC(S^{(k)}\mathcal{C}, X; p)\}.$$

If A is a ring we let

$$\mathbf{TC}(A, X; p) = \mathbf{TC}(\mathcal{P}_A, X; p)$$

and we see that $\mathbf{TC}(A, X; p)$ is equivalent to $\underline{TC}(A, X; p)$.

6.4.2 *Some Structural Properties of $TC(-; p)$*

A priori, the category $\mathcal{RF}_p$ can seem slightly too big for comfort, but it turns out to be quite friendly, especially if we consider the F and R maps separately. This separation gives us good control over the homotopy limit defining $TC(-; p)$. For instance, we shall see that it implies that $\underline{TC}(-; p)$ is -2-connected, can be computed degreewise and almost preserves id-cartesian cubes (see Definition 3.3.1.6 and more thoroughly, Sect. A.8 for terminology), and hence is "determined" by its value on discrete rings.

6.4.2.1 Calculating Homotopy Limits over $\mathcal{RF}_p$

Consider the two subcategories $\mathcal{F}_p$ and $\mathcal{R}_p$ of the category $\mathcal{RF}_p$ of Definition 6.2.3.2, namely the ones with only the $F^{p^k} = f_{1,p^k}$ (Frobenius = inclusion of fixed points) maps or only the $R_{p^k} = f_{p^k,1}$ (restriction) maps. We will typically let

$$TR(A, X; p) = \underset{p^n \in \mathcal{R}_p}{\underleftarrow{\operatorname{holim}}}\, T\langle p^n\rangle(A, X) \quad \text{and} \quad TF(A, X; p) = \underset{p^n \in \mathcal{F}_p}{\underleftarrow{\operatorname{holim}}}\, T\langle p^n\rangle(A, X)$$

and similarly for the spectra and the related functors of $\Gamma\mathcal{S}_*$-categories.

Let $\langle x, y, \dots\rangle$ be the free symmetric monoid generated by the letters $x, y, \dots$. If $\langle x\rangle$ acts on a space Y, we write $\operatorname{holim}_{\overleftarrow{\langle x\rangle}} Y$ as Y^{hx}, in analogy with the group case, and it may be calculated as the homotopy pullback

$$Y^{hx} \simeq \underleftarrow{\operatorname{holim}} \begin{pmatrix} & Map_*(I_+, Y) \\ & \Big\downarrow{\scriptstyle f\mapsto(f(0),f(1))} \\ Y \xrightarrow{y\mapsto(y,xy)} & Y \times Y \end{pmatrix}$$

Let L be any functor from $\mathcal{RF}_p$ to spaces, and suppose it has fibrant values so that we may suppress some fibrant replacements below. We see that $\langle R, F\rangle$ acts on $\prod_{p^n\in\mathcal{RF}_p} L(p^n)$, and writing out the cosimplicial replacement carefully, we get that

$$\begin{aligned}\underset{\overleftarrow{\mathcal{RF}_p}}{\operatorname{holim}} L &\cong Tot\left(q \mapsto \prod_{N_q\langle F,R\rangle}\left(\prod_{p^n\in\mathcal{RF}_p} L(p^n)\right)\right)\\ &\cong \underset{\overleftarrow{\langle R,F\rangle}}{\operatorname{holim}}\left(\prod_{p^n\in\mathcal{RF}_p} L(p^n)\right) \cong \underset{\overleftarrow{\langle R\rangle\times\langle F\rangle}}{\operatorname{holim}}\left(\prod_{p^n\in\mathcal{RF}_p} L(p^n)\right).\end{aligned}$$

We may take the homotopy limit over the product $\langle R\rangle \times \langle F\rangle$ in the order we choose. If we take the R map first we get

Lemma 6.4.2.1 *Let L be a functor from $\mathcal{RF}_p$ to spaces. Then*

$$\begin{aligned}\underset{\overleftarrow{\mathcal{RF}_p}}{\operatorname{holim}} L &\cong \underset{\overleftarrow{\langle F\rangle}}{\operatorname{holim}}\left(\underset{\overleftarrow{\langle R\rangle}}{\operatorname{holim}}\left(\prod_{p^n\in\mathcal{RF}_p} L(p^n)\right)\right)\\ &\cong \underset{\overleftarrow{\langle F\rangle}}{\operatorname{holim}}\ \underset{\overleftarrow{p^n\in\mathcal{R}_p}}{\operatorname{holim}} L(p^n) = \left(\underset{\overleftarrow{p^n\in\mathcal{R}_p}}{\operatorname{holim}} L(p^n)\right)^{hF}\end{aligned}$$

☺

Similarly we may take the F map first and get the same result with R and F interchanged.

For our applications we note that

$$TC(A,X;p) \cong TR(A,X;p)^{hF} \cong TF(A,X;p)^{hR}.$$

Lemma 6.4.2.2 *The spectrum $\underline{TC}(-;p)$ is -2 connected.*

Proof Consider the short exact sequence

$$0 \to \underset{\overleftarrow{p^n\in\mathcal{R}_p}}{\lim}{}^{(1)} \pi_{k+1}\underline{T}\langle p^n\rangle(\mathcal{C},X) \to \pi_k\underline{TR}(\mathcal{C},X;p) \to \underset{\overleftarrow{p^n\in\mathcal{R}_p}}{\lim} \pi_k\underline{T}\langle p^n\rangle(\mathcal{C},X) \to 0$$

of the tower defining TR. Since $\pi_0\underline{T}\langle p^n\rangle(\mathcal{C},X) \to \pi_0\underline{T}\langle p^{n-1}\rangle(\mathcal{C},X)$ is always surjective (its cokernel is $\pi_{-1}sd_{p^n}\underline{T}\langle p^n\rangle(\mathcal{C},X)_{hC_{p^n}} = 0$), the $\lim_{\overleftarrow{R}}{}^{(1)}\pi_{k+1}$-term vanishes for $k<0$, and $\underline{TR}$ is always -1-connected (alternatively, look at the spectral sequence of the R tower, and note that all the homotopy fibers are -1-connected). Hence the pullback $\underline{TC}(\mathcal{C};p) \simeq TR(\mathcal{C};p)^{hF}$ is at least -2-connected. □

An $\mathbf{S}$-algebra has a simplicial direction (as all Γ-spaces do), and if A is a *simplicial* $\mathbf{S}$-algebra, diag^*A is the $\mathbf{S}$-algebra you get by precomposing A with the diagonal $\operatorname{diag}\colon \Delta^o \to \Delta^o \times \Delta^o$.

Lemma 6.4.2.3 *If A is a simplicial* **S***-algebra, then $\underline{TC}(A, X; p)$, may be calculated degreewise in the sense that*

$$\mathrm{diag}^*\big\{[q] \mapsto \underline{TC}(A_q, X; p)\big\} \simeq \underline{TC}\big(\mathrm{diag}^* A, X; p\big).$$

Proof This is true for THH (Lemma 4.2.3.1), and so, by the fundamental cofibration sequence Proposition 6.2.4.2 it is true for all $sd_{p^n} THH(A, X)^{C_{p^n}}$. By Corollary A.8.2.7, homotopy limits of towers of connective simplicial spectra may always be computed degreewise, so

$$\underline{TR}(A; p) = \operatorname{holim}_{\overleftarrow{R}} sd_{p^n} \underline{T}(A)^{C_{p^n}}$$

is naturally equivalent to $\mathrm{diag}^*\{[q] \mapsto \underline{TR}(A_q; p)\}$. Now, $\underline{TC}(A; p) \simeq \underline{TR}(A; p)^{hF}$, a homotopy pullback construction which may be calculated degreewise. □

Lemma 6.4.2.4 *Let $f: A \to B$ be a k-connected map of* **S***-algebras and X an l-connected space. Then*

$$\underline{TC}(A, X; p) \to \underline{TC}(B, X; p)$$

is $k + l - 1$-connected.

Proof Since $THH(-, X)$, and hence the homotopy orbits of $THH(-, X)$, rise connectivity by l, we get by the tower defining TR that $\underline{TR}(-, X; p)$ also rises connectivity by l. We may loose one when taking the fixed points under the F-action to get $\underline{TC}(-, X; p)$. □

When restricted to simplicial rings, there is a cute alternative to this proof using the fact that any functor from simplicial rings to n-connected spectra which preserves equivalences and may be computed degreewise, sends $k \geq 0$-connected maps to $n + k + 1$-connected maps.

Lemma 6.4.2.5 *Assume $\mathcal{A}$ is a cube of* **S***-algebras such that $\underline{T}(\mathcal{A}, X)$ is id-cartesian. Then $\underline{TR}(\mathcal{A}, X; p)$ is also id-cartesian.*

Proof Choose a big k such that $THH(\mathcal{A}, S^k \wedge X)$ is $id + k$ cartesian. Let $\mathcal{X}$ be any m subcube and $\mathcal{X}^l = sd_{p^l} \mathcal{X}^{C_{p^l}}$. We are done if we can show that $\operatorname{holim}_{\overleftarrow{R}} \mathcal{X}^l$ is $(m + k)$-cartesian. Let Z^l be the iterated homotopy fiber of $\mathcal{X}^l$ (i.e., the homotopy fiber of $\mathcal{X}^l_\emptyset \to \operatorname{holim}_{\overleftarrow{S \neq \emptyset}} \mathcal{X}^l_S$). Then $Z = \operatorname{holim}_{\overleftarrow{R}} Z^l$ is the iterated homotopy fiber of $\operatorname{holim}_{\overleftarrow{R}} \mathcal{X}^l$, and we must show that Z is $m+k-1$ connected. Since homotopy orbits preserve connectivity and homotopy colimits, $THH(\mathcal{A}, S^k \wedge X)_{hC_{p^l}}$ must be $id + k$ cartesian, and so the homotopy fiber of $R: \mathcal{X}^l \to \mathcal{X}^{l-1}$ is $id + k$ cartesian. Hence $\pi_q Z^l \to \pi_q Z^{l-1}$ is surjective for $q = m + k$ and an isomorphism for $q < m + k$, and so $\pi_q Z \cong \lim^{(1)}_{\overleftarrow{R}} \pi_{q+1} Z^l \times \lim_{\overleftarrow{R}} \pi_q Z^l = 0$ for $q < m + k$. □

Proposition 6.4.2.6 *Assume $\mathcal{A}$ is a cube of* **S***-algebras such that* $\underline{T}(\mathcal{A}, X)$ *is* id*-cartesian. Then* $\underline{TC}(\mathcal{A}, X; p)$ *is* $id - 1$ *cartesian.*

Proof This follows from the lemma, plus the interpretation of $\underline{TC}(-; p) \simeq \underline{TR}(-; p)^{hF}$ as a homotopy pullback. □

When applying this to the canonical resolution of **S**-algebras by $H\mathbf{Z}$-algebras of Definition 3.3.1.8, we get the result saying essentially that TC is determined by its value on simplicial rings:

Theorem 6.4.2.7 *Let A be an* **S***-algebra and X a space. Let $S \mapsto (A)_S$ be the cubical diagram of Definition* 3.3.1.8. *Then*

$$\underline{TC}(A, X; p) \xrightarrow{\sim} \operatorname{holim}_{\overleftarrow{S \neq \emptyset}} \underline{TC}(\mathcal{A}_S, X; p)$$ ☺

6.4.2.2 The Frobenius Maps

The reason the map F, given by the inclusion of fixed points, is now often called the *Frobenius map*, is that Hesselholt and Madsen [129] have shown that if A is a commutative ring, then $\pi_0 TR(A; p)$ is canonically isomorphic to the p-typical Witt vectors $W(A) = W(A; p)$, and that the F-map corresponds to the Frobenius map.

Even better, they prove that for $n \geq 0$ there is an isomorphism

$$\pi_0 THH(A)^{C_{p^n}} \cong W_{n+1}(A),$$

where $W_n(A)$, is the ring of truncated p-typical Witt vectors, i.e., it is A^n as set, but with addition and multiplication defined by requiring that the "ghost map"

$$w\colon W_n(A) \to A^n, \qquad (a_0, \ldots, a_{n-1}) \mapsto (w_0, \ldots, w_{n-1})$$

where

$$w_i = a_0^{p^i} + p a_1^{p^{i-1}} + \cdots + p^i a_i$$

is naturally a ring map. If A has no p-torsion the ghost map is injective.

The map

$$R\colon W_{n+1}(A) \to W_n(A), \qquad (a_0, \ldots, a_n) \mapsto (a_0, \ldots, a_{n-1}),$$

is called the restriction and the isomorphisms

$$\pi_0 THH(A)^{C_{p^n}} \cong W_n(A)$$

respect the restriction maps.

On the Witt vectors the Frobenius and Verschiebung are given by

$$F, V \colon W(A) \to W(A),$$

$$F(w_0, w_1, \dots) = (w_1, w_2, \dots),$$

$$V(a_0, a_1, \dots) = (0, a_0, a_1, \dots)$$

(the Frobenius map F is given in ghost coordinates, and it must be verified that this determines F uniquely) satisfying the relations

$$x \cdot V(y) = V\big(F(x) \cdot y\big), \qquad FV = p, \qquad VF = \text{multiplication by } V(1).$$

If A is an $\mathbf{F}_p$-algebra, then $V(1) = p$.

6.4.2.3 $TC(-;p)$ of Spherical Group Rings

Let G be a simplicial group. We briefly sketch the argument of [27] giving $TC(\mathbf{S}[G]; p)$ (see also [192] and [243]). Recall from Corollary 6.2.5.3 that

$$\big|THH\big(\mathbf{S}[G]\big)\big|^{C_{p^n}} \xrightarrow{\sim} \prod_{j=0}^{n} \big|THH\big(\mathbf{S}[G]\big)\big|_{hC_{p^j}},$$

and that the restriction map corresponds to the projection

$$\prod_{j=0}^{n} \big|THH\big(\mathbf{S}[G]\big)\big|_{hC_{p^j}} \to \prod_{j=0}^{n-1} \big|THH\big(\mathbf{S}[G]\big)\big|_{hC_{p^j}}.$$

What is the inclusion of fixed point map $|THH(\mathbf{S}[G])|^{C_{p^n}} \subseteq |THH(\mathbf{S}[G])|^{C_{p^{n-1}}}$ in this factorization? Write T as shorthand for $|THH(\mathbf{S}[G])|$, and consider the diagram

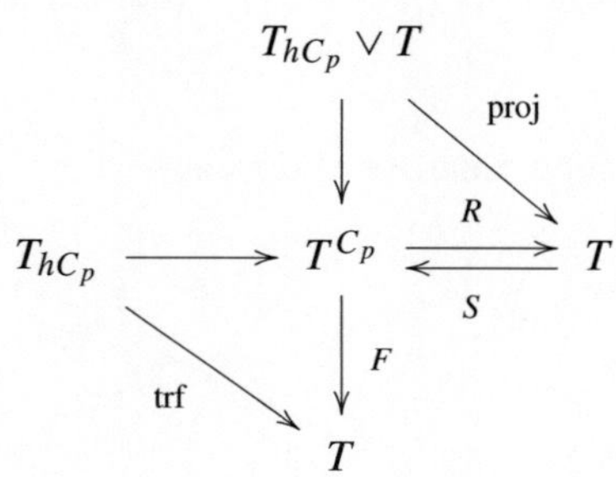

where S is the section of R defined in the proof of Lemma 6.2.5.1.

The trf in the diagram above is the composite

$$T_{hC_p} \longrightarrow T^{C_p} \xrightarrow{F} T$$

(where the first map is the homotopy fiber of the restriction map) and is called the *transfer*. Generally we will let the transfer be any (natural) map in the stable homotopy category making

$$\begin{array}{ccc} T_{hC_{p^n}} & \longrightarrow & T^{C_{p^n}} \\ {\scriptstyle \mathrm{trf}}\downarrow & & \downarrow{\scriptstyle F} \\ T_{hC_{p^{n-1}}} & \longrightarrow & T^{C_{p^{n-1}}} \end{array}$$

commute.

Hence, the inclusion of fixed points $F\colon T^{C_{p^{n+1}}} \to T^{C_{p^n}}$ acts as $FS\colon T \to T$ on the zeroth factor, and as $\mathrm{trf}\colon T_{hC_{p^n}} \to T_{hC_{p^{n-1}}}$ on the others:

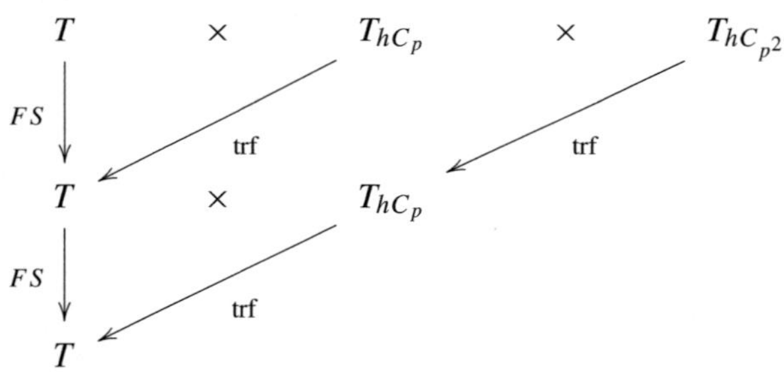

Using this, we see that

$$TC(\mathbf{S}[G]; p) = \underset{\overleftarrow{\mathcal{RF}_p}}{\mathrm{holim}}\, T\langle p^n\rangle(\mathbf{S}[G]) \simeq \big(\underset{\overleftarrow{R}}{\mathrm{holim}}\, T\langle p^n\rangle(\mathbf{S}[G])\big)^{hF}$$

is equivalent to the homotopy equalizer of the map

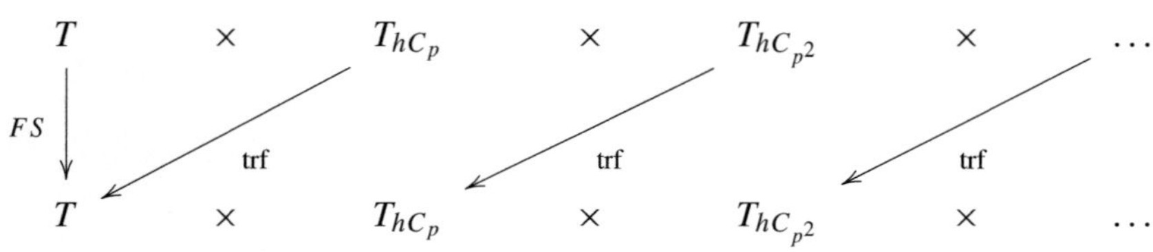

and the identity; or equivalently, the "diagram"

$$\begin{array}{ccc} TC(\mathbf{S}[G];p) & \longrightarrow & \underset{\overleftarrow{\mathrm{trf}}}{\mathrm{holim}}\big|THH(\mathbf{S}[G])\big|_{hC_{p^n}} \\ \downarrow & & \downarrow \\ \big|THH(\mathbf{S}[G])\big| & \xrightarrow{FS-1} & \big|THH(\mathbf{S}[G])\big| \end{array}$$

is homotopy cartesian (in order to make sense of this, one has to have chosen models for all the maps, see e.g., [243]).

We will identify these terms more closely in Sect. 7.3

6.4.3 The Definition and Properties of TC

Note that the topological Hochschild spectrum $\underline{T}$ is stably fibrant (in the sense of Appendix A.4.1(6): it is an Ω-spectrum and each of the spaces constituting the spectrum are Kan complexes), and so the fibrant replacements (Q and $\sin|-|$) in Definition 6.3.0.1 of homotopy fixed points are redundant. In the interest of having lean models, we write $\underline{T}^{hG}$ for the spectrum without these extra fibrant replacements (so that we actually get an honest map $\underline{T}^{hG} \to \underline{T}$, and not to some blown-up version of $\underline{T}$).

Definition 6.4.3.1 Define $\underline{TC}$ as the functor from $\mathbf{S}$-algebras, or more generally $\Gamma\mathcal{S}_*$-categories, to spectra obtained as the homotopy limit of

$$\begin{array}{ccc} & & \underline{T}(-)^{h\mathbf{S}^1} \\ & & \downarrow \\ \prod_{p \text{ prime}} \underline{TC}(-;p)\widehat{_p} & \longrightarrow & \prod_{p \text{ prime}} \underset{\overleftarrow{p^r \in \mathcal{F}_p}}{\mathrm{holim}} \underline{T}(-)^{hC_{p^r}}\widehat{_p} \end{array}$$

where the lower map is given by the projection onto $\mathcal{F}_p \subseteq \mathcal{RF}_p$

$$\underline{TC}(-;p) = \underset{\overleftarrow{p^r \in \mathcal{RF}_p}}{\mathrm{holim}}\ \underline{T}\langle p^r\rangle(-) \to \underset{\overleftarrow{p^r \in \mathcal{F}_p}}{\mathrm{holim}} \underline{T}\langle p^r\rangle(-)$$

followed by the map from the fixed points to the homotopy fixed points

$$\underline{T}\langle p^r\rangle(-) \cong \underline{T}(-)^{C_{p^r}} \to \underline{T}(-)^{hC_{p^r}}.$$

Likewise, the functor $\boldsymbol{TC}$ from discrete rings or more generally additive categories to bispectra is defined by the same square with $\mathbf{T}$ as in Definition 5.1.2.2 instead of $\underline{T}$.

In this definition we have used that the map from fixed points to homotopy fixed points is compatible with inclusion of fixed points.

More useful than the definition is the characterization given by the following lemma, where $X^{\wedge} = \underline{\mathcal{S}pt}(\Sigma^{-1}M\mathbf{Q}/\mathbf{Z}, X) \simeq \prod_{p \text{ prime}} X^{\wedge}_p$ signifies the profinite completion of Sect. A.2 of the spectrum X.

Lemma 6.4.3.2 *All the squares in*

$$\begin{array}{ccccc}
\underline{TC}(-) & \longrightarrow & \underline{T}(-)^{h\mathbf{S}^1} & \longrightarrow & \left(\underline{T}(-)_{(0)}\right)^{h\mathbf{S}^1}\\
\downarrow & & \downarrow & & \downarrow\\
\underline{TC}(-)^{\wedge} & \longrightarrow & \left(\underline{T}(-)^{\wedge}\right)^{h\mathbf{S}^1} & \longrightarrow & \left(\left(\underline{T}(-)^{\wedge}\right)_{(0)}\right)^{h\mathbf{S}^1}
\end{array}$$

are homotopy cartesian, where the upper left horizontal maps are given by the definition of TC, the lower is its profinite completion composed with the isomorphism $(\underline{T}(-)^{h\mathbf{S}^1})^{\wedge} \cong (\underline{T}(-)^{\wedge})^{h\mathbf{S}^1}$ and the right horizontal arrows are induced by rationalization.

Proof The rightmost square is cartesian as it is an arithmetic square Theorem A.7.6.1 to which $-^{h\mathbf{S}^1}$ is applied. The leftmost square is cartesian since (after commuting profinite completion and homotopy fixed points) it is the top square in

$$\begin{array}{ccc}
\underline{TC} & \longrightarrow & \underline{T}(-)^{h\mathbf{S}^1}\\
\downarrow & & \downarrow\\
\underline{TC}^{\wedge} & \longrightarrow & \left(\underline{T}(-)^{h\mathbf{S}^1}\right)^{\wedge}\\
\downarrow & & \simeq\big\downarrow 6.3.1.1\\
\prod_{p \text{ prime}} \underline{TC}(-;p)^{\wedge}_p & \longrightarrow & \prod_{p \text{ prime}} \underset{\overleftarrow{p^r \in \mathcal{F}_p}}{\operatorname{holim}} \underline{T}(-)^{hC_{p^r}\,\wedge}{}_p
\end{array}$$

where the outer and lower squares are cartesian by the definition of TC and the marked map is an equivalence by Lemma 6.3.1.1. □

Corollary 6.4.3.3 *Let A be a simplicial ring. Then*

$$\begin{array}{ccc}
\underline{TC}(A, X) & \longrightarrow & \left(\underline{HH}(A, X)_{(0)}\right)^{h\mathbf{S}^1}\\
\downarrow & & \downarrow\\
\underline{TC}(A, X)^{\wedge} & \longrightarrow & \left(\underline{HH}(A, X)^{\wedge}_{(0)}\right)^{h\mathbf{S}^1}
\end{array}$$

is homotopy cartesian.

Proof This follows from Lemma 6.4.3.2 by extending the square to the right with the $\mathbf{S}^1$-homotopy fixed point spectra of the square

$$\begin{array}{ccc} \underline{T}(-)_{(0)} & \xrightarrow{\simeq} & \underline{HH}(-)_{(0)} \\ \downarrow & & \downarrow \\ \underline{T}(-)^{\widehat{}}_{(0)} & \xrightarrow{\simeq} & \underline{HH}(-)^{\widehat{}}_{(0)} \end{array}$$

which is cartesian by Lemma 4.2.3.8 which says that the horizontal maps are equivalences. □

Theorem 6.4.3.4 *Let A be an* $\mathbf{S}$*-algebra and X a space. Let $\mathcal{A}$ be the cubical diagram of Definition* 3.3.1.8. *Then*

$$\underline{TC}(A, X) \xrightarrow{\sim} \operatorname{holim}_{\overleftarrow{S\neq\emptyset}} \underline{TC}(\mathcal{A}_S, X).$$

Proof By Theorem 6.4.2.7 this is true for $TC(-, X)^{\widehat{}}$ (products and completions of spectra commute with homotopy limits). Since $\underline{T}(\mathcal{A}, X)$ is id-cartesian, so are $\underline{T}(\mathcal{A}, X)_{(0)}$ and $\underline{T}(\mathcal{A}, X)^{\widehat{}}_{(0)}$, and hence

$$\underline{T}(A, X)_{(0)} \xrightarrow{\sim} \operatorname{holim}_{\overleftarrow{S\neq\emptyset}} \big(\underline{T}(\mathcal{A}_S, X)_{(0)}\big)$$

and

$$\underline{T}(A, X)^{\widehat{}}_{(0)} \xrightarrow{\sim} \operatorname{holim}_{\overleftarrow{S\neq\emptyset}} \big(\underline{T}(\mathcal{A}_S, X)^{\widehat{}}_{(0)}\big).$$

Since homotopy fixed points commute with homotopy limits we are done since we have proved the theorem for all the theories but TC in the outer homotopy cartesian square of Lemma 6.4.3.2. □

6.5 The Connection to Cyclic Homology of Simplicial Rings

Theorem 6.4.3.4 tells us that we can obtain much information about TC from our knowledge of simplicial rings. We have seen (Corollary 6.4.3.3) that, when applied to a simplicial ring A, TC fits into the cartesian square

$$\begin{array}{ccc} \underline{TC}(A, X) & \longrightarrow & \big(\underline{HH}(A, X)_{(0)}\big)^{h\mathbf{S}^1} \\ \downarrow & & \downarrow \\ \underline{TC}(A, X)^{\widehat{}} & \longrightarrow & \big(\underline{HH}(A, X)^{\widehat{}}_{(0)}\big)^{h\mathbf{S}^1} \end{array}$$

We can say something more about the right hand column, especially in some relative cases. As a matter of fact, it is calculated by *negative cyclic homology*, a theory which we will recall the basics about shortly.

To make the comparison to negative cyclic homology easier we first give some general results about (naïve) $\mathbf{S}^1$-spectra, and then describe spectral sequences computing the homotopy groups of the homotopy fixed and orbit spectra.

6.5.1 On the Spectral Sequences for the $\mathbb{T}$-Homotopy Fixed Point and Orbit Spectra

If V is an inner product space, let $S(V) = \{v \in V \mid |v| = 1\}$ be the unit sphere in V, $D(V) = \{v \in V \mid |v| \leq 1\}$ the unit disc, S^V the one point compactification of V (with base point at infinity) and $D^\infty(V) = \{v \in S^V \mid |v| \geq 1\}$. We consider $S(V)_+$, $D(V)_+$ and $D^\infty(V)$ as pointed subspaces of S^V. We recognize $S(\mathbf{C})$ as the circle, but we will continue to write $\mathbb{T}$ when we consider the circle as a group, and make the convention that when we write $S(\mathbf{C}^n)$ it is only a space (which may or may not be considered as a $\mathbb{T}$-space through the diagonal action).

Recall that the stable homotopy group $\pi_0(\Sigma^\infty S^0) \cong \mathbf{Z}$, generated by the identity $S^n = S^n$, whereas $\pi_1(\Sigma^\infty S^0) \cong \mathbf{Z}/2\mathbf{Z}$, generated by the suspensions of the Hopf map from the 3-sphere to the 2-sphere:

$$S(\mathbf{C}^2) \to S^{\mathbf{C}}, \qquad (z_0, z_1) \mapsto z_0/z_1$$

The collapse maps $S^1_+ \to S^1$ and $S^1_+ \to S^0$ give an isomorphism

$$\pi_*(\Sigma^\infty S^1_+) \cong \pi_*(\Sigma^\infty S^1) \oplus \pi_*(\Sigma^\infty S^0)$$

(this isomorphism is realized after a single suspension in view of the fact that collapsing contractible arcs give homotopy equivalences between the reduced and unreduced suspensions of $|S^1|_+$ and from the unreduced suspension to $|S^1| \vee S^{\mathbf{C}^1}$), and let σ be the element in $\pi_1(\Sigma^\infty S^1_+)$ projecting down to the identity class in $\pi_1(\Sigma^\infty S^1) \cong \pi_0(\Sigma^\infty S^0)$ and η be the element in $\pi_1(\Sigma^\infty \mathbf{S}^1{}_+)$ projecting down to the stable Hopf map in $\pi_1(\Sigma^\infty S^0)$.

The spectral sequences of Lemma 6.3.0.3 for the circle group are of particular importance to us, and so we must analyze the skeleton filtration in this particular case. The one sided bar construction $E\mathbb{T}$ for the circle group $\mathbb{T} = S(\mathbf{C})$ may be identified with $S(\mathbf{C}^\infty)$, and the skeleton filtration of $E(\mathbb{T})_+$ consists of the inclusions

$$S(\mathbf{C})_+ \subseteq S(\mathbf{C}^2)_+ \subseteq \cdots$$

induced by the standard inclusions $\mathbf{C}^n \subseteq \mathbf{C}^{n+1}$ onto the first factors. The $\mathbb{T}$-action is the diagonal one: $z \cdot (z_1, \ldots, z_n) = (zz_1, \ldots, zz_n)$. The $\mathbb{T}$-attaching maps are given by the pushout

$$\begin{array}{ccc} \mathbb{T} \times S(\mathbf{C}^n) & \longrightarrow & \mathbb{T} \times D(\mathbf{C}^n) \\ \downarrow & & \downarrow \\ S(\mathbf{C}^n) & \longrightarrow & S(\mathbf{C}^{n+1}), \end{array}$$

where the horizontal maps are the inclusions, the $\mathbb{T}$-action on the top row is on the first factor only, the left vertical map is the action map (which, with the given structure is a $\mathbb{T}$-map) and the right vertical map sends $(z, w) = (z, (w_1, \ldots, w_n)) \in \mathbb{T} \times D(\mathbf{C}^n)$ to $(zw_1, \ldots, zw_n, z\sqrt{1 - |w|^2}) \in S(\mathbf{C}^{n+1})$. The cofiber of the inclusion $S(\mathbf{C}^n)_+ \subseteq S(\mathbf{C}^{n+1})_+$ is thus $\mathbb{T}_+ \wedge D(\mathbf{C}^n)/S(\mathbf{C}^n)$. Since there are only cells in even degrees, the spectral sequences have terms only in the even columns, and the d^2 differentials are induced by the composite

$$\begin{array}{ccc} \mathbb{T}_+ \wedge D(\mathbf{C}^n)/S(\mathbf{C}^n) \longrightarrow & \Sigma(\mathbb{T} \times S(\mathbf{C}^n))_+ & \\ & \downarrow \text{action} & \\ & \Sigma S(\mathbf{C}^n)_+ & \longrightarrow \Sigma(\mathbb{T}_+ \wedge D(\mathbf{C}^{n-1})/S(\mathbf{C}^{n-1})), \end{array}$$

where the first map is the boundary map associated with the cofibration $(\mathbb{T} \times S(\mathbf{C}^n))_+ \subseteq (\mathbb{T} \times D(\mathbf{C}^n))_+$ and the last map the projection onto the cofiber.

A geometric analysis of this map as in [123, 1.4.2] makes it possible to identify the differential

Lemma 6.5.1.1 (Hesselholt) *Let* $\mathbf{S}^1 = \sin \mathbb{T}$ *be the circle group and let* X *be an* $\mathbf{S}^1$*-spectrum. The* E^2*-sheet of the spectral sequences for* $X_{h\mathbf{S}^1}$ *and* $X^{h\mathbf{S}^1}$ *of Lemma* 6.3.0.3 *are concentrated in the even columns with* $E^2_{2s,t} \cong \pi_t X$*, and the differential*

$$d^2 : E^2_{2s,t} \to E^2_{2s-2,t+1}$$

is induced by the map $\sigma + s \cdot \eta : \pi_t X \to \pi_{t+1}(\mathbf{S}^1{}_+ \wedge X)$ *composed with* π_{t+1} *of the* $\mathbf{S}^1$*-action* $\mathbf{S}^1{}_+ \wedge X \to X$. ☺

Note that for rational $\mathbf{S}^1$-spectra—like the ones we are talking about in connection with TC—or more generally, Eilenberg–Mac Lane $\mathbf{S}^1$-spectra, the Hopf map η is trivial, and so the differentials are simply given by the $\mathbf{S}^1$-action:

Corollary 6.5.1.2 *Let* X *be an* $\mathbf{S}^1$*-spectrum such that* $\eta : \pi_* X \to \pi_{*-1} X$ *is trivial, then the differential*

$$d^2 : E^2_{2s,t} \to E^2_{2s-2,t+1}$$

is the composite

$$\pi_t X \xrightarrow{\sigma} \pi_{t+1}(\mathbf{S}^1{}_+ \wedge X) \xrightarrow{\text{action}} \pi_{t+1} X.$$

☺

The homotopy orbits and fixed point spectra fit together to form the so-called *Tate spectrum* $X^{t\mathbf{S}^1} = [\widetilde{E\mathbf{S}^1} \wedge Map_*(E\mathbf{S}^1{}_+, X)]^{\mathbf{S}^1}$ of [116] where $\widetilde{E\mathbf{S}^1}$ is the homotopy cofiber of the nontrivial map $E\mathbf{S}^1{}_+ \to S^0$. From the cofiber sequence $E\mathbf{S}^1{}_+ \to$

$S^0 \to \widetilde{E\mathbf{S}^1}$ (which may be modelled concretely by $S(\mathbf{C}^\infty)_+ \subseteq D(\mathbf{C}^\infty)_+ \to S^{\mathbf{C}^\infty}$) we get a cofiber sequence [116]

$$\Sigma X_{h\mathbf{S}^1} \to X^{h\mathbf{S}^1} \to X^{t\mathbf{S}^1}.$$

There is a corresponding "Greenlees filtration" of $\widetilde{E\mathbf{S}^1}$ which is positive degrees correspond to the filtration for the homotopy orbit spectral sequence and in negative degrees to the filtration for the homotopy fixed point spectral sequence (see e.g., [28, Lemma 2.12]). The resulting *Tate spectral sequence*

$$\hat{E}^2_{s,t} = E^2_{s,t}(X^{t\mathbf{S}^1}) = \hat{H}^{-s}(\mathbf{S}^1, \pi_t X) \Rightarrow \pi_{s+t} X^{t\mathbf{S}^1},$$

(where $\hat{H}$ is the so-called Tate cohomology) is an upper half plane spectral sequence with vanishing odd columns and with even columns given by $\hat{E}^2_{2s,t} = \pi_t X$. Of course, this construction is not particular to the circle group, and the reader may want to compare the current discussion with the concrete examples relating to finite subgroups reviewed in Sect. 7.3.1.4.

The cofibration sequence relating the orbit, fixed point and Tate spectra is reflected in a short exact sequence of E^2-terms:

$$0 \to E^2_{s,t}(X^{h\mathbb{T}}) \to E^2_{s,t}(X^{t\mathbf{S}^1}) \to E^2_{s-1,t}(X_{h\mathbb{T}}) \to 0.$$

The norm map $S^1 \wedge X_{h\mathbf{S}^1} \to X^{h\mathbf{S}^1}$ induces the edge homomorphism. This has the following consequence:

Lemma 6.5.1.3 *Let X be an $\mathbf{S}^1$-spectrum. If the Tate spectral sequence converges strongly to zero, then the norm map*

$$S^1 \wedge X_{h\mathbf{S}^1} \to X^{h\mathbf{S}^1}$$

is a stable equivalence. ☺

6.5.2 Cyclic Homology and Its Relatives

Let Z be a cyclic module, i.e., a functor from Λ^o to simplicial abelian groups. Let $B\colon Z_q \to Z_{q+1}$ be Connes' operator

$$Z_q \xrightarrow{N=\sum(-1)^{qj}t^j} Z_q \xrightarrow{(-1)^q s_q} Z_{q+1} \xrightarrow{(1+(-1)^q t)} Z_{q+1}$$

satisfying $B \circ B = 0$ and $B \circ b + b \circ B = 0$ where $b = \sum(-1)^j d_j$. Due to these relations, the B-operator defines a complex

$$(\pi_* Z, B) = \left(\pi_0 Z \xrightarrow{B} \pi_1 Z \xrightarrow{B} \cdots \xrightarrow{B} \pi_q Z \xrightarrow{B} \cdots\right)$$

whose homology $H_*^{dR}(Z) = H_*(\pi_* Z, B)$ we call the *de Rham cohomology* of Z.

Using the b and the B, and their relations one can form bicomplexes (called (b, B)-bicomplexes, see e.g., [181, 5.1.7] for more detail) with (s,t)-entry Z_{t-s} connected by the bs vertically and the Bs horizontally (the relations guarantee that this becomes a bicomplex).

$$
\begin{array}{ccccccc}
 & & \vdots & & \vdots & & \\
 & & {\scriptstyle b}\big\downarrow & & {\scriptstyle b}\big\downarrow & & \\
\cdots & \xleftarrow{B} & Z_2 & \xleftarrow{B} & Z_1 & \xleftarrow{B} & \cdots \\
 & & {\scriptstyle b}\big\downarrow & & {\scriptstyle b}\big\downarrow & & \\
\cdots & \xleftarrow{B} & Z_1 & \xleftarrow{B} & Z_0 & & \\
 & & {\scriptstyle b}\big\downarrow & & & & \\
\cdots & \xleftarrow{B} & Z_0 & & & &
\end{array}
$$

If you allow all $t \geq s$ you get the so-called *periodic* (b, B)*-bicomplex* $B^{per}(Z)$ (called $\mathcal{B}Z^{\mathrm{per}}$ in [181]), if you allow $t \geq s \geq 0$ you get the *positive* (b, B)*-bicomplex* $B^+(Z)$ and if you allow $t \geq s \leq 0$ you get the *negative* (b, B)*-complex* $B^-(Z)$.

If $-\infty \leq m \leq n \leq \infty$ we let $T^{m,n}Z$ be the total complex of the part of the normalized (b, B)–bicomplex which is between the mth and nth column: $T_q^{m,n}Z = \prod_{k=m}^{n} C_{k,q-k}^{\mathrm{norm}}(Z)$ (where C^{norm} denotes the normalized chains defined in Sect. A.3.1.4). The associated homologies $H_*(T^{-\infty,0}Z)$, $H_*(T^{-\infty,\infty}Z)$ and $H_*(T^{0,\infty}Z)$ are called *periodic*, *negative* and (simply) *cyclic* homology, and often denoted $HP_*(Z)$, $HC^-(Z)$ and $HC(Z)$. The associated short exact sequence of complexes

$$0 \to T^{-\infty,0}Z \to T^{-\infty,\infty}Z \to T^{1,\infty}Z \to 0$$

together with the isomorphism $T_q^{1,\infty}Z \cong T_{q-2}^{0,\infty}Z$ gives rise to the well-known long exact sequence

$$\cdots \longrightarrow HC_{q-1}(Z) \longrightarrow HC_q^-(Z) \longrightarrow HP_q(Z) \longrightarrow HC_{q-2}(Z) \longrightarrow \cdots,$$

and similarly one obtains the sequence

$$\cdots \longrightarrow HC_{q-1}(Z) \xrightarrow{B} HH_q(Z) \xrightarrow{I} HC_q(Z) \xrightarrow{S} HC_{q-2}(Z) \longrightarrow \cdots$$

(the given names of the maps are the traditional ones, and we will have occasion to discuss the S-map a bit further).

Notice that $T^{-\infty,n}Z = \lim_{\overleftarrow{m}} T^{m,n}Z$, and so if $\cdots \twoheadrightarrow Z^{k+1} \twoheadrightarrow Z^k \twoheadrightarrow \ldots$ is a sequence of surjections of cyclic modules with $Z = \lim_{\overleftarrow{k}} Z^k$, then $T^{-\infty,n}Z \cong$

$\lim_{\overleftarrow{k}} T^{-\infty,n}Z^k$, and you have $\lim_{\leftarrow}{}^{(1)}$-$\lim_{\leftarrow}$ exact sequences, e.g.,

$$0 \to \lim_{\overleftarrow{k}}{}^{(1)} HC^-_{q+1}(Z^k) \to HC^-_q(Z) \to \lim_{\overleftarrow{k}} HC^-_q(Z^k) \to 0.$$

Also, from the description in terms of bicomplexes we see that we have a short exact sequence

$$0 \to \lim_{\overleftarrow{k}}{}^{(1)} HC_{q+1+2k}(Z) \to HP_q(Z) \to \lim_{\overleftarrow{k}} HC_{q+2k}(Z) \to 0$$

describing the periodic homology of a cyclic complex Z in terms of the cyclic homology and the S-maps connecting them.

We see that filtering $B^{\text{per}}(Z)$ by columns we get a spectral sequence for $HP_*(Z)$ with E^2-term given by $H^{dR}_*(Z)$.

6.5.2.1 The Upper Half Plane Bicomplex

There is another representation of these theories, which Loday and Quillen [182] attribute to Tsygan, by means of an upper half plane bicomplex, and which is closer to the spectral sequences associated to spectra with $\mathbf{S}^1$-actions we have discussed (the difference essentially boiling down to an indexation issue in the skeletal filtrations). If Z is a cyclic module, we let $CC(Z) = \{CC_{s,t}(Z)\}$ be the bicomplex with $CC_{s,t}(Z) = Z_t$ and with vertical differentials given the alternating sum $b = \sum_{i=0}^{t}(-1)^i d_i : Z_t \to Z_{t-1}$ of all the face maps in the even columns and the alternating sum b' of all *but the last* face map in the odd columns. The horizontal differentials leaving odd columns are given by $1 - (-1)t^j : Z_j \to Z_j$ and those leaving even columns by $N = \sum_{i=1}^{j+1}((-1)^j t)^i : Z_j \to Z_j$.

The odd columns are contractible, with $u = (-1)^j s_j : Z_j \to Z_{j+1}$ as a contraction ($b'u + ub' = 1$). This contraction makes the odd columns inessential, and the (b, B) is exactly what you get by systematically erasing them, see [182] or [100]. Thus the total complex of $CC(Z)$ is quasi isomorphic to $T^{-\infty,\infty}$, and similarly, if you restrict to non-negative (resp. non-positive) horizontal degrees, you get a complex which is quasi isomorphic to $T^{0,\infty}$ (resp. $T^{-\infty,0}$).

Notice that the presence of degeneracies is crucial for the contractibility of the odd columns, which is a pivotal point when extending Hochschild homology to rings without unit elements: you may still define the bicomplex, but the odd columns may no longer be contractible. In this situation, the homology of the odd columns are of particular importance, giving obstructions for excision, see e.g., [279].

Filtering $CC(Z)$ by columns, you get a spectral sequence with $E^1_{s,t} = E^2_{s,t}$ being $\pi_t Z$ if s is even and 0 otherwise. This may remind the reader of the Tate spectral sequence, and that this is no coincidence is the contents of the next section.

6.5.2.2 Geometrical Interpretation

These cyclic homology theories have clear geometrical meaning in terms of orbit and fixed point spectra as is apparent from Theorem 6.5.2.1 below.

By a *spectrum* M of simplicial abelian groups, we mean a sequence $\{n \mapsto M^n\}$ of simplicial abelian groups together with homomorphisms $\tilde{\mathbf{Z}}[S^1] \otimes M^n \to M^{n+1}$ (according to the notion of spectra in any simplicial model category). A map $f\colon M \to N$ between spectra is, as usual, a sequence of homomorphisms $f^n\colon M^n \to N^n$ respecting the structure maps.

In view of the identity $C_*^{\mathrm{norm}}(\tilde{\mathbf{Z}}[S^1]) = \mathbf{Z}[-1]$, the normalized chain complex transforms spectra of simplicial abelian groups to spectra of chain complexes (concentrated in non-negative dimensions):

$$\begin{aligned} C_*^{\mathrm{norm}}(M^n)[-1] = C_*^{\mathrm{norm}}(\tilde{\mathbf{Z}}[S^1]) \otimes C_*^{\mathrm{norm}}(M^n) &\to C_*^{\mathrm{norm}}(\tilde{\mathbf{Z}}[S^1] \otimes M^n) \\ &\to C_*^{\mathrm{norm}}(M^{n+1}), \end{aligned}$$

where the first map is the Eilenberg-Zilber (shuffle) map and the last is induced by the structure map on M. In turn, a spectrum $C = \{C^n[-1] \to C^{n+1}\}$ of chain complexes gives rise to a single (unbounded) chain complex $\lim_{\overrightarrow{n}} C^n[n]$, and we get a functor

$$M \mapsto C_*^{\mathrm{spt}} M = \lim_{\overrightarrow{n}} C_*^{\mathrm{norm}}(M^n)[n]$$

from spectra of simplicial abelian groups to chain complexes, sending stable equivalences to quasi isomorphisms (i.e., maps of chain complexes inducing isomorphism in homology). A more refined approach gives rise to (suitably monoidal) Quillen equivalences between $H\mathbf{Z}$-modules, spectra of simplicial abelian groups and chain complexes, see Schwede and Shipley [254].

Theorem 6.5.2.1 *Let $M\colon \Lambda^o \to sAb$ be a cyclic simplicial abelian group. There are natural chains of quasi isomorphisms*

$$C_*^{\mathrm{spt}} \sin|HM|_{h\mathbf{S}^1} \simeq T^{0,\infty} M,$$

$$C_*^{\mathrm{spt}} \sin|HM|^{h\mathbf{S}^1} \simeq T^{-\infty,0} M,$$

$$C_*^{\mathrm{spt}} \sin|HM|^{t\mathbf{S}^1} \simeq T^{-\infty,\infty} M.$$

Proof The first statement follows e.g., from the corresponding statement in Loday's book [181] which shows that there is a natural chain of quasi isomorphisms between $C^{\mathrm{norm}} \sin|M|_{h\mathbf{S}^1}$ and $T^{0,\infty} M$ and the fact that $\sin|HM|_{h\mathbf{S}^1}$ is a connected Ω-spectrum.

All the statements can be proved hands on by the standard filtration, giving chains of quasi isomorphisms in some range $T^{s,t}$ and then extending by (co)limits. We do the case $T^{-\infty,0}$. Choose the colimit $S(\mathbf{C}^\infty)$ of

$$S(\mathbf{C}) \subseteq S(\mathbf{C}^2) \subseteq \cdots \subseteq S(\mathbf{C}^n) \subseteq \cdots$$

as your model for the contractible free $\mathbb{T}$-space $E\mathbb{T}$. Then we have a natural equivalence

$$\sin |HM|^{h\mathbf{S}^1}(X) \xleftarrow{\sim} \lim_{\overleftarrow{n}} (\mathbb{T}\text{-}Top_*)\big(S(\mathbf{C}^n)_+, |M \otimes \tilde{\mathbf{Z}}[X]|\big)$$

(this is only a natural equivalence and not an isomorphism since the definition of the homotopy fixed points of a topological spectrum involves a fibrant replacement, which is unnecessary since HM is an Ω-spectrum).

Hence we are done once we have shown that there is a natural (in n and M) chain of maps connecting

$$P_M\big(S(\mathbf{C}^{n+1})_+\big) = C_*^{\text{norm}}(\mathbb{T}\text{-}Top_*)\big(S(\mathbf{C}^{n+1})_+, |M|\big), \quad \text{and} \quad T^{-n,0}M$$

inducing an isomorphism in homology in positive dimensions.

This is done by induction on n. For $n = 0$ this is obvious since $P_M(S(\mathbf{C})_+) \cong C_*^{\text{norm}} \sin |M|$ whereas $T^{-0,0}M = C_*^{\text{norm}} M$. Let $n > 0$ and consider the two pullback diagrams

$$\begin{array}{ccc}
\mathbb{T}_+ \wedge S(\mathbf{C}^n) & \xrightarrow{id \times \text{inclusion}} & \mathbb{T}_+ \wedge D(\mathbf{C}^n) \\
\uparrow & & \uparrow \\
\mathbb{T}_+ \wedge S(\mathbf{C}^n)_+ & \xrightarrow{id \times \text{inclusion}} & \mathbb{T}_+ \wedge D(\mathbf{C}^n)_+ \\
{\scriptstyle \text{action}}\downarrow & & \downarrow \\
S(\mathbf{C}^n)_+ & \longrightarrow & S(\mathbf{C}^{n+1})_+
\end{array}$$

where the upwards pointing arrows are the natural projections, while the lower square is the $\mathbb{T}$-cell attachment. Applying $P_M = C_*^{\text{norm}}(\mathbb{T}\text{-}Top_*)(-, |M|)$ to this diagram we get that $P_M(S(\mathbf{C}^{n+1})_+)[1]$ can be described by means of the mapping cylinder of the composite J_n

$$\begin{array}{ccccc}
P_M\big(S(\mathbf{C}^n)_+\big) & \xrightarrow{P_M(\text{action})} & P_M\big(\mathbb{T}_+ \wedge S(\mathbf{C}^n)_+\big) & \xrightarrow{\cong} & C_*^{\text{norm}} \sin Top_*\big(S(\mathbf{C}^n)_+, |M|\big) \\
 & & & & {\scriptstyle \text{proj.}}\downarrow \\
 & & C_*^{\text{norm}} \sin |M|[2n-1] & \xrightarrow{\cong} & C_*^{\text{norm}} \sin Top_*\big(S(\mathbf{C}^n), |M|\big),
\end{array}$$

where the first isomorphism is the adjunction isomorphism of the free/forgetful pair between pointed $\mathbb{T}$-spaces and pointed spaces, proj. is induced by the section to the inclusion of the pointed maps into the free maps given by sending a function to the function minus its value value at the basepoint (when considered as a based space, $S(\mathbf{C}^n)$ is based at $(1, 0, \ldots, 0)$), and the last isomorphism is given by the singular/realization adjunction, the identification $|S^{2n-1}| = S(\mathbf{C}^n)$ and the correspondence between shift and loop. For simplicity, we have assumed that M is highly connected, so that we don't have to worry about boundary effects and have suppressed truncations.

The commutativity of

$$\begin{array}{ccc} \mathbb{T}_+\wedge S(\mathbf{C}^n)_+ & \longrightarrow & \mathbb{T}_+\wedge\mathbb{T}_+\wedge D(\mathbf{C}^{n-1})/S(\mathbf{C}^{n-1}) \\ \text{action}\downarrow & & \downarrow\text{action}\wedge\text{id} \\ S(\mathbf{C}^n)_+ & \longrightarrow & \mathbb{T}_+\wedge D(\mathbf{C}^{n-1})/S(\mathbf{C}^{n-1}), \end{array}$$

where the horizontal maps are the cofibers of the $\mathbb{T}$-filtration, shows that on the top cell, J_n is nothing but a shifted copy of J_1, and since we are evaluating at Eilenberg-Mac Lane spaces there is no further attachment.

Hence, by induction $P_M(S(\mathbf{C}^{n+1})_+)$ is represented by a bicomplex of exactly the same shape as $T^{-n,0}$, the only question is the identification of the attaching maps.

Hence, we need to identify $J_1\colon P_M(\mathbb{T}_+) \to C^{\text{norm}} \sin|M|[1]$. The action $P_M(\mathbb{T}_+) \to P_M(\mathbb{T}_+\wedge\mathbb{T}_+)$ can be factored as $C^{\text{norm}} \sin$ applied to the composite

$$\begin{aligned} (\mathbb{T}\text{-}Top_*)\big(\mathbb{T}_+, |M|\big) &\xrightarrow{\mathbb{T}_+\wedge-} (\mathbb{T}\text{-}Top_*)\big(\mathbb{T}_+\wedge\mathbb{T}_+, \mathbb{T}_+\wedge|M|\big) \\ &\xrightarrow{\text{action}} (\mathbb{T}\text{-}Top_*)\big(\mathbb{T}_+\wedge\mathbb{T}_+, |M|\big) \end{aligned}.$$

Recall that the $\mathbb{T}$-action on $|M|$ is obtained through a certain isomorphism $\mathbb{T}_+\wedge|M| \cong |j_*j^*UM|$, where j_*j^*UM is what you get if you let UM be the underlying cyclic *pointed set* on M and use the free/forgetful pair (j_*, j^*) between simplicial and cyclic pointed sets (recall that $j\colon \Delta^o \subseteq \Lambda^o$ is the inclusion: $S^1_+\wedge UM$ and j_*j^*UM give different triangularizations of $\mathbb{T}_+\wedge|M|$). We could equally well have used the free/forgetful pair (j_*^{Ab}, j_{Ab}^*) between simplicial and cyclic *abelian groups*, and we get a canonical map $j_*j^*UM \to Uj_*^{Ab}j_{Ab}^*M$ (induced by inclusion of wedges into sums, and so is a stable equivalence). The $\mathbb{T}$-action can then be factored

$$\mathbb{T}_+\wedge|M| \cong \big|j_*j^*UM\big| \to \big|j_*^{Ab}j_{Ab}^*M\big| \to |M|,$$

where the last map is the unit of adjunction.

Using the $\mathbb{T}$-Top_*/Top_* adjunction, we get that J_1 is given by $C^{\text{norm}} \sin$ applied to

$$|M| \longrightarrow Top_*\big(\mathbb{T}, \big|j_*^{Ab}j_{Ab}^*M\big|\big) \longrightarrow Top_*\big(\mathbb{T}, |M|\big),$$

where the first map is given by $\mathbb{T}_+\wedge-$ and projecting to the pointed maps followed by postcomposing with the map $\mathbb{T}_+\wedge|M| \cong |j_*j^*UM| \to |j_*^{Ab}j_{Ab}^*M|$, and the last map is induced by the unit of adjunction. Hence, up to shift, the attaching map is given by the map

$$C_*^{\text{norm}}\big(\tilde{\mathbf{Z}}[S^1]\big) \otimes C_*^{\text{norm}}(M) \to C_*^{\text{norm}}\big(j_*^{Ab}j_{Ab}^*M\big) \to C_*^{\text{norm}}(M)$$

analyzed in [151, p. 415] and [123, p. 15], showing that the attaching map is the map (Connes' B-map) needed to express $T^{-1,0}$ as the mapping cone of $B\colon C^{\text{norm}}_* M[1] \to C^{\text{norm}}_* M[2]$ (recall that the sign of the differentials in a chain complex are changed by shifts, so that B is indeed a chain map). □

An important distinction for our purposes between homotopy orbits and fixed points is that homotopy orbits may be calculated degreewise. This is false for the homotopy fixed points.

Lemma 6.5.2.2 *Let X be a simplicial $\mathbf{S}^1$-spectrum. Then* $\operatorname{diag}^*(X_{h\mathbf{S}^1})$ *is naturally equivalent to* $(\operatorname{diag}^* X)_{h\mathbf{S}^1}$. *In particular, if A is a simplicial ring, then $HC(A)$ can be calculated degreewise.*

Proof True since homotopy colimits commute, and the diagonal may be calculated as $\operatorname{holim}_{\overrightarrow{[q]\in\Delta}} X_q$. □

Notice that if we filter by columns, the proof of Theorem 6.5.2.1 says that the resulting spectral sequences for periodic, cyclic and negative cyclic agree with the Tate, orbit and fixed point spectrum spectral sequences of Lemma 6.5.1.1 (with shape depending on the indexation of the filtration: either with vanishing odd columns and concentrated in the upper half plane, or as the for the (b, B)-bicomplex with no columns vanishing).

6.5.2.3 Derivations

The following is lifted from [100] (see also [181, 4.1]), and we skip the gory calculations. Let A be a simplicial ring. A *derivation* is a simplicial map $D\colon A \to A$ satisfying the Leibniz relation $D(ab) = D(a)b + aD(b)$. A derivation $D\colon A \to A$ induces an endomorphism of cyclic modules $L_D\colon HH(A) \to HH(A)$ by sending $a = a_0 \otimes \cdots \otimes a_q \in A_p^{\otimes q+1}$ to

$$L_D(a) = \sum_{i=0}^{q} a_0 \otimes \cdots \otimes a_{i-1} \otimes D(a_i) \otimes a_{i+1} \otimes \cdots \otimes a_q.$$

If A is a simplicial ring, let $(C_*(A), b)$ be the chain (normalized) complex associated to the bisimplicial abelian group $HH(A)$.

One then constructs maps

$$e_D\colon C_q(A) \to C_{q-1}(A), \text{ and } E_D\colon C_q(A) \to \bar{C}_{q+1}(A)$$

satisfying

Lemma 6.5.2.3 *Let $D\colon A \to A$ be a derivation. Then*

$$e_D b + b e_D = 0,$$

$$e_D B + B e_D + E_D b + b E_D = L_D$$

and

$$E_D B + B E_D$$

is degenerate. ☺

To be explicit, the maps are given by sending $a = a_0 \otimes \cdots \otimes a_q \in A_p^{\otimes q+1}$ to

$$e_D(a) = (-1)^{q+1} D(a_q) a_0 \otimes a_1 \otimes \cdots \otimes a_{q-1}$$

and

$$E_D(a) = \sum_{1 \le i \le j \le q} (-1)^{iq+1} \otimes a_i \otimes a_{i+1} \otimes \cdots \otimes a_{j-1} \otimes D(a_j) \otimes a_{j+1}$$
$$\otimes \cdots \otimes a_q \otimes a_0 \otimes a_{i-1}$$

The first equation of Lemma 6.5.2.3 is then a straightforward calculation, but the second is more intricate (see [100] or [181]).

Corollary 6.5.2.4 [100] *Let D be a derivation on a flat ring A. Then*

$$L_D S\colon HC_* A \to HC_{*-2} A$$

is the zero map.

Proof Collecting the relations in Lemma 6.5.2.3 we get that $(E_D + e_D)(B + b) + (B + b)(E_D + e_D) = L_D$ on the periodic complex. However, this does not respect the truncation to the positive part of the complex. Hence we shift once and get the formula $((E_D + e_D)(B + b) + (B + b)(E_D + e_D))S = L_D S$ which gives the desired result. □

Corollary 6.5.2.5 *Let $f\colon A \to B$ be a map of simplicial rings inducing a surjection $\pi_0 A \to \pi_0 B$ with nilpotent kernel. Let X be the homotopy fiber (in the category of cyclic abelian groups) of $HH(A) \to HH(B)$. Then $HP_*(X_{(0)}) = HP_*(\widehat{X_{(0)}}) = 0$.*

Proof By considering the square

$$\begin{array}{ccc} A & \longrightarrow & B \\ \downarrow & & \downarrow \\ \pi_0 A & \longrightarrow & \pi_0 B \end{array}$$

we see that it is enough to prove the case where f is a surjection with nilpotent kernel and f is a surjection with connected kernel separately.

Let P be completion followed by rationalization or just rationalization. The important thing is that P is an exact functor with rational values.

The basic part of the proof, which is given in [100, II.5], is the same for the connected and the nilpotent cases. In both situations we end up by proving that the shift map S is nilpotent on the relative part, or more precisely: for every q and every $k > q$ the map

$$S^k: HC_{q+2k}Y \to HC_qY$$

is zero, where Y is the homotopy fiber of $P(HH(A)) \to P(HH(B))$ (actually in this formulation we have assumed that the kernel was square zero in the nilpotent case, but we will see that this suffices for giving the proof). From this, and from the fact that periodic homology sits in a $\lim_{\overleftarrow{S}}{}^{(1)}$-$\lim_{\overleftarrow{S}}$ short exact sequence, we conclude the vanishing of periodic homology.

The main difference between our situation and the rational situation of [100] is that we can not assume that our rings are flat. That means that HH is not necessarily calculated by the Hochschild complex.

In the connected case, this is not a big problem, since the property of being connected is a homotopy notion, and so we can replace everything in sight by degreewise free rings and we are in business as explained in [100, IV.2.1]. Being nilpotent is not a homotopy notion, and so must be handled with a bit more care. First, by considering

$$A \to \cdots \to A/I^n \to A/I^{n-1} \to \cdots \to A/I^2 \to A/I = B$$

we see that it is enough to do the square zero case. Let $X \overset{\sim}{\twoheadrightarrow} B$ be a free resolution of B and consider the pullback

$$\begin{array}{ccc} Q & \longrightarrow & X \\ \downarrow & & \downarrow \\ A & \longrightarrow & B \end{array}.$$

Since the vertical maps are equivalences (using "properness" of simplicial rings: pullbacks of maps that are fibrations and weak equivalences—on underlying simplicial sets—are fibrations and weak equivalences), we have reduced to the case where $A \to B$ is a surjection of simplicial rings with discrete square zero kernel I and where B is free in every degree. But since cyclic homology can be calculated degreewise by Lemma 6.5.2.2, it is enough to prove this in every degree, but since B is free in every degree it is enough to prove it when $A \to B$ is a split surjection of discrete rings with square zero kernel I. Choosing a splitting we can write $A \cong B \ltimes I$, where I is a B-bimodule with square zero multiplication. Let $J \overset{\sim}{\twoheadrightarrow} I$ be a free resolution of I as B-bimodules. Then we have an equivalence $B \ltimes J \overset{\sim}{\to} B \ltimes I$, and again since cyclic homology can be calculated degreewise we have reduced to the case $B \ltimes I \to B$ where B is free and I is a free B-bimodule.

Hence we are in the flat case with $A = B \ltimes I \to B$ with $I^2 = 0$, and can prove our result in this setting. Then the distributive law provides a decomposition of $A^{\otimes q+1} \cong (B \oplus I)^{\otimes q+1}$, and if we let F_q^k consist of the summands with k or more I-factors we get a filtration

$$0 = F^\infty = \bigcap_n F^n \subset \cdots \subset F^2 \subset F^1 \subset F^0 = HH(A)$$

(it is of finite length in each degree, in fact $F_k^n = 0$ for all $n - 1 > k$). Note that we have isomorphisms of cyclic modules $HH(B) = F^0/F^1$, $HH(A) = F^0 \cong \bigoplus_{k\geq 0} F^k/F^{k+1}$.

We must show that for every q and every $k > q$ the map

$$S^k\colon HC_{q+2k}\big(P(F^1)\big) \to HC_q\big(P(F^1)\big)$$

is zero. Since $F_k^n = 0$ for all $n - 1 > k$ we have that $HC_q(P(F^n)) = 0$ for all $q < n - 1$. Hence it is enough to show that for every q

$$S^k\colon HC_{q+2k}\big(P\big(F^1/F^{k+1}\big)\big) \to HC_q\big(P\big(F^1/F^{k+1}\big)\big)$$

is zero.

Since I is square zero, the projection followed by inclusion $D\colon B \ltimes I \to I \subseteq B \ltimes I$ is a derivation ($D((b,i)\cdot(b',i')) = D((bb', bi' + ib')) = (0, bi' + ib') = (b,i)\cdot(0,i') + (0,i)\cdot(b',i') = (b,i)\cdot D(b',i') + D(b,i)\cdot(b',i')$), and it acts as multiplication by m on F^m/F^{m+1}. Therefore we have by Corollary 6.5.2.4 (whose proof is not affected by the insertion of P) that

$$m \cdot S = L_D S = 0$$

on $HC_*(P(F^m/F^{m+1}))$.

Since $m \geq 0$ is invertible in $\mathbf{Q}$, we get that $S = 0$ on $HC_*(P(F^m/F^{m+1}))$, and by induction $S^k = 0$ on $HC_*(P(F^1/F^{k+1}))$.

The proof of the connected case is similar: first assume that I is *reduced* (has just one zero-simplex: this is obtained by the Lemma 6.5.2.6 we have cited below). Use the "same" filtration as above (it no longer splits), and the fact that F^k is zero in degrees less than k since I is reduced.

Filter A by the powers of I:

$$\cdots \subseteq I^m \subseteq \cdots \subseteq I^1 \subseteq I^0 = A.$$

This gives rise to a filtration of the Hochschild homology

$$0 = F^\infty = \bigcap_n F^n \subset \cdots \subset F^2 \subset F^1 \subset F^0 = HH(A)$$

by defining

$$F_q^k = im\left\{ \bigoplus_{\sum k_i = k} \bigotimes_{i=0}^{q} I^{k_i} \to HH(A)_q \right\}.$$

Consider the associated graded ring $gr(A)$ with $gr_k A = I^k/I^{k+1}$. Note that we have isomorphisms of cyclic modules $HH(B) = F^0/F^1$, $HH(A) = F^0$ and $HH(grA) \cong \oplus_{k\geq 0} F^k/F^{k+1}$.

We define a derivation D on grA by letting it be multiplication by k in degree k. Note that L_D respects the filtration and acts like k on F^k/F^{k+1}. The proof then proceeds as in the nilpotent case. □

In the above proof we used the following result of Goodwillie [102, I.1.7]:

Lemma 6.5.2.6 *Let $f\colon A \twoheadrightarrow B$ be a k-connected surjection of simplicial rings. Then there is a diagram*

$$\begin{array}{ccc} R & \xrightarrow{\sim} & A \\ {\scriptstyle g}\downarrow & & \downarrow{\scriptstyle f} \\ S & \xrightarrow{\sim} & B \end{array}$$

of simplicial rings such that the horizontal maps are equivalences, the vertical maps surjections, and the kernel of g is k-reduced (i.e., its $(k-1)$-skeleton is a point). If A and B are flat in every degree, then we may choose R and S flat too. ☺

Proposition 6.5.2.7 *Let $f\colon A \to B$ be a map of simplicial rings inducing a surjection $\pi_0 A \to \pi_0 B$ with nilpotent kernel, then the diagrams induced by the norm map*

$$\begin{array}{ccc} S^1 \wedge \big(THH(A)_{(0)}\big)_{h\mathbf{S}^1} & \longrightarrow & \big(THH(A)_{(0)}\big)^{h\mathbf{S}^1} \\ \downarrow & & \downarrow \\ S^1 \wedge \big(THH(B)_{(0)}\big)_{h\mathbf{S}^1} & \longrightarrow & \big(THH(B)_{(0)}\big)^{h\mathbf{S}^1} \end{array}$$

and

$$\begin{array}{ccc} S^1 \wedge \big(THH(A)\widehat{\ }_{(0)}\big)_{h\mathbf{S}^1} & \longrightarrow & \big(THH(A)\widehat{\ }_{(0)}\big)^{h\mathbf{S}^1} \\ \downarrow & & \downarrow \\ S^1 \wedge \big(THH(B)\widehat{\ }_{(0)}\big)_{h\mathbf{S}^1} & \longrightarrow & \big(THH(B)\widehat{\ }_{(0)}\big)^{h\mathbf{S}^1} \end{array}$$

are homotopy cartesian.

Proof Recall that by Lemma 4.2.3.8 *THH* is equivalent to *HH* after rationalization, or profinite completion followed by rationalization, and so can be regarded as the Eilenberg–Mac Lane spectrum associated with a cyclic module.

By Theorem 6.5.2.1 and Lemma 6.5.1.3 we are done if the corresponding periodic cyclic homology groups vanish, and this is exactly the contents of Corollary 6.5.2.5. □

Remark 6.5.2.8 A priori $(\underline{T}_{(0)})^{h\mathbf{S}^1}$ should not preserve connectivity, and does not do so (look e.g., at the zero-connected map $\mathbf{Z} \twoheadrightarrow \mathbf{Z}/p\mathbf{Z}$: $(\underline{T}(\mathbf{Z})_{(0)})^{h\mathbf{S}^1}$ is not connective (its homotopy groups are the same as rational negative cyclic homology of the integers and so have a $\mathbf{Q}$ in in every even non-positive dimension), but $(\underline{T}(\mathbf{Z}/p\mathbf{Z})_{(0)})^{h\mathbf{S}^1}$ vanishes.

However, since homotopy colimits preserve connectivity Proposition 6.5.2.7 gives that we do have the following result.

Corollary 6.5.2.9 *Let $A \to B$ be a $k > 0$-connected map of simplicial rings, and let X be either $THH_{(0)}{}^{h\mathbf{S}^1}$ or $THH\hat{}_{(0)}{}^{h\mathbf{S}^1}$ considered as a functor from simplicial rings to spectra. Then $X(A) \to X(B)$ is $k+1$ connected. If $A \to B$ induces a surjection $\pi_0 A \to \pi_0 B$ with nilpotent kernel, then $X(A) \to X(B)$ is -1-connected.* ☺

6.5.3 Structural Properties for Integral TC

As remarked earlier, the importance of the results about the $\mathbf{S}^1$-homotopy fixed point spectra in Sect. 6.5.2.2 above comes from the homotopy cartesian square of Lemma 6.4.3.2

$$\begin{array}{ccc} \underline{TC}(-) & \longrightarrow & \left(\underline{T}(-)_{(0)}\right)^{h\mathbf{S}^1} \\ \downarrow & & \downarrow \\ \underline{TC}(-)\hat{} & \longrightarrow & \left(\underline{T}(-)\hat{}_{(0)}\right)^{h\mathbf{S}^1} \end{array}.$$

So combining these facts with the properties of $TC(-; p)$ exposed in Sect. 6.4 we get several results on TC quite for free.

Proposition 6.5.3.1 *If $A \to B$ is $k > 0$-connected map of $\mathbf{S}$-algebras, then $TC(A) \to TC(B)$ is $(k-1)$-connected. If $A \to B$ induces a surjection $\pi_0 A \to \pi_0 B$ with nilpotent kernel, then $TC(A) \to TC(B)$ is -1-connected.*

Proof Consider the cubical approximation in Definition 3.3.1.8. In this construction the conditions on the maps of $\mathbf{S}$-algebras are converted to conditions on homomorphisms of simplicial rings (that the maps in the cubes are not themselves homomorphisms does not affect the argument). Hence by Theorem 6.4.3.4 the result follows from the homotopy cartesian square, Corollary 6.5.2.9 and Lemma 6.4.2.4. □

In fact, for the same reason this applies equally well to higher dimensional cubes:

Proposition 6.5.3.2 *Let $\mathcal{A}$ be cubical diagram (of positive dimension) of $\mathbf{S}$-algebras, and assume that all maps are k-connected and induce surjections with nilpotent kernel on π_0. Assume that we have shown that $\underline{T}(\mathcal{A})$ is $id - k$ cartesian. Then $\underline{TC}(\mathcal{A})$ is $id - k - 1$ cartesian.*

Proof Again we do the proof for each of the vertices in the cartesian square giving TC. For $TC(-)\hat{} \simeq \prod_{p \text{ prime}} TC(-; p)\hat{}_p$ this is Proposition 6.4.2.6. For the two other vertices we again appeal to Theorem 6.4.3.4 which allow us to prove it only for simplicial rings, and then to Proposition 6.5.2.7 which tells us that the cubes involving $(\underline{T}_{(0)})^{h\mathbf{S}^1}$ and $(\underline{T}\widehat{_{p}}_{(0)})^{h\mathbf{S}^1}$ are as (co)cartesian as the corresponding cubes, $\Sigma(\underline{HH}(\mathcal{A})_{(0)})_{h\mathbf{S}^1}$ and $\Sigma(\underline{HH}(\mathcal{A})\widehat{}_{(0)})_{h\mathbf{S}^1}$. Thus we are done since homotopy colimits preserve cocartesianness. □

Notice that this is slightly stronger than what we used in Theorem 6.4.3.4 to establish the approximation property for TC: There we went all the way in the limit, obtaining stable equivalences before taking the homotopy fixed point construction. Here we actually establish that the connectivity grows as expected in the tower, not just that it converges.

Proposition 6.5.3.3 *Topological cyclic homology of simplicial rings can locally be calculated degreewise. That is, given a map of simplicial connective* **S**-*algebras* $A \to B$ *inducing a surjection with nilpotent kernel on* π_0 *in every degree, let* $\underline{TC}^{\delta}(A) = \mathrm{diag}^*\{[q] \mapsto \underline{TC}(A_q)\}$ *and* $\underline{TC}(A) = \underline{TC}(\mathrm{diag}^* A)$. *Then there is a natural equivalence between the homotopy fibers of* $\underline{TC}^{\delta}(A) \to \underline{TC}^{\delta}(B)$ *and* $\underline{TC}(A) \to \underline{TC}(B)$.

Proof Consider the homotopy cartesian square of Lemma 6.4.3.2

$$\begin{array}{ccc} \underline{TC}(-) & \longrightarrow & \left(\underline{T}(-)_{(0)}\right)^{h\mathbf{S}^1} \\ \downarrow & & \downarrow \\ \underline{TC}(-)\widehat{} & \longrightarrow & \left(\underline{T}(-)\widehat{}_{(0)}\right)^{h\mathbf{S}^1} \end{array}.$$

If we can show that each of the three other vertices can be calculated degreewise, so can $\underline{TC}$ (homotopy pullbacks of simplicial spectra can be performed degreewise). That $\underline{TC}(-)\widehat{}$ has this property follows since profinite completion of simplicial spectra can be performed degreewise, since $TC \simeq TC(-;p)$ after p-completion and since Lemma 6.4.2.3 gave us that $TC(-;p)$ could be calculated degreewise. Hence the statement that TC can be calculated degreewise is equivalent to the statement that the last two vertices can be calculated degreewise locally.

Now, by inspection we see that $\mathrm{diag}^* A \to \mathrm{diag}^* B$ itself induces a surjection with nilpotent kernel on π_0. Let X be either $\underline{T}_{(0)}$ or $\underline{T}\widehat{}_{(0)}$, and apply X to $\mathrm{diag}^* A \to \mathrm{diag}^* B$. By Lemma 6.5.2.7 the homotopy fiber of $X(\mathrm{diag}^* A)^{h\mathbf{S}^1} \to X(\mathrm{diag}^* B)^{h\mathbf{S}^1}$ is equivalent to the homotopy fiber of $S^1 \wedge X(\mathrm{diag}^* A)_{h\mathbf{S}^1} \to S^1 \wedge X(\mathrm{diag}^* B)_{h\mathbf{S}^1}$ which can be calculated degreewise (homotopy colimits can be calculated degreewise) and since the condition on $A \to B$ was satisfied in every degree we can translate back to the homotopy $\mathbf{S}^1$-fixed points in each degree. □

6.5.3.1 Summary of Results

In addition to the above results depending on the careful analysis of the homotopy fixed points of topological Hochschild homology we have the following more trivial results following from our previous analyses of $TC(-;p)$ and the general properties of homotopy fixed points:

- TC is an Ω-spectrum.
- TC can be calculated degreewise in certain relative situations
- TC preserves ΓS_*-equivalences
- TC is Morita-equivariant
- TC preserves finite products
- TC of triangular matrices give the same result as products.
- TC satisfies strict cofinality
- TC of **S**-algebras "depends only" on its values on simplicial rings.

Chapter 7
The Comparison of K-Theory and TC

At long last we come to the comparison between algebraic K-theory and topological cyclic homology. In Definition 7.1.1.2 below we define the *cyclotomic trace* as a factorization of the Dennis trace for symmetric monoidal $\Gamma\mathcal{S}_*$-categories of Definition 5.2.3.2, and with the simplified notation introduced immediately below Definition 7.1.1.2 the statement reads as follows:

Theorem 7.0.0.1 *Let A be an* **S***-algebra. Then the Dennis trace map $K(A) \to THH(A)$ factors naturally as*

$$K(A) \xrightarrow{trc} TC(A) \longrightarrow THH(A).$$

This theorem is proved in Sect. 7.1.1 and presupposes using proper models for all the functors, as made precise in the statement and proof of Lemma 7.1.1.1 and permanently frozen by the definition of the cyclotomic trace Definition 7.1.1.2 below when applied to the $\Gamma\mathcal{S}_*$-category of finite free A-modules $\mathcal{F}_A$ of Definition 3.2.4.1. In the interest of readability we have used undecorated symbols for these precise models.

The Dennis trace map was originally a map from algebraic K-theory to Hochschild homology. When Connes used cyclic homology to produce an analogue of de Rham cohomology for non-commutative rings, he also indicated how the classical Chern character from algebraic K-theory to de Rham cohomology for commutative rings could, at least rationally, be obtained by a factorization of the Dennis trace map through cyclic homology. Several others, perhaps most notably Jones and Goodwillie, gave an integral factorization of the Dennis trace map through negative cyclic homology, which rationally recovered Connes earlier constructions. In an influential letter [103] to Waldhausen, Goodwillie showed how one could recover the Jones-Goodwillie Chern character to negative cyclic homology using a cyclic bar construction in conjunction with the S-construction Sect. 1.2.2 which factored though actual fixed points for every finite subgroup of the natural circle action. Using methods of Illusie in [145], Bökstedt constructed the conjectured topological Hochschild homology and the factorization of the Dennis trace

B.I. Dundas et al., *The Local Structure of Algebraic K-Theory*,
Algebra and Applications 18,
DOI 10.1007/978-1-4471-4393-2_7, © Springer-Verlag London 2013

map through the linearization map Definition 4.2.3.4 $THH \to HH$. Using edgewise subdivision, Bökstedt, Hsiang and Madsen factored the Dennis trace in [27] compatibly though the fixed points of topological Hochschild homology. In his ICM lecture [104], Goodwillie further indicated how to map algebraic K-theory to TC, essentially unifying the various character and trace constructions obtained previously as factorizations of the original Dennis trace map.

The reader should be aware, that even though we propose topological cyclic homology as an approximation to algebraic K-theory, there are marked differences between the two functors. This is exposed by a number of different formal properties, as well as the fact that in most cases they give radically different output.

However, the local structure is the same. We immediately get that this is the case if we use the myopic view of stabilizing (see Corollary 7.1.2.6, which is stated for simplicial rings only, but obviously extends to all **S**-algebras by denseness), but we will see that algebraic K-theory and topological cyclic homology have the same local structure even with the eyes of deformation theory.

More precisely, we prove

Theorem 7.0.0.2 *Let $B \to A$ be a map of* **S***-algebras inducing a surjection $\pi_0 B \to \pi_0 A$ with nilpotent kernel, then the square induced by the naturality of the cyclotomic trace*

$$\begin{array}{ccc} K(B) & \xrightarrow{trc} & TC(B) \\ \downarrow & & \downarrow \\ K(B) & \xrightarrow{trc} & TC(A) \end{array}$$

is homotopy cartesian.

The version where the map $B \to A$ is 1-connected was proposed as a conjecture by Goodwillie at the ICM in Kyoto 1990, [104].

The study of algebraic K-theory of nilpotent extensions has a long history. Already in [17] Bloch studies the "tangent space" of algebraic K-theory and compares it with the Kähler differentials. In the 1980's there were several rational results. Soulé calculated the rational K-groups of the dual numbers of a ring of algebraic integers in [266], and Dwyer, Hsiang and Staffeldt calculated the rational homotopy groups of the homotopy fiber of $K(\mathbf{S}[G]) \to K(\mathbf{S}[\pi_0 G])$ in [72, 73] when G is a simplicial group. In the following years, many papers focused on finding the rational K-groups of $\mathbf{S}[G]$ for connected groups G, see for instance [47, 49, 50, 142, 143], and finally, in [102] the rational version of Theorem 7.0.0.2 appears.

A variant of Theorem 7.0.0.2 can be found in Bökstedt, Carlsson, Cohen, Goodwillie, Hsiang and Madsen's paper [26], for the case $\mathbf{S}[G] \to \mathbf{S}$ when G is a connected simplicial group. See also the forerunner [51]. In [104] the conjecture was made for 1-connected maps of ring spectra, since there was some concern about commuting a homotopy colimit with a homotopy inverse limit for the more general case of Theorem 7.0.0.2 as this case is at the boundary of the radius of analyticity for the constructions in terms of their calculus as functors.

The profinite statement was proved by McCarthy in 1993 for simplicial rings in [205], and extended to connective **S**-algebras by Dundas in 1995 in [63]. The full statement is somewhat more than the sum of the rational and the profinite statements, mostly concerned with technicalities as to when homotopy limits and colimits commute. Most of the fine points have already been covered in Chap. 6. The proof for the full statement was found in 1996, but depended on quite a lot that was known to the experts, but not documented elsewhere in the literature. We apologize for the long delay in the publication.

We prove Theorem 7.0.0.2 in two steps. In Sect. 7.2.1 we prove the result for the case where $B \to A$ is a split surjection of simplicial rings with square zero kernel. This case is possible to attack by means of a concrete cosimplicial resolution calculating the loops of the classifying space of the kernel of $B \to A$. Some connectedness bookkeeping then gives the result. In Sect. 7.2.2 we get rid of the square zero condition and the condition that $B \to A$ is split. This last point requires some delicate handling made possible by the fact that we know that in the relative situations both K-theory and TC can be calculated degreewise. Using the "denseness" of simplicial rings in **S**-algebras, and the "continuity" of K and TC we are finished.

In the last section we give an overview of instances where the cyclotomic trace map has been used to prove theorems about algebraic K-theory. Apart from the original application of Bökstedt, Hsiang and Madsen to the algebraic K-theory Novikov conjecture in Sect. 7.3.6, we give a brief overview in Sect. 7.3.1 of Bökstedt and Madsen's setup for calculating topological cyclic homology, made concrete by three central examples. We also make some inadequate comments about the connection to the Lichtenbaum–Quillen conjecture and the redshift conjecture in Sect. 7.3.2. One of the striking structures emerging from Hesselholt and Madsen's work is the de Rham–Witt complex discussed in Sect. 7.3.4, which has been crucial in many situations, but most prominently appears in their calculation of the K-theory of local number fields, cited as Theorem 7.3.3.1 below.

Several other applications are discussed, giving a picture of the rich collection of calculations of algebraic K-theory that has become available through trace methods. The reader will find them most easily by consulting the index under "K-theory of". The section ends by giving an insultingly weak presentation of the interplay with algebraic geometry. It is the nature of the choices made in the previous chapters that issues about commutative algebra have been downplayed. This is deplorable for several reasons, not the least because motivic cohomology has provided some of the most striking tools and interesting avenues for algebraic K-theory, but also because of the recent interest in the arithmetical properties of commutative ring spectra.

7.1 Lifting the Trace and Square Zero Extensions

The purpose of this section is twofold. First and foremost, we lift the Dennis trace map to a map to cyclic homology, and so proving Theorem 7.0.0.1. The use of categorical input makes the checking that our construction of the cyclotomic trace is well defined fairly straightforward at the price of having ridiculously complicated models for the source and target.

Secondly, we investigate the special case where the input is a split zero extension of **S**-algebras. This is a special case of Theorem 7.0.0.2 and is used in its proof.

7.1.1 The Cyclotomic Trace

Lemma 7.1.1.1 *Let $\mathcal{C}$ be a $\Gamma\mathcal{S}_*$-category. Then the Dennis trace map*

$$ob\mathcal{C} \to THH(\mathcal{C}, S^0) \to \sin\left|THH(\mathcal{C}, S^0)\right| = \underline{T}(\mathcal{C})_0$$

of Sect. 4.3.2 *factors through the projection* $\underline{TC}(\mathcal{C})_0 \to \underline{T}(\mathcal{C})_0$.

Proof Remember that the first map in the Dennis trace map was defined as the composite

$$ob\mathcal{C} \to \bigvee_{c\in ob\mathcal{C}} \mathcal{C}(c,c)\left(S^0\right) \to THH\left(\mathcal{C}; S^0\right)_0 \to THH\left(\mathcal{C}, S^0\right)$$

where the first map assigns to every object its identity map, and the last map is the inclusion by degeneracies. This map lands in the fixed points of the cyclic action in the following sense. Recall from Lemma 6.2.1.4 that if X is a pointed cyclic set, then $\lim_{\overleftarrow{\Lambda^o}} X \cong \{x \in X_0 | s_0x = ts_0x\} \cong |X|^{\mathbb{T}}_{\Lambda}$. The Dennis trace factors over the cyclic nerve

$$\left\{[q] \mapsto \bigvee_{c_0,\dots,c_q} \mathcal{C}(c_0, c_q)\left(S^0\right) \wedge \bigwedge_{i=1}^{q} \mathcal{C}(c_i, c_{i-1})\left(S^0\right)\right\},$$

and in the cyclic nerve both s_0 and ts_0 send the zero simplex given by the identity on $c \in ob\mathcal{C}$ to $(c = c = c)$ in $\bigvee_{c_0,c_1} \mathcal{C}(c_0, c_1) \wedge \mathcal{C}(c_1, c_0)$ (and conversely, the equalizer of s_0 and ts_0 is exactly $ob\mathcal{C}$). Hence the Dennis trace map factors over the circle fixed points of $|THH(\mathcal{C}, S^0)|$ (as a matter of fact, $ob\mathcal{C}$ is mapped isomorphically onto the fixed points in our particular model), and also, when varying r we get compatible factorizations

$$ob\mathcal{C} \to sd_r THH\left(\mathcal{C}, S^0\right)^{C_r} \subseteq sd_r THH\left(\mathcal{C}, S^0\right).$$

To see that it commutes with the restriction maps, one chases an object $c \in ob\mathcal{C}$ through

$$ob\mathcal{C} \to sd_{rs} THH\left(\mathcal{C}, S^0\right)^{C_{rs}} \xrightarrow{R} sd_r THH\left(\mathcal{C}, S^0\right)^{C_r},$$

and sees that it coincides with its image under $ob\mathcal{C} \to sd_r THH(\mathcal{C}, S^0)^{C_r}$. □

By naturality this applies equally well if we take as input any of our K-theory machines producing diagrams of $\Gamma\mathcal{S}_*$-categories. In particular, if $\mathcal{C}$ is a symmetric monoidal $\Gamma\mathcal{S}_*$-category we may apply the cyclotomic trace to the $\mathcal{K}(\mathcal{C})$-construction of Sect. 5.2.1.4. The letter combination $\underline{TC}(\mathcal{K}(\mathcal{C}))$ is supposed to signify the bispectrum $m, n \mapsto \underline{TC}(\mathcal{K}(\mathcal{C}, w)(S^n), S^m)$. In order to have the following definition well defined, we consider K-theory as a bispectrum in the trivial way: $\Sigma^\infty ob\mathcal{K}(\mathcal{C}) = \{(m,n) \mapsto ob\mathcal{K}(\mathcal{C})(S^n) \wedge S^m\}$.

Definition 7.1.1.2 Let $\mathcal{C}$ be a symmetric monoidal $\Gamma\mathcal{S}_*$-category. Then the *cyclotomic trace* is the lifting of the Dennis trace for symmetric monoidal $\Gamma\mathcal{S}_*$-categories (Definition 5.2.3.2)

$$\Sigma^\infty ob\mathcal{K}(\mathcal{C}) \to \underline{TC}\big(\mathcal{K}(\mathcal{C})\big) \to \underline{T}\big(\mathcal{K}(\mathcal{C})\big)$$

considered as a map of bispectra.

With this we have established Theorem 7.0.0.1 which uses the simplified notation

$$K(A) = \Sigma^\infty ob\mathcal{K}(\mathcal{F}_A) \to \underline{TC}\big(\mathcal{K}(\mathcal{F}_A)\big) = TC(A),$$

where $\mathcal{F}_A$ is the category of finitely generated free A-modules of Definition 3.2.4.1 (its objects were natural numbers and its morphisms were matrices).

That the source and target actually model algebraic K-theory and topological cyclic homology follows: for the K-theory side this is Theorem 5.2.2.2, and for the TC-side the proof of Theorem 5.2.3.5 establishes that $\underline{T}(\mathcal{K}(C))$ is naturally equivalent to $\Sigma^\infty \underline{T}(C)$ through a chain of equivalences that immediately lifts by Lemma 6.4.1.2 and Definition 6.4.3.1 of TC to a chain of equivalences between $\underline{TC}(\mathcal{K}(C))$ and $\Sigma^\infty \underline{TC}(C)$. Lastly, Morita equivalence Lemma 6.4.1.2 gives that the map $\underline{TC}(\mathcal{F}_A) \leftarrow \underline{TC}(A)$ induced by the inclusion of the rank one module is an equivalence.

7.1.2 Split Square Zero Extensions and the Trace

Let A be an $\mathbf{S}$-algebra and P an A-bimodule. We define $A \vee P$ as in Sect. 5.3.2, and recall that, for every $\mathbf{x} \in \mathcal{I}^{q+1}$, we have a decomposition $V(A \vee P)(\mathbf{x}) \cong \bigvee_{j \geq 0} V^{(j)}(A, P)(\mathbf{x})$, with

$$V^{(j)}(A,P)(\mathbf{x}) = \bigvee_{\phi \in \Delta_m([j-1],[q])} \bigwedge_{0 \leq i \leq q} F_{i,\phi}(x_i)$$

where

$$F_{i,\phi}(x) = \begin{cases} A(S^x) & \text{if } i \notin im\,\phi, \\ P(S^x) & \text{if } i \in im\,\phi. \end{cases}$$

This gives an equivalence $THH(A \vee P; X) \xrightarrow{\sim} \prod_{j \geq 0} THH^{(j)}(A, P; X)$ of cyclic spaces, where

$$THH^{(j)}(A,P;X)_q = \underrightarrow{\operatorname{holim}}_{\mathbf{x} \in \mathcal{I}^{q+1}} \Omega^{\sqcup \mathbf{x}}\big(X \wedge V^{(j)}(A,P)(\mathbf{x})\big).$$

In order to get a description of $TC(A \vee P)$ we investigate the effect of the restriction maps on this decomposition.

Lemma 7.1.2.1 *For every positive integer a the canonical map*

$$sd_a THH(A \vee P; X)^{C_a} \xrightarrow{\sim} \prod_{j \geq 0} sd_a THH^{(j)}(A, P; X)^{C_a}$$

is an equivalence.

The restriction maps respects this decomposition, sending $sd_a THH^{(j)}(A, P; X)^{C_a}$ *to* $sd_{a/p} THH^{(j/p)}(A, P; X)^{C_{a/p}}$ *(which is defined to be trivial if p does not divide j).*

Proof This follows by analysis of the proof of the fundamental cofiber sequence Proposition 6.2.4.2. Note that $X \wedge V(A \vee P)(\mathbf{x}^a) \cong \bigvee_{j \geq 0} (X \wedge V^{(j)}(A, P)(\mathbf{x}^a))$ is a C_a-isomorphism, and furthermore that

$$V^{(j)}(A, P)(\mathbf{x}^a)^{C_a} \cong \begin{cases} V^{(j/a)}(A, P)(\mathbf{x}) & \text{if } j \equiv 0 \bmod a, \\ * & \text{otherwise.} \end{cases}$$

We have maps of fibrations (we have deleted "(A, P)" from the V's to fit the line width)

$$\begin{array}{ccc} Map_*(S^{\sqcup \mathbf{x}^a}, X \wedge \bigvee_{j \geq 0} V^{(j)}(\mathbf{x}^a))^{C_a} & \longrightarrow & Map_*(\bigcup_{1 \neq b | a} (S^{\sqcup \mathbf{x}^a})^{C_b}, X \wedge \bigvee_{j \geq 0} V^{(j)}(\mathbf{x}^a))^{C_a} \\ \downarrow & & \downarrow \\ \prod_{j \geq 0} Map_*(S^{\sqcup \mathbf{x}^a}, X \wedge V^{(j)}(\mathbf{x}^a))^{C_a} & \longrightarrow & \prod_{j \geq 0} Map_*(\bigcup_{1 \neq b | a} (S^{\sqcup \mathbf{x}^a})^{C_b}, X \wedge V^{(j)}(\mathbf{x}^a))^{C_a} \end{array}$$

whose map of homotopy homotopy fibers is, by the proof of Proposition 6.2.4.2, after stabilization given by

$$\begin{array}{c} \operatorname{holim}_{\overrightarrow{k}} \Omega^k Map_*(S^{\sqcup \mathbf{x}^a}, S^k \wedge X \wedge \bigvee_{j \geq 0} V^{(j)}(A, P)(\mathbf{x}^a))_{hC_a} \\ \downarrow \\ \prod_{j \geq 0} \operatorname{holim}_{\overrightarrow{k}} \Omega^k Map_*(S^{\sqcup \mathbf{x}^a}, S^k \wedge X \wedge V^{(j)}(A, P)(\mathbf{x}^{ap}))_{hC_a} \end{array}.$$

The map of homotopy fibers factors through

$$\operatorname{holim}_{\overrightarrow{k}} \Omega^k \left(\prod_{j \geq 0} Map_*(S^{\sqcup \mathbf{x}^a}, S^k \wedge X \wedge V^{(j)}(A, P)(\mathbf{x}^a)) \right)_{hC_a}.$$

By Blakers–Massey, the map into this space is an equivalence, and also the map out of this space (virtually exchange the product for a wedge to tunnel it through the orbits, which is possible as the connectivity goes to infinity with j).

Hence

$$\begin{array}{ccc} sd_a THH(A \vee P; X)^{C_a} & \xrightarrow{R} & \underset{\overleftarrow{1\neq b|a}}{\operatorname{holim}} sd_{a/b} THH(A \vee P; X)^{C_{a/b}} \\ \downarrow & & \downarrow \\ \prod_{j\geq 0} sd_a THH^{(j)}(A \vee P; X)^{C_a} & \xrightarrow{R} & \underset{\overleftarrow{1\neq b|a}}{\operatorname{holim}} \prod_{j\geq 0} sd_{a/b} THH^{(j/b)}(A \vee P; X)^{C_{a/b}} \end{array}$$

is homotopy cartesian. Since the map in question is an equivalence when $a = 1$, this homotopy cartesian square gives the proposition by induction over a. □

Proposition 7.1.2.2 *Let A be an* **S***-algebra, P an $n-1$ connected A-bimodule, X an $m-1$ connected space and p a prime. Then the projection to the terminal pieces of the R-towers*

$$\begin{aligned} TR(A \vee P; X; p) &\to TR(A, X; p) \times \prod_{r\geq 0} \underset{\overleftarrow{p^t \in \mathcal{R}_p}}{\operatorname{holim}} sd_{p^{t+r}} THH^{(p^t)}(A, P; X)^{C_{p^{t+r}}} \\ &\to TR(A, X; p) \times \prod_{r\geq 0} sd_{p^r} THH^{(1)}(A, P; X)^{C_{p^r}} \end{aligned}$$

is $2n + m$-connected.

Proof Consider the restriction map

$$sd_{ap} THH^{(jp)}(A, P; X)^{C_{ap}} \xrightarrow{R} sd_a THH^{(j)}(A, P; X)^{C_a},$$

where a is a power of p. The homotopy fiber is

$$\underset{\overrightarrow{k}}{\operatorname{holim}} \left(\Omega^k sd_{ap} THH^{(jp)}\left(A, P, S^k \wedge X\right)_{hC_{ap}}\right)$$

and is by assumption $jpn + m - 1$ connected. If p does not divide j, we get an equivalence

$$sd_a THH^{(j)}(A, P; X)^{C_a} \simeq \underset{\overrightarrow{k}}{\operatorname{holim}} \left(\Omega^k sd_a THH^{(j)}\left(A, P, S^k \wedge X\right)_{hC_a}\right) \tag{7.1}$$

(since the target of the restriction map is contractible), and since homotopy orbits preserve connectivity, this is $jn + m - 1$ connected. So, consider

$$sd_{p^r} THH^{(lp^s)}(A, P; X)^{C_{p^r}}$$

where p does not divide l. If $s \geq r$ the R maps will compose to an $lp^{s-r+1}n + m$ connected map to $THH^{(lp^{s-r})}(A, P; X)$, which is $lp^{s-r}n + m - 1$ connected. If $r \geq s$ the R-maps will compose to an $lpn + m$ connected map to $sd_{p^{r-s}} THH^{(l)}(A, P; X)^{C_{p^{r-s}}} \simeq \operatorname{holim}_{\overrightarrow{k}} \Omega^k (sd_{p^{r-s}} THH^{(l)}(A, P, S^k \wedge X))_{hC_{p^{r-s}}}$, which is $ln + m - 1$ connected.

Hence

$$\underset{\overleftarrow{p^t \in \mathcal{R}_p}}{\operatorname{holim}}\, sd_{p^{r+t}} THH^{(lp^{s+t})}(A, P; X)^{C_{p^{r+t}}}$$

will be $\max(ln + m - 1, lp^{s-r}n + m - 1)$ connected. This means that there is a $2n + m$ connected map (first discarding all terms with $l \geq 2$, then projecting down to the terminal term in each tower)

$$\begin{aligned} TR(A \vee P; X; p) &\to TR(A, X; p) \times \prod_{r \geq 0} \underset{\overleftarrow{p^t \in \mathcal{R}_p}}{\operatorname{holim}}\, sd_{p^{t+r}} THH^{(p^t)}(A, P; X)^{C_{p^{t+r}}} \\ &\to TR(A, X; p) \times \prod_{r \geq 0} sd_{p^r} THH^{(1)}(A, P; X)^{C_{p^r}} \end{aligned}$$

(the last map is $pn + m \geq 2n + m$ connected as all maps in the homotopy limit on the first line are $pn + m$ connected). □

Lemma 7.1.2.3 *Let A, P, X and a be as before. Then the map of fixed points into homotopy fixed points*

$$\left|THH^{(1)}(A, P; X)\right|^{C_a} \longrightarrow \left|THH^{(1)}(A, P; X)\right|^{hC_a}$$

is an equivalence.

Proof Recall from Lemma 5.3.3.2 that $j_* \underline{T}(A, P; X) \to \underline{T}^{(1)}(A, P; X)$ is a map of cyclic spectra and an equivalence, where j_* is the left adjoint to the forgetful functor from cyclic to simplicial sets. From Lemma 6.2.1.3 we have a $\mathbb{T}$-equivariant homeomorphism $|j_* \underline{T}(A, P; X)| \cong \mathbb{T}_+ \wedge |\underline{T}(A, P; X)|$. Let G be a finite subgroup of the circle group and consider the commuting diagram (in the homotopy category)

$$\begin{array}{ccccc} \left|T^{(1)}(A, P, X)\right|_{hG} & \xrightarrow{\simeq} & \left|T^{(1)}(A, P, X)\right|^{G} & \longrightarrow & \left|T^{(1)}(A, P, X)\right|^{hG} \\ \simeq \uparrow & & & & \uparrow \simeq \\ \left|j_* T(A, P; X)\right|_{hG} & & \xrightarrow[\simeq]{\text{norm}} & & \left|j_* T(A, P; X)\right|^{hG} \end{array}$$

where the first upper horizontal map is the map from the homotopy fiber of the restriction map as in the proof of Proposition 7.1.2.2 and the second is the map from the fixed point to the homotopy fixed point spectra. The top composite is then homotopic to the norm of Sect. 6.3.2. Once we have shown that the marked arrows are weak equivalences we are done.

By Proposition 6.3.2.5, the G-norm map for the G-free spectrum $|j_* T(A, P; X)|$ is an equivalence, so the bottom arrow is an equivalence. Since homotopy orbits and homotopy fixed points preserve naïve G-equivalences, the two vertical maps in the diagram are equivalences as well. The map $|T^{(1)}(A, P, X)|_{hG} \to |T^{(1)}(A, P, X)|^{G}$

is an equivalence as shown in (7.1) in the proof of Proposition 7.1.2.2. Hence the second map $|T^{(1)}(A,P,X)|^G \to |T^{(1)}(A,P,X)^{hG}$ is an equivalence, as claimed. □

Collecting the information so far, and recalling from Lemma 5.3.3.2 that $\underline{T}^{(1)}(A,P;X) \leftarrow j_*\underline{T}(A,P;X) \to S^1_+\wedge\underline{T}(A,P;X)$ are equivalences, where j_* is the free cyclic functor, we get

Lemma 7.1.2.4 *There is a $(2n+m)$-connected map*

$$\begin{array}{ccc} \underline{TC}(A\vee P;X;p) & \longrightarrow & \underline{TC}(A,X;p)\times \underset{p^r\in\mathcal{F}_p}{\underleftarrow{\operatorname{holim}}}\left|\underline{T}^{(1)}(A,P;X)\right|^{hC_{p^r}} \\ & & \uparrow\simeq \\ & & \underline{TC}(A,X;p)\times \underset{p^r\in\mathcal{F}_p}{\underleftarrow{\operatorname{holim}}}\left|j_*\underline{T}(A,P;X)\right|^{hC_{p^r}}. \end{array}$$

Proof Take $-^{h\langle F\rangle}$ of the TR expression, and insert Lemma 7.1.2.3 to get the desired connectivity of the horizontal map. □

Theorem 7.1.2.5 (Hesselholt) *Let A, P, X and p be as above. The "composite"*

$$\begin{array}{ccccccc} \widetilde{\underline{TC}}(A\vee P;X;p) & \longrightarrow & \tilde{\underline{T}}(A\vee P;X) & \longleftarrow & j_*\underline{T}(A,P;X) & \xrightarrow{\sim} & S^1_+\wedge\underline{T}(A,P;X) \\ &&&&&& \downarrow \\ &&&&&& S^1\wedge\underline{T}(A,P;X) \end{array}$$

is $2n+m-1$ connected after p completion.

Proof In a $2n+m-1$ range, the "composite" looks like

$$\begin{array}{ccccccc} \underset{p^r\in\mathcal{F}_p}{\underleftarrow{\operatorname{holim}}}\left(j_*\underline{T}(A,P;X)\right)^{hC_{p^r}} & \longrightarrow & Qj_*\underline{T}(A,P;X) & \xleftarrow{\sim} & j_*\underline{T}(A,P;X) & \xrightarrow{\sim} & S^1_+\wedge\underline{T}(A,P;X) \\ &&&&&& \downarrow \\ &&&&&& S^1\wedge\underline{T}(A,P;X) \end{array}$$

but the diagram

$$\begin{array}{ccccc} \underset{p^r\in\mathcal{F}_p}{\underleftarrow{\operatorname{holim}}}\left(j_*\underline{T}(A,P;X)\right)^{hC_{p^r}} & \xrightarrow{\sim} & \underset{p^r\in\mathcal{F}_p}{\underleftarrow{\operatorname{holim}}}\left(S^1_+\wedge\underline{T}(A,P;X)\right)^{hC_{p^r}} & \longleftarrow & \left(S^1_+\wedge\underline{T}(A,P;X)\right)^{hS^1} \\ \downarrow & & \downarrow & & \simeq\downarrow \\ Qj_*\underline{T}(A,P;X) & \xrightarrow{\sim} & Q\left(S^1_+\wedge\underline{T}(A,P;X)\right) & \longrightarrow & Q\left(S^1\wedge\underline{T}(A,P;X)\right) \end{array}$$

gives the result as the left hand maps are the equivalence from Lemma 6.2.1.2 and the upper right hand map is an equivalence after p-completion by Lemma 6.3.1.1. The right hand vertical map is an equivalence by Corollary 6.3.2.3. □

Corollary 7.1.2.6 *Let A be a simplicial ring and P a simplicial A-bimodule. The trace induces an equivalence*

$$D_1\mathbf{K}(A \ltimes -)(P) \to D_1\mathbf{TC}(A \ltimes -)(P).$$

Proof Comparing Proposition 5.3.4.3 and Theorem 7.1.2.5, we see that the map of differentials $D_1\mathbf{K}(A \ltimes -)(P) \to D_1\mathbf{TC}(A \ltimes -; p)(P)$ is an equivalence after p-completion, and so by the definition of topological cyclic homology, the cyclotomic trace induces an equivalence on differentials after profinite completion. Hence we must study what happens for the other corners in the definition of $\underline{TC}$. But here we may replace the S^1 homotopy fixed points by the negative cyclic homology, and as we are talking about square zero extensions, even by shifted cyclic homology. But as cyclic homology respects connectivity we see that the horizontal maps in

$$\begin{array}{ccc}
\left(j_*\underline{T}(A, P; X)_{(0)}\right)^{hS^1} & \longrightarrow & \left(\tilde{\underline{T}}(A \ltimes P; X)_{(0)}\right)^{hS^1} \\
\downarrow & & \downarrow \\
\left(j_*\underline{T}(A, P; X)^{\wedge}_{(0)}\right)^{hS^1} & \longrightarrow & \left(\tilde{\underline{T}}(A \ltimes P; X)^{\wedge}_{(0)}\right)^{hS^1}
\end{array}$$

are both $2k + m$ connected if P is $k - 1$ connected and X is $m - 1$ connected.

Summing up: both maps going right to left in

$$\begin{array}{ccccccc}
\tilde{\mathbf{K}}(A \ltimes P) & \longrightarrow & \widetilde{\mathbf{TC}}(A \ltimes P) & \longleftarrow & \left(j_*\mathbf{T}(A, P)\right)^{hS^1} & \xrightarrow{\sim} & \left(S^1_+ \wedge \mathbf{T}(A, P)\right)^{hS^1} \\
 & & \downarrow & & \downarrow & & \downarrow \\
 & & \tilde{\mathbf{T}}(A \ltimes P) & \longleftarrow & Q j_*\mathbf{T}(A, P) & \xrightarrow{\sim} & Q\left(S^1_+ \wedge \mathbf{T}(A, P)\right) \\
 & & & & \simeq\uparrow & & \simeq\uparrow \\
 & & & & j_*\mathbf{T}(A, P) & \xrightarrow{\sim} & S^1_+ \wedge \mathbf{T}(A, P) \\
 & & & & & & \downarrow \\
 & & & & & & S^1 \wedge \mathbf{T}(A, P)
\end{array}$$

are $2k$-connected, and all composites from the top row to the bottom are $2k$-connected. □

7.2 The Difference Between K-Theory and TC is Locally Constant

7.2.1 The Split Algebraic Case

In this subsection we prove Theorem 7.0.0.2 in the special case of a split square zero extension of simplicial rings. The original proof in [205] used calculus of functors.

Here we offer a more elementary approach by disassembling the cobar construction. We start by reviewing the few facts we will need.

7.2.1.1 The Cobar Construction

Let A be a simplicial ring and X a connected A-bimodule. Model the loops ΩX by means of the cosimplicial object

$$\omega(0, X, 0) = \left\{[q] \mapsto \underline{\mathcal{S}_*}\left(S^1_q, X\right) \cong X^{\times q}\right\}.$$

More precisely, the weak equivalence $\underset{\overrightarrow{[q]\in\Delta^o}}{\operatorname{holim}} S^1_q \xrightarrow{\sim} S^1$ of Lemma A.7.1.1 induces a weak equivalence

$$\begin{aligned}\Omega X &= \underline{\mathcal{S}_*}\left(S^1, \sin|X|\right) \xrightarrow{\sim} \underline{\mathcal{S}_*}\Big(\underset{\overrightarrow{[q]\in\Delta^o}}{\operatorname{holim}} S^1_q, \sin|X|\Big) \cong \underset{\overleftarrow{[q]\in\Delta}}{\operatorname{holim}} \underline{\mathcal{S}_*}\left(S^1_q, X\right) \\ &= \underset{\overleftarrow{\Delta}}{\operatorname{holim}}\, \omega(0, X, 0).\end{aligned}$$

This is a special case of the *cobar construction* of coalgebras and cobimodules, dual to Hochschild homology. We don't need the full generality of this construction and write out the slight extension we need explicitly. Let $p\colon E \twoheadrightarrow X$ be a surjection of A-bimodules. Then $\omega(0, X, E)$ is the cosimplicial simplicial A-bimodule which in codegree q is $\omega(0, X, E)_q = X^{\times q} \times E$, with coface maps

$$d^i(x_1, \dots, x_q, e) = \begin{cases} (0, x_1, \dots, x_q, e) & \text{if } i = 0, \\ (x_1, \dots, x_{i-1}, x_i, x_i, x_{i+1}, \dots, x_q, e) & \text{if } 0 < i \leq q, \\ (x_1, \dots, x_q, pe, e) & \text{if } i = q+1, \end{cases}$$

and codegeneracy maps by removing appropriate factors. Another way of casting $\omega(0, X, E)$ making the cosimplicial structure clear is to observe that $\omega(0, X, E)_q$ can be identified with the set of diagrams of the sort

$$\begin{array}{ccccc} \Delta[0]_q & \xrightarrow{d^1} & \Delta[1]_q & \xleftarrow{d^0} & \Delta[0]_q \\ \downarrow & & \downarrow & & \downarrow \\ * & \longrightarrow & X & \xleftarrow{p} & E. \end{array}$$

We notice that if E is contractible then the map $\omega(0, X, 0) \to \omega(0, X, E)$ is a pointwise equivalence of cosimplicial objects, and so induces an equivalence of homotopy limits:

$$\underset{\overleftarrow{[q]\in\Delta}}{\operatorname{holim}}\, \omega(0, X, 0) \xrightarrow{\sim} \underset{\overleftarrow{[q]\in\Delta}}{\operatorname{holim}}\, \omega(0, X, E).$$

The latter model for ΩX has the benefit of coming with the coaugmentation $P = \omega(0,0,P) \subseteq \omega(0,X,E)$ where P is the kernel of $E \twoheadrightarrow X$, modelling the equivalence $P \xrightarrow{\sim} \Omega X$.

7.2.1.2 Split Square Zero Extensions

Recall that if A is a ring and P is an A-bimodule we write $A \ltimes P$ for the ring whose underlying abelian group is $A \oplus P$ and whose multiplication is defined by $(a_1, p_1) \cdot (a_2, p_2) = (a_1 a_2, a_1 p_2 + p_1 a_2)$. Then the projection $A \ltimes P \to A$ is a surjection of rings whose kernel is a the square zero ideal which we identify with P.

Definition 7.2.1.1 For A a simplicial ring and P an A-bimodule, let $\mathbf{F}_A P$ be the iterated homotopy fiber (Definition A.8.0.3) of

$$\begin{array}{ccc} \mathbf{K}(A \ltimes P) & \longrightarrow & TC(A \ltimes P) \\ \downarrow & & \downarrow \\ \mathbf{K}(A) & \longrightarrow & TC(A) \end{array}$$

regarded as a functor from A-bimodules to spectra.

Theorem 7.2.1.2 *Let A be a simplicial ring and P an A-bimodule. Then $\mathbf{F}_A P$ is contractible. That is, the diagram in Definition* 7.2.1.1 *is homotopy cartesian.*

The proof of this theorem will occupy the rest of this section. In the next, we will show how the theorem extends to $\mathbf{S}$-algebras to prove Theorem 7.0.0.2.

Proof In view of Propositions 3.1.4.2 and 6.5.3.3, we may, without loss of generality, assume that A is discrete. We know by Corollary 7.1.2.6 (or more precisely, by the diagram that ends the proof of Corollary 7.1.2.6) that if P is k-connected, then $\mathbf{F}_A P$ is $2k$-connected; so for general P it is natural to study $\Omega^k \mathbf{F}_A(B^k P)$ (whose connectivity goes to infinity with k), or more precisely, the map

$$\mathbf{F}_A P \xrightarrow{\eta_P^k} \Omega^k \mathbf{F}_A\big(B^k P\big).$$

The map $\eta\colon \mathbf{F}_A P \to \Omega \mathbf{F}_A BP$ appears naturally as the map of homotopy fibers of $\mathbf{F}_A$ applied to the homotopy cartesian square

$$\begin{array}{ccc} P & \longrightarrow & 0 \\ \downarrow & & \downarrow \\ 0 & \longrightarrow & BP \end{array}.$$

Since $\mathbf{F}_A$ does not a priori preserve homotopy cartesian diagrams, we don't know that η_P is an equivalence, but we will show that η_P is as connected as $\mathbf{F}_A(BP)$ is. This means that $\mathbf{F}_A P$ is as connected as $\Omega\mathbf{F}_A(BP)$, and by induction $\mathbf{F}_A P$ must be arbitrarily connected, and we are done.

Recall the cobar construction $\omega(0, BP, EP)$ discussed in 7.2.1.1 at the beginning of the section and the augmentation $P = \omega(0, 0, P) \xrightarrow{\sim} \omega(0, BP, EP)$, thought of as a functor from the category $\Delta \cup \emptyset$. Here $EP = P \otimes \tilde{\mathbf{Z}}[\Delta[1]]$, and we identify the cokernel EP/P by the zero skeleton with $BP = B \otimes \tilde{\mathbf{Z}}[S^1]$.

By a choice of equivalence of categories between $\Delta \cup \emptyset$ and the category Ord of ordered finite sets, we think of $\omega(0, BP, EP)$ as a functor from Ord to A-bimodules.

Let $\mathcal{P} \subseteq Ord$ be the category of finite sets of positive integers and inclusions (inclusions are order preserving). Fix, for the moment, $n \geq 0$, and let $\mathcal{P}\mathbf{n} \subseteq \mathcal{P}$ be the full subcategory of subsets of $\mathbf{n} = \{1, \dots, n\}$. Let $S \mapsto \mathcal{P}^n_S$ be the composite

$$\mathcal{P}\mathbf{n} \xrightarrow{\subset} Ord \xrightarrow{\omega(0,B,E)\cup P} \text{simplicial } A\text{-bimodules}.$$

Notice that this becomes a strongly cocartesian n-cube (in the sense that all subsquares are homotopy pushouts of simplicial abelian groups, cf. Sect. A.8.2).

For any n-cube $\mathcal{X}$ and $1 \leq j \leq n$, consider the subcube you get by restricting $\mathcal{X}$ to $\mathcal{P}\mathbf{j} \subseteq \mathcal{P}\mathbf{n}$. For instance $\mathcal{X}|\mathcal{P}\emptyset$ is the object $\mathcal{X}_\emptyset$, $\mathcal{X}|\mathcal{P}\mathbf{1}$ is the map $\mathcal{X}_{\{1\}} \to \mathcal{X}_\emptyset$, and $\mathcal{X}|\mathcal{P}\mathbf{2}$ is the square

$$\begin{array}{ccc} \mathcal{X}_{\{1\}} & \longrightarrow & \mathcal{X}_\emptyset \\ \downarrow & & \downarrow \\ \mathcal{X}_{\{2\}} & \longrightarrow & \mathcal{X}_{\{1,2\}} \end{array} .$$

Let F_j be the iterated homotopy fiber of $\mathcal{X}|\mathcal{P}\mathbf{j}$, and consider the resulting sequence

$$F_n \to F_{n-1} \to \cdots \to F_1 \to F_0 = \mathcal{X}_\emptyset.$$

This is nothing more than a specific choice of path for computing the iterated homotopy fiber (Definition A.8.0.3).

Let $\overline{\mathcal{P}\mathbf{j}} = \{S \in \mathcal{P}\mathbf{j} | j \in S\} \subseteq \mathcal{P}\mathbf{n}$ (the set of all subsets of $\{1, \dots, j\}$ that actually contain j; for instance $\overline{\mathcal{P}\mathbf{2}} = \{\{2\} \subseteq \{1, 2\}\}$), and notice that $\mathcal{X}|\mathcal{P}\mathbf{j}$ can be viewed as a map

$$\mathcal{X}|\mathcal{P}(\mathbf{j}-\mathbf{1}) \to \mathcal{X}|\overline{\mathcal{P}\mathbf{j}}$$

of $j-1$-cubes. Hence, if we define $\Phi_1 = \mathcal{X}_{\{1\}}$ and Φ_j for $1 < j \leq n$ as the iterated homotopy fiber of $\mathcal{X}|\overline{\mathcal{P}\mathbf{j}}$, we get fiber sequences

$$F_j \to F_{j-1} \to \Phi_j$$

for all $1 \leq j \leq n$.

In the case $\mathcal{X} = \mathbf{F}_A\mathcal{P}^n$ we get $F_1 \xrightarrow{\sim} F_0 = \mathbf{F}_A P$ and $\Phi_2 \simeq \mathrm{hofib}\{* \simeq F_A(E) \to F_A(B)\} \simeq \Omega\mathbf{F}_A BP$. Theorem 7.2.1.2 follows from the claim that F_2 is as connected as $\Omega \mathbf{F}_A BP$ is. This will again follow if we know this to be true for the Φ_is and for F_{n+1}.

We first consider the question for the Φ_js. Note that the maps in $\overline{\mathcal{P}\mathbf{j}}$ always preserve j. Translated to Δ, for all $0 < l < j$, it has all the inclusions $d^i : [l] \to [l+1]$ but the one omitting $l+1$. This leaves some room for a change of base isomorphism of j-cubes $\mathcal{P}^n|\overline{\mathcal{P}\mathbf{j}} \cong \mathcal{Q}^j$ given by sending d^i to δ^i which omits the $(i+1)$st coordinate, and is the identity on the vertices of cardinality ≤ 1. Here we have used that the cubes are strongly (co)cartesian. The important outcome is that $\mathcal{Q}^j$ can be constructed iteratively by taking products with BP (no diagonals).

Thus Φ_j is the iterated homotopy fiber of $\mathbf{F}_A \mathcal{Q}^j$, which can be analyzed as follows. Let $P_0, \dots P_n$ be A-bimodules, and define

$$\mathbf{F}_A(P_0; P_1 \dots, P_j)$$

inductively by letting $\mathbf{F}_A(P_0)$ be as before, and setting

$$\begin{aligned}\mathbf{F}_A(P_0; P_1, \dots, P_j) = \operatorname{hofib}\bigl\{&\mathbf{F}_A(P_0; P_1, \dots, P_{j-1}) \\ &\longrightarrow \mathbf{F}_A(P_0 \times P_j; P_1, \dots, P_{j-1})\bigr\}.\end{aligned}$$

We see that

$$\mathbf{F}_A(*; BP, \dots, BP) \simeq \mathbf{F}_A(EP; BP, \dots, BP) \simeq \Phi_j.$$

Now, assume that we know that $\mathbf{F}_A(*; BP) \simeq \Omega \mathbf{F}_A BP$ is m-connected for all A and P.

We will show that $\Phi_j \simeq \mathbf{F}_A(*; BP, \dots, BP)$ is also m-connected. This will follow from the more general statement, that if all the P_i are 1-reduced (their zero-skeleta are trivial), then $\mathbf{F}_A(-; P_1, \dots, P_j)$ is m-connected. For $j = 1$ this is immediate as

$$\mathbf{F}_A(P_0; P_1) = \operatorname{hofib}\bigl\{\mathbf{F}_A(P_0) \to \mathbf{F}_A(P_0 \times P_1)\bigr\} \simeq \Omega \mathbf{F}_{A \ltimes P_0}(P_1)$$

is m-connected by assumption. So assume inductively that $\mathbf{F}_A(-; P_1, \dots, P_{j-1})$ is m-connected. In particular $\mathbf{F}_A(P_0; P_1, \dots, P_{j-1})$ and $\mathbf{F}_A(P_0 \times (P_j)_q; P_1, \dots, P_{j-1})$ are m-connected, and using that $\mathbf{F}_A(-)$ may be calculated degreewise we see that

$$\mathbf{F}_A\bigl(P_0; P_1, \dots, (P_j)_q\bigr) \text{ is } \begin{cases} 0 & \text{if } q = 0, \\ m-1\text{-connected} & \text{if } q > 0 \end{cases}$$

and hence the conclusion follows.

We are left with showing that the iterated homotopy fiber ("F_{n+1}") of $\mathbf{F}_A \mathcal{P}^n$ is as highly connected as we need. In fact, we will show that $\mathbf{F}_A \mathcal{P}^n$ is $(n-3)$-cartesian, and so choosing n large enough we are done. In order to prove this—and so to prove the triviality of $\mathbf{F}_A P$—it is enough to prove Lemmas 7.2.1.3 and 7.2.1.4 below. □

Lemma 7.2.1.3 *The n-cube $K(A \ltimes \mathcal{P}^n)$ is n-cartesian.*

Proof This follows from Lemma 1.2.5.6 with $k = 0$. □

Lemma 7.2.1.4 *The n-cube $TC(A \ltimes \mathcal{P}^n)$ is $n-3$ cartesian.*

Proof By Proposition 6.5.3.2 with $k = -2$, this follows from the corresponding statement about topological Hochschild homology. This is proved in Lemma 7.2.1.7 below. □

For each $0 \le i \le n$, let P_i be an A-bimodule, and let $\underline{T}(A; P_0, \dots, P_n) = \{k \mapsto \underline{T}(A; P_0, \dots, P_n, S^k)\}$ be the n-reduced simplicial spectrum given by

$$[q] \mapsto \underset{\mathbf{x} \in I^{q+1}}{\underrightarrow{\operatorname{holim}}}\, \Omega^{\sqcup \mathbf{x}}\left(S^k \wedge \bigvee_{\phi \in \Delta_m([n],[q])} \bigwedge_{0 \le i \le q} F^j \otimes \tilde{\mathbf{Z}} S^{x_i}\right)$$

for $q \ge n$, where $\Delta_m([n],[q]) \subseteq \Delta([n],[q])$ is the set of *injective* and order preserving functions $\phi\colon [n] \to [q]$, and where $F^j = A$ if $j \notin im\phi$ and $F^j = P_{\phi^{-1}(j)}$ otherwise. The simplicial operations are the ordinary Hochschild ones, where the Ps multiply trivially. This is a functor from simplicial A-bimodules to spectra, and restricted to each factor it preserves homotopy cartesian diagrams. We let $\underline{T}^{(n+1)}(A, P) = \underline{T}(A; P, \dots, P)$ be the composite with the diagonal. We see that this agrees with our earlier definition.

Lemma 7.2.1.5 *Let $\mathcal{M}$ be a strongly (co)cartesian S-cube of simplicial A-bimodules. Then $\underline{T}^{(n)}(A, \mathcal{M})$ is cartesian if $|S| > n$.*

Proof We define a new S-cube $\mathcal{Z}$ as follows. If $T \subseteq S$, let $S_T \subseteq \mathcal{P}S$ be the full subcategory with objects U containing T and with $|S - U| \le 1$, let

$$\mathcal{Z}_T = \underset{U_1 \in S_T}{\underleftarrow{\operatorname{holim}}} \dots \underset{U_n \in S_T}{\underleftarrow{\operatorname{holim}}}\, \underline{T}\big(A, \{\mathcal{M}_{U_i}\}\big).$$

As $\mathcal{M}$ is strongly cartesian $\mathcal{M}|S_T$ is cartesian, and so the map $\underline{T}^{(n)}(A, \mathcal{M}_T) \to \mathcal{Z}_T$ is an equivalence for each T. The homotopy limits may be collected to be over $S_T^{\times n}$ which may be written as $\bigcap_{s \in T} \mathcal{A}_s$ where $\mathcal{A}_s$ is the full subcategory of $\mathcal{A} = S_\emptyset^{\times n}$ such that s is in every factor. As $|S| > n$ the $\mathcal{A}_s$ cover $\mathcal{A}$ as in the hypothesis of [106, Lemma 1.9], and so $\mathcal{Z}$ is cartesian. □

Lemma 7.2.1.6 *If P is $(k-1)$-connected, then*

$$\underline{T}(A \ltimes P) \xleftarrow{\sim} \bigvee_{0 \le j < \infty} \underline{T}^{(j)}(A, P) \longrightarrow \bigvee_{0 \le j < n} \underline{T}^{(j)}(A, P)$$

is $(n(k+1)-1)$-connected.

Proof In view of the equivalence $HA \vee HP \xrightarrow{\sim} H(A \ltimes P)$ Corollary 5.3.2.2 gives that

$$\underline{T}(A \ltimes P) \xleftarrow{\sim} \bigvee_{0 \le j < \infty} \underline{T}^{(j)}(A, P).$$

Since P is $(k-1)$-connected and each simplicial dimension contains a smash product of j copies of P, we get that

$$\underline{T}^{(j)}(A,P)_q \text{ is } \begin{cases} 0 & \text{if } q < j-1, \\ kj-1 \text{ connected} & \text{if } q \geq j-1 \end{cases}$$

and so $\underline{T}^{(j)}(A,P)$ is $j-1+kj-1=(k+1)j-2$ connected. □

Lemma 7.2.1.7 *$\underline{T}(A \ltimes \mathcal{P}^n)$ is $id-2$ cartesian.*

Proof Consider

$$\begin{array}{ccc} \underline{T}(A \ltimes P) & \xrightarrow{a} & \prod_{0\leq j<n} \underline{T}^{(j)}(A,P) \\ {\scriptstyle d}\downarrow & & {\scriptstyle b}\downarrow \\ \underset{\overleftarrow{S\neq\emptyset}}{\operatorname{holim}}\, \underline{T}(A \ltimes \mathcal{P}^n_S) & \xrightarrow{c} & \underset{\overleftarrow{S\neq\emptyset}}{\operatorname{holim}} \prod_{0\leq j<n} \underline{T}^{(j)}(A,\mathcal{P}^n_S). \end{array}$$

By Lemma 7.2.1.6, the map a is $(n-1)$-connected. By Lemma 7.2.1.5 $\underline{T}^{(j)}(A,-)$ is n-excisive for $j<n$, and so

$$\bigvee_{0\leq j<n} \underline{T}^{(j)}(A,P) \xrightarrow{\sim} \bigvee_{0\leq j<n} \operatorname{holim}_{\overleftarrow{S\neq\emptyset}} \underline{T}^{(j)}(A,\mathcal{P}^n_S).$$

Since the map from finite wedges to products of spectra is a stable equivalence, this implies that b is an equivalence.

Again, by Lemma 7.2.1.6

$$\underline{T}(A \ltimes \mathcal{P}^n_S) \xrightarrow{a} \bigvee_{0\leq j<n} \underline{T}^{(j)}(A,\mathcal{P}^n_S)$$

is $(2n-1)$-connected for $S \neq \emptyset$, with homotopy fiber, say $\mathcal{F}_S$, $(2n-2)$-connected. The fiber of c equals $\operatorname{holim}_{\overleftarrow{S\neq\emptyset}} \mathcal{F}_S$, and must then be $2n-2-n+1=n-1$ connected (an n-cube consisting of l-connected spaces must have $(l-n)$-connected iterated fiber: by induction). Hence c is n-connected.

This means that d must be $(n-2)$-connected. Likewise for all subcubes (some are id-cartesian). □

7.2.2 The General Case

In this section we will finally prove Theorem 7.0.0.2:

Theorem 7.2.2.1 *Let $B \to A$ be a map of connective* **S***-algebras inducing a surjection $\pi_0 B \to \pi_0 A$ with nilpotent kernel, then the square induced by the naturality of* *the cyclotomic trace*

$$\begin{array}{ccc} K(B) & \longrightarrow & TC(B) \\ \downarrow & & \downarrow \\ K(A) & \longrightarrow & TC(A) \end{array}$$

is homotopy cartesian.

Following the procedure of [102] we first prove the theorem in the case where $B \to A$ is a map of simplicial ring, and then use the density argument to extend it to (connective) **S**-algebras.

We start up with some consequences of the split square zero case.

Lemma 7.2.2.2 *Let $f\colon B \to A$ be a simplicial ring map such that each f_q is an epimorphism with nilpotent kernel. Then*

$$\begin{array}{ccc} K(B) & \longrightarrow & TC(B) \\ \downarrow & & \downarrow \\ K(A) & \longrightarrow & TC(A) \end{array}$$

is homotopy cartesian.

Proof Let $I = \ker(f)$. As we may calculate the relative K-theory and topological cyclic homology of a simplicial radical extension degreewise (Proposition 3.1.4.2 and Corollary 6.5.3.3), the statement will follow if for each q we can prove it for the map $f_q\colon B_q \to A_q$. That is, we may assume that B and A are discrete, and that $I = \ker(f)$ satisfies $I^n = 0$. Note that each of the maps

$$B = B/I^n \to B/I^{n-1} \to \cdots B/I^2 \to B/I = A$$

are square zero extensions, so it will be enough to show the lemma when $I^2 = 0$.

Let $F \overset{\sim}{\twoheadrightarrow} A$ be a free resolution of A; in particular F is a degreewise free simplicial ring, $F \twoheadrightarrow A$ is a surjective homomorphism and a weak equivalence. Consider the pullback

$$\begin{array}{ccc} P & \longrightarrow & F \\ \simeq\downarrow & & \simeq\downarrow \\ B & \xrightarrow{f} & A \end{array}$$

of simplicial rings. Using, again, that we may calculate the K-theory and TC of a simplicial radical extension degreewise, the result will follow for $P \to F$ (and hence for f) if we can prove the statement for $P_q \to F_q$ for each q. Since the ring F_q is free, the surjection $P_q \to F_q$ must be a split square zero extension, for which the theorem is guaranteed by Theorem 7.2.1.2. □

Lemma 7.2.2.3 *Let $f: B \to A$ be a 1-connected epimorphism of simplicial rings, then*

$$\begin{array}{ccc} K(B) & \longrightarrow & TC(B) \\ \downarrow & & \downarrow \\ K(A) & \longrightarrow & TC(A) \end{array}$$

is homotopy cartesian.

Proof Note that if $R \to S$ is a k-connected map of simplicial rings, then $K(R) \to K(S)$ and $TC(R) \to TC(S)$ will be (at least) $(k-1)$-connected. We will clearly be done if we can show that any $k \geq 1$ connected map $f: B \to A$ has a diagram of the following sort

$$\begin{array}{ccccc} B & \xleftarrow{\simeq} & B' & & \\ & & & \searrow^{g} & \\ \downarrow f & & \downarrow & & C \\ & & & \swarrow_{h} & \\ A & \xleftarrow{\simeq} & A' & & \end{array}$$

where g is a $(k+1)$-connected epimorphism and h is a square zero extension. The horizontal maps are simply the replacement of I by a k-reduced ideal $I' \subseteq B'$ described in [102, I.1.7]. We set g to be the projection $B' \to B'/(I')^2 = C$. We have a short exact sequence of simplicial abelian groups

$$0 \longrightarrow \ker(m) \longrightarrow I' \otimes_{\mathbf{Z}} I' \xrightarrow{m} (I')^2 \longrightarrow 0.$$

As I' is k-reduced, so is $\ker(m)$, and $I' \otimes_{\mathbf{Z}} I'$ is $(2k-1)$-connected, and accordingly $\ker(g) = (I')^2$ must be at least k-connected. □

Proposition 7.2.2.4 *Let $f: B \to A$ be a map of simplicial rings inducing a surjection $\pi_0 B \to \pi_0 A$ with nilpotent kernel, then*

$$\begin{array}{ccc} K(B) & \longrightarrow & TC(B) \\ \downarrow & & \downarrow \\ K(A) & \longrightarrow & TC(A) \end{array}$$

is homotopy cartesian.

Proof Consider the diagram

$$\begin{array}{ccc} B & \xrightarrow{f} & A \\ \downarrow & & \downarrow \\ \pi_0 B & \xrightarrow{\pi_0 f} & \pi_0 A \end{array}$$

The proposition holds for f if it is true for the three other maps. This follows for the vertical maps by Lemma 7.2.2.3, and for $\pi_0 f$ by Lemma 7.2.2.2. □

Proof of Theorem 7.2.2.1 As in the preceding proof, it is enough to consider the maps $A \to H\pi_0 A$. Consider the resolution $S \mapsto A_S$ of Definition 3.3.1.8. We know that for every $S \in ob\mathcal{P}$, the commutative diagram

$$\begin{array}{ccc} K(A_S) & \longrightarrow & TC(A_S) \\ \downarrow & & \downarrow \\ K(\pi_0 A) & \longrightarrow & TC(\pi_0 A) \end{array}$$

induced by the naturality of the cyclotomic trace is cartesian, and furthermore, by Theorem 3.3.2.2 and Theorem 6.4.3.4, that

$$K(A) \stackrel{\sim}{\to} \underset{\overleftarrow{S \in \mathcal{P} - \emptyset}}{\operatorname{holim}} K(A_S)$$

and

$$TC(A) \stackrel{\sim}{\to} \underset{\overleftarrow{S \in \mathcal{P} - \emptyset}}{\operatorname{holim}} TC(A_S),$$

and the result follows. □

7.3 Some Hard Calculations and Applications

In this section we give a short presentation of the many calculations and structural results about topological cyclic homology. The calculations are the result of the Herculean efforts of first and foremost Ib Madsen, Marcel Bökstedt, Lars Hesselholt, John Rognes and Christian Ausoni, but many more have played an active and important rôle.

We refer away most details to the original sources, which in most cases are very carefully written, but explain enough so that the general idea behind the strategies might be possible to grasp. Also, we will freely use the most convenient technology, deviating sharply from our general aim of being fairly self contained. We hope the reader will follow up the references for notions and theories that are not covered elsewhere in these notes.

7.3.1 General Framework for Calculating $TC(A; p)$

In their tour de force [28], Bökstedt and Madsen give a general procedure for calculating topological cyclic homology, which has been followed in next to all calculations to date. We give a very brief account of the procedure, exposing the results at the various stages for the prime field $\mathbf{F}_p$, for the integers $\mathbf{Z}$ and for the p-complete Adams summand ℓ_p of complex K-theory with homotopy groups $\pi_*(\ell_p) = \mathbf{Z}_p[v_1]$ with v_1 in degree $2p-2$. The Adams summand can most conveniently be realized as an $\mathbf{S}$-algebra by setting $\ell_p = K(k)_p$ [230], where $k = \bigcup_{n>0} \mathbf{F}_{g^n} \subset \bar{\mathbf{F}}_p$ where g is a topological generator of the p-adic units (or equivalently, g is an integer generating the units in $\mathbf{Z}/p^2$). These three examples ($\mathbf{F}_p$, $\mathbf{Z}$ and ℓ_p) together give rise to a tantalizing picture which we hope will whet the reader's interest in the underlying sources for these calculations.

For simplicity, we will fix an odd prime p. We leave the case of the "macho prime" 2 to the reader/the references, with a few exceptions.

We use the notation

$$P(x_1, x_2, \dots) = \mathbf{F}_p[x_1, x_2, \dots]$$

for the polynomial algebra and

$$E(x_1, x_2, \dots) = \mathbf{F}_p[x_1, x_2, \dots]/\left(x_i^2\right)$$

for the exterior algebra on a set of generators $x_1, x_2, \dots$, and we write $|x|$ for the degree of an element x.

If X is a spectrum, we write $H_*(X) = \pi_*(H\mathbf{F}_p \wedge X)$ for the mod p spectrum homology. Let $\mathcal{A}_* = H_*(H\mathbf{F}_p) = \pi_*(H\mathbf{F}_p \wedge H\mathbf{F}_p)$ be the dual Steenrod algebra. Using that

$$\begin{aligned}\pi_*(H\mathbf{F}_p \wedge H\mathbf{F}_p \wedge H\mathbf{F}_p) &\cong \pi_*\left((H\mathbf{F}_p \wedge H\mathbf{F}_p) \wedge_{H\mathbf{F}_p} (H\mathbf{F}_p \wedge H\mathbf{F}_p)\right) \\ &\cong \pi_*(H\mathbf{F}_p \wedge H\mathbf{F}_p) \otimes_{\mathbf{F}_p} \pi_*(H\mathbf{F}_p \wedge H\mathbf{F}_p)\end{aligned}$$

we get that $\mathcal{A}_*$ is a graded commutative Hopf algebra: the algebra structure is inherited by the fact that $H\mathbf{F}_p \wedge H\mathbf{F}_p$ is an $H\mathbf{F}_p$-algebra, the comultiplication $\mathcal{A}_* \to \mathcal{A}_* \otimes_{\mathbf{F}_p} \mathcal{A}_*$ is given by the unit $\mathbf{S} \to H\mathbf{F}_p$ via (the homomorphism of homotopy groups induced by the map)

$$H\mathbf{F}_p \wedge H\mathbf{F}_p \cong H\mathbf{F}_p \wedge \mathbf{S} \wedge H\mathbf{F}_p \to H\mathbf{F}_p \wedge H\mathbf{F}_p \wedge H\mathbf{F}_p,$$

the counit $\mathcal{A}_* \to \mathbf{F}_p$ by multiplication $H\mathbf{F}_p \wedge H\mathbf{F}_p \to H\mathbf{F}_p$, and conjugation $\chi : \mathcal{A}_* \to \mathcal{A}_*$ given by reversing the order of the $H\mathbf{F}_p$-factors.

For $p > 2$ Milnor [211] shows that as an algebra

$$\begin{aligned}&\mathcal{A}_* = \pi_*(H\mathbf{F}_p \wedge H\mathbf{F}_p) = P(\xi_k | k > 0) \otimes E(\tau_k | k \geq 0), \\ &|\xi_k| = 2p^k - 2, \ |\tau_k| = 2p^k - 1.\end{aligned}$$

With the convention $\xi_0 = 1$, the coproduct $\psi\colon \mathcal{A}_* \to \mathcal{A}_* \otimes \mathcal{A}_*$ is given by

$$\Delta(\xi_k) = \sum_{i+j=k} \xi_i^{p^j} \otimes \xi_j,$$

$$\Delta(\tau_k) = \tau_k \otimes 1 + \sum_{i+j=k} \xi_i^{p^j} \otimes \tau_j,$$

the unit and counit are isomorphisms in degree 0 and the conjugation is given recursively by

$$\chi(\xi_0) = 1,$$

$$\sum_{i+j=k} \xi_i^{p^j} \chi(\xi_j) = 0 \quad \text{for } n > 0,$$

$$\tau_k + \sum_{i+j=k} \xi_i^{p^j} \chi(\tau_j) = 0.$$

Many formulas are easier to formulate using the conjugate of Milnor's generators, and we write $\bar{\xi}_i = \chi(\xi_i)$ and $\bar{\tau}_i = \chi(\tau_i)$.

If X is a spectrum, then the spectrum homology $H_*(X) = \pi_*(H\mathbf{F}_p \wedge X)$ is a comodule over the dual Steenrod algebra with coaction $\Delta\colon H_*(X) \to \mathcal{A}_* \otimes_{\mathbf{F}_p} H_*(X)$ induced by the map $H\mathbf{F}_p \wedge X \cong H\mathbf{F}_p \wedge \mathbf{S} \wedge X \to H\mathbf{F}_p \wedge H\mathbf{F}_p \wedge X \cong (H\mathbf{F}_p \wedge H\mathbf{F}_p) \wedge_{H\mathbf{F}_p} (H\mathbf{F}_p \wedge X)$. Recall that an element x in an $\mathcal{A}_*$-comodule M is $\mathcal{A}_*$-comodule *primitive* if $\Delta(x) = 1 \otimes x$. Note that if $M = \mathcal{A}_* \otimes_{\mathbf{F}_p} V$ for some $\mathbf{F}_p$-vector space V, then the primitive elements are all of the form $1 \otimes v$ with $v \in V$.

Let $V(0) = \mathbf{S}/p$ be the cofiber of the map $p\colon \mathbf{S} \to \mathbf{S}$ given by multiplication by p. The mod p homotopy group of a spectrum X is the graded group $V(0)_*(X) = \pi_*(V(0) \wedge X)$, and is often (confusingly) denoted $\pi_*(X; \mathbf{F}_p)$. Note that we get a long exact sequence

$$\cdots \longrightarrow \pi_* X \xrightarrow{p} \pi_* X \longrightarrow V(0)_*(X) \longrightarrow \pi_{*-1}(X) \longrightarrow \cdots.$$

We identify $H_* V(0) = V(0)_* H\mathbf{F}_p = E(\tau_0)$ as an $\mathcal{A}_*$-comodule subalgebra of $\mathcal{A}_*$.

7.3.1.1 Commutative S-Algebras vs. Γ-Spaces

It is an unfortunate fact that the category of Γ-spaces, although modelling all connective spectra and connective ring spectra, does not support a good theory for E_∞-ring spectra (see Lawson [170]). This means that in order to exploit the extra structure on topological Hochschild homology in the commutative case, one must base the theory on alternative frameworks, such as symmetric spectra [141] where the E_∞-ring spectra are modelled by the strictly commutative **S**-algebras (see [196] for a comparison between the different alternatives). This poses no real difficulty if we restrict ourselves to the connective case (and allowing some fibrant replacements that were conveniently unnecessary in the Γ-space case, see also [261] and [262] for the non-connective situation), since all the constructions are based on the associated simplicial functors evaluate on spheres. In the following we will hence tacitly refer to this framework when we talk about commutative **S**-algebras.

Recall that if A is a E_∞-ring spectrum, its homology supports the so-called Dyer-Lashof operations, see [46, III.1.1]

$$Q^k\colon H_*(A) \to H_{*+2k(p-1)}(A)$$

coming from certain classes in $H_*(S^{tp}_{h\Sigma_p})$. Explicitly, represent a class in $H_t(A)$ by a map $f\colon S^t \to H\mathbf{F}_p\wedge A$ (or really a fibrant replacement thereof), consider the composite

$$\big(S^t\wedge\cdots\wedge S^t\big)_{h\Sigma_p} \to \big((H\mathbf{F}_p\wedge A)\wedge_{H\mathbf{F}_p}\cdots\wedge_{H\mathbf{F}_p}(H\mathbf{F}_p\wedge A)\big)_{h\Sigma_p} \to H\mathbf{F}_p\wedge A,$$

giving a map $H_*((S^t\wedge\cdots\wedge S^t)_{h\Sigma_p}) \to H_*(A)$. These operations satisfy the Cartan formulas, Adem relations and Nishida relations [46]. For $p>2$, and $x\in H_t(A)$, then $Q^k(x)=0$ for $k<2t$ and $Q^{2t}(x)=x^p$. For $A=H\mathbf{F}_p$, we have the Dyer-Lashof operations on $\mathcal{A}_*$, with $Q^{p^k}(\bar{\xi}_k)=\bar{\xi}_{k+1}$ and $Q^{p^k}(\bar{\tau}_k)=\bar{\tau}_{k+1}$.

That the cyclotomic trace is multiplicative for discrete rings is shown in the appendix of [92]. The more general situation follows by the methods of [65] or [248], and a proper reference will hopefully soon appear.

7.3.1.2 The Bökstedt Spectral Sequence

Given an **S**-algebra A, $H\mathbf{F}_p\wedge\underline{T}(A)\simeq \underline{HH}^{\mathbf{F}_p}(\tilde{\mathbf{F}}_p[A])$. This can be realized as a bisimplicial $\mathbf{F}_p$-vector space. The resulting spectral sequence is called the Bökstedt spectral sequence, and takes the form

$$E^2_{**}=HH_*\big(H_*(A)\big)\Rightarrow H_*\big(\underline{T}(A)\big).$$

This is an $\mathcal{A}_*$-comodule spectral sequence. If A is commutative, this has more structure: Angeltveit and Rognes [7, 4.2] prove that it is an augmented commutative $\mathcal{A}_*$-comodule $H_*(A)$-algebra spectral sequence.

The suspension map $\sigma\colon S^1\wedge A\to S^1_+\wedge A\to \underline{T}(A)$ (where the first map is induced by the stable splitting of $S^0\to S^1_+\to S^1$ Sect. 6.5.1) corresponds to the usual suspension in Hochschild homology in the sense that if $x\in H_*A$, then the class $\sigma x\in H_*(\underline{T}(A))$ is represented by $1\otimes x$ in the normalized chain complex calculating $HH^{\mathbf{F}_p}_*(H_*(A))$.

There are important extensions in the Bökstedt spectral sequence. Most notably, in the case where A is commutative we have extensions given by the Dyer-Lashof operations

$$Q^k(\sigma x)=\sigma\big(Q^k x\big),$$

see [30, 2.9] or [7, 5.9].

The calculations below are due to Bökstedt (unpublished) and to McClure and Staffeldt [207].

Theorem 7.3.1.1 *There are isomorphisms of $\mathcal{A}_*$-algebras*

1. $H_*(\underline{T}(H\mathbf{F}_p)) \cong \mathcal{A}_* \otimes P(\sigma\bar{\tau}_0)$,
2. $H_*(\underline{T}(\mathbf{Z})) \cong H_*(H\mathbf{Z}) \otimes E(\sigma\bar{\xi}_1) \otimes P(\sigma\bar{\tau}_1)$,
3. $H_*(\underline{T}(\ell_p)) \cong H_*(\ell_p) \otimes E(\sigma\bar{\xi}_1, \sigma\bar{\xi}_2) \otimes P(\sigma\bar{\tau}_2)$.

Here $H_*(H\mathbf{Z})$ (resp. $H_*(\ell_p)$) is the $\mathcal{A}_*$-comodule subalgebra of $\mathcal{A}_*$ generated by all the $\bar{\xi}_i$ and all the $\bar{\tau}_j$ but $\bar{\tau}_0$ (resp. all but $\bar{\tau}_0$ and $\bar{\tau}_1$). In fact, the second isomorphism in the theorem above is as $H_*(H\mathbf{Z})$ algebras, and the third as $H_*(\ell_p)$-algebras.

McClure and Staffeldt [207] also calculate $H_*(\underline{T}(A))$ when A is complex cobordism spectrum MU and the Brown-Peterson spectrum BP.

For the first two cases, the mod-p-homotopy now follows by taking the $\mathcal{A}_*$-comodule primitives, but in the third case McClure and Staffeldt needed to enlist the help of the entire Adams spectral sequence and the answer gets rather complicated. For the sake of exposition, we therefore follow Ausoni and Rognes by giving its $V(1)$-homotopy, where $V(1)$ is the Smith-Toda spectrum given as the cofiber of a periodic self map $v_1 \colon \Sigma^{2p-2}V(0) \to V(0)$ (note that $V(1)_*(\ell_p) = V(0)_*(H\mathbf{Z}) = \mathbf{F}_p$, and we identify $H_*(V(1)) = E(\tau_0, \tau_1)$ as an $\mathcal{A}_*$-comodule subalgebra of $\mathcal{A}_*$). By [219], $V(1)$ is a commutative ring spectrum (in the weakest sense) when our fixed odd prime p is greater than 3, and in the following we restrict ourselves to these primes when discussing results involving $V(1)$.

Corollary 7.3.1.2 *There are algebra isomorphisms*

1. $\pi_*(\underline{T}(H\mathbf{F}_p)) \cong P(\mu_0)$,
2. $V(0)_*(\underline{T}(H\mathbf{Z})) \cong E(\lambda_1) \otimes P(\mu_1)$
3. $V(1)_*(\underline{T}(\ell_p)) \cong E(\lambda_1, \lambda_2) \otimes P(\mu_2)$.

with $|\mu_i| = 2^i$ *and* $|\lambda_i| = 2^i - 1$.

Here μ_i corresponds to $1 \otimes \sigma\bar{\tau}_i + \tau_0 \otimes \sigma\bar{\xi}_i \in H_*(V(i)) \otimes H_*(\underline{T}(B)) \cong H_*(V(i) \wedge \underline{T}(B))$ and λ_i to $1 \otimes \sigma\bar{\xi}_i$.

7.3.1.3 Calculating the Homotopy Fixed Points

In order to calculate the fixed point spectra of topological Hochschild homology, one compares the fundamental cofibration sequence Proposition 6.2.4.2 with a similar sequence involving the homotopy fixed point and the Tate spectrum. We now give a brief sketch the procedure.

Let G be a finite group, and consider the homotopy cofiber sequence

$$EG_+ \to S^0 \to \widetilde{EG}.$$

Thus, $\widetilde{EG}$ is a contractible space with G-action, and its fixed points satisfy $\widetilde{EG}^H \simeq S^0$ for all nontrivial subgroups $H \subseteq G$. When G is a subgroup of the circle, we can choose explicit representatives, with $EG = S(\mathbf{C}^\infty)$ the unit circle in $\mathbf{C}^\infty = \lim_\rightarrow \mathbf{C}^n$ and $EG_+ \to S^0$ modelled by the inclusion $S(\mathbf{C}^\infty)_+ \subseteq D(\mathbf{C}^\infty)_+$ into the unit disc, whose quotient $\widetilde{EG}$ is equal to $S^{\mathbf{C}^\infty}$, the one-point compactification of $\mathbf{C}^\infty$. When H is a nontrivial finite subgroup of the circle, we see that the only H-fixed points in $S^{\mathbf{C}^\infty}$ are 0 and ∞.

In what follows we take the notion of "G-spectra" not in the naïve sense used in the main body of the text, but rather as "genuine G-spectra" indexed on a complete universe of G-representations, see e.g., [173, 195, 197] or [80]. The homotopy category of G-spectra in this sense has a very rich structure. In particular, one should note the presence of transfers $X^H \to X^G$ for inclusions $H \subseteq G$ of finite groups making $H \mapsto \pi_* X^H$ a so-called Mackey functor (the inclusion of fixed points making up the other half of the structure). For our applications to topological Hochschild homology, these transfers provide the Verschiebung maps considered in connection with the de Rham–Witt complex in Sect. 7.3.4 and which is apparent in the Witt-vector descriptions of the homotopy groups of TC and algebraic K-theory (see e.g., Sect. 6.1.4, Theorem 7.3.7.2 and Sect. 6.4.2.2). The intimacy between the actual fixed points used in the definition of $TC(-;p)$ and the homotopy-notions discussed here comes through the fact that topological Hochschild homology is a cyclotomic spectrum, see Sect. 7.3.1.5 below.

If X is a G-spectrum let the homotopy orbit spectrum be defined as $X_{hG} = EG_+ \wedge_G X$, the homotopy fixed point spectrum as $X^{hG} = Map_*(EG_+, X)^G$ and the *Tate spectrum* as $X^{tG} = [\widetilde{EG} \wedge Map_*(EG_+, X)]^G$. Consider the cofibration sequence of spectra [116]

$$X_{hG} \to X^{hG} \to X^{tG}$$

obtained by applying $[- \wedge Map_*(EG_+, X)]^G$ to the cofiber sequence $EG_+ \to S^0 \to \widetilde{EG}$ under the equivalence $[EG_+ \wedge Map_*(EG_+, X)]^G \simeq X_{hG}$. The map $X_{hG} \to X^{hG}$ may be identified with the norm map of Proposition 6.3.2.4.

Tate cohomology of a discrete group G is defined as follows: consider $\mathbf{Z}[G]$-projective resolutions of $\mathbf{Z}$

$$\cdots \leftarrow P_{-2} \leftarrow P_{-1} \leftarrow \mathbf{Z} \leftarrow 0 \quad \text{and} \quad 0 \leftarrow \mathbf{Z} \leftarrow P_0 \leftarrow P_1 \leftarrow \cdots,$$

splice these together to form the "complete" resolution

$$\cdots \leftarrow P_{-2} \leftarrow P_{-1} \leftarrow P_0 \leftarrow P_1 \leftarrow \cdots,$$

and apply $\mathrm{Hom}_{\mathbf{Z}[G]}(-, M)$ for some $\mathbf{Z}[G]$-module M. Tate cohomology $\hat{H}^*(G; M)$ of G with coefficients in M is then the homology of the resulting chain complex.

The skeleton filtration for EG gives spectral sequences for the homotopy orbit and fixed point spectra, and splicing these together to the Greenlees filtration [115] for $\widetilde{EG}$ gives a spectral sequence for the Tate spectrum. For details the reader may consult [131, Sect. 4], which treats the comparison with the classical case and the multiplicative structure. The filtration leads to a conditionally convergent upper half plane spectral sequence

$$\hat{E}^2_{s,t}(X^{tG}) = \hat{H}^{-s}(G;\pi_t X) \Rightarrow \pi_{s+t}(X^{tG}).$$

The inclusion of the non-positive columns gives a map of spectral sequences (which is a surjection $E^2_{s,t} \twoheadrightarrow \hat{E}^2_{s,t}$ for $s \le 0$) from the spectral sequence

$$E^2_{s,t}(X^{hG}) = H^{-s}(G;\pi_t X) \Rightarrow \pi_{s+t}(X^{hG})$$

calculating the homotopy of the homotopy fixed point spectrum. On the abutment this map is induced by the map $X^{hG} \to X^{tG}$.

Likewise, associated to the boundary map $X^{tG} \to \Sigma X_{hG}$, we have a map of spectral sequences $\hat{E}^2_{s+1,t}(X^{tG}) \to E^2_{s,t}(X_{hG}) = H_s(G;\pi_t X)$. This map is injective for $s \ge 0$ and $r \ge 2$.

The interrelationship between these spectral sequences is described in close detail in [28, Sect. 2], and in particular they give the highly useful description of elements in the kernel of $E^\infty_{-s,t}(X^{hG}) \to \hat{E}^\infty_{-s,t}(X^{tG})$: if α is in this kernel (for $s \ge 0$), then there exists an $r > s$ such that α is hit by the rth differential of an element $\beta \in \hat{E}^r_{r-s,t+r-1}(X^{tG}) \subseteq E^r_{r-s-1,t-r-1}(X_{hG})$, with β surviving to $E^\infty_{r-s-1,t-r-1}(X_{hG})$, and such that the image of β is sent to α under the norm map $X_{hG} \to X^{hG}$.

If X is a ring spectrum, then $E^r(X^{hG}) \to \hat{E}^r(X^{tG})$ is a map of algebra spectral sequences (in particular, the differentials are derivations).

7.3.1.4 The C_p-Tate Spectral Sequence for *THH*

Consider the cyclic group $C = C_m$. We identify $\mathbf{Z}[C]$ with $\mathbf{Z}[t]/t^m - 1$ and see that

$$\mathbf{Z}[C] \xleftarrow{1-t} \mathbf{Z}[C] \xleftarrow{1+t+\cdots+t^{m-1}} \mathbf{Z}[C] \xleftarrow{1-t} \cdots$$

is a free $\mathbf{Z}[C]$-resolution of $\mathbf{Z}$. If M is a trivial C-module, this becomes, after applying $\mathrm{Hom}_{\mathbf{Z}[C]}(-,M)$,

$$M \xrightarrow{0} M \xrightarrow{m} M \xrightarrow{0} \quad \xrightarrow{m} \cdots,$$

and so the cohomology $H^k(C,M)$ is M if $k=0$, ${}_mM = \ker\{m\colon M \to M\}$ if k is odd and M/mM if k is even and positive. In particular, if multiplication by m is trivial, then the cohomology is M in all degrees.

If m is a power of p and R is a graded commutative $\mathbf{F}_p$-algebra (considered as a C-module with trivial C-action), then the Tate homology is

$$\hat{H}^{-*}(C;R) = E(u) \otimes P(t,t^{-1}) \otimes R$$

with u in dimension -1 and t in dimension -2.

Hence, we may identify the E^2-terms of the spectral sequences for calculating the appropriate groups of the Tate spectra from Corollary 7.3.1.2 (recall that the C-action on the homotopy groups are trivial since the C action is a restriction of the $\mathbb{T}$-action):

Corollary 7.3.1.3 *Let $C = C_{p^n}$. There are strongly convergent upper half plane spectral sequences*

1. $\hat{E}^2(\underline{T}(H\mathbf{F}_p)) = E(u) \otimes P(t, t^{-1}) \otimes P(\mu_0) \Rightarrow \pi_*(\underline{T}(H\mathbf{F}_p)^{tC})$,
2. $\hat{E}^2(V(0)\wedge\underline{T}(H\mathbf{Z})) = E(u) \otimes P(t, t^{-1}) \otimes E(\lambda_1) \otimes P(\mu_1) \Rightarrow V(0)_*(\underline{T}(H\mathbf{Z})^{tC})$,
3. $\hat{E}^2(V(1)\wedge\underline{T}(\ell_p)) = E(u) \otimes P(t, t^{-1}) \otimes E(\lambda_1, \lambda_2) \otimes P(\mu_1) \Rightarrow V(1)_*(\underline{T}(\ell_p)^{tC})$,

where the bidegree of u is $(-1, 0)$, of t is $(-2, 0)$ of μ_i is $2p^i$ and of λ_i is $2p^i - 1$.

The structure of the differentials in these spectral sequences are increasingly complicated, however in all three cases they are completely determined by the differentials on the base line, i.e., on u and on the powers of t. The spectral sequences are conditionally convergent by construction, and since they are finite in each bidegree, Boardman's first conditional convergence theorem [25, 7.1] implies strong convergence.

In the $C = C_p$ and $H\mathbf{F}_p$-case, everything is killed off by $d^3(u) = t^2\mu_0$. This is an interesting differential in that its origin gives a first example of what seems to be a general phenomenon. In the mod p spectral sequence

$$\begin{aligned}\hat{H}^{-*}\big(C_p, V(0)_*\big(\underline{T}(H\mathbf{F}_p)\big)\big) &= E(u) \otimes P\big(t, t^{-1}\big) \otimes E(\epsilon_0) \otimes P(\sigma\epsilon_0)\\ &\Rightarrow V(0)_*\big(\underline{T}(H\mathbf{F}_p)^{tC_p}\big)\end{aligned}$$

there is a d^2-differential $d^2(\epsilon_0) = t\sigma\epsilon_0$ coming from the calculation of $\underline{T}(H\mathbf{F}_p)$. For dimension reasons, there can be no d^2-differentials in the integral spectral sequence, so $t\mu_0$ represents a class in total degree 0 in the homotopy fixed point spectral sequence which is killed mod p. Ultimately this gives rise to the said d^3-differential which expresses that the class of $t\mu_0$ in $\pi_0(\underline{T}(H\mathbf{F}_p)^{hC_p})$ comes from $\pi_0(\underline{T}(H\mathbf{F}_p)_{hC_p}) \cong \mathbf{Z}/p$ in the fundamental cofiber sequence

$$\underline{T}(H\mathbf{F}_p)_{hC_p} \to \underline{T}(H\mathbf{F}_p)^{C_p} \to \underline{T}(H\mathbf{F}_p),$$

reflecting that $\pi_0(\underline{T}(H\mathbf{F}_p)^{hC_p}) \cong \mathbf{Z}/p^2$.

The analysis of the $H\mathbf{F}_p$-case is carried through by Hesselholt and Madsen in [129], the $\mathbf{Z}$-case by Bökstedt and Madsen in [28] and [29] with input from [288], and for $p = 2$ by Rognes in [240], and finally the ℓ_p-case by Ausoni and Rognes in [11].

The outcome is particularly striking when $C = C_p$:

Proposition 7.3.1.4 *The spectral sequences $i = 1, 2, 3$ in Corollary* 7.3.1.3 *degenerate at the $2p^i + 2$nd page, leaving no room for extensions*:

1. $\pi_*(\underline{T}(H\mathbf{F}_p)^{tC_p}) \cong \hat{E}^\infty(\underline{T}(H\mathbf{F}_p)) = P(t, t^{-1})$
2. $V(0)_*(\underline{T}(\mathbf{Z})^{tC_p}) \cong \hat{E}^\infty(V(0) \wedge \underline{T}(H\mathbf{Z})) = E(\lambda_1) \otimes P(t^p, t^{-p})$
3. $V(1)_*(\underline{T}(\ell_p)^{tC_p}) \cong \hat{E}^\infty(V(1) \wedge \underline{T}(\ell_p)) = E(\lambda_1, \lambda_2) \otimes P(t^{p^2}, t^{-p^2})$

The reader may note that in non-negative degrees (but for the class $\lambda_1\lambda_2 t^{p^2}$ in degree $2p-2$) these groups are abstractly isomorphic to the corresponding topological Hochschild homology groups. Whether this is a coincidence or not depends on your point of view.

7.3.1.5 Comparison of Fixed Point and Homotopy Fixed Points

Recall the fundamental cofibration sequence Proposition 6.2.4.2. The map $EC_{p^n} \to *$ is a C_{p^n}-map, and induces a map Γ_n from fixed point spectra to homotopy spectra such that we get a map of cofiber sequences

$$\begin{array}{ccccc} \underline{T}(A)_{hC_{p^n}} & \longrightarrow & \underline{T}(A)^{C_{p^n}} & \longrightarrow & \underline{T}(A)^{C_{p^{n-1}}} \\ \| & & \Gamma_n \downarrow & & \hat{\Gamma}_n \downarrow \\ \underline{T}(A)_{hC_{p^n}} & \longrightarrow & \underline{T}(A)^{hC_{p^n}} & \longrightarrow & \underline{T}(A)^{tC_{p^n}} \end{array},$$

where the leftmost square commutes up to homotopy, as one sees from the constructions, forcing the existence of the slightly more mysterious map $\hat{\Gamma}_n$. In the fully equivariant world we are adopting, this map can be seen quite geometrically. In that framework, the upper sequence should be replaced by a sequence which exists for general (appropriately fibrant and cofibrant) G-spectra X formed by taking fixed points of the sequence one gets by smashing the cofibration sequence $EG_+ \to S^0 \to \widetilde{EG}$ with X (just as the lower sequence is obtained by taking the fixed point of the smash of this sequence with $Map_*(EG_+, X)$). The spectrum $[\widetilde{EG} \wedge X]^G$ is the so-called spectrum of *geometric fixed points* $\Phi^G X$ and enjoys numerable good properties (for instance the geometric fixed points of the suspension spectrum of a G-space is the suspension spectrum of the G-fixed point space, see e.g., [195] for an exposition of some basic facts), and the vertical maps in the diagram are induced by the projection $EG \to S^0$. The connection to the diagram as it appears above when $G = C_p$ comes through the fact that topological Hochschild homology is a *cyclotomic spectrum*: its C_p-geometric fixed point spectrum is equivalent to *THH* itself. The theory of cyclotomic spectra was introduced in [28], and expanded upon in [129], and is an important tool both for calculations and foundations.

Adapting ideas from Carlsson's proof of the Segal conjecture [52], Tsalidis proved [289] the following theorem starting an induction procedure for calculating the fixed point spectra of topological Hochschild homology.

Theorem 7.3.1.5 (Tsalidis) *If Γ_1 induces a p-adic equivalence after smashing with a finite CW-spectrum on some connected cover, then so does Γ_n for all $n \geq 1$.*

Using knowledge about certain classes with origin in K-theory, Hesselholt, Madsen, Bökstedt, Ausoni and Rognes consequently prove

Theorem 7.3.1.6 *The map $\hat{\Gamma}_1$ is a V-equivalence on k-connective covers for $A = H\mathbf{F}_p$ ($V = \mathbf{S}$, $k = -1$), $H\mathbf{Z}$ ($V = V(0)$, $k = -1$) and ℓ_p ($V = V(1)$, $k = 2p - 2$), sending λ_i to λ_i and μ_i to t^{-p^i} (up to multiplication by units in $\mathbf{F}_p$).*

Not only does this show that the C_{p^n}-fixed point spectra can be calculated from the homotopy fixed point spectral sequence, it also gives important feedback to these spectral sequences.

Except for the case $A = H\mathbf{F}_p$, where $\pi_*(\underline{T}(H\mathbf{F}_p)^{C_{p^n}}) \cong \mathbf{Z}/p^{n+1}(\mu_0(n))$ with $\mu_0(0) = \mu_0$, $R(\mu_0(n)) = p\mu_0(n-1)$ and $F(\mu_0(n)) = \mu_0(n-1)$, the explicit groups for the fixed points are rather messy, and we will not list them here. However, when taking the homotopy limit over the restriction and Frobenius maps things shape up a bit. In particular, given the formula for the restriction map R above, all higher groups vanish in the homotopy limit over R, leaving

$$\operatorname*{holim}_{\overleftarrow{R}} \underline{T}(H\mathbf{F}_p)^{C_{p^n}} \simeq H\mathbf{Z}_p,$$

and

$$TC(H\mathbf{F}_p; p) \simeq H\mathbf{Z}_p \vee \Sigma^{-1} H\mathbf{Z}_p.$$

The most remarkable feature is perhaps that by taking fixed points we have gone from a situation where $p = 0$ to a case where p acts injectively. A similar thing happens for the two other cases:

Theorem 7.3.1.7 *There are algebra isomorphism*

1. $\pi_* TC(H\mathbf{F}_p; p) \cong \mathbf{Z}_p[\partial]/\partial^2$,
2. $V(0)_* TC(Z; p) \cong (E(\partial, \lambda_1) \otimes \mathbf{F}_p\{\lambda_1 t^i | 0 < i < p\}) \otimes P(v_1)$,
3. $V(1)_* TC(\ell_p; p) \cong M \otimes P(v_2)$, *where*

$$M = E(\partial, \lambda_1, \lambda_2) \oplus E(\lambda_2) \otimes \mathbf{F}_p\{\lambda_1 t^i || 0 < i < p\} \oplus E(\lambda_1) \otimes \mathbf{F}_p\{\lambda_2 t^{ip} || 0 < i < p\}.$$

with $|\partial| = -1$, $|\lambda_i| = 2p^i - 1$, $|t| = -2$, $|v_i| = 2p^i - 2$.

The classes v_i has appeared above as $t\mu_i$, and are named thus since they are mapped all the way from the periodic maps in $\Sigma^{2p^{i-1}-2} V(i-1) \to V(i-1)$ of the same name.

7.3.1.6 Algebraic K-Theory

The cyclotomic trace gives that these results have direct bearings on the algebraic K-theory. The comparison goes by a sort of induction, with the start of the induction being the case $A = H\mathbf{F}_p$. Before stating the theorem it is perhaps appropriate to remind the reader that we look only at odd primes p and when discussing $V(1)$-homotopy we assume that $p > 3$.

Theorem 7.3.1.8 *Let A be an* **S**-*algebra with a surjective ring map* $\mathbf{Z}_p \twoheadrightarrow \pi_0 A$. *Then the cyclotomic trace fits in a cofiber sequence*

$$K(A)_p \xrightarrow{trc} TC(A)_p \longrightarrow \Sigma^{-1} H\mathbf{Z}_p.$$

In particular,

1. $\pi_* K(H\mathbf{F}_p)_p \cong \mathbf{Z}_p$
2. $V(0)_* K(\mathbf{Z}_p)_p \cong (E(\lambda_1) \otimes \mathbf{F}_p\{\lambda_1 t^i | 0 < i < p\}) \otimes P(v_1)$
3. $V(1)_* K(\ell_p)_p \cong \mathbf{F}_p\{a\} \oplus N \otimes P(v_2)$, *where*

$$N = E(\lambda_1, \lambda_2) \oplus e(\lambda_2) \otimes \mathbf{F}_p\{\lambda_1 t^i || 0 < i < p\} \oplus E(\lambda_1) \otimes \mathbf{F}_p\{\lambda_2 t^{ip} || 0 < i < p\}.$$

The class $a \in V(1)_{2p-3} K(\ell_p)_p$ *arises as the* $V(1)$-*Bockstein of* $1 \in \pi_0 H\mathbf{Z}_p$.

Proof By [230] the higher homotopy groups of $K(H\mathbf{F}_p)$ are all finite torsion, but there is no p-torsion, so $K(H\mathbf{F}_p)_p \xrightarrow{\sim} H\mathbf{Z}_p$. By the appendix of [92] the cyclotomic trace map is multiplicative, and so $trc\colon K(H\mathbf{F}_p)_p \to TC(H\mathbf{F}_p)_p$ must induce the identity on π_0 (since both source and target are copies of $\mathbf{Z}_p$), and the cofiber sequence is established in the case $A = H\mathbf{F}_p$. If $\pi_0 A$ is finite, it must be a nilpotent extension of $\mathbf{F}_p$, and so

$$\begin{array}{ccc} K(A)_p & \longrightarrow & TC(A)_p \\ \downarrow & & \downarrow \\ K(H\mathbf{F}_p)_p & \longrightarrow & TC(H\mathbf{F}_p)_p \end{array}$$

is homotopy cartesian, and we get the stated cofiber sequence. If $\pi_0 A = \mathbf{Z}_p$ we are done by using that $K(\mathbf{Z}_p)_p \xrightarrow{\sim} \operatorname{holim}_{\overleftarrow{n}} K(\mathbf{Z}/p^n)_p$ [221] and $TC(\mathbf{Z}_p)_p \xrightarrow{\sim} \operatorname{holim}_{\overleftarrow{n}} TC(\mathbf{Z}/p^n)_p$ [129]. □

Remark 7.3.1.9 Note that the comparison between algebraic K-theory and topological cyclic homology goes from being a relative statement, as in the main body of the book, to an absolute statement after p-completion, thanks to the fact that the higher K-groups of $H\mathbf{F}_p$ are finite torsion away from p.

This means that trace methods are particularly well suited to understanding K-theory at the characteristic, which should be thought of as the harder part - analyzing K-theory away from the characteristic is open to attack by a wider array of methods like comparison with étale K-theory and motivic cohomology. Another important line of reductions away from the characteristic arise through Gabber's rigidity theorem [88] which states that if (A, I) is a Hensel pair Sect. 1.1.3.4(5) with $1/p \in A$, then $K(A)_p \simeq K(A/I)_p$.

The good behavior of both algebraic K-theory and topological cyclic homology implies that the above statements extend to a wide array of situations. For instance, generalizing a result in [129] slightly, one gets

Theorem 7.3.1.10 *Let k be a perfect field of characteristic $p > 0$ (so, taking the pth power is an automorphism) and A a connective $\mathbf{S}$-algebra such that $\pi_0 A$ is a $W(k)$-algebra which is finitely generated as a $W(k)$-module. Then the cyclotomic trace fits in a cofiber sequence*

$$K(A)_p \to TC(A)_p \to \Sigma^{-1} H\bigl(\operatorname{coker}\{1 - F\}\bigr)$$

where $F\colon W(\pi_0 A) \to W(\pi_0 A)$ is the Frobenius endomorphism.

7.3.1.7 Space Level Descriptions and the Case $p = 2$

Since $TC(H\mathbf{F}_p; p)$ is the Eilenberg–Mac Lane spectrum $H\mathbf{Z}_p \vee \Sigma^{-1} H\mathbf{Z}_p$, we get that $TC(A; p)$ is an Eilenberg–Mac Lane spectrum whenever A is an $\mathbf{F}_p$-algebra, and so is determined completely by its homotopy groups.

For other rings this is definitely not true. For instance, Bökstedt and Madsen prove in [29] that

Theorem 7.3.1.11 (Bökstedt and Madsen) *Let p be an odd prime, let g be a topological generator of the units in $\mathbf{Z}_p$ and let (the image of) J be the homotopy fiber of $1 - \Psi^g : \mathbf{Z} \times BU \to BU$, where Ψ^g is the gth Adams operation. Then the algebraic K-theory space of $W(\mathbf{F}_{p^s})$ for $s \geq 1$ is equivalent after p-completion to*

$$J \times BJ \times SU \times U^{\times (s-1)}.$$

In particular $K(\mathbf{Z}_p)_p \simeq J_p \times BJ_p \times SU_p$.

This result is stated in terms of infinite loop spaces, and B signifies the associated bar-construction, shifting homotopy groups by one. We note that neither J nor the infinite unitary group $U = \bigcup_{n>0} U(n)$ are Eilenberg–Mac Lane spaces. Due to Bott periodicity, the special unitary group SU could alternatively be given as BBU.

For the homotopy groups, one notices that apart from the group $\pi_0(J_p) = \mathbf{Z}_p$, the nonzero homotopy groups of J_p and U are $\pi_{2k(p-1)-1}(J_p) = \mathbf{Z}/p^{\nu_p(k)+1}$ and $\pi_{2k-1}(U) = \mathbf{Z}$ for $k > 0$, where ν_p is the p-adic valuation: $\nu_p(np^m) = m$ if $\gcd(n, p) = 1$. In order to get to this space level description, one of course has to reach beyond the mod p homotopy groups we have listed. In the case of the Eilenberg–Mac Lane spectrum $H\mathbf{Z}_p$, Bökstedt and Madsen [28] compare directly to the more computable topological cyclic homology of $\mathbf{S}_p$, and uses that the spectra involved are in a sense very rigid. In particular, there are canonical maps from J_p and BJ_p to $K(W(\mathbf{F}_{p^s}))_p$, but the identification of the factors involving the unitary groups relies on structure theorems in the K-local category, and so is highly non-canonical. As far as the authors know, the multiplicative structure on $K(W(\mathbf{F}_{p^s}))_p$ is unknown.

The case $p = 2$ is in many ways quite different. For one thing the algebraic properties (or lack thereof) of the mod p Moore spectra $\mathbf{S}/p = V(0)$ are bad for small primes p. In particular, $\mathbf{S}/2$ is not a homotopy commutative ring spectrum, even in a weak sense. In a series of papers culminating in [240], Rognes resolves this by carefully comparing with mod 4 homotopy, and also replacing Bökstedt and Madsen's comparison with $K(\mathbf{S}_p)$ with a comparison through a "Galois reduction map" $K(\mathbf{Z}_2) \to K(\mathbf{F}_3)$. Here $K(\mathbf{F}_3)$ plays the rôle as the (complex!) image of J-space, but the splitting results of the odd primary case fails, giving

Theorem 7.3.1.12 (Rognes) *After 2-completion, there are (non-split) fiber sequences*

$$\begin{array}{ccccc} B(K\mathbf{F}_3) & \longrightarrow & K^{red}(\mathbf{Z}_2) & \longrightarrow & BBU \\ & & \downarrow & & \\ & & K(\mathbf{Z}_2) & & \\ & & \downarrow & & \\ & & K(\mathbf{F}_3) & & \end{array}$$

7.3.2 The Lichtenbaum–Quillen Conjecture, the Milnor Conjecture and the Redshift Conjecture

Since we are mainly concerned with phenomena in algebraic K-theory which can be understood from trace methods, we will do the Lichtenbaum–Quillen conjecture grave injustice. This conjecture and its relatives have led to vast amounts of deep mathematics, and the final solution in the original cases of interest comes through motivic cohomology (although trace methods played an interesting part in the early identification). For a nice exposition, containing a chronological overview of this and related results and conjecture, see the papers of B. Kahn [153], Weibel [309] or Gajda [90].

Quillen proved that for extensions of finite fields $k_1 \subseteq k_2$, the map

$$K(k_1) \to K(k_2)^{hGal(k_2/k_1)}$$

is an equivalence. Here the Galois group $Gal(k_2/k_1)$ acts on the category of finite dimensional k_2-vector spaces and hence on $K(k_2)$. In most models of algebraic K-theory we may identify $K(k_1)$ with the actual fixed point spectrum $K(k_2)^{Gal(k_2/k_1)}$.

So, for a group G of ring-automorphisms of a given ring A, one may ask about the relationship between the algebraic K-theory $K(A^G)$ of the fixed ring A^G and the homotopy fixed point spectrum $K(A)^{hG}$.

Lichtenbaum conjectured [175] a relationship between the values of the Dedekind zeta function for a number field and the order of the (higher) K-groups of the ring of integers and Quillen conjectured [233] that there should exist a sort of analog of the

Atiyah–Hirzebruch spectral sequence for algebraic K-theory built out of étale cohomology. These two conjectures are closely related and have been refined over the years in various directions. Dwyer and Friedlander [77] refined Soulé's work [267], and defined a surjective map from algebraic K-theory to something called *étale K-theory* which is the abutment of the Atiyah–Hirzebruch spectral sequence mentioned above, so that the Lichtenbaum–Quillen conjecture amounted to the claim that the map was injective. The step towards stable homotopy theory was taken when it was realized that under favourable circumstances étale K-theory essentially was what you got when you "invert the Bott element" in algebraic K-theory [76, 283].

More precisely, in Waldhausen's description [300] of Snaith's setup [265], let $V(0) = \mathbf{S}/p$ be the homotopy cofiber of a degree p self-map $\mathbf{S} \to \mathbf{S}$. There is a self-map $v_1 \colon \Sigma^{2p-2}V(0) \to V(0)$, and the Bott inverted algebraic K-theory $K(A; \mathbf{Z}/p)[\beta^{-1}]$ is equivalent to the homotopy colimit of

$$K(A)\wedge V(0) \xrightarrow{v_1} K(A)\wedge\Sigma^{-2p+2}V(0) \xrightarrow{v_1} K(A)\wedge\Sigma^{-24+4}V(0) \xrightarrow{v_1} \cdots,$$

which on the other hand (by formal properties of Bousfield's L_1-localization with respect to Morava K-theory $K(1)$, see also the discussion about the redshift conjecture below) is the same as $L_1K(A)\wedge\mathbf{S}/p$. Eventually, the Lichtenbaum–Quillen conjecture then can be formulated to say that for suitable discrete rings A the localization map

$$K(A)_{(p)} \to L_1K(A)_{(p)}$$

is an equivalence in sufficiently high degrees.

In this formulation, one sees that the calculations of Bökstedt, Madsen and Rognes confirm the Lichtenbaum–Quillen conjectures in the cases they cover: the answers they provide are clearly equal to their L_1-localizations in high degrees: v_1 acts injectively. Also the calculations of Hesselholt and Madsen on finite extensions of local number fields (see Theorem 7.3.3.1 below) is phrased so that this becomes apparent.

For algebraically closed fields, Suslin proved the Lichtenbaum–Quillen conjecture, in that he proved that if $F \subseteq E$ is an extension of algebraically closed fields, then $K(F) \to K(E)$ is an equivalence after profinite completion, and both spectra have the value predicted by étale K-theory. In particular, if F is an algebraically closed field of characteristic 0, then there is an equivalence

$$K(F)_p \simeq ku_p,$$

and the values for the positive characteristic case is given by Quillen's calculation of the K-theory of the algebraic closure of finite fields in Sect. 1.1.7. Suslin's result could be stated as saying that the proposed spectral sequence in the Quillen part of the conjecture states that the algebraic K-theory of a field k is, in fact, the homotopy fixed set of the absolute Galois group action on the K-theory of its algebraic closure.

The ultimate goal of calculating the algebraic K-theory of the integers is beyond the scope of trace methods, since topological cyclic homology is notoriously bad at distinguishing between a ring and its completions: $TC(\mathbf{Z}; p)_p \xrightarrow{\sim} TC(\mathbf{Z}_p; p)_p$, whereas K-theory sees a huge difference.

Voevodsky's proof of the Milnor conjecture [293] made it possible for Rognes and Weibel [239] to complete the 2-primary piece of the Lichtenbaum–Quillen Conjecture (see also the paper by Weibel [308] and the influential preprint [152] by Kahn where the authors obtain the result upon relying on a multiplicative structure of the Bloch-Lichtenbaum spectral sequence). The result is most elegantly summarized by stating that a certain 2-completed homotopy commutative square proposed by Bökstedt

$$\begin{array}{ccc} K(\mathbf{Z})_2 & \longrightarrow & BO_2 \\ \downarrow & & \downarrow \\ K(\mathbf{F}_3)_2 & \longrightarrow & BU_2 \end{array}$$

is homotopy cartesian. For a fuller discussion see Hodkin and Østvær [137]. See also the table in Sect. 1.1 which is lifted from Weibel [308].

The situation for odd primes was much more painful, in that the odd companion of the Milnor conjecture, called the Bloch–Kato conjecture, turned out to be hard to prove (a statement which in no way is meant to belittle the monumental achievements leading to the results for $p = 2$). Eventually this is now a theorem, the *norm residue isomorphism theorem*, due to Rost and Voevodsky, see [294, 303]. The link between the Bloch–Kato conjecture and the Beilinson–Lichtenbaum conjectures was proved by Suslin and Voevodsky assuming resolution of singularities (this condition is removed by Geisser and Levine in [95]). The relevant spectral sequence was first constructed by Grayson [114] with the last pieces being laid by Suslin [274]. Voevodsky and Levine also have constructions of the spectral sequence. See also [149]. We are indebted to Weibel and a careful referee for making the status of various conjectures clearer to us.

7.3.2.1 Redshift

In another direction, there is ongoing work trying to understand algebraic K-theory of $\mathbf{S}$-algebras intermediate between the Eilenberg–Mac Lane spectrum of rings and the sphere spectrum itself. Much of this activity draws its motivation from Waldhausen's paper [300] in which he extends beyond the Lichtenbaum–Quillen conjecture to speculate about the homotopy theoretic significance of a filtration of the linearization map $K(\mathbf{S}) \to K(\mathbf{Z})$ through the K-theory of intermediate rings ideally extracting the "arithmetic properties" of ring spectra.

Given that we now understand the linearization map, to some degree, through a blend of motivic and trace method information (see Sect. 7.3.8.1 below), this idea is especially tantalizing.

Slightly reinterpreted, let $L_n = L_{E(n)}$ be Bousfield localization with respect to the nth Johnson–Wilson spectra $E(n)$ with $\pi_* E(n) = \mathbf{Z}_p[v_1, \ldots, v_n][v_n^{-1}]$ and consider the tower

$$X \to \cdots \to L_n X \to \cdots \to L_1 X \to L_0 X \simeq H\mathbf{Q}_p \wedge X$$

approximating the finite p-local CW-spectrum X in the sense of chromatic convergence: $X \xrightarrow{\sim} \operatorname{holim}_{\overleftarrow{n}} L_n X$. See Ravenel's orange book [237] for background. In [300], Waldhausen considered a somewhat different localization functor L_n^f, characterized by its behavior on finite spectra, which turned out to be different due to the failure of the so-called telescope conjecture (it fails for $n = 2$, but is true for $n = 1$, so $L_1 = L_1^f$), see [238] and [206]. McClure and Staffeldt [206] prove that one has chromatic convergence for algebraic K-theory, in the sense that $K(\mathbf{S}_p) \simeq \operatorname{holim}_{\overleftarrow{n}} K(L_n \mathbf{S}_p)$, but it is unknown whether this holds for the L^f-version. Given the close connection between algebraic K-theory and topological cyclic homology, one is tempted to believe that this should be possible to establish through trace methods.

As touched upon above, Thomason [283] reinterpreted the Lichtenbaum–Quillen conjecture to say that under favourable circumstances you had something resembling Bott periodicity. This was again reinterpreted by Waldhausen [300] and further developed by Mitchell [216] to say that the localization map $K(A) \to L_1 K(A)$ was an equivalence in high dimensions. As a matter of fact, Mitchell proved that for any discrete ring A, the map $L_n K(A) \to L_1 K(A)$ is an equivalence, and hence [238], so is $L_n^f K(A) \to L_1 K(A)$.

This interpretation puts the discoveries of Bökstedt and Madsen in an interesting context, and already in their first papers they emphasize that their trace calculations give exactly this sort of behavior. In [28, Sect. 9] and [29, Sect. 5] they show that algebraic K-theory of unramified number rings is v_1-periodic in the sense that it agrees with its L_1-localization in high degrees, cf. the formula for $V(0)_* K(\mathbf{Z}_p)$ in Theorem 7.3.1.8 for a typical example. Later, in [131] more general local number fields are included into the picture (see Theorem 7.3.3.1 below).

The "redshift conjecture" of Rognes is an offspring of this line of ideas, generalizing this behavior to higher chromatic filtration. As presented in [2], Ausoni and Rognes connect this up with speculations on the interplay between algebraic K-theory and Galois theory for commutative $\mathbf{S}$-algebras, much as for the Lichtenbaum–Quillen conjecture.

Redshift is clearly also present in their calculation of the algebraic K-theory of the Adams summand ℓ_p above in Theorem 7.3.1.8, lending credibility to the speculation that this may be a more general phenomenon.

Also Ausoni's calculation of the K-theory of all of ku_p for $p > 3$, [10, Theorem 8.1], supports the idea:

$$V(1)_* K(ku_p) \cong P(b) \otimes M \oplus \mathbf{F}_s\{s\}$$

where M is some finite $\mathbf{F}_p$-vector space (given explicitly by Ausoni, see also Sect. 7.3.3.1 below), b satisfies $b^{p-1} = -v_2$ and the degree of s is $2p - 3$.

A conceptual explanation for *why* one would expect red shift for algebraic K-theory is still missing, but there is hope that the cyclotomic trace may shed some light on the phenomenon, through the rich algebraic structure on the fixed point spectra of topological Hochschild homology.

For instance, one notices that for the commutative ring spectra $A = \mathbf{S}$ and $A = MU$, one has a sort of Segal conjecture in the narrowest sense, in that $\underline{T}(\mathbf{S})^{C_n} \to \underline{T}(\mathbf{S})^{hC_n}$ and $\underline{T}(MU)^{C_n} \to \underline{T}(MU)^{hC_n}$ are both equivalences [52, 185]. In some way this reflects that both $\mathbf{S}$ and MU have "infinite chromatic height", in contrast to the more algebraic rings A where $\underline{T}(A)^{C_n} \to \underline{T}(A)^{hC_n}$ may be an equivalence in high dimension, but where the homotopy fixed point spectrum exhibits periodic phenomena in negative dimensions, starting in dimensions that somehow correspond to the chromatic height of the input. For instance, we note that in the spectral sequence Sect. 6.5.1 for the homotopy C_p-fixed points for $H\mathbf{F}_p$, $H\mathbf{Z}$ and ℓ_p, we have infinite cycles t, t^p and t^{p^2}, where t is a generator in $H^2(BS^1)$.

This should be contrasted with the case where one looks at the fixed point spectra of topological Hochschild homology for one degree at a time: according to Lunøe-Nilsen and Rognes [184]

$$THH_{jp-1}(A)^{C_p} \to THH_{jp-1}(A)^{hC_p}$$

is an equivalence if for each i the ith spectrum homology group $\pi_i(H\mathbf{F}_p \wedge A)$ is finite (recall that $THH_{jp-1}(A)$ is just a particular model for the jp-fold smash power of A. As a matter of fact, the result is not dependent on the multiplicative structure of A and is valid for any spectrum A which is bounded below and satisfies the said finiteness condition). Hence the phenomenon that fixed points and homotopy fixed points differ for topological Hochschild homology is a consequence of the fact that homotopy fixed points do not commute with realization.

It seems likely that the fixed point spectra of topological Hochschild homology might shed more light on the algebraic side of redshift. The very first step, moving from a situation where the prime $v_0 = p$ is zero to where it acts injectively is encoded in the Witt vectors, which we recognize as the path components of TR: $\pi_0 TR(A) \simeq W(\pi_0 A)$ for a commutative $\mathbf{S}$-algebra A, and for the next step much information is encoded in the de Rham–Witt complex Sect. 7.3.4. In [44] it is shown that for commutative $\mathbf{S}$-algebras, one can extend the definition of topological Hochschild homology to a functor $X \mapsto \Lambda_X A$ such that $THH(A) \simeq \Lambda_{S^1} A$, and such that one retains full control over the equivariant structure, and in [53] this is used to study the interplay between the various fixed points of iterated topological Hochschild homology. This gives us access to classes that conjecturally detect higher chromatic phenomena, but we are still very far from understanding the interrelationship between the commutativity of the $\mathbf{S}$-algebra and the chromatic structure of the spectrum underlying the fixed point spectra.

7.3.3 Topological Cyclic Homology of Local Number Fields

Algebraic K-theory has a localization sequence, making the connection between examples like the algebraic K-theory of a number field and its ring of integers quite transparent. For instance, there is a fiber sequence

$$K(\mathbf{F}_p) \to K(\mathbf{Z}_p) \to K(\mathbf{Q}_p),$$

so that the p-torsion in the K-groups of $\mathbf{Z}_p$ and of $\mathbf{Q}_p$ agree. The "transfer" $K(\mathbf{F}_p) \to K(\mathbf{Z}_p)$ is induced from the inclusion of the torsion $\mathbf{Z}_p$-modules in the category of finitely generated $\mathbf{Z}_p$-modules. By the resolution Theorem 1.2.7.6, the K-theory of the former category is equivalent to $K(\mathbf{F}_p)$ and the K-theory of the latter is equivalent to $K(\mathbf{Z}_p)$.

For topological cyclic homology the situation is quite different. Since p is invertible in $\mathbf{Q}_p$, the C_{p^n}-fixed points of topological cyclic homology becomes quite uninteresting: $H\mathbf{Q}_p \simeq \underline{T}(\mathbf{Q}_p) \simeq \underline{T}(\mathbf{Q}_p)_{hC_{p^n}}$, so $\underline{T}(\mathbf{Q}_p)^{C_{p^n}} \simeq \mathbf{Q}_p \times \cdots \times \mathbf{Q}_p$ and $TR(\mathbf{Q}_p; p) \simeq HW\mathbf{Q}_p$, an infinite product of copies of $H\mathbf{Q}_p$. The Frobenius action cuts this down to size so that $TC(\mathbf{Q}_p; p) \simeq H\mathbf{Q}_p$.

Hesselholt and Madsen handle this problem in [131] by forcing localization on topological cyclic homology and get a map of fiber sequences

$$\begin{array}{ccccc} K(k) & \longrightarrow & K(A) & \longrightarrow & K(K) \\ \downarrow & & \downarrow & & \downarrow \\ TC(k; p) & \longrightarrow & TC(A; p) & \longrightarrow & TC(A|K; p) \end{array} \tag{7.2}$$

where K is a complete discrete valuation field of characteristic zero with valuation ring A and perfect residue field k of characteristic $p > 2$.

Explicitly, they introduce three categories with cofibrations and weak equivalences, which all are full subcategories of the category $C^b(\mathcal{P}_A)$ of bounded complexes of finitely generated projective A-modules

1. $C^b_z(\mathcal{P}_A)$: all objects and the weak equivalences are the homology isomorphisms,
2. $C^b_q(\mathcal{P}_A)$: all objects and the weak equivalences are the rational homology isomorphisms, and
3. $C^b_z(\mathcal{P}_A)^q$: the objects are the complexes whose homology is torsion and the weak equivalences are the homology isomorphisms.

We have maps of categories with cofibrations and weak equivalences

$$C^b_z(\mathcal{P}_A)^q \xrightarrow{\subseteq} C^b_z(\mathcal{P}_A) \longrightarrow C^b_q(\mathcal{P}_A),$$

and localization Theorem 1.2.7.4 implies that this gives a fibration sequence in K-theory, and Hesselholt and Madsen use that one has a corresponding fiber sequence for topological cyclic homology, resulting in a map of fiber sequences

$$\begin{array}{ccccc} K\big(C^b_z(\mathcal{P}_A)^q\big) & \longrightarrow & K\big(C^b_z(\mathcal{P}_A)\big) & \longrightarrow & K\big(C^b_q(\mathcal{P}_A)\big) \\ \downarrow & & \downarrow & & \downarrow \\ TC\big(C^b_z(\mathcal{P}_A)^q; p\big) & \longrightarrow & TC\big(C^b_z(\mathcal{P}_A); p\big) & \longrightarrow & TC\big(C^b_q(\mathcal{P}_A); p\big). \end{array} \tag{7.3}$$

Hesselholt and Madsen now define $TC(A|K;p) = TC(C_q^b(\mathcal{P}_A);p)$, and variations on the approximation Theorem 1.2.7.3 can be used to identify diagram (7.3) with diagram (7.2).

Combined with the technology they develop for the de Rham–Witt complex with log poles, briefly discussed in Sect. 7.3.4 below, Hesselholt and Madsen then prove in [131] that

Theorem 7.3.3.1 *Let K be a finite extension of $\mathbf{Q}_p$ where p is an odd prime. For an integer r let $F\Psi^r$ be the homotopy fiber of $1-\Psi^r : \mathbf{Z}\times BU \to BU$, where Ψ^r is the rth Adams operation. Then the algebraic K-theory space is equivalent after p-completion to*

$$F\Psi^{g^{p^{a-1}d}} \times BF\Psi^{g^{p^{a-1}d}} \times U^{|K:\mathbf{Q}_p|},$$

where $d = (p-1)/|K(\mu_p):K|$, $a = \max\{v|\mu_{p^v} \subseteq K(\mu_p)\}$ and g is a topological generator of the units in $\mathbf{Z}_p$ (or equivalently, an integer which generate the units in Z/p^2).

7.3.3.1 Quotient Fields in a More General Framework

It should be noted that the localization idea have been extended in some cases beyond discrete rings. In particular, in [22] Blumberg and Mandell show that there is a fiber sequence

$$K(H\mathbf{Z}) \to K(ku) \to K(KU)$$

where the first map is a transfer-type homomorphism similar to the inclusion of the category of finite abelian groups into the category of finitely generated abelian groups, and where the last map is induced by the map from connective complex K-theory, ku, to (periodic) complex K-theory, KU, given by inverting the Bott class. Similar sequences hold for the Adams summand, and Hesselholt has observed that the calculations of Ausoni and Rognes could be efficiently codified if one extends the techniques from the local field case, see e.g., the discussion in [10, Remark 8.4], and according to Ausoni would give the calculation [10, Theorem 8.3]

$$\begin{aligned} V(1)_*K(KU_p) &\overset{?}{\cong} P(b)\otimes E(\lambda_1,d)\oplus P(b)\otimes \mathbf{F}_p\{\partial\lambda_1,\partial b,\partial a_1,\partial\lambda_1 d\} \\ &\oplus P(b)\otimes E(d)\otimes \mathbf{F}_p\{t^k\lambda_1 \mid 0<k<p\} \\ &\oplus P(b)\otimes E(\lambda_1)\otimes \mathbf{F}_p\{\sigma_n, \lambda_2 t^{p^2-p} \mid 0<n<p-1\} \end{aligned}$$

where b satisfies $b^{p-1} = -v_2$, and the degrees of the other generators are $|\partial| = -1$, $|\lambda_1| = 2p-1$, $|\lambda_2| = 2p^2-1$, $|a_1| = 2p+3$, $|\sigma_n| = 2n+1$ and $|t| = -2$.

This has been taken further by Ausoni and Rognes into a speculation on the rôle of quotient fields for more general **S**-algebras, and interpretations in the vein of the redshift conjecture.

7.3.4 The de Rham–Witt Complex

In [18] Spencer Bloch outlined a connection between algebraic K-theory and the crystalline cohomology of Berthelot–Grothendieck. In [123] Hesselholt used trace methods to confirm Bloch's ideas for any smooth algebra A over a perfect field k of positive characteristic (see also Sect. 7.3.5.2 below). Hesselholt's result was accomplished by showing that for any such ring A, the homotopy data in $TR(A; p)$ assembles into a pro-complex isomorphic to the de Rham–Witt complex of Bloch–Deligne–Illusie. This observation has lead Hesselholt and his collaborators (most notably Madsen and Geisser) towards a sequence of remarkable calculations of algebraic K-theory as well as new purely algebraic generalizations of the Witt vectors [124] and the de Rham–Witt complex itself (see [127] for a very readable, purely algebraic construction of these).

Two important base computations suggest the connection of $TR(A; p)$ with the de Rham–Witt complex. The first is that for A a commutative ring, $\pi_0 TR_n(A; p) \cong W_n(A)$ where $W_n(A)$ is the ring of p-typical Witt vectors of length n in A (the p is assumed, see Sect. 6.4.2.2 for a brief outline of the algebraic structure). Moreover, this isomorphism can be chosen (naturally in A) so that π_0 of the restriction, Frobenius and Verschiebung maps for the fixed points of $T(A)$ correspond to the classical restriction, Frobenius and Verschiebung maps of the Witt rings (note that, since we are working with fully equivariant spectra, we have transfers, and in particular in the homotopy category we have the *Verschiebung* $T(A)^{C_{rs}} \to T(A)^{C_s}$ on the fixed points of topological Hochschild homology).

The second observation is that by the Hochschild–Kostant–Rosenberg theorem [136], for A a smooth k-algebra, the de Rham complex $(\Omega^*_{A/k}, d)$ is isomorphic to $(HH_*(A/k), B)$ where B is Connes B-operator (see Sect. 6.5.2) and $HH(A/k) = \pi_* HH^k(A)$ is the Hochschild homology of the k-algebra A. If we let δ be the map from $\pi_n T(A)$ to $\pi_{n+1} T(A)$ induced by the $\mathbb{T}$-action (cf. Corollary 6.5.1.2), then $\delta \circ \delta = 0$ and the linearization map $T(A) \to HH(A/k)$ takes δ to B. We obtain a limit system of differential graded algebras

$$\begin{array}{ccccccc}
\vdots & & \vdots & & \vdots & & \\
\downarrow R & & \downarrow R & & \downarrow R & & \\
\pi_0 T(A)^{C_{p^2}} & \xrightarrow{\delta} & \pi_1 T(A)^{C_{p^2}} & \xrightarrow{\delta} & \pi_2 T(A)^{C_{p^2}} & \xrightarrow{\delta} & \cdots \\
\downarrow R & & \downarrow R & & \downarrow R & & \\
\pi_0 T(A)^{C_p} & \xrightarrow{\delta} & \pi_1 T(A)^{C_p} & \xrightarrow{\delta} & \pi_2 T(A)^{C_p} & \xrightarrow{\delta} & \cdots \\
\downarrow R & & \downarrow R & & \downarrow R & & \\
\pi_0 T(A) & \xrightarrow{\delta} & \pi_1 T(A) & \xrightarrow{\delta} & \pi_2 T(A) & \xrightarrow{\delta} & \cdots
\end{array} \qquad (7.4)$$

The first column is the limit system of the p-typical Witt ring of A, and after linearization to $HH_*(A/k)$ the bottom row is the de Rham cohomology of A (cf. Sect. 6.5.2) when A is smooth as a k-algebra. Now, since $\pi_1 T(A) \cong HH_1(A/k)$ is isomorphic to the first Kähler differentials $\Omega^1_{A/k}$, and $\pi_* T(A)$ is a graded commutative algebra, one obtains a map of differential graded algebras from $\Omega^*_{A/k}$ to $\pi_* TH(A)$. After checking additional relations about how the restriction, Frobenius, and Verschiebung behave on higher homotopy groups, this implies by the universal properties of the de Rham–Witt complex shown by Illusie in [146] that one has a map of pro-complexes from the de Rham–Witt complex

$$\begin{array}{ccccccc}
\vdots & & \vdots & & \vdots & & \\
\downarrow R & & \downarrow R & & \downarrow R & & \\
W_3\Omega^0_A & \xrightarrow{\delta} & W_3\Omega^1_A & \xrightarrow{\delta} & W_3\Omega^2_A & \xrightarrow{\delta} & \cdots \\
\downarrow R & & \downarrow R & & \downarrow R & & \\
W_2\Omega^0_A & \xrightarrow{\delta} & W_2\Omega^1_A & \xrightarrow{\delta} & W_2\Omega^2_A & \xrightarrow{\delta} & \cdots \\
\downarrow R & & \downarrow R & & \downarrow R & & \\
W_1\Omega^0_A & \xrightarrow{\delta} & W_1\Omega^1_A & \xrightarrow{\delta} & W_1\Omega^2_A & \xrightarrow{\delta} & \cdots
\end{array}$$

to the limit system (7.4). The map of limit systems of differential graded algebras is an isomorphism when A is smooth over a perfect field k. Hesselholt obtains these results by showing that the isomorphism first holds for polynomial algebras $k[x_1, \ldots, x_n]$, and then uses that both complexes behave well for étale maps (see [123, 2.4.3]. A referee points out that by van der Kallen's Theorem [291, Theorem 2.4] the requirement that we are over $\mathbf{F}_p$ may be removed):

Proposition 7.3.4.1 *If $f : A \to B$ is an étale map of $\mathbf{F}_p$-algebras then the canonical map*

$$W_r(B) \otimes_{W_r(A)} \pi_* T(A)^{C_{p^{r-1}}} \to \pi_* T(B)^{C_{p^{r-1}}}$$

is an isomorphism.

In fact, a sharper result is obtained, namely

Theorem 7.3.4.2 *Suppose that A is a smooth k-algebra. Then there is an isomorphism*

$$W_n\Omega^*_A \otimes_{W_n(k)} S_{W_n}\{\sigma_n\} \to \pi_* T(A)^{C_{p^{n-1}}}, \quad \deg \sigma_n = 2$$

Moreover, $F(\sigma_n) = \sigma_{n-1}$, $V(\sigma_n) = p\sigma_{n+1}$ and $R(\sigma_n) = p\lambda_n\sigma_{n-1}$, where λ_n is a unit of $W_n(F_p)$.

One of our referees points out that the conclusion of Theorem 7.3.4.2 holds more generally for every regular Noetherian $\mathbf{F}_p$-algebra A, since by generalized Néron desingularization [226] A is isomorphic to a filtered colimit of smooth $\mathbf{F}_p$-algebras.

Let V be a complete discrete valuation ring of mixed characteristic $(0, p)$ with quotient field K and perfect residue field k. For A a smooth V-algebra one has a localization sequence in K-theory

$$\cdots \to \pi_q K(A \otimes_V k) \to \pi_q K(A) \to \pi_q K(A \otimes_V K) \to \cdots.$$

In [131], a corresponding sequence related by trace maps is constructed

$$\cdots \to \pi_q TR(A \otimes_V k) \to \pi_q TR(A) \to \pi_q TR(A|A_K; p) \to \cdots,$$

cf. the discussion in Sect. 7.3.3. The term $\pi_q TR(A|A_K; p)$ is calculated in [132] when p is odd and $\mu_{p^v} \subset K$, by an isomorphism of pro-abelian groups

$$W.\Omega^*_{(A,M_A)} \otimes_{\mathbf{Z}} S_{\mathbf{Z}/p^v}(\mu_{p^v}) \to \pi_* TR^{\cdot}\big(A|A_K; p, \mathbf{Z}/p^v\big).$$

Here $\pi_* TR^{\cdot}(A|A_K; p, \mathbf{Z}/p^v)$ denotes the graded pro-group with coefficients in $\mathbf{Z}/p^v$ coming from the system of the restriction maps (not just the homotopy limit) and $W.\Omega^*_{(A,M_A)}$ is a universal Witt complex over the log-ring (A, M_A) with the map

$$d\log_n : M_A \to TR^n_1(A|A_K; p)$$

given by the composite

$$M_A = A \cap (A \otimes_V K)^\times \longrightarrow (A \otimes_V K)^\times \longrightarrow K_1(A \otimes_V K) \xrightarrow{\mathrm{tr}} TR^n_1(A|A_K; p).$$

In order to describe $\Omega^*_{(A,M_A)}$ we recall that a pre-log structure on a ring R is a map of monoids α from a symmetric monoid M to R considered a monoid via multiplication. A log ring (R, M) is a ring with a pre-log structure and a derivation to an R-module E is a pair of maps $(D, D\log)$ with $D : R \to E$ a derivation and $D\log : M \to E$ a map of monoids such that $\alpha(a) D\log a = D\alpha(a)$. There is a universal derivation of a log ring (R, M) given by

$$\Omega^1_{(R,M)} = \big(\Omega^1_R \oplus \big(R \otimes_{\mathbf{Z}} M^{gp}\big)\big)/\big\langle d\alpha(a) - \alpha(a) \otimes a | a \in M\big\rangle$$

where M^{gp} is the group completion of M and $\langle \cdots \rangle$ is the R-submodule generated by the given set. One defines $\Omega^*_{(R,M)}$ to be the usual differential graded ring $\Lambda^*_R(\Omega^1_{(R,M)})$ generated by

$$d : R \to \Omega^1_{(R,M)}, \qquad da = da \oplus 0,$$

$$d\log : M \to \Omega^1_{(R,M)}, \qquad d\log a = 0 \oplus (1 \otimes a).$$

In [131], the relation between algebraic K-theory and the de Rham–Witt complex with log poles in this situation (p is odd) is nicely expressed as a sequence

$$\begin{array}{llll} \cdots \longrightarrow & \pi_* K(K, \mathbf{Z}/p^v) & \longrightarrow W\Omega^*_{(A,M)} \otimes S_{\mathbf{Z}/p^v}(\mu_{p^v}) \\ \xrightarrow{1-F} & W\Omega^*_{(A,M)} \otimes S_{\mathbf{Z}/p^v}(\mu_{p^v}) \xrightarrow{\partial} & \cdots \end{array}$$

which is exact in degrees ≥ 1.

7.3.5 *Curves and Nil Terms*

If A is a ring, there is a close connection between finitely generated modules over A and over the polynomial ring $A[t]$. For instance, Serre's problem asks whether finitely generated projective $k[t_0, \ldots, t_n]$-modules are free when k is a field (that the answer is "yes" is the Quillen–Suslin theorem, [234, 275]). Consequently, there is a close connection between $K(A)$ and $K(A[t])$, and the map $K(A) \to K(A[t])$ is an equivalence if A is regular (finitely generated modules have finite projective dimension) and $A[s,t]$ is coherent (every finitely generated module is finitely presented), see e.g., [97] or [297] which cover a wide range of related situations.

7.3.5.1 The Algebraic K-Theory of the Polynomial Algebra

In the general case, $K(A) \to K(A[t])$ is not an equivalence, and one can ask questions about the cofiber $NK(A)$. By extending from the commutative case, we might think of $A[t]$ as the affine line on A, and so $NK(A)$ measures to what extent algebraic K-theory fails to be "homotopy invariant" over A. In the regular Noetherian case we have that $NK = 0$, which is essential for the comparison with the motivic literature which is based on homotopy invariant definitions, as those of Karoubi–Villamayor [157, 158] or Weibel [307].

The situation for topological Hochschild and cyclic homology is worse, in that $TC(A) \to TC(A[t])$ and $THH(A) \to THH(A[t])$ are rarely equivalences, regardless of good regularity conditions on A ($THH(A[t])$ is accessible through the methods of Sect. 7.3.7 below). That said, we still can get information about the K-theory nil-term $NK(A)$. Let Nil_A be the category of nilpotent endomorphisms of finitely generated projective A-modules. That is, an object of Nil_A is a pair (P, f) where P is a finitely generated module and $f\colon P \to P$ is an A-module homomorphism for which there exist an n such that the nth iterate is trivial, $f^n = 0$. The zero endomorphisms split off, giving an equivalence $K(\mathrm{Nil}_A) \simeq K(A) \vee Nil(A)$. By [108, p. 236] there is a natural equivalence $NK(A) \xrightarrow{\sim} \Sigma Nil(A)$, and it is the latter spectrum which is accessible through trace methods.

In particular, if A is a regular Noetherian $\mathbf{F}_p$-algebra, Hesselholt and Madsen [130] give a description of $Nil(A[t]/t^n)$ in terms of the big de Rham–Witt complex of Sect. 7.3.4.

7.3.5.2 Curves on K-Theory

In [18] Bloch defined a notion of p-typical curves built on the algebraic K-theory of truncated polynomial rings $k[t]/t^n$ as n varied, and established a connection to crystalline cohomology. With the connection between K-theory and topological cyclic homology of nilpotent extensions, this allows for a reinterpretation in terms of topological cyclic homology. Hesselholt redefines in [123] the *curves* $C(A)$ on $K(A)$ of a commutative ring A to be the homotopy fiber of the canonical map

$$\operatorname{holim}_{\overleftarrow{n}} \Sigma^{-1} K\big(A[t]/t^n\big) \to \Sigma^{-1} K(A).$$

When A is a $\mathbf{Z}_{(p)}$-algebra, there is a splitting

$$C(A) \simeq \prod_{\gcd(k,p)=1} C(A;\, p)$$

of $C(A)$ into copies of a spectrum $C(A;\, p)$, the *p-typical curves* on $K(A)$.

Theorem 7.3.5.1 (Hesselholt) *If A is a commutative $\mathbf{Z}/p^j$-algebra, then there is a natural equivalence*

$$C(A;\, p) \simeq TR(A;\, p).$$

If k is a perfect field of characteristic p and A is a smooth A-algebra, then the p-typical curves split as an Eilenberg–Mac Lane spectrum with homotopy groups

$$\pi_* C(A;\, p) \cong W\Omega^*_A$$

where $W\Omega^*_A$ is the de Rham–Witt complex of Sect. 7.3.4.

Theorem 7.3.5.1 completes Bloch's program and puts it in a new and more structured context. For the same reason as was noted after Theorem 7.3.4.2, Theorem 7.3.5.1 holds for regular Noetherian $\mathbf{F}_p$-algebras A.

7.3.6 The Algebraic K-Theory Novikov Conjecture

As mentioned before, topological cyclic homology was developed by Bökstedt, Hsiang and Madsen [27] in order to prove the the analog in algebraic K-theory of the Novikov conjecture on the invariance of higher signatures. Novikov's original conjecture was reformulated by Quinn in his thesis from 1970 into a question of whether a certain map

$$L(\mathbf{Z}) \wedge BG_+ \to L\big(\mathbf{Z}[G]\big),$$

called the *L-theory assembly map*, was rationally injective. Here L is a certain functor, called *L-theory* which plays a central rôle in surgery theory, and G is a group such that BG has the homotopy type of a compact manifold. See for instance the survey [82] for an overview of the Novikov conjecture.

The K-theoretic analog of the Novikov conjecture was originally proposed by Hsiang in 1983 [144], and in [27] the following is proved:

Theorem 7.3.6.1 (Bökstedt, Hsiang and Madsen) *Let G be a discrete group G whose homology is finitely generated in every degree. Then the "K-theory assembly map"*

$$K(\mathbf{Z}) \wedge BG_+ \to K\big(\mathbf{Z}[G]\big)$$

is injective on rational homotopy groups.

From this they deduce that the there is an inclusion

$$H_i(G;\mathbf{Q}) \oplus \bigoplus_{k\geq 1} H_{i-4k-1}(G;\mathbf{Q}) \subseteq K_i\big(\mathbf{Z}[G]\big) \otimes \mathbf{Q}.$$

The K-theory assembly map can be described in many ways, but in essence boils down to the obvious map

$$\begin{aligned} &Map_*\big(m_+, m_+ \wedge A\big(S^n\big)\big) \wedge G_+ \to Map_*\big(m_+, m_+ \wedge A\big(S^n\big) \wedge G_+\big) \\ &\quad = Map_*\big(m_+, m_+ \wedge A[G]\big(S^n\big)\big), \end{aligned}$$

"assembling" the A-matrix $M = (m_{ij})$ and the group element g to the $A[G]$-matrix $Mg = (m_{ij}g)$.

Bökstedt, Hsiang and Madsen's original argument is simplified in [192, 4.5], to a statement of Soulé's [267] comparing the K-theory of the integers and the p-adic integers and the rational equivalence $K(\mathbf{S}_p) \to K(\mathbf{Z}_p)$ combined with the following lemma:

Lemma 7.3.6.2 *For any discrete group G the p-completion of the assembly map*

$$K(\mathbf{S}_p) \wedge BG_+ \to K\big(\mathbf{S}_p[G]\big)$$

is split injective in the homotopy category.

This last fact in turn follows, since by Theorem 7.3.1.10 the cyclotomic trace $K(\mathbf{S}_p)_p \to TC(\mathbf{S}_p; p)_p$ is an equivalence in non-negative dimensions, and from a direct analysis of the TC-assembly map $TC(\mathbf{S}_p; p) \wedge BG_+ \to TC(\mathbf{S}_p[G])$ which shows that it is split injective after p-completion. We refer to Madsen's survey [192, 4.5] for details.

A further simplification of Bökstedt, Hsiang and Madsen's result has been given by Holger Reich (unpublished) who considers the commutative diagram

$$\begin{array}{ccccc}
K(\mathbf{Z})\wedge BG_+ & & \longrightarrow & & K\big(\mathbf{Z}[G]\big) \\
\downarrow & & & & \downarrow \\
K(\mathbf{Z})_p\wedge BG_+ & \longrightarrow & \big(K(\mathbf{Z})\wedge BG_+\big)_p & \longrightarrow & K\big(\mathbf{Z}[G]\big)_p \\
\downarrow & & \downarrow & & \downarrow \\
K(\mathbf{Z}_p)_p\wedge BG_+ & \longrightarrow & \big(K(\mathbf{Z}_p)\wedge BG_+\big)_p & \longrightarrow & K\big(\mathbf{Z}_p[G]\big)_p \\
\downarrow & & \downarrow & & \downarrow \\
TC(\mathbf{Z}_p)_p\wedge BG_+ & \longrightarrow & \big(TC(\mathbf{Z}_p)\wedge BG_+\big)_p & \longrightarrow & TC\big(\mathbf{Z}_p[G]\big)_p,
\end{array}$$

where the horizontal maps are either p-completion or assembly maps, the upper and middle vertical maps are induced by p-completion and the lower vertical maps by the cyclotomic trace. All the maps along the left hand side and the bottom are rational injections for the following reasons (in order from top to bottom): the K-groups of the integers are finitely generated [231]; the above mentioned theorem of Soulé [267]; Theorem 7.3.1.10; by the assumption that the homology groups are finitely generated in every dimension; the direct splitting alluded to above.

One should mention that the Novikov conjecture is a very sharp version of much more general conjectures that purport to give the K-theory of group rings of wider classes of groups. The isomorphism conjecture of Farrell and Jones, allows for deeper knowledge about the subgroup lattice than what you get by simply smashing the K-theory of the ring with the classifying space of the group. These set-ups have analogs in topological cyclic homology and a joint effort by Lück, Reich, Rognes and Varisco will hopefully give new insight. See also [183].

7.3.7 Pointed Monoids and Truncated Polynomial Rings

Definition 7.3.7.1 A *pointed monoid* is a monoid in $(\mathcal{S}_*, \wedge, S^0)$, or in other words, a pointed space M, a "unit map" $S^0 \to M$ and a "multiplication" $M\wedge M \to M$ satisfying unitality and associativity.

If A is an $\mathbf{S}$-algebra, the *pointed monoid ring* $A[M]$ is given by $X \mapsto A[M](X) = A(X)\wedge M$, with unit and multiplication given by the obvious maps $X = X\wedge S^0 \to A(X)\wedge M$ and

$$A[M](X)\wedge A[M](Y) = \big(A(X)\wedge M\big)\wedge\big(A(Y)\wedge M\big) \cong \big(A(X)\wedge A(Y)\big)\wedge(M\wedge M)$$

$$\to A(X\wedge Y)\wedge M = A[M](X\wedge Y).$$

If G is a simplicial group, we may consider G_+ as a pointed monoid, but we write $A[G]$ instead of $A[G_+]$.

We define the *cyclic bar construction* of a pointed monoid M as before: $B^{cy}(M) = \{[q] \mapsto M^{\wedge(q+1)}\}$ with multiplication defining the face maps and the unit giving the degeneracy maps. This is a cyclic space, and we note the natural isomorphisms

$$sd_n B^{cy}(M)^{C_n} \cong B^{cy}(M)$$

given by skipping the repetitions that necessarily have to be present in the fixed points, cf. Example 6.2.3.3. Under these isomorphisms the cyclic bar construction for pointed monoid becomes an epicyclic object, Definition 6.2.3.1. Now, writing out the definition of *THH*, we see that we have an isomorphism of cyclic spectra

$$THH\bigl(A[M], X\bigr) \cong THH\bigl(A, B^{cy} M\wedge X\bigr),$$

(the latter should be thought of as the diagonal of a bicyclic object) and under this isomorphism the restriction map $sd_n THH(A[M], X)^{C_n} \to THH(A[M], X)$ corresponds to the composite

$$\begin{aligned} sd_n THH\bigl(A, sd_n B^{cy} M\wedge X\bigr)^{C_n} &\to THH\bigl(A, sd_n B^{cy} M^{C_n}\wedge X\bigr)\\ &\cong THH\bigl(A, B^{cy} M\wedge X\bigr), \end{aligned}$$

where the first map is the obvious variant of the restriction map when there is an action on the coefficient space (in this case, the cyclic action on $sd_n B^{cy} M\wedge X$). As a matter of fact, in the equivariant framework in which much of the literature on the subject is written, the coefficients **do** come equipped with an action. This is a convenient framework, simplifying much notation, for instance, we obviously get a map of cyclic spectra

$$\underline{T}(A)\wedge B^{cy}(M) \to \underline{T}\bigl(A[M]\bigr)$$

which is an stable equivalence of underlying spectra, but in order for this to be an equivalence of fixed points one should take care to work in an equivariant setting so that deloopings with respect to non-trivial representations is implicit.

The equivalence $\underline{T}(A)\wedge B^{cy}(M) \to \underline{T}(A[M])$ makes it easy to calculate $\pi_*\underline{T}(A[M])$ when A is a simplicial ring, for then $\underline{T}(A)$ is an Eilenberg-Mac Lane spectrum and $\underline{T}(A) \simeq \bigvee_{n=0}^{\infty} \Sigma^n H(\pi_n\underline{T}(A))$, so that

$$\underline{T}\bigl(A[M]\bigr) \simeq \bigvee_{n=0}^{\infty} \Sigma^n H\bigl(\pi_n\underline{T}(A)\bigr)\wedge B^{cy} M,$$

and so $\pi_*\underline{T}(A[M]) \cong \bigoplus_{n=0}^{\infty} H_{*-n}(B^{cy} M; \pi_n\underline{T}(A))$.

If R is a commutative ring, then $\pi_*(HR \wedge B^{cy}M) \cong H_*(B^{cy}M; R) \cong HH_*^R(R[M])$ (rewrite $R[M \times M]$ as $R[M] \otimes_R R[M]$), and so for $A = H\mathbf{F}_p$ or $A = H\mathbf{Z}$ where we have Bökstedt's explicit calculations of $\underline{T}(A)$ we can rewrite the above in terms of Hochschild homology:

$$\pi_*\underline{T}\big(H\mathbf{Z}[M]\big) \cong HH_*\big(\mathbf{Z}[M]\big) \oplus \bigoplus_{n=1}^{\infty} HH_{*-2n+1}\big(\mathbf{Z}/i[M]\big)$$

and

$$\pi_*\underline{T}\big(H\mathbf{F}_p[M]\big) \cong \bigoplus_{n=0}^{\infty} HH_{*-2n}\big(\mathbf{F}_p[M]\big).$$

In order to calculate the topological cyclic homology of the truncated polynomial algebras $k[t]/t^{n+1}$ over a perfect field k of characteristic p, Hesselholt and Madsen do a thorough investigation of the equivariant structure on $\underline{T}(Hk[\Pi_n])$ and $N^{cy}(\Pi_n)$, where $\Pi_n = \{0, 1, t, t^2, \dots, t^n\}$ with $t^{n+1} = 0$, so that $k[t]/t^{n+1} \cong k[\Pi_n]$, see [129, Sect. 8 and 9] and [128], and obtain

Theorem 7.3.7.2 (Hesselholt and Madsen) *There is an isomorphism*

$$\underline{TC}\big(k[t]/(t^n)\big)\widehat{_p} \cong \Sigma^{-1}HW(k)_F \vee H\mathbf{Z}_p \vee \bigvee_{m>0} \Sigma^{2m-1}H\big(\mathbf{W}_{nm-1}(k)/V_n\mathbf{W}_{m-1}(k)\big)$$

where $\mathbf{W}_j(k) = (1 + tk[[t]])^\times/(1 + t^{j+1}k[[t]])^\times$ *is the group of truncated* big Witt vectors, *and* $V_n : \mathbf{W}_{m-1}(k) \to \mathbf{W}_{nm-1}(k)$ *is the Verschiebung map sending* $f(t) = 1 + t\sum_{i=1}^{\infty} a_i t^i$ *to* $f(t^n)$.

The result is extended to smooth (or better, regular Noetherian) $\mathbf{F}_p$-algebras in [130].

7.3.7.1 K-Theory of $\mathbf{Z}/p^n$

Non-split nilpotent extensions can not in any good way be encoded by means of pointed monoids, and are also less well understood than, say, truncated polynomial algebras. Ironically enough, we know the profinitely completed K-theory of the p-adic integers and of the prime field, but we have only partial knowledge about the K-groups of the intermediate rings $\mathbf{Z}/p^n$. Using that the first p-torsion in topological Hochschild homology of the integers appears in dimension $2p - 1$ (the class λ_1 appearing in Corollary 7.3.1.2) and a comparison to Hochschild homology through a spectral sequence like Lemma 4.2.3.7, Brun [43, Theorem 6.1] overcomes part of the problem with the extension being non-split through filtration techniques.

Theorem 7.3.7.3 (Brun) *Let A be a simplicial ring with an ideal I satisfying $I^m = 0$ and with both A and A/I flat. Then the square*

$$\begin{array}{ccc} K(A) & \longrightarrow & HH(A)^{h\mathbb{T}} \\ \downarrow & & \downarrow \\ K(A/I) & \longrightarrow & HH(A/I)^{h\mathbb{T}} \end{array}$$

is $p/(m-1)-1$-cartesian after p-completion. Here $HH(A)^{h\mathbb{T}}$ is the homotopy circle fixed point spectrum of the Eilenberg–Mac Lane spectrum associated with Hochschild homology.

Brun states this in terms of (shifted) cyclic homology groups, and concludes after a calculation of cyclic homology groups that

Corollary 7.3.7.4 *For $0 < i < p-2$ the K-groups of $\mathbf{Z}/p^n$ are zero in even dimensions, and the odd groups are given by $K_i(\mathbf{Z}/p^n) \cong \mathbf{Z}/p^{j(n-1)}(p^j-1)$ when $i = 2j-1$.*

In a recent preprint, V. Angeltveit [6] goes significantly further (also by filtration methods):

Theorem 7.3.7.5 (Angeltveit) *Let $n > 0$. The K-groups of $\mathbf{Z}/p^n$ are finite in positive dimensions. Furthermore the order of the groups satisfy*

$$\frac{|K_{2i-1}(\mathbf{Z}/p^n)|}{|K_{2i-2}(\mathbf{Z}/p^n)|} = p^{(n-1)i}\left(p^i - 1\right)$$

for all $i \geq 2$.

Angeltveit also determines the groups $K_q(\mathbf{Z}/p^n)$ for $q \leq 2p-2$.

7.3.8 Spherical Group Rings and Thom Spectra

Recall the identification of the topological cyclic homology of spherical group rings from Sect. 6.4.2.3. There we saw that if G is a simplicial group, then the restriction map $\underline{T}(\mathbf{S}[G])^{C_{p^n}} \to \underline{T}(\mathbf{S}[G])^{C_{p^{n-1}}}$ had a splitting S, and eventually that there was a homotopy cartesian square

$$\begin{array}{ccc} \underline{TC}(\mathbf{S}[G];p) & \longrightarrow & \underset{\overleftarrow{\mathrm{trf}}}{\mathrm{holim}}\,\underline{T}(\mathbf{S}[G])_{hC_{p^n}} \\ \downarrow & & \downarrow \\ \underline{T}(\mathbf{S}[G]) & \xrightarrow{FS-1} & \underline{T}(\mathbf{S}[G]) \end{array}$$

in the homotopy category, where the homotopy limit is over the transfer maps. Just as for Example A.7.6.4, we get that $BC_{p^\infty} \simeq_p BS^1$ implies that we may (after p-completion) exchange the upper right corner with the S^1-homotopy orbit spectrum $S^1 \wedge \underline{T}(\mathbf{S}[G])_{hS^1} \simeq (\Sigma^\infty S^1 \wedge \Lambda BG_+)_{hS^1}$, and that we get a homotopy cartesian diagram

$$\begin{array}{ccc} \underline{TC}(\mathbf{S}[G]; p) & \longrightarrow & (\Sigma^\infty S^1 \wedge \Lambda BG_+)_{hS^1} \\ \downarrow & & \downarrow \\ \Sigma^\infty \Lambda BG_+ & \xrightarrow{\Delta_p - 1} & \Sigma^\infty \Lambda BG_+ \end{array} \tag{7.5}$$

after p-completion, where $\Delta_p \colon \Lambda BG \to \Lambda BG$ is precomposition with the p-th power $S^1 \to S^1$ (an argument is presented in [192, 4.4.9 and 4.4.11]. See also [242, 1.12]).

7.3.8.1 The Whitehead Spectrum of a Point

Given that the topological cyclic homology of both $\mathbf{S}$ and $\mathbf{Z}$ has been calculated, one knows the "difference" between the K-theory of $\mathbf{S}$ and $\mathbf{Z}$ if one understands the linearization map $TC(\mathbf{S}) \to TC(\mathbf{Z})$ and the cyclotomic trace $K(\mathbf{Z}) \to TC(\mathbf{Z})$.

Let G be a simplicial group and let $X = BG$. By Waldhausen [302], there is a splitting $K(\mathbf{S}[G]) \simeq \mathrm{Wh}^{\mathrm{Diff}}(X) \vee \Sigma^\infty X_+$ (the map to the $\Sigma^\infty X_+$-piece is related to the trace to topological Hochschild homology), where the so-called smooth Whitehead spectrum $\mathrm{Wh}^{\mathrm{Diff}}(X)$ is strongly related to smooth pseudo-isotopies of manifolds.

If one likewise splits off $\mathbf{S}$ from $TC(\mathbf{S})$, so that $TC(\mathbf{S}) \simeq \widetilde{TC}(\mathbf{S}) \vee \mathbf{S}$, one obtains a homotopy cartesian diagram

$$\begin{array}{ccc} \mathrm{Wh}^{\mathrm{Diff}}(*) & \longrightarrow & K(\mathbf{S}) \\ \downarrow & & \downarrow \mathrm{trc} \\ \widetilde{TC}(\mathbf{S}) & \longrightarrow & TC(\mathbf{S}) \end{array}.$$

Letting $\mathbf{CP}^\infty_{k-1}$ be the truncated complex projective space with one cell in each even dimension greater than $2k$, we get a stable equivalence $\Sigma^\infty \mathbf{CP}^\infty_k \simeq Th(k\gamma^1)$ to the Thom spectrum of k times the canonical line bundle over $\mathbf{CP}^\infty$. The right hand side makes sense for negative k as well, and it is customary to write $\mathbf{CP}^\infty_k$ even for negative k. Knapp identifies $\Sigma\mathbf{CP}^\infty_{-1}$ with the homotopy fiber of the "S^1-transfer" (the right vertical map in the homotopy cartesian diagram (7.5) giving $\underline{TC}(\mathbf{S}; p)$ above), and so there is an equivalence

$$\Sigma\mathbf{CP}^\infty_{-1} \simeq_p \widetilde{TC}(\mathbf{S})$$

after p-completion.

Rognes analyzes the cohomology of the Whitehead spectrum in two papers, [242] which gives the 2-primary information and [243] which gives the information at odd regular primes. For simplicity (and for a change) we focus on the 2-primary results.

In [242] trace information is combined with information about the K-theory of the integers from the Milnor conjecture, giving

Theorem 7.3.8.1 (Rognes) *Let* hofib(trc) *be the homotopy fiber of the cyclotomic trace map $K(\mathbf{S})_2 \to TC(\mathbf{S})_2$ completed at 2. In positive dimensions it has homotopy groups given by the table*

j	0	1	2	3	4	5	6	7
$\pi_{8k+j}(\text{hofib(trc)})$	0	0	$\mathbf{Z}_2$	$\mathbf{Z}/16$	$\mathbf{Z}/2$	$\mathbf{Z}/2$	$\mathbf{Z}_2$	$\mathbf{Z}/2^{v_2(k+1)+4}$

The only other nonzero homotopy groups are in dimension -2 *and* 0, *where there is a copy of* $\mathbf{Z}_2$. *After* 2*-completion there is a cofiber sequence*

$$\mathbf{CP}^{\infty}_{-1} \to \text{hofib(trc)} \to \text{Wh}^{\text{Diff}}(*).$$

Hence, calculating $\text{Wh}^{\text{Diff}}(*)$ is dependent upon understanding $\mathbf{CP}^{\infty}_{-1}$. The homotopy groups are hard to calculate, but Rognes does obtain the 2-primary part of the homotopy groups of the smooth Whitehead spectrum in dimensions up to 20. Instead, he calculates the mod 2 spectrum cohomology

Theorem 7.3.8.2 (Rognes) *The mod* 2 *cohomology of the smooth Whitehead spectrum fits in an extension of left modules over the Steenrod algebra* $\mathcal{A}$

$$\Sigma^{-2}C/\mathcal{A}(Sq^1, Sq^3) \longrightarrow H^*(\text{Wh}^{\text{Diff}}(*)) \longrightarrow \Sigma^3\mathcal{A}/\mathcal{A}(Sq^1, Sq^2),$$

where $C \subseteq \mathcal{A}$ is the annihilator ideal of the generator for $H^*(\mathbf{CP}^{\infty}_{-1})$. There exists just two extensions of $\mathcal{A}$-modules of this sort, and $H^*(\text{Wh}^{\text{Diff}}(*))$ fits in the nontrivial extension.

7.3.8.2 Thom Spectra

Another line of development generalizing the case of spherical group rings is the investigation by Blumberg, Cohen and Schlichtkrull [20] of topological Hochschild homology of Thom spectra. This has many refinements, but the case simplest to state is the following [20, Corollary 1.1]:

Theorem 7.3.8.3 (Blumberg, Cohen and Schlichtkrull) *Let G be either of the infinite dimensional Lie groups O, SO, $Spin$, U or Sp. Then there is a stable equivalence*

$$\underline{T}(MG) \simeq MG \wedge BBG_+.$$

These equivalences arise as chains of natural equivalences respecting the E_∞-structure (see references to preprints in [20]). The multiplicative structure of topological Hochschild homology needed for the more refined versions of the theorem above, rely on the fact [45] that if A is an $\mathbf{S}$-algebra with a so-called E_n-structure (that is, the multiplication comes from an action by an operad equivalent to the little n-cubes operad; roughly saying that A is homotopy commutative with homotopy coherency down to the nth level), then $THH(A)$ has an E_{n-1}-structure.

Unfortunately, a priori Theorem 7.3.8.3 does not tell us much about topological cyclic homology (we do not know of a circle action on $MG\wedge BBG_+$ making the equivalence of Theorem 7.3.8.3 equivariant), and more work is needed in this direction. However, the results are strong enough to give information about homotopy fixed point spectra, and in many applications this is enough. In particular, in the preprint [185] the following is proved:

Theorem 7.3.8.4 (Lunøe-Nilsen and Rognes) *The canonical map*

$$THH(MU)^{C_{p^n}} \to THH(MU)^{hC_{p^n}}$$

is a stable equivalence after p-completion.

7.3.9 Topological Cyclic Homology of Schemes and Excision

In [92], Geisser and Hesselholt extend the definition of topological cyclic homology to schemes by applying Thomason's extension of the Godement construction. This definition is not the same as what you get by applying topological cyclic homology to (Waldhausen's S-construction on) the category of vector bundles on X, however a preprint [23] of Blumberg and Mandell show that Geisser and Hesselholt' definition agrees with the topological cyclic homology of a certain category of perfect complexes over X enriched in symmetric spectra.

Geisser and Hesselholt prove that if $A \to B$ is an étale map of commutative rings, then the induced map

$$HB\wedge_{HA}\underline{T}(A) \to \underline{T}(B)$$

is an equivalence, from which it follows that their construction agrees with the original definition in the affine case:

$$TC(A;p) \xrightarrow{\sim} TC\big((\mathrm{Spec})_{\text{ét}};p\big).$$

In many cases, the topology is not really important: if X is quasi-compact and quasi separated, then [92, Corollary 3.3.4] states that $TC(X_{\text{ét}};p)$ is equivalent to $TC(X_\tau;p)$ for any topology τ coarser than the étale topology.

The cyclotomic trace extends to Thomason's definition [284] of the algebraic K-theory

$$K(X) \to TC(X;p).$$

From the construction, one obtains a descent spectral sequence, which is the basis for a comparison to the étale K-theory mentioned above. Knowledge about the topological cyclic homology for fields of finite characteristic yields [94, Theorem A]:

Theorem 7.3.9.1 (Geisser and Hesselholt) *Let X be a smooth and proper scheme over a Henselian discrete valuation ring of mixed characteristic. If the residue characteristic is p, then for all integers q and $v \geq 1$, the cyclotomic trace induces an isomorphism*

$$K^{\acute{e}t}(X, \mathbf{Z}/p^v) \xrightarrow{\sim} TC(X; p, \mathbf{Z}/p^v).$$

These results are stated with "finite coefficients", i.e., after smashing with the Moore spectrum $\mathbf{S}/p^v$.

If the schemes in question are not required to be smooth, the situation is very different, unless one focuses on situation where one is working with coefficients that avoid troublesome primes, like in [305] and [304]. General excision for closed embeddings is covered rationally by Cortiñas in [57]. The line of argument is interesting in the context of these notes, in that Cortiñas approximates by means of nilpotent extensions, using ideas of among others Cuntz and Quillen [58] and thereby getting a comparison with cyclic homology. To tackle the difference between the nilpotent approximations and the problem at hand, Cortiñas adapts the technique of Suslin and Wodzicki [279]. See also [278].

The result was shown to hold also after p-completion by Geisser and Hesselholt [93]. The results extend to $\mathbf{S}$-algebras by techniques very similar to those discussed in the main body of the text, and this extension was undertaken in [67] and [68] resulting in

Theorem 7.3.9.2 *Let*

$$\mathcal{A} = \left\{ \begin{array}{ccc} A^0 & \longrightarrow & A^1 \\ \downarrow & & \downarrow f^1 \\ A^2 & \xrightarrow[f^2]{} & A^{12} \end{array} \right\}$$

be a homotopy cartesian square of connective $\mathbf{S}$-algebras and 0-connected maps. Then the resulting cube

$$trc_{\mathcal{A}} \colon K(\mathcal{A}) \to TC(\mathcal{A})$$

is homotopy cartesian.

Notice that there are no commutativity requirements. The geometric counterpart of the requirement that the maps are 0-connected is that this is excision for closed embeddings, and in all but the full integral statement it is known that it suffices that just one of the maps are 0-connected.

As an example of how this closed excision property is useful for concrete calculations, Hesselholt [126] uses the excision result to calculate the K-theory of the coordinate axes in the following sense

Theorem 7.3.9.3 (Hesselholt) *Let k be a regular $\mathbf{F}_p$-algebra. Then there is a canonical isomorphism between $K_q(k[x,y]/(xy))$ and the group $K_q(k) \oplus \bigoplus_{m\geqslant 1} \mathbf{W}_m\Omega_k^{q-2m}$, where $\mathbf{W}_m\Omega_k^j$ is the group of big de Rham–Witt j-forms of k.*

Appendix
Homotopical Foundations

Part of the reason for the existence of this book is that, when writing down the proof of the local correspondence between algebraic K-theory and topological cyclic homology (Theorem 7.0.0.2), we found that we needed quite a number of results that were either not in the literature (but still probably well known), or else appearing in a context that was just *similar* to the one we needed. Though not adding mathematical content, the effort to fit all these pieces together turned out to be more formidable than we had anticipated.

Much of this effort has made it into the preceding chapters, but certain topics—i.e., those included in this appendix—were needed at places where the flow of ideas would be severely disrupted if one were to digress into them, and yet others are used at several places that are logically independent of each other.

We collect these results in this appendix, along with as much background as is convenient for readability and for setting the notation. Most standard results are referred away (but stated for reference in the form we need them), and we only provide proofs when no convenient reference was available, or when the proofs have some independent interest.

For a general background on simplicial techniques the reader may consult the books of May [201], Gabriel and Zisman [89], Bousfield and Kan [40] or Goerss and Jardine [99]. For model categories, the books of Quillen [235], Hovey [139], Hirschhorn [135] and Dwyer, Hirschhorn and Kan [78] are all warmly recommended. For questions pertaining to algebraic topology, one may consult Spanier [269], Hatcher [120] or May [199] (the two latter available online). Finally, the basics of category theory are nicely summed up by Mac Lane [191] and Borceux [31–33].

A.1 Foundations

A.1.1 The Category Δ

Let Δ be the category consisting of the finite ordered sets $[n] = \{0 < 1 < 2 < \dots < n\}$ for every non-negative integer n, and monotone (non-decreasing) maps.

B.I. Dundas et al., *The Local Structure of Algebraic K-Theory*,
Algebra and Applications 18,
DOI 10.1007/978-1-4471-4393-2,

In particular, for $0 \le i \le n$ we have the maps

$$d^i : [n-1] \to [n], \qquad d^i(j) = \begin{cases} j & j < i, \\ j+1 & i \le j, \end{cases} \quad \text{“skips } i\text{”},$$

$$s^i : [n+1] \to [n], \qquad s^i(j) = \begin{cases} j & j \le i, \\ j-1 & i < j, \end{cases} \quad \text{“hits } i \text{ twice”}.$$

Every map in Δ has a factorization in terms of these maps. Given $\phi \in \Delta([n],[m])$, let $\{i_1 < i_2 < \cdots < i_k\} = [m] - im(\phi)$, and $\{j_1 < j_2 < \cdots < j_l\} = \{j \in [n] | \phi(j) = \phi(j+1)\}$. Then

$$\phi(j) = d^{i_k} d^{i_{k-1}} \cdots d^{i_1} s^{j_1} s^{j_2} \cdots s^{j_l}(j).$$

This factorization is unique, and hence we could describe Δ as being generated by the maps d^i and s^i subject to the “cosimplicial identities”:

$$d^j d^i = d^i d^{j-1} \quad \text{for } i < j,$$

$$s^j s^i = s^{i-1} s^j \quad \text{for } i > j$$

and

$$s^j d^i = \begin{cases} d^i s^{j-1} & \text{for } i < j, \\ id & \text{for } i = j, j+1, \\ d^{i-1} s^j & \text{for } i > j+1. \end{cases}$$

A.1.2 Simplicial and Cosimplicial Objects

If $\mathcal{C}$ is a category, the *opposite* category, $\mathcal{C}^o$, is the category you get by letting $ob\mathcal{C}^o = ob\mathcal{C}$, but where you have reversed the direction of all arrows: $\mathcal{C}^o(c, c') = \mathcal{C}(c', c)$ (with the obvious units and compositions). A functor from $\mathcal{C}^o$ is sometimes called a *contravariant functor*.

If $\mathcal{C}$ is any category, a *simplicial $\mathcal{C}$-object* (or *simplicial object in $\mathcal{C}$*) is a functor $\Delta^o \to \mathcal{C}$, and a *cosimplicial $\mathcal{C}$-object* is a functor $\Delta \to \mathcal{C}$.

If X is a simplicial object, we let X_n be the image of $[n]$, and for a map $\phi \in \Delta$ we will often write ϕ^* for $X(\phi)$. For the particular maps d^i and s^i, we write simply d_i and s_i for $X(d^i)$ and $X(s^i)$, and call them *face* and *degeneracy maps*. Note that the face and degeneracy maps satisfy the “simplicial identities” which are the duals of the cosimplicial identities. Hence a simplicial object is often defined in the literature to be a sequence of objects X_n and maps d_i and s_i in $\mathcal{C}$ satisfying these identities.

Dually, for a cosimplicial object X, we let $X^n = X([n])$, $\phi_* = X(\phi)$, and the coface and codegeneracy maps are written d^i and s^i.

A map between two (co)simplicial $\mathcal{C}$-objects is a natural transformation. Generally, we let $s\mathcal{C}$ and $c\mathcal{C}$ be the categories of simplicial and co-simplicial $\mathcal{C}$-objects.

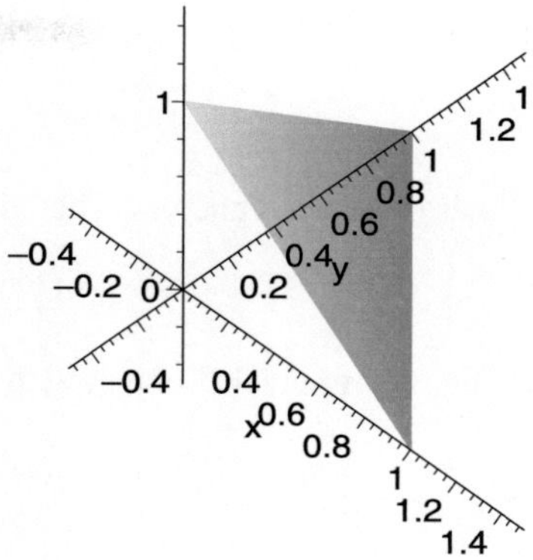

Fig. A.1 The standard topological 2-simplex $\Delta^2 \in \mathbf{R}$

Functor categories like $s\mathcal{C}$ and $c\mathcal{C}$ inherit limits and colimits from $\mathcal{C}$ (and in particular sums and products), when these exist. We say that (co)limits are formed degreewise.

Example A.1.2.1 (The Topological Standard Simplices) There is an important cosimplicial topological space $[n] \mapsto \Delta^n$, where Δ^n is the *standard topological n-simplex*

$$\Delta^n = \left\{(x_0, \dots, x_n) \in \mathbf{R}^{n+1} \,\middle|\, \sum x_i = 1, x_i \geq 0\right\}$$

see Fig. A.1 for the case $n = 2$, with

$$d^i(x_0, \dots, x_{n-1}) = (x_0, \dots, x_i, 0, x_{i+1}, \dots, x_{n-1}),$$

$$s^i(x_0, \dots, x_{n+1}) = (x_0, \dots, x_{i-1}, x_i + x_{i+1}, x_{i+2}, \dots, x_{n+1}).$$

A.1.3 Resolutions from Adjoint Functors

Adjunctions are an important source of (co)simplicial objects. Recall that an *adjunction* is a pair of functors

$$\mathcal{D} \underset{U}{\overset{F}{\rightleftarrows}} \mathcal{C}$$

together with a natural bijection of morphism sets

$$\mathcal{C}\big(F(d), c\big) \cong \mathcal{D}\big(d, U(c)\big)$$

(the bijection is part of the data, but when there is little chance for confusion we often refer to an adjunction as an *adjoint pair* and list only the functors). The bijection can alternatively be given by declaring the *unit* $\sigma_d \colon d \to UF(d)$ (corresponding to $id_{F(d)} \in \mathcal{C}(F(d), F(d))$) and *counit* $\delta_c \colon FU(c) \to c$ (corresponding to $id_{U(c)} \in \mathcal{D}(U(c), U(c))$).

Let

$$\mathcal{D} \underset{U}{\overset{F}{\rightleftarrows}} \mathcal{C}$$

be a pair of adjoint functors. Then the assignment

$$[q] \mapsto (FU)^{q+1}(c)$$

underlies a simplicial $\mathcal{C}$-object with structure maps defined by

$$d_i = (FU)^i \delta_{(FU)^{q-i+1}} : (FU)^{q+2}(c) \to (FU)^{q+1}(c)$$

and

$$s_i = (FU)^i F\sigma_{U(FU)^{q-i}} : (FU)^q(c) \to (FU)^{q+1}(c).$$

Dually, $[q] \mapsto (UF)^{q+1}(d)$ defines a cosimplicial $\mathcal{D}$-object. These (co)simplicial objects are called the *(co)simplicial resolutions associated with the adjoint pair.*

The composite $T = UF$ (together with the associated natural transformations $1 \to T$ and $TT \to T$) is occasionally referred to as a *triple* or *monad* (probably short for "monoid in the monoidal category of endofunctors and composition"), and likewise FU a *cotriple* or *comonad*, but never mind: the important thing to us are the associated (co)simplicial resolutions.

A.2 Simplicial Sets

Let *Ens* be the category of sets (when we say "sets" they are supposed to be small in some fixed universe, see e.g., [162, 2.6 and 3.11] for some comments on enlargements of universes). Let $\mathcal{S} = s\mathit{Ens}$ the category of simplicial sets. Since all (co)limits exist in *Ens*, all (co)limits exist in $\mathcal{S}$. The category of simplicial sets has close connections with the category *Top* of topological spaces. In particular, the realization and singular functors (see Sect. A.2.1) induce equivalences between their respective "homotopy categories" (see Sect. A.4.3 below).

In view of this equivalence, we let a "*space*" mean a simplicial set (unless explicitly called a *topological* space). We also have a pointed version. A *pointed set* is a set with a preferred element, called the base point, and a pointed map is a map respecting base points. The category of pointed spaces (= pointed simplicial sets = simplicial pointed sets) is denoted $\mathcal{S}_*$. Being a category of functors to sets, the category $\mathcal{S}_*$ has (co)limits. In particular the coproduct is the wedge

$$X \vee Y = X \coprod_* Y$$

and we define the *smash* by

$$X \wedge Y = X \times Y / X \vee Y.$$

If $X \in \mathcal{S}$ we can add a disjoint basepoint and get the pointed simplicial set

$$X_+ = X \coprod *.$$

Clearly, the assignment $X \mapsto X_+$ underlies a functor $\mathcal{S} \to \mathcal{S}_*$ left adjoint to the forgetful functor $\mathcal{S}_* \to \mathcal{S}$.

A.2.1 Simplicial Sets vs. Topological Spaces

There are adjoint functors

$$Top \underset{\sin}{\overset{|-|}{\leftrightarrows}} \mathcal{S}$$

defined as follows. For $Y \in Top$, the *singular functor* is defined as

$$\sin Y = \{[n] \mapsto Top(\Delta^n, Y)\}$$

(the set of unbased continuous functions from the topological standard simplex to Y). As $[n] \mapsto \Delta^n$ is a cosimplicial space, this becomes a simplicial set. For $X \in \mathcal{S}$, the *realization functor* is defined as

$$|X| = \left(\coprod_n X_n \times \Delta^n\right) / (\phi^* x, u) \sim (x, \phi_* u)$$

(for $\phi \in \Delta([m],[n])$, $u \in \Delta^m$ and $x \in X_n$). The realization functor is left adjoint to the singular functor via the bijection

$$Top(|X|, Y) \cong \mathcal{S}(X, \sin Y)$$

induced by the maps

$$X \longrightarrow \sin|X|$$

$$x \in X_n \mapsto \left(\Delta^n \overset{u \mapsto (x,u)}{\longrightarrow} X_n \times \Delta^n \to |X|\right) \in \sin|X|_n$$

and

$$|\sin Y| \longrightarrow Y$$

$$(y, u) \in \sin(Y)_n \times \Delta^n \mapsto y(u) \in Y.$$

From the adjointness we see that the singular functor preserves all limits and the realization functor preserves all colimits.

What is not formal, but very useful, is the following result. See e.g., [139, 3.1.8], or alternatively see [61] and its references, where the geometric realization is written as a filtered colimit of metric spaces.

Proposition A.2.1.1 *The geometric realization of a simplicial set is compactly generated and Hausdorff. As a functor to the category of compactly generated (weak Hausdorff) spaces, the geometric realization preserves finite limits.* ☺

The singular and realization functors also define adjoint functors between the category of simplicial pointed sets, $\mathcal{S}_*$ and the category of pointed topological spaces, Top_*.

Definition A.2.1.2 If $x \in X \in \mathcal{S}$, we define the homotopy groups to be those of the realization:

$$\pi_*(X, x) = \pi_*\big(|X|, x\big).$$

In the based situation we simply write $\pi_*(X)$.

Definition A.2.1.3 A space X is 0-*connected* (or simply *connected*) if $\pi_0 X$ is a point, and if it is connected it is k-*connected* for a $k > 0$ if for all vertices $x \in X_0$ we have that $\pi_q(X, x) = 0$ for $0 \le q \le k$. A space is -1-connected by definition if it is nonempty. A space X is k-*reduced* if $X_j = *$ for all $j < k$. A space is *reduced* if it is 1-reduced.

A map $X \to Y$ is k-*connected* if its homotopy fiber (defined in Sect. A.5 below) over each 0-simplex is $(k-1)$-connected.

So a k-reduced space is $(k-1)$-connected.

A.2.2 The Standard Simplices, and Homotopies

We define a cosimplicial space (cosimplicial simplicial set)

$$[n] \mapsto \Delta[n] = \big\{[q] \mapsto \Delta\big([q], [n]\big)\big\}.$$

The spaces $\Delta[n]$ are referred to as the *standard simplices*. Note that the realization $|\Delta[n]|$ of the standard simplex may be identified with Δ^n, the *topological* standard simplex. The standard simplices are in a precise way the building blocks (representing objects) for all simplicial sets: if X is a simplicial set, then there is a functorial isomorphism (an instance of the Yoneda lemma)

$$\mathcal{S}\big(\Delta[n], X\big) \cong X_n, \qquad f \mapsto f\big([n] = [n]\big).$$

We let $S^1 = \Delta[1]/\partial\Delta[1]$, where $\partial\Delta[1]$ is the discrete subspace $\{0, 1\}$ of the vertices of $\Delta[1]$ (the "endpoints of the (simplicial) interval"). To us, the q-sphere is the q-fold smash $S^q = S^1 \wedge \cdots \wedge S^1$. This is not equal to the competing model $\Delta[q]/\partial\Delta[q]$ used in other texts, but their geometric realizations are homeomorphic.

A (simplicial) *homotopy* between two maps $f_0, f_1\colon X \to Y \in \mathcal{S}$ is a map $H\colon X \times \Delta[1] \to Y$ such that the composites

$$X \cong X \times \Delta[0] \xrightarrow{id \times d^i} X \times \Delta[1] \xrightarrow{H} Y, \quad i = 0, 1$$

are f_0 and f_1. Since $|X \times \Delta[1]| \cong |X| \times |\Delta[1]|$, we see that the realization of a homotopy gives a homotopy in *Top*. The pointed version of a homotopy is a map

$$H\colon X \wedge \Delta[1]_+ \to Y$$

(the subscript $_+$ means a disjoint basepoint added).

We say that f_0 and f_1 are *strictly* (or *simplicially*) *homotopic* if there is a homotopy between them, and *homotopic* if there is a finite chain of homotopies which connect f_0 and f_1. In this way, "homotopic" forms an equivalence relation.

Another way to say this is that two maps $f_0, f_1\colon X \to Y$ are homotopic if there is a map

$$H\colon X \times I \to Y, \text{ or in the pointed case } H\colon X \wedge I_+ \to Y$$

which is equal to f_0 and f_1 at the "ends" of I, where I is a finite number of $\Delta[1]$s glued together at the endpoints, i.e., for some sequence of numbers $i_j \in \{0, 1\}$, $1 \leq j \leq n$, I is the colimit of

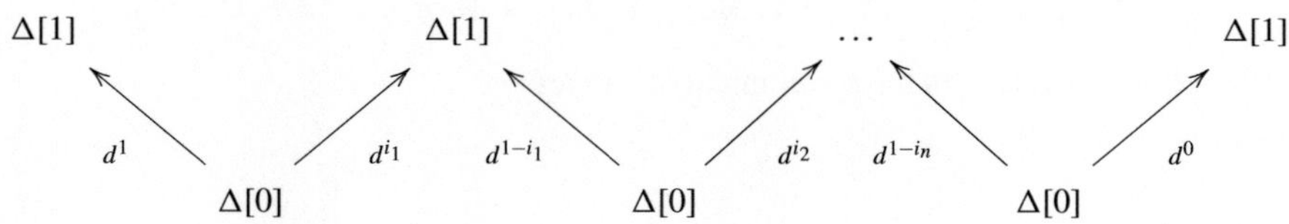

We still denote the two end inclusions $d^0, d^1\colon * = \Delta[0] \to I$.

We note that elements in $\pi_1(X)$ can be represented by maps $\alpha\colon I \to X$ such that $\alpha d^0 = \alpha d^1 = 0$.

A.2.3 Function Spaces

We define the simplicial function space of maps from X to Y to be the simplicial set

$$\underline{\mathcal{S}}(X, Y) = \{[q] \mapsto \mathcal{S}(X \times \Delta[q], Y)\},$$

where the cosimplicial structure of the standard simplices $[q] \to \Delta[q]$ makes this into a simplicial set. In the pointed case we set

$$\underline{\mathcal{S}_*}(X, Y) = \{[q] \mapsto \mathcal{S}_*(X \wedge \Delta[q]_+, Y)\}.$$

We reserve the symbol Y^X for the pointed case: $Y^X = \underline{\mathcal{S}_*}(X, Y)$, and so $Y^{X_+} = \underline{\mathcal{S}}(X, Y)$. Unfortunately, these definitions are not homotopy invariant; for instance, the weak equivalence $B\mathbf{N} \to \sin|B\mathbf{N}|$ does not induce an equivalence $\underline{\mathcal{S}_*}(S^1, B\mathbf{N}) \to \underline{\mathcal{S}_*}(S^1, \sin|B\mathbf{N}|)$ (on π_0 it is the inclusion $\mathbf{N} \subset \mathbf{Z}$). To remedy this we define

$$Map(X, Y) = \underline{\mathcal{S}}\big(X, \sin|Y|\big)$$

and

$$Map_*(X, Y) = \underline{\mathcal{S}_*}\big(X, \sin|Y|\big).$$

In fact, using the adjointness of the singular and realization functors we see that

$$\begin{aligned} Map(X, Y) &\cong \big\{[q] \mapsto Top\big(|X| \times |\Delta[q]|, |Y|\big)\big\} \cong \big\{[q] \mapsto Top\big(\Delta^q, Top(|X|, |Y|)\big)\big\} \\ &= \sin\big(Top\big(|X|, |Y|\big)\big) \end{aligned}$$

and likewise in the pointed case. These function spaces still have some sort of adjointness properties, in that

$$Map(X \times Y, Z) \cong \mathcal{S}\big(X, Map(Y, Z)\big) \xrightarrow{\sim} Map\big(X, Map(Y, Z)\big)$$

and

$$Map_*(X \wedge Y, Z) \cong \underline{\mathcal{S}_*}\big(X, Map_*(Y, Z)\big) \xrightarrow{\sim} Map_*\big(X, Map_*(Y, Z)\big)$$

where the equivalences have canonical left inverses.

A.2.4 The Nerve of a Category

For every $n \geq 0$, regard $[n] = \{0 < 1 < \cdots < n\}$ as a category (if $a \leq b$ there is a unique map $a \leftarrow b$: beware that many authors let the arrow point in the other direction. The choice of convention does not matter to the theory). Furthermore, we identify the maps in Δ with the corresponding functors, so that Δ sits as a full subcategory of the category of (small) categories.

Definition A.2.4.1 The *nerve* $\mathbf{N}\mathcal{C}$ of the small category $\mathcal{C}$ is the simplicial category

$$[q] \mapsto \mathbf{N}_q\mathcal{C} = \big\{\text{category of functors } [q] \to \mathcal{C}\big\}.$$

The nerve is a functor from the category of small categories to simplicial categories.

We see that the set of objects $B_q\mathcal{C} = ob\mathbf{N}_q\mathcal{C}$, is the set of all chains $c_0 \leftarrow c_1 \leftarrow \cdots \leftarrow c_q$ in $\mathcal{C}$, and in particular $B_0\mathcal{C} = ob\mathcal{C}$. Frequently, the underlying simplicial set $B\mathcal{C} = ob\mathbf{N}\mathcal{C}$ is also referred to as the nerve or *classifying space* of $\mathcal{C}$. Note that there is a unique isomorphism $B[q] \cong \Delta[q]$.

The classifying space functor B, as a functor from small categories to spaces, has a left adjoint given by sending a simplicial set X to the category CX defined as follows. The set of objects is X_0. The set of morphisms is generated by X_1, where $y \in X_1$ is regarded as an arrow $y\colon d_0y \to d_1y$, subject to the relations that $s_0x = 1_x$ for every $x \in X_0$, and for every $z \in X_2$

$$\begin{array}{ccc} d_0d_0z & \xrightarrow{\;d_0z\;} & d_1d_0z \\ & {\scriptstyle d_1z}\searrow \quad \swarrow{\scriptstyle d_2z} & \\ & d_1d_1z & \end{array}$$

commutes. The obvious functor $CB\mathcal{C} \to \mathcal{C}$ is an isomorphism, and the induced function from $\mathcal{S}(X, B\mathcal{C})$ to the set of functors $CX \to \mathcal{C}$ (sending f to $CX \stackrel{Cf}{\to} CB\mathcal{C} \cong \mathcal{C}$) is an isomorphism too, giving the adjunction. Notice the similarity between the classifying space $B_q\mathcal{C} = \mathcal{C}at([q], \mathcal{C})$ and the singular functor $\sin(Y)_q = Top(\Delta^q, Y)$. Their left adjoints can both be expressed as left Kan extensions along the Yoneda embedding $[q] \mapsto \Delta[q]$:

$$\begin{array}{ccc} \Delta & \xrightarrow{[q]\mapsto\Delta^q} & Top, \\ {\scriptstyle [q]\mapsto\Delta[q]}\downarrow & \nearrow{\scriptstyle X\mapsto|X|} & \\ \mathcal{S} & & \end{array} \qquad \begin{array}{ccc} \Delta & \xrightarrow{[q]\mapsto[q]} & \mathcal{C}at \\ {\scriptstyle [q]\mapsto\Delta[q]}\downarrow & \nearrow{\scriptstyle C} & \\ \mathcal{S}. & & \end{array}$$

The classifying space B is a full and faithful functor (recall that a functor $F\colon \mathcal{C} \to \mathcal{D}$ is *full* (resp. *faithful*) if it induces a surjection (resp. injection) $\mathcal{C}(c, c') \to \mathcal{D}(F(c), F(c'))$ of morphism sets for $c, c' \in ob\mathcal{C}$). The reader should be warned that the functor $\mathcal{C} \mapsto |B\mathcal{C}|$ to topological spaces is not full.

A.2.4.1 Natural Transformations and Homotopies

The classifying space takes natural transformations to homotopies: if $\eta\colon F_1 \to F_0$ is a natural transformation of functors $\mathcal{C} \to \mathcal{D}$, regard it as a functor $\eta\colon \mathcal{C} \times [1] \to \mathcal{D}$ with $\eta(c, k) = F_k(c)$ for $c \in ob\mathcal{C}$ and $k \in \{0, 1\}$, sending the morphism $(c \stackrel{f}{\leftarrow} c',\ 0 < 1)$ to the composites around

$$\begin{array}{ccc} F_1(c) & \xleftarrow{\;F_1(f)\;} & F_1(c') \\ \downarrow{\scriptstyle \eta_c} & & \downarrow{\scriptstyle \eta_{c'}} \\ F_0(c) & \xleftarrow{\;F_0(f)\;} & F_0(c'). \end{array}$$

Thus we have defined a homotopy between F_0 and F_1:

$$BC \times \Delta[1] \cong BC \times B[1] \cong B\big(\mathcal{C} \times [1]\big) \to B\mathcal{D},$$

where the isomorphisms are the canonical ones.

In view of this, and of the isomorphisms between the category of small categories and its image in $\mathcal{S}$ (resulting from the fact that the classifying space is full and faithful and injective on objects), it is customary to use language that normally refers to spaces to categories. For instance, a functor may be said to be a weak equivalence if the induced map of classifying spaces is.

A.2.4.2 Over and Under Categories

If $\mathcal{C}$ is a category and c an object in $\mathcal{C}$, the category over c, written $\mathcal{C}/c$, is the category whose objects are maps $f: d \to c \in \mathcal{C}$, and a morphism from f to g is a factorization $f = g\alpha$ in $\mathcal{C}$. More generally, if $F: \mathcal{C} \to \mathcal{D}$ is a functor and d is an object of $\mathcal{D}$, the *over category* F/d is the category whose objects are pairs (c, f), where $c \in ob\mathcal{C}$ and $f: F(c) \to d \in \mathcal{D}$. A morphism from (c, f) to (c', f') is a morphism $\alpha: c \to c' \in \mathcal{C}$ such that $f = f'F(\alpha)$. The *under categories* $c/\mathcal{C}$ and d/F are defined dually. Over and under categories are frequently referred to as *comma categories* in the literature.

The over category can be used to define simplicial homotopies for simplicial objects in an arbitrary category $\mathcal{C}$ as follows. A homotopy between two maps in $\mathcal{S}$ is a map $X \times \Delta[1] \to Y$. In dimension q, this is simply a function $\coprod_{\phi \in \Delta([q],[1])} X_q \cong X_q \times \Delta[1]_q \to Y_q$, or in other word a collection of functions $\eta_\phi: X_q \to Y_q$ indexed over $\phi \in \Delta([q],[1])$, satisfying compatibility conditions. This can be summarized and generalized as follows:

Definition A.2.4.2 Let X and Y be simplicial objects in a category $\mathcal{C}$, and consider two maps $F_0, F_1: X \to Y$. Let $S: \Delta/[1] \to \Delta$ be the forgetful functor, sending $\phi: [q] \to [1]$ to $[q]$. A *simplicial homotopy* from F_0 to F_1 is a natural transformation $H: X \circ S \to Y \circ S$ such that for $i = 0$ and $i = 1$, $F_i = H(\phi_i)$, where $\phi_i[q] \to [1]$ is the constant order preserving function with value i.

This makes it clear that

Lemma A.2.4.3 *Any functor $\mathcal{C} \to \mathcal{D}$, when applied degreewise to simplicial objects, takes simplicial homotopies to simplicial homotopies.* ☺

A.2.5 Filtered Colimits in $\mathcal{S}_*$

A.2.5.1 Subdivisions and Kan's Ex^∞

Consider the subcategory $\Delta_m \subset \Delta$ with all objects, but just monomorphisms. For any $n \geq 0$ we consider the (barycentric) *subdivision* of the standard n-simplex $\Delta[n]$.

To be precise, it is $B(\Delta_m/[n])$, the classifying space of the category of order preserving monomorphisms into $[n]$. For every $\phi\colon [n] \to [m] \in \Delta$ we get a functor $\phi_*\colon \Delta_m/[n] \to \Delta_m/[m]$ sending a monomorphism $\alpha \in \Delta_m([q],[n])$ to the unique monomorphism $\phi_*(\alpha)$ such that $\phi\alpha = \phi_*(\alpha)\psi$ where ψ is an epimorphism (see Sect. A.1.1). This means that $B(\Delta_m/-)$ is a cosimplicial space, and the functor $\Delta_m/[n] \to [n]$ sending $\alpha\colon [p] \to [n]$ to $\alpha(p) \in [n]$ defines a cosimplicial map to the standard simplices $\{[n] \mapsto \Delta[n] = B[n]\}$.

For any simplicial set X, Kan then defines

$$Ex(X) = \big\{[q] \mapsto \mathcal{S}\big(B\big(\Delta_m/[q]\big), X\big)\big\}.$$

This is a simplicial set, and $B(\Delta_m/[q]) \to \Delta[q]$, defines an inclusion $X \subseteq Ex(X)$. Set

$$Ex^\infty X = \lim_{\overrightarrow{k}} Ex^{(k)}(X).$$

The inclusion $X \subseteq Ex^\infty X$ is a weak equivalence, and $Ex^\infty X$ is always a *Kan complex*, that is a *fibrant* object in the sense of Sect. A.4. These constructions restrict to give a fibrant replacement also in the pointed case, and give the possibility of defining the homotopy groups without reference to topological spaces via

$$\pi_q X = \underline{\mathcal{S}_*}\big(S^q, Ex^\infty X\big)/\text{homotopy}.$$

A.2.5.2 Filtered Colimits in $\mathcal{S}_*$

Recall that a *filtered category* J is a nonempty category such that for any $j, j' \in ob J$ there are morphisms $j \to k$, $j' \to k$ in J to a common object, and such that given two morphisms $f, g\colon j \to j'$ in J there is a morphism $h\colon j' \to k$ in J such that $hf = hg$.

A *filtered colimit* is a colimit over a *filtered category* (see [191, p. 207]). Filtered colimits of sets are especially nice because they commute with finite limits (see [191, p. 211]). It immediately follows that filtered colimits commute with finite limits also for simplicial sets, and this is one of the many places we should be happy for not considering general topological spaces.

Given a space Y, its N-*skeleton* is the subspace $sk_N Y \subseteq Y$ generated by simplices in dimension less than or equal to N.

A space Y is *finite* if it has only finitely many non degenerate simplices. Alternatively finiteness can be spelled out as, $Y = sk_N Y$ for some N, Y_0 is finite, and its q-skeleton for $q \leq N$ is formed by iterated pushouts over finite sets D_q

$$\begin{array}{ccc} \bigvee_{D_q} \partial\Delta[q]_+ & \longrightarrow & \bigvee_{D_q} \Delta[q]_+ \\ \downarrow & & \downarrow \\ sk_{q-1}Y & \longrightarrow & sk_q Y \end{array}$$

(in view of the applications below we have displayed the pointed case. For the unpointed case, remove the extra base points and substitute disjoint unions for the wedges).

Lemma A.2.5.1 *Let J be a filtered category, $X\colon J \to \mathcal{S}_*$ a functor and Y a finite space. Then the canonical map*

$$\varinjlim_J \underline{\mathcal{S}_*}(Y, X) \to \underline{\mathcal{S}_*}\big(Y, \varinjlim_J X\big)$$

is an isomorphism.

Proof Since $\underline{\mathcal{S}_*}(Y, -)_q = \mathcal{S}_*(Y \wedge \Delta[q]_+, -)$ and $Y \wedge \Delta[q]_+$ is finite, it is clearly enough to prove that

$$\varinjlim_J \mathcal{S}_*(Y, X) \cong \mathcal{S}_*\big(Y, \varinjlim_J X\big).$$

Remember that filtered colimits commute with finite limits. Since Y is a finite colimit of diagrams made out of $\Delta[q]$'s, this means that it is enough to prove the lemma for $Y = \Delta[q]$, which is trivial since $\mathcal{S}_*(\Delta[q], X) = X_q$ and colimits are formed degreewise. □

Lemma A.2.5.2 *If J is a filtered category, then the canonical map*

$$\varinjlim_J Ex^\infty X \to Ex^\infty \varinjlim_J X$$

is an isomorphism.

Proof Since colimits commute with colimits, it is enough to prove that Ex commutes with filtered colimits, but this is clear since $Ex(X)_n = \mathcal{S}_*(B(\Delta_m/[n]), X)$ and $B(\Delta_m/[n])$ is a simplicial finite set equal to its n-skeleton. □

Proposition A.2.5.3 *Homotopy groups commute with filtered colimits.*

Proof Let $X \in ob\mathcal{S}_*$ and J be a filtered category. First note that π_0, being a colimit itself, commutes with arbitrary colimits. For $q \geq 0$ we have isomorphisms

$$\begin{aligned}\pi_q \varinjlim_J X &\cong \pi_0 \underline{\mathcal{S}_*}\big(S^q, Ex^\infty \varinjlim_J X\big) \cong \pi_0 \underline{\mathcal{S}_*}\big(S^q, \varinjlim_J Ex^\infty X\big) \\ &\cong \varinjlim_J \pi_0 \underline{\mathcal{S}_*}\big(S^q, Ex^\infty X\big) \cong \varinjlim_J \pi_q X.\end{aligned}$$

□

A.2.6 The Classifying Space of a Group

Let G be a (discrete) group, and regard it as a one point category whose morphisms are the group elements. Then the classifying space Sect. A.2.4 takes the simple form

$B_q G = G^{\times q}$, and BG is called the *classifying space* of the group G (which makes a lot of sense, since homotopy classes of maps into BG are in bijective correspondence with isomorphism classes of principal G-bundles). The homotopy groups of BG are given by

$$\pi_i(BG) = \begin{cases} G & \text{if } i = 1, \\ 0 & \text{otherwise,} \end{cases}$$

a fact that characterizes the space BG up to weak equivalence.

A.2.6.1 The $\overline{W}$-Construction

If G is a simplicial group there is an alternative construction for the homotopy type of $\operatorname{diag}^* BG$, called $\overline{W}G$, which is most easily described as follows: We have a functor $\sqcup \colon \Delta \times \Delta \to \Delta$ sending two ordered sets S and T to the naturally ordered disjoint union $S \sqcup T$ (elements in S are smaller than elements in T). For any simplicial set X we may consider the bisimplicial set $sd_2 X$ obtained by precomposing X with $\sqcup$ (so that the $(sd_2 X)_{p,q} = X_{p+q+1}$: the diagonal of this construction was called the (second) *edgewise subdivision* in Sect. 6.2.2). We define $\overline{W}G$ to be space with q-simplices

$$\overline{W}_q G = \{\text{bisimplicial maps } sd_2 \Delta[q] \to BG\},$$

where the simplicial structure is induced by the cosimplicial structure of $[q] \mapsto \Delta[q]$. This description is isomorphic to the one given in [201] (with reversed orientation).

A.2.6.2 Kan's Loop Group

The classifying space BG of a group G is a reduced space (i.e., it only has one zero simplex). On the category of reduced spaces X there is a particularly nice model GX for the loop functor due to Kan [155], see [201, p. 118] or [99]. If $q \geq 0$ we have that $G_q X$ is the free group generated by X_{q+1} modulo contracting the image of s_0 to the base point. The degeneracy and face maps are induced from X except the extreme face map (which extreme depends on your choice of orientation, see [201, Definition 26.3] for one choice). The Kan loop group is adjoint to the $\overline{W}$-construction described above. As a matter of fact, the adjoint pair $(G, \overline{W})$ defines something called a Quillen equivalence, which among other things implies that the homotopy category of reduced spaces is equivalent to the homotopy category of simplicial groups.

A.2.7 Path Objects

Let Y be a simplicial object in a category $\mathcal{C}$. There is a convenient combinatorial model mimicking the path space Y^I. Let $\sqcup\colon \Delta \times \Delta \to \Delta$ be the ordered disjoint union.

Definition A.2.7.1 Let Y be a simplicial object in a category $\mathcal{C}$. Then the *path object* is the simplicial object PY given by precomposing Y with $[0]\sqcup ?\colon \Delta \longrightarrow \Delta$.

Hence $P_q Y = Y_{q+1}$. The map $PY \to Y$ corresponding to evaluation is given by the natural transformation $d^0\colon id \to [0] \sqcup id$ (concretely: it is $P_q Y = Y_{q+1} \xrightarrow{d_0} Y_q$).

Lemma A.2.7.2 *The maps $Y_0 \to PY \to Y_0$ induced by the natural maps $[0] \to [0] \sqcup [q] \to [0]$ are simplicial homotopy equivalences.*

Proof That $PY \to Y_0 \to PY$ is simplicially homotopic to the identity follows, using the formulation in Definition A.2.4.2, by considering the natural transformation of functors $(\Delta/[1])^o \to \Delta^o$ sending $\phi\colon [q] \to [1]$ to $\phi_*\colon [0] \sqcup [q] \to [0] \sqcup [q]$ with $\phi_*(0) = 0$ and

$$\phi_*(j+1) = \begin{cases} 0 & \text{if } \phi(j) = 0, \\ j+1 & \text{if } \phi(j) = 1. \end{cases}$$ □

The connection to the path-space is the following: considering Δ as a subcategory of the category of small categories in the usual way, there is a projection $[1] \times [q] \to [0] \sqcup [q]$ sending $(0, j)$ to $0 \in [0] \sqcup [q]$ and $(1, j)$ to $1 + j \in [1+q] = [0] \sqcup [q]$. If X is a simplicial set, the usual path space is

$$\underline{\mathcal{S}}(\Delta[1], X) = \{[q] \mapsto \mathcal{S}(\Delta[1] \times \Delta[q], X) = \mathcal{S}(N([1] \times [q]), X)\}$$

whereas $PX = \{[q] \mapsto \mathcal{S}(N([0] \sqcup [q]), X)\}$, and the injection $PX \subseteq \underline{\mathcal{S}}(\Delta[1], X)$ is induced by the above projection.

A.2.8 Cosimplicial Spaces

Recall that a cosimplicial space is a functor $X\colon \Delta \to \mathcal{S}$. The category of cosimplicial spaces is a "simplicial category". If Z is a space and X is a cosimplicial space, then $Z \times X$ is the cosimplicial space whose value on $[q] \in \Delta$ is $Z \times X^q$. The function space

$$c\mathcal{S}(X, Y) \in \mathcal{S}$$

of maps from the cosimplicial space X to the cosimplicial space Y has q-simplices the set of maps (natural transformation of functors from $\Delta^o \times \Delta$ to sets)

$$\Delta[q] \times X \to Y.$$

The *total space* of a cosimplicial space X is the space

$$\operatorname{Tot} X = c\mathcal{S}\big(\Delta[-], X\big)$$

(where $\Delta[-]$ is the cosimplicial space whose value on $[q] \in \Delta$ is $\Delta[q] = \Delta(-, [q])$). The q-simplices are cosimplicial maps $\Delta[q] \times \Delta[-] \to X$.

A.2.8.1 The Pointed Case

In the pointed case we make the usual modifications: A pointed cosimplicial space is a functor $X\colon \Delta \to \mathcal{S}_*$, the function space $c\mathcal{S}_*(X, Y)$ has q-simplices the set of maps $\Delta[q]_+ \wedge X \to Y$, and the total space $\operatorname{Tot} X = c\mathcal{S}_*(\Delta[-]_+, X)$ is isomorphic (as an unbased space) to what you get if you forget the basepoint before taking Tot.

A.3 Spectra and Simplicial Abelian Groups

A.3.1 Simplicial Abelian Groups

Let *Ab* be the category of abelian groups. Consider the free/forgetful adjoint pair

$$Ab \underset{U}{\overset{\mathbf{Z}[-]}{\leftrightarrows}} Ens$$

where $\mathbf{Z}[X]$ is the free abelian group on the set X (i.e., the coproduct of $\mathbf{Z}$ with itself indexed over the set X). We extend this to an adjoint pair between the category $A = sAb$ of simplicial abelian groups and the category $\mathcal{S}$ of spaces. If M is a simplicial abelian group, then its homotopy groups are defined as $\pi_*(M) = \pi_*(UM, 0)$. The *homology* of a simplicial set X defined to be $H_*(X) = \pi_*(\mathbf{Z}[X])$, and this definition is naturally isomorphic to the singular homology of the realization.

In the pointed case, the adjoint pair

$$Ab \underset{U}{\overset{\tilde{\mathbf{Z}}[-]}{\leftrightarrows}} Ens_*,$$

where Ens_* is the category of pointed sets and $\tilde{\mathbf{Z}}[X] = \mathbf{Z}[X]/\mathbf{Z}[*]$, gives rise to an adjoint pair between A and $\mathcal{S}_*$, and $\tilde{\mathbf{Z}}[X]$ represent the *reduced homology*: $\tilde{H}_*(X) = \pi_*(\tilde{\mathbf{Z}}[X])$. The unit of adjunction $X = 1 \cdot X \subseteq U\tilde{\mathbf{Z}}[X]$ induces the *Hurewicz map* $\pi_*(X) \to \tilde{H}_*(X)$ on homotopy groups.

A.3.1.1 Closed Structure

The category of simplicial abelian groups has the structure of a closed category in the sense of Definition A.10.1.1. That is, we have a tensor product

$$M \otimes N = \{[q] \mapsto M_q \otimes N_q\},$$

and morphism objects

$$\underline{A}(M, N) = \{[q] \mapsto A(M \otimes \mathbf{Z}[\Delta[q]], N)\}$$

with natural isomorphisms $\underline{A}(M \otimes N, P) \cong \underline{A}(M\underline{A}(N, P))$ satisfying the necessary conditions (see Definition A.10.1.1).

A.3.1.2 Eilenberg–Mac Lane Spaces

If G is a (discrete) abelian group, then the classifying space BG of Sect. A.2.6 becomes a simplicial abelian group. Hence we may apply the construction again in every degree, and get a bisimplicial abelian group $B^2G = BBG$, and so on. Taking the diagonal, we get a sequence G, BG, $\mathrm{diag}^* BBG, \ldots$. The nth term, $\mathrm{diag}^* B^n G$, is naturally isomorphic to $\tilde{\mathbf{Z}}[S^n] \otimes G$, and is often written $H(G, n)$, and is characterized up to homotopy by having only one nonzero homotopy group $\pi_n = G$, and such spaces are called *Eilenberg–Mac Lane spaces*. We call any space (weakly) equivalent to a simplicial abelian group, an Eilenberg–Mac Lane space (some call these spaces generalized Eilenberg–Mac Lane spaces, reserving the term Eilenberg–Mac Lane space to spaces with only one nontrivial homotopy group).

Note that there is a map $S^1 \wedge H(G, n) \to \tilde{\mathbf{Z}}[S^1] \otimes H(G, n) \cong H(G, n+1)$ (given by the inclusion $\vee \subseteq \oplus$), and so the Eilenberg–Mac Lane spaces give examples of spectra (see Sect. A.3.2 below).

A.3.1.3 Chain Complexes

A *chain complex* is a sequence of abelian groups

$$C_* = \{\cdots \stackrel{d}{\leftarrow} C_{q-1} \stackrel{d}{\leftarrow} C_q \stackrel{d}{\leftarrow} C_{q+1} \stackrel{d}{\leftarrow} \cdots\}$$

such that any composite is zero. A map of chain complexes $f_*\colon C_* \to D_*$ is a collection of maps $f_q\colon C_q \to D_q$ such that the diagrams

$$\begin{array}{ccc} C_q & \xrightarrow{f_q} & D_q \\ d\downarrow & & d\downarrow \\ C_{q-1} & \xrightarrow{f_{q-1}} & D_{q-1} \end{array}$$

commute. We let Ch be the category of chain complexes, and $Ch^{\geq 0}$ be the full subcategory of chain complexes C_* such that $C_q = 0$ if $q < 0$.

If C_* is a chain complex, we let $Z_q C = \ker\{C_q \to C_{q-1}\}$ (cycles), $B_q C = \mathrm{im}\{C_{q+1} \to C_q\}$ (boundaries) and $H_q C_* = Z_q C / B_q C$ (homology).

A.3.1.4 The Normalized Chain Complex

The isomorphism between simplicial abelian groups and chain complexes concentrated in non-negative degrees is given by the *normalized chain complex*: If M is a simplicial abelian group, then $C_*^{\mathrm{norm}}(M)$ (which is usually called N_*M, an option unpalatable to us since this notation is already occupied by the nerve) is the chain complex given by

$$C_q^{\mathrm{norm}}(M) = \bigcap_{i=0}^{q-1} \ker\{d_i : M_q \to M_{q-1}\}$$

and boundary map $C_q^{\mathrm{norm}} M \to C_{q-1}^{\mathrm{norm}} M$ given by the remaining face map d_q. As commented earlier, this defines an isomorphism of categories between A and $Ch^{\geq 0}$, see [201, 22.4]. This isomorphism sends homotopies to chain homotopies (and conversely) and interacts well with tensor products and shifts. Note that $C_*^{\mathrm{norm}}(\tilde{\mathbf{Z}}[S^1])$ is isomorphic to the chain complex $\mathbf{Z}[-1] = \{\cdots = 0 = 0 \to \mathbf{Z} \to 0\}$ (the "-1" signifies—confusingly—that the copy of the integers is in degree 1).

Also, using a combinatorial description of the homotopy groups as in [201] (valid for "Kan complexes") one gets that, for a simplicial abelian group M, there is a natural identification $\pi_* M \cong H_* C^{\mathrm{norm}}(M)$ between the homotopy groups of the underlying simplicial set of M and the homology groups of the normalized chain complex.

A.3.1.5 The Moore Complex

Associated to a simplicial abelian group M there is another chain complex, the *Moore complex* $C_* M$, defined by $C_q M = M_q$ with boundary map given by the alternating sum $\delta = \sum_{j=0}^{q} (-1)^j d_j : M_q \to M_{q-1}$. The inclusion of the normalized complex into the Moore complex $C_*^{\mathrm{norm}}(M) \subseteq C_*(M)$ is a homotopy equivalence (see e.g., [201, 22.1]), and so one has a chain of natural isomorphisms

$$\pi_* M \cong H_* C^{\mathrm{norm}}(M) \cong H C_*(M).$$

It should also be mentioned that the Moore complex is the direct sum of the normalized complex and the subcomplex generated by the images of the degeneracy maps. Hence you will often see the normalized complex defined as the quotient of the Moore complex by the degenerate chains.

A.3.2 Spectra

There are many different models for spectra, each having their own merits (but all have equivalent homotopy categories with respect to the stable equivalences—the *stable category*). We will only need the very simplest version.

A *spectrum* is a sequence of spaces $X = \{X^0, X^1, X^2, \dots\}$ together with (structure) maps $S^1 \wedge X^k \to X^{k+1}$ for $k \geq 0$. A map of spectra $f\colon X \to Y$ is a sequence of maps $f^k\colon X^k \to Y^k$ compatible with the structure maps: the diagrams

$$\begin{array}{ccc} S^1 \wedge X^k & \longrightarrow & X^{k+1} \\ \downarrow{\scriptstyle 1_{S^1} \wedge f^k} & & \downarrow{\scriptstyle f^{k+1}} \\ S^1 \wedge Y^k & \longrightarrow & Y^{k+1} \end{array}$$

commute.

We let $\mathcal{S}pt$ be the resulting category of spectra. This category is enriched in $\mathcal{S}_*$, and also tensored and cotensored in the sense of Definition A.10.2.2. If X is a spectrum and K is a pointed space, then $X \wedge K = \{n \mapsto X^n \wedge K\}$ and the space of maps from K to X is $\{n \mapsto \underline{\mathcal{S}_*}(K, X^n)\}$. The morphism spaces are given by

$$\underline{\mathcal{S}pt}^0(X, Y) = \bigl\{[q] \mapsto \mathcal{S}pt\bigl(X \wedge \Delta[q]_+, Y\bigr)\bigr\}.$$

In fact, this is the zero space of a function spectrum

$$\underline{\mathcal{S}pt}(X, Y) = \bigl\{k \mapsto \underline{\mathcal{S}pt}^0\bigl(X, Y^{(k+?)}\bigr)\bigr\}.$$

There is a specially important spectrum, namely the *sphere spectrum*

$$\underline{\mathbf{S}} = \bigl\{k \mapsto S^k = S^1 \wedge \cdots \wedge S^1\bigr\}$$

whose structure maps are the identity. Note that there is a natural isomorphism between the function spectrum $\underline{\mathcal{S}pt}(\underline{\mathbf{S}}, X)$ and X.

The Eilenberg–Mac Lane spaces of Sect. A.3.1.2 give a rich supply of important spectra—the *Eilenberg–Mac Lane spectra*: if M is a simplicial abelian group, then $HM = \{n \mapsto M \otimes \tilde{\mathbf{Z}}[S^n]\}$ is the Eilenberg–Mac Lane spectrum associated with M (cf. the Γ-space version in Example 2.1.2.1(1)).

Recall that the loop space of a pointed space Y is defined by $\Omega Y = \underline{\mathcal{S}_*}(S^1, \sin|Y|)$.

Definition A.3.2.1 Let X be a spectrum and q an integer. The qth *homotopy group* of X is defined to be the group

$$\pi_q X = \varinjlim_k \pi_{q+k} X^k$$

where the colimit is over the maps $\pi_{q+k} X^k \to \pi_{q+k} \Omega X^{k+1} \cong \pi_{q+k+1} X^{k+1}$ induced by the map $X^k \to \underline{\mathcal{S}_*}(S^1, X^{n+1}) \to \underline{\mathcal{S}_*}(S^1, \sin|X^{n+1}|) = \Omega X^{n+1}$ given by the adjoint of the structure maps.

We note that π_q defines a functor from the category of spectra to abelian groups.

Definition A.3.2.2 A map of spectra $f\colon X\to Y$ is a *levelwise equivalence* if for each integer n the map $f^n\colon X^n\to Y^n$ is a weak equivalence of spaces, and f is a *stable equivalence* if it induces an isomorphism $\pi_* f\colon \pi_* X\to\pi_* Y$ on homotopy groups.

A.3.2.1 Ω-Spectra

We say that a spectrum X is an *Ω-spectrum* if for all n the map $X^n\to \underline{\mathcal{S}_*}(S^1,X^{n+1})\to\underline{\mathcal{S}_*}(S^1,\sin|X^{n+1}|)=\Omega X^{n+1}$ induced by the adjoint of the structure map is a weak equivalence.

For a spectrum X, we define its levelwise fibrant replacement fX by declaring that $(fX)^n=\sin|X^n|$ with structure maps

$$\begin{aligned}S^1\wedge(fX)^n=S^1\wedge\sin\left|X^n\right|&\to\sin\left|S^1\right|\wedge\sin\left|X^n\right|\to\sin\left|S^1\wedge X^n\right|\\&\to\sin\left|X^{n+1}\right|=(fX)^{n+1}\end{aligned}$$

induced by the structure maps of X (and the singular/realization adjoint pair). The map $X\to fX$ is a levelwise equivalence. We see that X is an Ω-spectrum if and only if for all n the adjoint of the structure maps $(fX)^n\to\underline{\mathcal{S}_*}(S^1,(fX)^{n+1})$ are weak equivalences.

If X is a spectrum, define the spectrum ωX by setting $(\omega X)^n=\underline{\mathcal{S}_*}(S^1,X^{n+1})$ with structure maps adjoint to the maps

$$(\omega X)^n=\underline{\mathcal{S}_*}\big(S^1,X^{n+1}\big)\to\underline{\mathcal{S}_*}\big(S^1,\underline{\mathcal{S}_*}\big(S^1,X^{n+2}\big)\big)=\underline{\mathcal{S}_*}\big(S^1,(\omega X)^{n+1}\big)$$

induced by the adjoints $\tilde\sigma\colon X^{n+1}\to\underline{\mathcal{S}_*}(S^1,X^{n+2})$ of the structure maps of X. The adjoints of the structure maps induce a map $\tilde\sigma\colon X\to\omega X$ of spectra and we define the spectrum

$$\omega^\infty X=\varinjlim\big\{X\xrightarrow{\tilde\sigma}\omega X\xrightarrow{\omega\tilde\sigma}\omega\omega X\xrightarrow{\omega\omega\tilde\sigma}\cdots\big\}.$$

We define the spectra $\Omega X=\omega(fX)$ and $QX=\omega^\infty(fX)$. Note that QX is an Ω-spectrum and the canonical map $X\to fX\to QX$ is a levelwise equivalence (even better: $QX\to\omega QX$ is an isomorphism and $\omega QX\to\Omega QX$ is a levelwise equivalence). For further discussions on stabilizations of this kind, the reader may consult [39, 140, 252] or [71].

Note that a map of spectra $X\to Y$ is a stable equivalence if and only if the induced map $QX\to QY$ is a levelwise equivalence.

A.3.3 Cofibrant Spectra

We say that a spectrum X is *cofibrant* if all the structure maps $S^1\wedge X^k\to X^{k+1}$ are cofibrations (i.e., inclusions). We say that a spectrum X is *n-connected* if $\pi_q X=0$

for $q \leq n$ and *connective* if it is -1-connected. A spectrum is *bounded below* if it is n-connected for some integer n.

We then get the trivial, but important, observation that any spectrum is the direct colimit of bounded below spectra

Lemma A.3.3.1 *Let X be a spectrum. Then there is a canonical levelwise equivalence $C(X) \xrightarrow{\sim} X$ where $C(X)$ is a cofibrant spectrum, and a natural filtration*

$$C_0(X) \subseteq C_1(X) \subseteq \cdots \subseteq \lim_{\overrightarrow{n}} C_n(X) = C(X)$$

such that $C_n(X)$ is a cofibrant $-(n+1)$-connected spectrum.

Proof Let $C(X)^0 = X^0$, and define $C(X)^1$ to be the mapping cone of $S^1 \wedge X^0 \to X^1$. Assuming $C(X)^n \xrightarrow{\sim} X^n$ has been constructed, let $C(X)^{n+1}$ be the mapping cone of the composite $S^1 \wedge C(X)^n \xrightarrow{\sim} S^1 \wedge X^n \to X^{n+1}$. This gives us a levelwise equivalence $C(X) \xrightarrow{\sim} X$, and by construction $C(X)$ is cofibrant. Let $C_n(X)$ be the spectrum with kth level

$$C_n(X)^k = \begin{cases} C(X)^k & \text{for } k < n, \\ S^{k-n} \wedge C(X)^n & \text{for } k \geq n, \end{cases}$$

with the obvious structure maps, and we see that $C(X) = \lim_{\overrightarrow{n}} C_n(X)$. □

Example A.3.3.2 Given an abelian group G and integer $n > 0$, the nth *Moore space* is a (choice of a) space $M(G,n)$ such that $\tilde{H}_k(M(G,n)) = 0$ if $k \neq n$ and $\tilde{H}_n(M(G,n)) = G$ (e.g., build it from some representation of G in terms of generators and relations). Note that we may choose $M(G \oplus H, n) = M(G,n) \vee M(H,n)$. The *Moore spectrum* MG is the associated suspension spectrum $n \mapsto \Sigma^{n-1} M(G,1)$ (interpreted as a point if $n = 0$).

A.4 Homotopical Algebra

In 1967, Quillen [235] provided a setup that summarized much of the formal structure seen in homotopy theory, with a view to applying it to situations that were not captured classically by homological algebra, in particular to the category of simplicial rings (the homomorphisms from one ring to another only form a set, not an abelian group). This theory has proved to be useful in a wide variety of situations, although considered to be rather on the abstract side until the techniques saw a dramatic renaissance in the 1990's. We summarize the little of the theory of *(simplicial closed) model categories* that is needed for our purposes. For fuller accounts, see any one of [135, 139, 235] or [99].

In homotopy theory there are three important concepts: *fibrations, cofibrations* and *weak equivalences*. The important thing is to know how these concepts relate to each other: Consider the (solid) commuting diagram

$$\begin{array}{ccc} A & \longrightarrow & E \\ {\scriptstyle i}\big\downarrow & \overset{s}{\nearrow} & \big\downarrow{\scriptstyle f} \\ X & \longrightarrow & B \end{array}$$

where i is a cofibration and f is a fibration. If either i or f is a weak equivalence, then there exists a (dotted) map $s\colon X \to E$ making the resulting diagram commutative. The map s will in general only be unique up to homotopy (there is a general rule in this game which says that "existence implies uniqueness", meaning that the existence property also can be used to prove that there is a homotopy between different liftings).

Note that there may be many meaningful choices of weak equivalences, fibrations and cofibrations on a given category.

A.4.1 Examples

1. **Spaces**. In $\mathcal{S}$ the weak equivalences are the maps $f\colon X \to Y$ which induce isomorphisms on path components and on all homotopy groups $\pi_*(X, x) \to \pi_*(Y, f(x))$ for all $x \in X_0$. The cofibrations are simply the injective maps, and the fibrations are all maps which have the lifting property described above with respect to the cofibrations which are weak equivalences. These are classically called *Kan fibrations*.

   These notions also pass over to the subcategory $\mathcal{S}_*$ of pointed simplicial sets. The inclusion of the basepoint is always a cofibration (i.e., all spaces are *cofibrant*), but the projection onto a one point space is not necessarily a fibration (i.e., not all spaces are *fibrant*). The fibrant spaces are also called *Kan spaces* (or *Kan complexes*).
2. **Topological Spaces**. In *Top* and *Top*$_*$, a weak equivalences is still a map $f\colon X \to Y$ which induces a bijection $\pi_0(X) \to \pi_0(Y)$ of path components and for every $x \in X$ and positive n induces an isomorphism $\pi_n(X, x) \to \pi(Y, f(x))$ of homotopy groups. The fibrations are the *Serre fibrations*, and the cofibrations are those which satisfy the lifting property with respect to the Serre fibrations which are weak equivalences. All topological spaces are fibrant, but not all are cofibrant. CW-complexes are cofibrant. Both the realization functor and the singular functor preserve weak equivalences, fibrations and cofibrations. The fact that the realization of a Kan fibration is a Serre fibration is a cornerstone in the theory, due to Gabriel–Zisman and Quillen, see e.g., [99, I.10.10].

3. **Simplicial Groups, Rings, Monoids, Abelian Groups**. In $s\mathcal{G}$, the category of simplicial groups, a map is a weak equivalence or a fibration if it is in $\mathcal{S}_*$, and the cofibrations are the maps which have the lifting property with respect to the fibrations which are weak equivalences. Note that this is much more restrictive than just requiring it to be a cofibration (inclusion) in $\mathcal{S}_*$ (the lifting is measured in different categories). However, if $X \to Y \in \mathcal{S}_*$ is a cofibration, then $F(X) \to F(Y)$ is also a cofibration, where $F\colon Ens_* \to \mathcal{G}$ is the free functor, which sends a pointed set X to the free group on X modulo the basepoint.

   Likewise in A, the category of simplicial abelian groups, and $sRing$, the category of simplicial rings.

   In these categories fibrations are easily recognized: a map $G \to H$ of simplicial groups is a fibration if and only if the induced map $G \to H \times_{\pi_0 H} \pi_0 G$ is a surjection, where $\pi_0 G$ is considered as a constant simplicial group. In particular, all simplicial groups are fibrant.
4. **Functor Categories**. Let I be any small category, and let $[I, \mathcal{S}_*]$ be the category of functors from I to $\mathcal{S}_*$. This is a closed simplicial model category in the "pointwise" structure: a map $X \to Y$ (natural transformation) is a weak equivalence (resp. fibration) if $X(i) \to Y(i)$ is a weak equivalence (resp. fibration) of simplicial sets, and it is a cofibration if it has the left lifting property with respect to all maps that are both weak equivalences and fibrations. Important examples are the pointwise structure on Γ-spaces (see Sect. 2.2.1.4) and G-spaces (see below).

   Generally, it is the pointwise structure which is used for the construction of homotopy (co)limits (see Sect. A.7 below).
5. **The Levelwise Structure on Spectra**. A map $X \to Y$ of spectra is a *levelwise equivalence* (resp. *levelwise fibration*) if for every k the map $X^k \to Y^k$ is a weak equivalence (resp. fibration) of pointed simplicial sets. A map is a *cofibration* if it has the lifting property with respect to maps that are both levelwise fibrations and levelwise equivalences. A spectrum X is cofibrant if all the structure maps $S^1 \wedge X^k \to X^{k+1}$ are cofibrations (i.e., inclusions).
6. **The Stable Structure on Spectra**. A map $X \to Y$ of spectra is a *stable equivalence* if it induces an isomorphism on homotopy groups (see Definition A.3.2.1), and a (stable) *cofibration* if it is a cofibration in the levelwise structure. The map is a *stable fibration* if it has the lifting property with respect to maps that are both cofibrations and stable equivalences. So, in particular, a spectrum is stably fibrant if it is a levelwise fibrant Ω-spectrum.
7. ***G*-Spaces and *G*-Spectra**. Let G be a simplicial monoid. The category of G-spaces (see Sect. A.9) is a closed simplicial model category with the following structure: a map is a G-equivalence (resp. G-fibration) if it is an equivalence (resp. fibration) of spaces, and a cofibration if it has the left lifting property with respect to all maps that are both G-equivalences and G-fibrations. Also, the category of *G-spectra* (see Sect. 6.3) has a levelwise and a stable structure giving closed simplicial model categories. *Levelwise fibrations* and *levelwise equivalences* (resp. *stable fibrations* and *stable equivalences*) are given by forgetting down to spectra, and *levelwise* (resp. *stable*) *cofibrations* are given by the left lifting property.

Many authors refer to G-spectra in this stable structure as *naïve G-spectra* or *spectra with G-action* to distinguish them from versions, often called *genuine G-spectra* or the *equivariant structure*, where representations of the group are built into the structure maps and where a weak equivalence is a G-map that induces a stable equivalence on fixed point spectra of all subgroups of G. Note that the more sophisticated theory is used in the calculations in Sect. 7.3.

Examples A.4.1(1)–A.4.1(3) can be summarized as follows: Consider the diagram

$$Top_* \underset{\sin}{\overset{|-|}{\leftrightarrows}} \mathcal{S}_* \underset{U}{\overset{F}{\rightleftarrows}} s\mathcal{G} \underset{U}{\overset{H_1(-)}{\rightleftarrows}} \mathcal{A} \underset{U}{\overset{T_{\mathbf{Z}}(-)}{\rightleftarrows}} sRing$$

where the U are forgetful functors, $T_{\mathbf{Z}}(A)$ the tensor ring on an abelian group A, and $H_1(-) = -/[-,-]\colon \mathcal{G} \to Ab$ the first homology (applied degreewise). We know what weak equivalences in Top_* are, and we define them everywhere else to be the maps which are sent to weak equivalences in Top_*. We know what cofibrations are in $\mathcal{S}_*$, and use the axiom to define the fibrations. We then define fibrations everywhere else to be the maps that are sent to fibrations in $\mathcal{S}_*$, and use the axioms to define the cofibrations.

These adjunctions also provide "morphism spaces". For instance, for simplicial groups G and G' we may define the *space* of group homomorphisms from G to G' to be $\underline{s\mathcal{G}}(G, G') = \{[q] \mapsto s\mathcal{G}(G * F(\Delta[q]_+), G')\}$, where $*$ is the coproduct in the category of groups (for obscure reasons sometimes referred to as the free product). For a more general view on morphism spaces in categories of simplicial objects, see Example A.10.2.2(4).

The proof that Examples A.4.1(1)–A.4.1(4) define closed simplicial model categories is contained in [235, II4], and the proof A.4.1(5) and A.4.1(6) is in [39]. None of these proofs state explicitly the functoriality of the factorizations of the axiom CM5 below, but for each of the cases it may be easily reconstructed from a *small object* kind of argument [139, Theorem 2.1.14]. For a discrete group G, the case of G-spaces is a special case of Example A.4.1(4) but a direct proof in the general case is fairly straight-forward, and the same proof works for the levelwise structure on G-spectra. The proof for the stable structure then follows from the levelwise structure by the same proof as in [39] for the case $G = *$.

A.4.2 The Axioms

For convenience we list the axioms for a closed simplicial model category $\mathcal{C}$. It is a category enriched in $\mathcal{S}$, and it is tensored and cotensored (see Definition A.10.2.2). We call the function spaces $\underline{\mathcal{C}}(-,-)$. Furthermore $\mathcal{C}$ has three classes of maps called *fibrations*, *cofibrations* and *weak equivalences* satisfying the following axioms

CM1 $\mathcal{C}$ has all (small) limits and colimits.

CM2 (Two-out-of-three) For two composable morphisms

$$b \xrightarrow{f} c \xrightarrow{g} d \in \mathcal{C},$$

if any two of f, g and gf are weak equivalences, then so is the third.

CM3 (Closed under retracts) If a map f is a retract of g (in the arrow category), and g is a weak equivalence, a fibration or a cofibration, then so is f.

CM4 Given a solid diagram

$$\begin{array}{ccc} A & \longrightarrow & E \\ {\scriptstyle i}\big\downarrow & \overset{s}{\nearrow} & \big\downarrow \\ X & \longrightarrow & B \end{array}$$

where i is a cofibration and f is a fibration. If either i or f are weak equivalences, then there exists a (dotted) map $s\colon X \to E$ making the resulting diagram commutative.

CM5 (Functorial factorization axiom) Any map f may be functorially factored as $f = pi = qj$ where i is a cofibration, p a fibration and a weak equivalence, j a cofibration and a weak equivalence, and q a fibration.

SM7 If $i\colon A \rightarrowtail B$ is a cofibration and $p\colon X \to Y$ is a fibration, then the canonical map

$$\underline{\mathcal{C}}(B, X) \xrightarrow{(i^*, p_*)} \underline{\mathcal{C}}(A, X) \times_{\underline{\mathcal{C}}(A,Y)} \underline{\mathcal{C}}(B, Y)$$

is a fibration of simplicial sets. If either i or p are weak equivalences, then (i^*, p_*) is also a weak equivalence.

An object X is a retract of Y if there are maps $X \to Y \to X$ whose composite is id_X. Note that the demand that the factorizations in CM5 should be functorial is not a part of Quillen's original setup, but is true in all examples we will encounter, and is sometimes extremely useful (Quillen also only demanded the existence of *finite* limits and colimits).

With the exception of $\mathcal{S}$ and *Top*, all our categories will be $\mathcal{S}_*$-categories, that is the function spaces have preferred basepoints.

A.4.3 The Homotopy Category

It makes sense to talk about the homotopy category $Ho(\mathcal{C})$ of a closed simplicial model category $\mathcal{C}$. These are the categories where the weak equivalences are formally inverted (see e.g., [235]).

The realization and singular functor induce equivalences of categories

$$Ho(\mathcal{S}_*) \simeq Ho(Top_*).$$

This has the consequence that for all practical purposes we can choose whether we rather want to work with simplicial sets or topological spaces. Both categories have their drawbacks, and it is useful to know that all theorems which are proven for either homotopy category holds for the other.

A.5 Fibrations in $\mathcal{S}_*$ and Actions on the Fiber

Let $f: E \to B \in \mathcal{S}_*$ be a fibration. We call $F = * \times_B E$ the *fiber* of f. Recall [99, I.3.7] that we get a long exact sequence

$$\cdots \to \pi_{q+1}E \to \pi_{q+1}B \to \pi_q F \to \pi_q E \to \pi_q B \to \cdots \to \pi_0 B$$

in the sense that kernel equals image everywhere. The π_is are groups for $i > 0$ and abelian groups for $i > 1$, $\pi_2 E$ maps into the center of $\pi_1 F$, all maps to the left of $\pi_1 B$ are group homomorphisms and two elements in $\pi_0 F$ are mapped to the same element in $\pi_0 E$ if and only if they are in the same orbit under the action of $\pi_1 B$ on $\pi_0 F$ (see Sect. A.5.1 below for details on the action on the fiber).

More generally, if $f: E \to B \in \mathcal{S}_*$ is any map we may define the *homotopy fiber* by the pullback diagram

$$\begin{array}{ccccc}
\mathrm{hofib}(f) & \longrightarrow & & & \underline{\mathcal{S}_*}(\Delta[1], \sin|B|) \\
\downarrow & & & & \downarrow \\
E & \xrightarrow{f} & B & \xrightarrow{\sim} & \sin|B|,
\end{array}$$

where the right vertical map is induced by $d^1: \Delta[0] \to \Delta[1]$ and the right lower horizontal map is the unit of adjunction. That this definition does what it is supposed to follows since $\sin|B|$ is fibrant, $\Delta[0] \to \Delta[1]$ is a cofibration, SM7 and since $\mathcal{S}_*$ is *right proper*: weak equivalences are preserved by pulling them back along fibrations, see e.g., [135, 13.1.13].

Note that there is a preferred map $\Omega B = \underline{\mathcal{S}_*}(S^1, \sin|B|) \to \mathrm{hofib}(f)$. If a sequence of spaces $\Omega B \to F \to X \to B$ is connected by a chain of weak equivalences to $\Omega B \to \mathrm{hofib}(f) \to E \to B$ we say that it is a (homotopy) *fiber sequence*.

An alternative functorial choice for the definition of the homotopy fiber is to factor $E \to B$ through a weak equivalence and cofibration followed by a fibration as in CM5 and let the homotopy fiber be the fiber of the fibration.

A.5.1 Actions on the Fiber

If π and G are groups and $\pi \to \mathrm{Aut}(G)$ is a group homomorphism from π to the group of automorphisms on G we say that π *acts* on G. If $H \subset G$ is a normal subgroup we have an action $G \to \mathrm{Aut}(H)$ via $g \mapsto \{h \mapsto g^{-1}hg\}$. In particular, any group acts on itself in this fashion, and these automorphisms are called the *inner automorphisms*.

Let $f\colon E \twoheadrightarrow B$ be a fibration and assume B is fibrant. Let $i\colon F = E \times_B * \subseteq E$ be the inclusion of the fiber. Then there are group actions

$$\pi_1 E \to \mathrm{Aut}(\pi_* F)$$

and

$$\pi_1 B \to \mathrm{Aut}(H_* F)$$

and the actions are compatible in the sense that the obvious diagram

$$\begin{array}{ccc} \pi_* F \times \pi_1 E & \longrightarrow & \pi_* F \\ \downarrow & & \downarrow \\ H_* F \times \pi_1 B & \longrightarrow & H_* F \end{array}$$

commutes. For reference, we review the construction.

The spaces F, E, and B are fibrant, so function spaces into these spaces are homotopy invariant. For instance, $B^{S^1} = \underline{\mathcal{S}}_*(S^1, B)$ is a model for the loops on B. We write X^I for the free path space $\underline{\mathcal{S}}_*(\Delta[1]_+, X)$.

Consider the map $p\colon X \to F \times B^{S^1}$ defined by

$$\begin{array}{ccccccccccc} X & = & \lim_{\leftarrow}\{F & \xrightarrow{i} & E & \xleftarrow{d_0} & E^I & \xrightarrow{d_1} & E & \xleftarrow{i} & F\} \\ {\scriptstyle p}\downarrow & & \| & & {\scriptstyle f}\downarrow & & {\scriptstyle f^I}\downarrow & & {\scriptstyle f}\downarrow & & \downarrow \\ F \times B^{S^1} & = & \lim_{\leftarrow}\{F & \xrightarrow{*} & B & \xleftarrow{d_0} & B^I & \xrightarrow{d_1} & B & \longleftarrow & *\}, \end{array}$$

where the upper "equality" is a definition, and the lower is the canonical isomorphism. We see that p is both a fibration and a weak equivalence.

Hence there exists splittings $F \times B^{S^1} \to X$, unique up to homotopy, which by adjointness give rise to an unbased homotopy class of maps $B^{S^1} \to \underline{\mathcal{S}}(F, X)$. Via the projection onto the last factor

$$X = F \times_E E^I \times_E F \xrightarrow{pr_3} F,$$

this gives rise to a homotopy class of maps $B^{S^1} \to \underline{\mathcal{S}}(F, F)$. For every such we have a commuting diagram

$$\begin{array}{ccccc} B^{S^1} & \longrightarrow & \underline{\mathcal{S}}(F, F) & & \\ \downarrow & & \downarrow & & \\ \pi_0 B^{S^1} = \pi_1 B & \longrightarrow & \pi_0 \underline{\mathcal{S}}(F, F) & \longrightarrow & \mathrm{End}\big(H_*(F)\big), \end{array}$$

and the lower map does not depend on the choice of the upper map. As F is fibrant, $\pi_0\underline{\mathcal{S}}(F, F)$ is the monoid of homotopy classes of unbased self maps. Any homotopy class of unbased self maps of F defines an element in $\mathrm{End}(H_*(F))$, and the map from $\pi_1 B$ is a monoid map, and giving rise to the desired group action $\pi_1 B \to \mathrm{Aut}(H_*(F))$.

For the pointed situation, consider the (solid) diagram

$$\begin{array}{ccccc} F \vee E^{S^1} & \xrightarrow{in_1+j} & X & \xrightarrow{pr_3} & F \\ \big\downarrow & \nearrow & \big\downarrow p & & \\ F \times E^{S^1} & \longrightarrow & F \times B^{S^1} & & \end{array} \tag{A.1}$$

where $in_1\colon F \to X = F \times_E E^I \times_E F$ is inclusion of the first factor, and j is the inclusion $E^{S^1} = * \times_E E^I \times_E * \subseteq F \times_E E^I \times_E F = X$. Again there is a homotopy class of liftings, and since the top row in the diagram is trivial, the composites

$$F \times E^{S^1} \to X \xrightarrow{pr_3} F$$

all factor through $F \wedge E^{S^1}$. So, this time the adjoints are pointed: $E^{S^1} \to \underline{\mathcal{S}}_*(F, F)$, giving rise to a unique

$$\pi_1 E = \pi_0 E^{S^1} \to \pi_0\big(\underline{\mathcal{S}}_*(F, F)\big) \to \mathrm{End}\big(\pi_*(F)\big).$$

Again the map is a map of monoids, and so factors through the automorphisms, and we get the desired group action $\pi_1 E \to \mathrm{Aut}(\pi_*(F))$, compatible with the homology operation.

A.5.2 Actions for Maps of Grouplike Simplicial Monoids

If $j\colon G \subseteq M$ is the inclusion of a subgroup in a monoid, then $j/1$ is the *over category* Sect. A.2.4.2 of j considered as a functor of categories. Explicitly, it has the elements of M as objects, and a map from m to m' is a $g \in G$ such that $m'g = m$.

We have an isomorphism

$$M \times G^{\times q} \xrightarrow{\cong} B_q(j/1)$$

given by

$$(m, g_1, \ldots, g_q) \mapsto \big(m \xleftarrow{g_1} mg_1 \xleftarrow{g_2} \cdots \xleftarrow{g_q} mg_1g_2\cdots g_q\big)$$

and $B(M, G, *) = \{[q] \mapsto M \times G^{\times q}\}$ with the induced simplicial structure, is called the *one-sided bar construction*. The projection $M \times G^{\times q} \to G^{\times q}$ away from M gives a map $B(j/1) \to BG$.

Theorem A.5.2.1 *Let M be a group-like simplicial monoid, and $j \colon G \subseteq M$ a (simplicial) subgroup. Then*

$$B(j/1) \to BG \to BM$$

is a fiber sequence, and the action

$$\Omega BG \times B(j/1) \to B(j/1) \in Ho(\mathcal{S}_*)$$

may be identified with the conjugation action

$$G \times B(j/1) \longrightarrow B(j/1),$$

$$\big(g, (m, g_1, \ldots, g_q)\big) \in G_q \times B_q(j/1) \mapsto \big(gmg^{-1}, gg_1g^{-1}, \ldots, gg_qg^{-1}\big).$$

Proof That the sequence is a fiber sequence follows from Corollary A.6.1.4. As to the action, replace the fiber sequence with the equivalent fiber sequence

$$F \xrightarrow{\ i\ } E \xrightarrow{\ f\ } B$$

defined by $B = \sin|BM|$, and the pullback diagrams

$$\begin{array}{ccc} E & \longrightarrow & BG \\ \downarrow & & \downarrow \\ B^I & \xrightarrow{d_0} & B \end{array} \quad \text{and} \quad \begin{array}{ccc} F & \xrightarrow[i]{} & E \\ \downarrow & & \downarrow{\scriptstyle f} \\ * & \longrightarrow & B \end{array}$$

where f is the composite

$$E = BG \times_B B^I \xrightarrow{pr_2} B^I \xrightarrow{d_1} B.$$

To describe the action, consider the diagram (the maps will be described below)

$$\begin{array}{ccccccc} F \vee G & \overset{\sim}{\rightarrowtail} & F \vee (BG)^{S^1} & \overset{\sim}{\rightarrowtail} & F \vee E^{S^1} & \xrightarrow{in_1 + j} & X \\ \downarrow & & \downarrow & & \downarrow & & \simeq \downarrow p \\ F \times G & \overset{\sim}{\rightarrowtail} & F \times (BG)^{S^1} & \overset{\sim}{\rightarrowtail} & F \times E^{S^1} & \longrightarrow & B^{S^1} \times F. \end{array}$$

The rightmost square is the same as the lifting square in (A.1). The leftmost horizontal weak equivalences are induced by the weak equivalence $G \overset{\sim}{\to} (BG)^{S^1} =$

$\underline{\mathcal{S}_*}(S^1, BG)$ adjoint to the canonical inclusion $S^1 \wedge G \subseteq BG$, and the middle horizontal weak equivalences in the diagram are induced by the weak equivalence $BG \to E = BG \times_B B^I$ given by $x \mapsto (x, f(x))$ (the constant map at $f(x)$). By the uniqueness of liftings (also homotopies lift in our diagram), any lifting $F \times G \to X$ is homotopic to $F \times G \xrightarrow{\sim} F \times E^{S^1}$ composed with a lifting $F \times E^{S^1} \to X$. Hence we may equally well consider liftings $F \times G \to X$. We will now proceed to construct such a lifting by hand, and then show that the constructed lifting corresponds to the conjugation action.

We define a map $BG \times G \times \Delta[1] \to (BG)$ by sending $(x, g, \phi) = (g, (x_1, \dots, x_q), \phi) \in G_q \times B_q G \times \Delta([q], [1])$ to

$$H^g(x)(\phi) = \left(g^{\phi(0)} x_1 g^{-\phi(1)}, g^{\phi(1)} x_2 g^{-\phi(2)}, \dots, g^{\phi(q-1)} x_q g^{-\phi(q)}\right)$$

(where $g^0 = 1$ and $g^1 = g$). Note that, if 1 is the constant map $[q] \to [1]$ sending everything to 1, then $H^g(x)(1) = (g x_1 g^{-1}, \dots, g x_q g^{-1})$. We let $H \colon BG \times G \to BG^I$ be the adjoint, and by the same formula we have a diagram

$$\begin{array}{ccc} BG \times G & \xrightarrow{H} & (BG)^I \\ \downarrow & & \downarrow \\ BM \times G & \xrightarrow{\bar{H}} & (BM)^I \\ \downarrow & & \downarrow \\ B \times G & \xrightarrow{\bar{H}} & B^I. \end{array}$$

This extends to a map $E \times G \to E^I$ by sending $(x, \alpha) \in BG \times_B B^I = E$ to $g \mapsto H^g(x, \alpha) = (H^g(x), \bar{H}^g(\alpha))$. Since

$$G \subset B \times G \xrightarrow{\bar{H}} B^I \xrightarrow{d_i} B$$

is trivial for $i = 0, 1$ (and so, if $(x, \alpha) \in F$, we have $H^g(x, \alpha)(i) = (H^g(x)(i), \bar{H}^g(\alpha)(i)) \in F$ for $i = 0, 1$), we get that, upon restricting to $F \times G$ this gives a lifting

$$F \times G \to F \times_E E^I \times_E F = X.$$

Composing with

$$X = F \times_E E^I \times_E F \xrightarrow{pr_3} F$$

we have the conjugation action

$$F \times G \to F, \qquad \left(g, (x, \alpha)\right) \mapsto c^g(x, \alpha) = H^g(x, \alpha)(1) = \left(H^g(x)(1), \bar{H}^g(\alpha)(1)\right)$$

is equivalent to the action of G on the fiber in the fiber sequence of the statement of the theorem.

Let $C = \sin|B(M/1)| \times_B B^I$ and $\tilde{F} = C \times_B E$. Since C is contractible, $F \xrightarrow{\sim} \tilde{F}$ is an equivalence. We define a conjugation action on C using the same formulas, such that $C \to B$ is a G-map, and this defines an action on $\tilde{F}$ such that

$$\begin{array}{ccc} F \times G & \longrightarrow & F \\ \downarrow & & \downarrow \\ \tilde{F} \times G & \longrightarrow & \tilde{F} \\ \uparrow & & \uparrow \\ B(j/1) \times G & \longrightarrow & B(j/1) \end{array}$$

commutes, where the lower map is the action in the theorem. As the vertical maps are equivalences by the first part of the theorem, this proves the result. □

A.6 Bisimplicial Sets

A *bisimplicial set* is a simplicial object in $\mathcal{S}$, that is, a simplicial space. From the diagonal and projection functors

$$\Delta \xrightarrow{\text{diag}} \Delta \times \Delta \underset{pr_2}{\overset{pr_1}{\rightrightarrows}} \Delta$$

we get functors

$$\mathcal{S} \xleftarrow{\text{diag}^*} s\mathcal{S} \underset{pr_2^*}{\overset{pr_1^*}{\leftleftarrows}} \mathcal{S}$$

where the leftmost is called the diagonal and sends X to $\text{diag}^*(X) = \{[q] \mapsto X_{q,q}\}$, and the two maps to the right reinterpret a simplicial space X as a bisimplicial set by letting it be constant in one direction (e.g., $pr_1^*(X) = \{([p],[q]) \mapsto X_p\}$).

There are important criteria for when information about each X_p may be sufficient to conclude something about $\text{diag}^* X$. We cite some useful facts. Proofs may be found either in the Appendix of [39] or in [99, Chap. IV].

Theorem A.6.0.1 (See e.g., [99, Proposition IV 1.9]) *Let $X \to Y$ be a map of simplicial spaces inducing a weak equivalence $X_q \xrightarrow{\sim} Y_q$ for every $q \geq 0$. Then $\text{diag}^* X \to \text{diag}^* Y$ is a weak equivalence.* ☺

Definition A.6.0.2 (The π_*-Kan Condition, [39]) Let

$$X = \{[q] \mapsto X_q\} = \{([p],[q]) \mapsto X_{p,q}\}$$

be a simplicial space. For a vertex a in X_q, consider the maps

$$d_i : \pi_p(X_q, a) \to \pi_p(X_{q-1}, d_i a), \quad 0 \le i \le q.$$

We say that X satisfies the *π_*-Kan condition at $a \in X_q$* if for every tuple of elements

$$(x_0, \dots, x_{k-1}, x_{k+1}, \dots x_q) \in \prod_{\substack{0 \le i \le m \\ i \ne k}} \pi_p(X_{q-1}, d_i a)$$

such that $d_i x_j = d_{j-1} x_i$ for $k \ne i < j \ne k$, there is an

$$x \in \pi_p(X_q, a)$$

such that $d_i x = x_i$ for $i \ne k$.

For an alternative description of the π_*-Kan condition see [99, Sect. IV.4].

Examples of simplicial spaces which satisfies the π_*-Kan condition are bisimplicial groups and simplicial spaces $\{[q] \mapsto X_q\}$ where each X_q is connected, see [39].

Recall that a square is *homotopy cartesian* if it is equivalent to a categorically cartesian square of fibrations. More precisely, a commutative square

$$\begin{array}{ccc} A & \longrightarrow & B \\ \downarrow & & \downarrow \in \mathcal{S} \\ C & \longrightarrow & D \end{array}$$

is homotopy cartesian if there is a factorization $B \xrightarrow{\sim} X \twoheadrightarrow D$ such that the resulting map $A \to C \times_D X$ is a weak equivalence. Note that if the condition is true for one factorization, it is true for all. This definition gives the same result as the more general one given in Definition A.8.0.2.

Theorem A.6.0.3 (Bousfield–Friedlander's Theorem B.4 [39]) *Let*

$$\begin{array}{ccc} V & \longrightarrow & X \\ \downarrow & & \downarrow \\ W & \longrightarrow & Y \end{array}$$

be a commutative diagram of simplicial spaces, such that

$$\begin{array}{ccc} V_p & \longrightarrow & X_p \\ \downarrow & & \downarrow \\ W_p & \longrightarrow & Y_p \end{array}$$

is homotopy cartesian for every p. If X and Y satisfy the π_*-Kan condition and if $\{[q] \mapsto \pi_0(X_q)\} \to \{[q] \mapsto \pi_0(Y_q)\}$ is a fibration, then

$$\begin{array}{ccc} \operatorname{diag}^* V & \longrightarrow & \operatorname{diag}^* X \\ \downarrow & & \downarrow \\ \operatorname{diag}^* W & \longrightarrow & \operatorname{diag}^* Y \end{array}$$

is homotopy cartesian. ☺

As an immediate corollary we have the important result that loops can often be calculated degreewise. Recall that if X is a space, then the loop space of Y is $\Omega X = \underline{\mathcal{S}_*}(S^1, \sin|X|)$.

Corollary A.6.0.4 *Let X be a simplicial space such that X_p is connected for every $p \geq 0$. Then there is a natural chain of weak equivalences between $\Omega \operatorname{diag}^* X$ and $\operatorname{diag}^*\{[p] \mapsto \Omega X_p\}$.*

Proof Let $Y_p = \sin|X_p|$, and consider the homotopy cartesian square

$$\begin{array}{ccc} \underline{\mathcal{S}_*}(S^1, Y_p) & \longrightarrow & \underline{\mathcal{S}_*}(\Delta[1], Y_p) \\ \downarrow & & \downarrow \\ * & \longrightarrow & Y_p. \end{array}$$

Now, since each Y_p is connected, this diagram satisfies the conditions of Theorem A.6.0.3, and so

$$\begin{array}{ccc} \operatorname{diag}^*\{[p] \mapsto \underline{\mathcal{S}_*}(S^1, Y_p)\} & \longrightarrow & \operatorname{diag}^*\{[p] \mapsto \underline{\mathcal{S}_*}(\Delta[1], Y_p)\} \\ \downarrow & & \downarrow \\ * & \longrightarrow & \operatorname{diag}^*\{[p] \mapsto Y_p\} \end{array}$$

is homotopy cartesian. The right vertical map factors canonically as

$$\operatorname{diag}^*\{[p] \mapsto \underline{\mathcal{S}_*}(\Delta[1], Y_p)\} \longrightarrow \operatorname{diag}^*\{[p] \mapsto \underline{\mathcal{S}_*}(\Delta[1], Y_p)\} \wedge \Delta[1] \longrightarrow \operatorname{diag}^* Y$$

where the first map is the non-basepoint inclusion and the second map sends $f \wedge \phi \in \mathcal{S}_*(\Delta[p]_+ \wedge \Delta[1], Y_p) \wedge \Delta([p],[1])$ to $f_p(id_{[p]} \wedge \phi) \in Y_{p,p}$. This gives a canonical map $\operatorname{diag}^*\{[p] \mapsto \underline{\mathcal{S}_*}(\Delta[1], Y_p)\} \to \underline{\mathcal{S}_*}(\Delta[1], \operatorname{diag}^* Y)$ of contractible spaces over $\operatorname{diag}^* Y$, and ultimately a canonical equivalence $\operatorname{diag}^* \Omega X = \operatorname{diag}^*\{[p] \mapsto \underline{\mathcal{S}_*}(S^1, Y_p)\} \to \underline{\mathcal{S}_*}(S^1, \operatorname{diag}^* Y) \to \Omega \operatorname{diag}^* Y$. Combining this with the equivalence $\operatorname{diag}^* Y \xleftarrow{\sim} \operatorname{diag}^* X$, we get the desired canonical chain of equivalences. □

In this proof, the attentive reader will notice the appearance of a useful map

$$\operatorname{diag}^* \underline{\mathcal{S}_*}(A, Y) \to \underline{\mathcal{S}_*}(A, \operatorname{diag}^* Y)$$

(for A a space and Y a simplicial space) which in degree q is given by sending $f\colon A\wedge\Delta[q]_+\to Y_q$ to the composite

$$A\wedge\Delta[q]_+ \xrightarrow{f} Y_q\wedge\Delta[q]_+ \longrightarrow \operatorname{diag}^*Y,$$

where the last map is given in degree p by sending $y\wedge\phi\in Y_{pq}\wedge\Delta([p],[q])_+$ to $\phi^*y\in Y_{pp}$. With $Y_p=\sin|X_p|$ as above, this gives a weak map

$$\operatorname{diag}^*Map_*(A,X) \longrightarrow \mathcal{S}_*(A,\operatorname{diag}^*Y) \longrightarrow Map_*(A,\operatorname{diag}^*Y) \xleftarrow{\sim} Map_*(A,\operatorname{diag}^*X),$$

where weak equivalence pointing the wrong way is induced by $\operatorname{diag}^*Y\xleftarrow{\sim}\operatorname{diag}^*X$.

By induction we get that if $A=S^{n+1}$ and X_p is n-connected for every $p\geq 0$, the corollary says that

$$\operatorname{diag}^*\Omega^{n+1}X\xrightarrow{\sim} Map_*\big(S^{n+1},\operatorname{diag}^*\{[q]\mapsto\sin|X_q|\}\big)\xleftarrow{\sim}\Omega^{n+1}\operatorname{diag}^*X$$

is a natural chain of weak equivalences.

Bousfield and Friedlander's setup [39] gives a nice proof of the spectral sequence associated to a simplicial space, generalizing that of Artin and Mazur [8] (see also Quillen's short proof for the case of a bisimplicial group [227] on which the published account of Artin and Mazur relies).

Theorem A.6.0.5 *Let X be a pointed simplicial space satisfying the π_*-Kan condition. Then there is a first quadrant convergent spectral sequence*

$$E^2_{pq}=\pi_p\big([n]\mapsto\pi_q(X_n)\big)\Rightarrow\pi_{p+q}\big(\operatorname{diag}^*X\big).$$ ☺

That E^2_{00} is just a set is not a problem: the spectral sequence at this stage just expresses the bijection $\pi_0([n]\to\pi_0(X_n))\cong\pi_0(\operatorname{diag}^*X)$; that E^2_{10} and E^2_{01} are not necessarily abelian means that the convergence of the spectral sequence at this stage expresses that there is an exact sequence

$$E^2_{20}\to E^2_{01}\to\pi_1\big(\operatorname{diag}^*X\big)\to E^2_{10}\to 1$$

of groups. Above total degree 1 all groups are abelian, and the spectral sequence has the usual meaning.

As an application we prove three corollaries, which is totally wrong historically, since the first result was used in [39] to prove Theorem A.6.0.5, and the second predates [39].

Corollary A.6.0.6 *Let G be a simplicial group, and let BG be the diagonal of $[n]\mapsto BG_n$. Then*

$$\pi_qBG\cong\pi_{q-1}G.$$

Proof Note that BG_n is connected for each n, and so BG satisfies the π_*-Kan condition. Now

$$E^2_{pq} = \pi_p\big([n] \mapsto \pi_q(BG_n)\big) = \begin{cases} 0 & \text{if } q \neq 1, \\ \pi_p G & \text{if } q = 1 \end{cases}$$

and the result follows. □

Corollary A.6.0.7 *Let X be a simplicial space. Then there is a convergent spectral sequence*

$$E^2_{pq} = H_p\big([n] \mapsto H_q(X_n)\big) \Rightarrow H_{p+q}\big(\operatorname{diag}^* X\big).$$

Proof Apply the spectral sequence of Theorem A.6.0.5 to the bisimplicial abelian group $\mathbf{Z}X$. □

Using the theorem repeatedly and using that the π_*-Kan condition is satisfied for simplicial spaces that are connected in every degree we get

Corollary A.6.0.8 *Let $X\colon (\Delta^o)^{\times n} \to Ens_*$ be a multi-simplicial set which is $k_i - 1$-connected in the ith direction for $i = 1, \dots, n$, and at least one of the $k_i - 1$s is positive. Then the diagonal space is $k_1 + k_2 + \cdots + k_n - 1$ connected.* ☺

A.6.1 Linear Simplicial Spaces

Definition A.6.1.1 A simplicial object X in a model category is *linear* if the natural maps

$$X_p \to X_1 \times_{X_0} X_1 \times_{X_0} \cdots \times_{X_0} X_1$$

are weak equivalences, where the ith component is induced from $[1] \cong \{i-1, i\} \subseteq [p]$ for $0 < i \leq p$.

This is inspired by categories, where a space X is the classifying space of a category, exactly if the said maps are isomorphisms. A slicker way of formulating this is to say that X is linear if it takes the pushouts of monomorphisms that exist in Δ to homotopy pullbacks.

Note that if $X_0 = *$, this gives a "weak multiplication" on X_1:

$$X_1 \times X_1 \xleftarrow[\sim]{(d_0, d_2)} X_2 \xrightarrow{d_1} X_1.$$

Saying that this weak multiplication has a *homotopy inverse* is the same as saying that all the diagrams

$$\begin{array}{ccc} X_p & \longrightarrow & X_1 \\ {\scriptstyle d_p}\big\downarrow & & \big\downarrow \\ X_{p-1} & \longrightarrow & * \end{array}$$

are homotopy cartesian, where the top map is induced by $[1] \cong \{0, p\} \subseteq [p]$.

The following proposition is proved in [257, p. 296] and is used several times in the text. The natural map in question is obtained as follows: you always have a map of simplicial spaces $\Delta[1] \times X_1 \to X$, but if $X_0 = *$ you may collapse the endpoints and get a pointed map $S^1 \wedge X_1 \to X$. Take the diagonal, and consider the adjoint map $X_1 \to \underline{S_*}(S^1, \mathrm{diag}^* X)$, which we map further to $\Omega X = \underline{S_*}(S^1, \sin|X|)$, where we have suppressed mention of the diagonal from the notation.

Proposition A.6.1.2 *Let X be a linear simplicial space with $X_0 = *$. Then the natural map*

$$X_1 \to \Omega X$$

is a weak equivalence if and only if the induced weak multiplication on X_1 has a homotopy inverse.

Proof Recall from Sect. A.2.7 the definition of the path space $PX = \{[q] \mapsto P_q X = X_{q+1}\}$. Since X is linear, we have that for each q the square

$$\begin{array}{ccccc} X_1 & \longrightarrow & P_q X & \xrightarrow{\sim} & X_q \times_{X_0} X_1 \\ \big\downarrow & & \big\downarrow & & \big\downarrow{\scriptstyle \text{proj.}} \\ X_0 = * & \longrightarrow & X_q & = & X_q \end{array}$$

is homotopy cartesian. That X_1 has a homotopy inverse implies that X and PX satisfy the π_*-Kan condition, and that $\{[q] \mapsto \pi_0(P_q X)\} \to \{[q] \mapsto \pi_0(X_q)\}$ is isomorphic to the classifying fibration $E(\pi_0 X_1) = B(\pi_0 X_1, \pi_0 X_1, *) \twoheadrightarrow B(\pi_0 X_1)$ (which is a fibration since $\pi_0 X_1$ is a group). Hence Theorem A.6.0.3 gives that

$$\begin{array}{ccc} X_1 & \longrightarrow & \mathrm{diag}^* PX \\ \big\downarrow & & \big\downarrow \\ * & \longrightarrow & \mathrm{diag}^* X \end{array}$$

is homotopy cartesian, and the result follows by the contractibility of PX. □

Applying this proposition to the bar construction of a group-like simplicial monoid, we get:

Corollary A.6.1.3 *Let M be a group-like simplicial monoid. Then the natural map $M \to \Omega BM$ is a weak equivalence.* ☺

Exactly the same argument as for Proposition A.6.1.2, but applied to the diagram

$$\begin{array}{ccc} B(*, N, M) & \longrightarrow & EM \\ \downarrow & & \downarrow \\ BN & \longrightarrow & BM \end{array}$$

yields

Corollary A.6.1.4 *Let $N \subseteq M$ be an inclusion of simplicial monoids where M is group-like. Then*

$$B(*, N, M) \to BN \to BM$$

is a fiber sequence. ☺

A.7 Homotopy Limits and Colimits

Let I be a small category, and $[I, \mathcal{S}_*]$ the category of functors from I to $\mathcal{S}_*$. This is a simplicial category in the sense that we have function spaces and "tensors" with pointed simplicial sets satisfying the usual properties (i.e., it is a tensored $\mathcal{S}$-category in the language of Definitions A.10.2.1 and A.10.2.2). If $F, G \in [I, \mathcal{S}_*]$ we define the function space to be the simplicial set $\underline{I\mathcal{S}_*}(F, G)$ whose set of q-simplices is

$$\underline{I\mathcal{S}_*}(F, G)_q = [I, \mathcal{S}_*]\big(F \wedge \Delta[q]_+, G\big)$$

i.e., the set of all pointed natural transformations $F(i) \wedge \Delta[q]_+ \to G(i)$, and whose simplicial structure comes from regarding $[q] \mapsto \Delta[q]$ as a cosimplicial object. If $F \in [I^o, \mathcal{S}_*]$ and $G \in [I, \mathcal{S}_*]$ we define

$$F \wedge G \in \mathcal{S}_*$$

to be the colimit of

$$\bigvee_{\gamma : i \to j \in I} F(j) \wedge G(i) \rightrightarrows \bigvee_{i \in I} F(i) \wedge G(i)$$

where the upper map sends the γ summand to the j summand via $1 \wedge G\gamma$, and the lower map sends the γ summand to the i summand via $F\gamma \wedge 1$ (in other words: it is the coend $\int^I F \wedge G$).

If $F \in [I^o, \mathcal{S}_*]$, $G \in [I, \mathcal{S}_*]$ and $X \in \mathcal{S}_*$, we get that

$$\underline{\mathcal{S}_*}(F \wedge G, X) \cong \underline{I^o\mathcal{S}_*}\big(F, \underline{\mathcal{S}_*}(G, X)\big) \cong \underline{I\mathcal{S}_*}\big(G, \underline{\mathcal{S}_*}(F, X)\big).$$

Recall the nerve Sect. A.2.4 and over category constructions Sect. A.2.4.2. Let $B(I/-)_+ \in [I, \mathcal{S}_*]$, be the functor which sends $i \in obI$ to $B(I/i)_+$.

Definition A.7.0.1 If $F \in [I, \mathcal{S}_*]$, then the *homotopy limit* is defined by

$$\operatorname*{holim}_{\overleftarrow{I}} F = \underline{I\mathcal{S}_*}\big(B(I/-)_+, \sin|F|\big)$$

and the *homotopy colimit* is defined by

$$\operatorname*{holim}_{\overrightarrow{I}} F = B\big(I^o/-\big)_+ \wedge F.$$

Note that according to the definitions, we get that

$$\underline{\mathcal{S}_*}\big(B\big(I^o/-\big)_+ \wedge F, X\big) \cong \underline{I^o\mathcal{S}_*}\big(B\big(I^o/-\big)_+, \underline{\mathcal{S}_*}(F, X)\big)$$

so many statements dualize. Most authors do not include the "$\sin| - |$" construction into their definition of the homotopy limit. This certainly has categorical advantages (i.e., the above duality becomes a duality on the nose between homotopy limits and colimits: $\underline{\mathcal{S}_*}(\operatorname{holim}_{\overrightarrow{I}} F, X) \cong \operatorname{holim}_{\overleftarrow{I^o}} \underline{\mathcal{S}_*}(F, X)$), but it has the disadvantage that whenever they encounter a problem in homotopy theory they have to assume that their functor has "fibrant values".

A.7.1 Connection to Categorical Notions

We can express the categorical notions in the same language using the constant functor $*: I \to \mathcal{S}_*$ with value the one-point space:

$$\lim_{\overleftarrow{I}} F = \underline{I\mathcal{S}_*}(*_+, F)$$

and

$$\lim_{\overrightarrow{I}} F = *_+ \wedge F.$$

The canonical maps $B(I/-) \to *$ and $B(I^o/-) \to *$ give natural maps (use in addition $F \to \sin|F|$ in the first map)

$$\lim_{\overleftarrow{I}} F \to \operatorname*{holim}_{\overleftarrow{I}} F, \text{ and } \operatorname*{holim}_{\overrightarrow{I}} F \to \lim_{\overrightarrow{I}} F.$$

A.7.1.1 (Co)simplicial Spaces

Let X be a cosimplicial space, e.g., a functor $X: \Delta \to \mathcal{S}_*$. The map $B(\Delta/-) \to \Delta[-]$ of cosimplicial spaces sending

$$[j] \xleftarrow{\sigma_0} [i_0] \xleftarrow{\sigma_1} \cdots \xleftarrow{\sigma_q} [i_q] \in B_q\big(\Delta/[j]\big)$$

to the map $t \mapsto \sigma_0\sigma_1\cdots\sigma_{q-t}(i_{q-t})$ in $\Delta([q],[j])$ defines a map $TotX = c\mathcal{S}_*(\Delta[-]_+, X) \to c\mathcal{S}_*(B(\Delta/-)_+, \sin|X|) = \operatorname{holim}_{\overleftarrow{\Delta}} X$. Likewise, if Y is a simplicial space (bisimplicial set), we get a map $\operatorname{holim}_{\overrightarrow{\Delta^o}} Y = B(\Delta/-)_+ \wedge Y \to \Delta[-]_+ \wedge Y \cong \operatorname{diag}^* Y$. This map is an equivalence:

Lemma A.7.1.1 [40, XII.4.3] or [135, 19.6.7] *If Y is a simplicial space, the induced map* $\operatorname{holim}_{\overrightarrow{\Delta^o}} Y \to \operatorname{diag}^* Y$ *is a weak equivalence.* ☺

The map $TotX \to \operatorname{holim}_{\overleftarrow{\Delta}} X$ is also a weak equivalence under the condition that X is a "Reedy fibrant" cosimplicial space, see e.g., [135, 19.6.4].

A.7.2 Functoriality

Let

$$I \xrightarrow{f} J \xrightarrow{F} \mathcal{S}_*$$

be functors between small categories. Then there are natural maps

$$f^*\colon \operatorname{holim}_{\overleftarrow{J}} F \to \operatorname{holim}_{\overleftarrow{I}} Ff$$

and

$$f_*\colon \operatorname{holim}_{\overrightarrow{I}} Ff \to \operatorname{holim}_{\overrightarrow{J}} F.$$

Under certain conditions these maps are equivalences.

Lemma A.7.2.1 (Cofinality Lemma, cf. [40, XI.9.2]) *Let I and J be small categories and let*

$$I \xrightarrow{f} J \xrightarrow{F} \mathcal{S}_*$$

be functors. Then

$$\operatorname{holim}_{\overrightarrow{I}} Ff \xrightarrow{f_*} \operatorname{holim}_{\overrightarrow{J}} F$$

is an equivalence if the under categories j/f are contractible for all $j \in obJ$ (f is "right cofinal"); and dually

$$\operatorname{holim}_{\overleftarrow{J}} F \xrightarrow{f^*} \operatorname{holim}_{\overleftarrow{I}} Ff$$

is an equivalence if the over categories f/j are contractible for all $j \in obJ$ (f is "left cofinal"). ☺

For a sketch of the proof, see the simplicial version, Lemma A.7.5.4 below.

The corresponding categorical statement to the cofinality lemma only uses the path components of I, and we list it here for comparison:

Lemma A.7.2.2 (Categorical Cofinality Lemma, cf. [191, p. 217]) *Let I and J be small categories and let*

$$I \xrightarrow{f} J \xrightarrow{F} \mathcal{S}_*$$

be functors. Then

$$\underset{\overrightarrow{I}}{\lim} Ff \xrightarrow{f_*} \underset{\overrightarrow{J}}{\lim} F$$

is an isomorphism if and only if the under categories j/f are connected for all $j \in \mathrm{ob}J$; and dually

$$\underset{\overleftarrow{J}}{\lim} F \xrightarrow{f^*} \underset{\overleftarrow{I}}{\lim} Ff$$

is an isomorphism if and only if the over categories f/j are connected for all $j \in \mathrm{ob}J$. ☺

Homotopy colimits are functors of "natural modules" (really of $\mathcal{S}_*$-natural modules, see Sect. A.10.4.1 for the general situation), that is the category of pairs (I, F) where I is a small category and $F: I \to \mathcal{S}_*$ is a functor. A morphism $(I, F) \to (J, G)$ is a functor $f: I \to J$ together with a natural transformation $F \to f^*G = G \circ f$ and induces the map

$$\underset{\overrightarrow{I}}{\mathrm{holim}}\, F \to \underset{\overrightarrow{I}}{\mathrm{holim}}\, f^*G \to \underset{\overrightarrow{J}}{\mathrm{holim}}\, G.$$

Homotopy limits should be thought of as a kind of cohomology. It is a functor of "natural comodules" (I, F) (really $\mathcal{S}_*$-natural comodules), that is, the category of pairs as above, but where a map $(I, F) \to (J, G)$ now is a functor $f: J \to I$ and a natural transformation $f^*F \to G$. Such a morphism induces a map

$$\underset{\overleftarrow{I}}{\mathrm{holim}}\, F \to \underset{\overleftarrow{J}}{\mathrm{holim}}\, f^*F \to \underset{\overleftarrow{J}}{\mathrm{holim}}\, G.$$

Lemma A.7.2.3 (Homotopy Lemma, cf. [40, XI.5.6 and XII.4.2]) *Let $\eta: F \to G \in [I, \mathcal{S}_*]$ be a pointwise weak equivalence (i.e., $\eta_i: F(i) \to G(i)$ is a weak equivalence for all $i \in \mathrm{ob}I$). Then*

$$\underset{\overleftarrow{I}}{\mathrm{holim}}\, F \xrightarrow{\sim} \underset{\overleftarrow{I}}{\mathrm{holim}}\, G$$

and

$$\underset{\overrightarrow{I}}{\mathrm{holim}}\, F \xrightarrow{\sim} \underset{\overrightarrow{I}}{\mathrm{holim}}\, G$$

are weak equivalences.

Proof The first statement follows from the fact that $B(I/-)$ is cofibrant and $\sin|F|$ and $\sin|G|$ are fibrant in the closed simplicial model category of $[I, \mathcal{S}_*]$ of Sect. A.4(4), and the second statement follows from duality. □

Lastly we have the following very useful observation. We do not know of any reference, but the first part is fairly obvious, and the second part follows by some work from the definition (remember that we take a functorial fibrant replacement when applying the homotopy limit):

Lemma A.7.2.4 *Let $f\colon I \subseteq J$ be an inclusion of small categories and $F\colon J \to \mathcal{S}_*$. Then the natural map*

$$f_*\colon \underset{\overrightarrow{I}}{\operatorname{holim}} Ff \to \underset{\overrightarrow{J}}{\operatorname{holim}} F$$

is a cofibration (i.e., an injection) and

$$f^*\colon \underset{\overleftarrow{J}}{\operatorname{holim}} F \to \underset{\overleftarrow{I}}{\operatorname{holim}} Ff$$

is a fibration. ☺

A.7.3 (Co)Simplicial Replacements

There is another way of writing out the definition of the homotopy (co)limit of a functor $F\colon I \to \mathcal{S}_*$. Note that

$$\underline{I\mathcal{S}_*}\big(B_q(I/-), F\big) = \prod_{i_0 \leftarrow \cdots \leftarrow i_q \in B_q(I)} F(i_0).$$

Using the simplicial structure of $B_q(I/-)$ this defines a cosimplicial space. This gives a functor

$$[I, \mathcal{S}_*] \xrightarrow{\prod^*} [\Delta, \mathcal{S}_*],$$

the so-called *cosimplicial replacement*, and the homotopy limit is exactly the composite

$$[I, \mathcal{S}_*] \xrightarrow{\sin|\ |} [I, \mathcal{S}_*] \xrightarrow{\prod^*} [\Delta, \mathcal{S}_*] \xrightarrow{\operatorname{Tot}} \mathcal{S}_*,$$

where Tot refers to the total complex of Sect. A.2.8.

Likewise, we note that

$$B_q\big(I^o/-\big)_+ \wedge F = \bigvee_{i_0 \leftarrow \cdots \leftarrow i_q \in B_q(I)} F(i_q)$$

defining a functor $\bigvee_*\colon [I,\mathcal{S}_*]\to[\Delta^o,\mathcal{S}_*]$, the so-called *simplicial replacement*, and the homotopy colimit is the composite

$$[I,\mathcal{S}_*]\xrightarrow{\bigvee_*}[\Delta^o,\mathcal{S}_*]\xrightarrow{\text{diag}^*}\mathcal{S}_*.$$

There is a strengthening of the homotopy lemma for colimits which does not dualize:

Lemma A.7.3.1 *Let $\eta\colon F\to G\in[I,\mathcal{S}_*]$ be a natural transformation such that*

$$\eta_i\colon F(i)\to G(i)$$

is n-connected for all $i\in \operatorname{ob} I$. Then

$$\operatorname*{holim}_{\overrightarrow{I}} F\to\operatorname*{holim}_{\overrightarrow{I}} G$$

is n-connected.

Proof Notice that, by the description above, the map $B_q(I^o/-)_+\wedge F\to B_q(I^o/-)_+\wedge G$ is n-connected for each q. The result then follows upon taking the diagonal. □

Lemma A.7.3.2 *Let $\cdots\twoheadrightarrow X_n\twoheadrightarrow X_{n-1}\twoheadrightarrow\cdots\twoheadrightarrow X_0\twoheadrightarrow *$ be a tower of fibrations. Then the canonical map*

$$\lim_{\overleftarrow{n}} X_n\to\operatorname*{holim}_{\overleftarrow{n}} X_n$$

is an equivalence. Dually, if $Y_0\to Y_0\to Y_2\to\cdots$ is a sequence of maps in $\mathcal{S}_*$, then the canonical map

$$\operatorname*{holim}_{\overrightarrow{n}} Y_n\to\lim_{\overrightarrow{n}} Y_n$$

is an equivalence.

Proof The limit part is covered by [40], and the colimit is easier, but we provide a pedestrian proof since many readers seem to believe that one has to assume that the maps $Y_i\to Y_{i+1}$ are inclusions. By Theorem A.6.0.1 and the simplicial replacement of the homotopy colimit it is enough to prove the claim when all the Y_k are discrete. Given $[x]\in\lim_{\overrightarrow{n}} Y_n$, there is a smallest number n such that $[x]$ is in the image of Y_n. call this number $n_{[x]}$ and choose an element $z_{[x]}\in Y_{n_{[x]}}$ mapped to $[x]$. This induces a map $\lim_{\overrightarrow{n}} Y_n\to\operatorname{holim}_{\overrightarrow{n}} Y_n$, sending $[x]$ to the vertex $z_{[x]}$ in the $n_{[x]}$-summand. The composite $\lim_{\overrightarrow{n}} Y_n\to\lim_{\overrightarrow{n}} Y_n$ is the identity, whereas the other sends the q-simplex $x\in Y_{n_q}$ in the $n_0\geq n_1\geq\cdots\geq n_q$-summand to $z_{[x]}$ in the $n_{[x]}=\cdots=n_{[x]}$-summand. The relation $n_q\geq n_{[x]}$ induces a homotopy from this composite to the identity. □

A.7.4 Homotopy (Co)Limits in Other Categories

Note that, when defining the homotopy (co)limit we only used the simplicial structure of $[I, \mathcal{S}_*]$, plus the possibility of functorially replacing any object by an equivalent (co)fibrant object. If $\mathcal{C}$ is any category with all (co)products (at least all those indexed by the various $B_q(I/i)$s etc.), we can define the (co)simplicial replacement functors for any $F \in [I, \mathcal{C}]$:

$$\prod{}^* F = \left\{ [q] \mapsto \prod_{i_0 \leftarrow \cdots \leftarrow i_q \in B_q(I)} F(i_0) \right\}$$

and

$$\coprod{}_* F = \left\{ [q] \mapsto \coprod_{i_0 \leftarrow \cdots \leftarrow i_q \in B_q(I)} F(i_q) \right\}.$$

In the special case of a closed simplicial model category, we can always precompose $\prod^*$ (resp. $\coprod_*$) with a functor assuring that $F(i)$ is (co)fibrant to get the right homotopy properties.

As an easy example, we could consider unbased spaces. For $F \in [I, \mathcal{S}]$ we let $\operatorname{holim}_{\overleftarrow{I}} F = Tot(\prod^* \sin|F|)$ and $\operatorname{holim}_{\overrightarrow{I}} F = \operatorname{diag}^* \coprod_* F$. Recall the adjoint functor pair

$$\mathcal{S} \underset{U}{\overset{X \mapsto X_+}{\rightleftarrows}} \mathcal{S}_*.$$

We get that if $F \in [I, \mathcal{S}]$ and $G \in [I, \mathcal{S}_*]$, then $\prod^* UG = U \prod^* G$, and $(\coprod_* F)_+ = \bigvee_* (F_+)$, so

$$U \operatorname{holim}_{\overleftarrow{I}} G \cong \operatorname{holim}_{\overleftarrow{I}} UG, \text{ and } \left(\operatorname{holim}_{\overrightarrow{I}} F\right)_+ \cong \operatorname{holim}_{\overrightarrow{I}} (F_+).$$

On the other hand, note that $U \operatorname{holim}_{\overrightarrow{I}} G \cong \operatorname{holim}_{\overrightarrow{I}} UG/BI$, so that

Lemma A.7.4.1 *Let $G\colon I \to \mathcal{S}_*$ and $U\colon \mathcal{S}_* \to \mathcal{S}$ be the forgetful functor to unbased spaces. If BI is contractible, then the projection* $\operatorname{holim}_{\overrightarrow{I}} UG \to U \operatorname{holim}_{\overrightarrow{I}} G$ *is a weak equivalence.*

Furthermore, the simplicial replacement of the homotopy colimit makes the following version of Quillen's theorem B accessible. The proof below is modelled on [99, IV 5.7].

Proposition A.7.4.2 *Let I be a small category, let λ be a natural number and let $X\colon I \to \mathcal{S}$ be such that for any $f\colon i \to j \in I$ the induced map $X(f)\colon X(i) \to X(j)$ is λ-connected. Let* $\operatorname{holim}_{\overrightarrow{I}} X \to \operatorname{holim}_{\overrightarrow{I}} * \cong BI$ *be induced by the natural transformation $X(i) \to *$.*

Given any object i in I, the map from $X(i)$ to the homotopy fiber of $\operatorname{holim}_{\overrightarrow{I}} X \to BI$ *over the vertex $i \in B_0 I$ is λ-connected.*

Proof Consider the map $* \to BI$ sending the single point to $i \in B_0 I$. Quillen's small object argument gives a factorization $* = Z_0 \overset{\sim}{\rightarrowtail} Z_1 \overset{\sim}{\rightarrowtail} Z_2 \overset{\sim}{\rightarrowtail} \cdots \overset{\sim}{\rightarrowtail} \lim_{\overrightarrow{n}} Z_n = Z \twoheadrightarrow BI$ through a fibration $Z \twoheadrightarrow BI$ where Z is contractible as follows: if Z_{n-1} has been constructed, let D_n be the set of all diagrams

$$\begin{array}{ccc} K & \rightarrowtail & \Delta[q] \\ \downarrow & & \downarrow \\ Z_{n-1} & \longrightarrow & BI \end{array}$$

where $K \subseteq \Delta[q]$ is any contractible subspace (it is enough to consider the "horns", but that only clutters up the notation and adds no clarity). Then we define Z_n as the pushout

$$\begin{array}{ccc} \coprod_{D_n} K & \rightarrowtail & \coprod_{D_n} \Delta[q] \\ \downarrow & & \downarrow \\ Z_{n-1} & \longrightarrow & Z_n . \end{array}$$

Clearly the inclusion $Z_{n-1} \to Z_n$ is a weak equivalence, and since the K's are "small" (i.e., the unbased version of Lemma A.2.5.1 is true with $Y = K$), the very construction shows that $Z \to BI$ is a fibration.

We must show that the induced map $X(i) = * \times_{BI} \operatorname{holim}_{\overrightarrow{I}} X \to Z \times_{BI} \operatorname{holim}_{\overrightarrow{I}} X$ is λ-connected, which follows if we can show that $Z_{n-1} \times_{BI} \operatorname{holim}_{\overrightarrow{I}} X \to Z_n \times_{BI} \operatorname{holim}_{\overrightarrow{I}} X$ is λ-connected for each n, which ultimately follows if we can show that each $K \times_{BI} \operatorname{holim}_{\overrightarrow{I}} X \to \Delta[q] \times_{BI} \operatorname{holim}_{\overrightarrow{I}} X$ is λ-connected. In level r the bisimplicial replacement of this map can be identified with the inclusion

$$\coprod_{s_0 \le \cdots \le s_r \in K_r} X\sigma(s_r) \subseteq \coprod_{s_0 \le \cdots \le s_r \in B_r[q]} X\sigma(s_r),$$

where we have written elements in $\Delta[q]_r = B_r[q]$ as sequences $0 \le s_0 \le \cdots \le s_r \le q$. Now, q is an initial object in $[q]$, giving the vertical maps in

$$\begin{array}{ccc} \coprod_{s_0 \le \cdots \le s_r \in K_r} X\sigma(q) & \longrightarrow & \coprod_{s_0 \le \cdots \le s_r \in B[q]} X\sigma(q) \\ \downarrow & & \downarrow \\ \coprod_{s_0 \le \cdots \le s_r \in K_r} X\sigma(s_r) & \longrightarrow & \coprod_{s_0 \le \cdots \le s_r \in B[q]} X\sigma(s_r), \end{array}$$

which by assumption are λ-connected. However, letting r vary, the diagonal of the top horizontal map is the weak equivalence $K \wedge X\sigma(q) \overset{\sim}{\rightarrowtail} \Delta[q] \wedge X\sigma(q)$. □

A.7.4.1 Simplicial Abelian Groups

In abelian groups, the product is the product of the underlying sets, whereas the coproduct is the direct sum. All simplicial abelian groups are fibrant, and we choose a functorial factorization $0 \rightarrowtail C(M) \stackrel{\sim}{\twoheadrightarrow} M$, for instance the one coming from the free/forgetful adjoint functor pair to sets. Note that the diagonal (total) of a (co)simplicial simplicial abelian group is a simplicial abelian group, and we define

$$\underset{\overleftarrow{I}}{\operatorname{holim}} F = Tot \prod{}^{*} F.$$

and

$$\underset{\overrightarrow{I}}{\operatorname{holim}} F = \operatorname{diag}^{*} \coprod{}_{*} F$$

Note that this last definition is "wrong" in that we have not replaced $F(i)$ by a cofibrant object, but this does not matter since

Lemma A.7.4.3 *Let $F\colon I \to A$ be a functor, and let $C\colon A \to A$ be a functorial cofibrant replacement. Then the map*

$$\underset{\overrightarrow{I}}{\operatorname{holim}} F \xleftarrow{\simeq} \underset{\overrightarrow{I}}{\operatorname{holim}} CF$$

is a weak equivalence.

Proof This follows by forgetting down to simplicial spaces, using that homotopy groups commute with filtered colimits Lemma A.2.5.3 and finite products Proposition A.2.1.1, and finally that a degreewise equivalence of bisimplicial sets induces an equivalence on the diagonal Theorem A.6.0.1. □

If F is a functor to abelian groups, the (co)homotopy groups of the (co)simplicial replacement functors above are known to algebraists as the derived functors of the (co)limit, i.e.,

$$\underset{\overleftarrow{I}}{\lim}{}^{(s)} F = H^{s}(I, F) = \pi^{s} \prod{}^{*} F, \quad \text{and} \quad \underset{\overrightarrow{I}}{\lim}{}_{(s)} F = H_{s}(I, F) = \pi_{s} \coprod{}_{*} F.$$

Definition A.7.4.4 A category J has *finite cohomological dimension* if there is some n such that for all functors $F\colon J \to Ab$ and $s > n$ the sth derived limit vanishes, $\lim_{\overleftarrow{J}}{}^{(s)} F = 0$.

For instance, the category $\mathbf{N} = \{\cdots \to 2 \to 1 \to 0\}$ of natural numbers has finite cohomological dimension ($n = 1$).

Finite cohomological dimension ensures convergence of Bousfield and Kan's spectral sequence [40, XI.7, XII.5] for homotopy limits.

Theorem A.7.4.5 *Let $X\colon J \to \mathcal{S}_*$ be a functor. If the $\pi_q X$ take values in abelian groups for all $q \geq 0$, then there is a spectral sequence with E^2 term*

$$E^2_{s,t} = \lim_{\overleftarrow{J}}{}^{(-s)} \pi_t X, \quad 0 \leq -s \leq t$$

which under favourable conditions converges (in Bousfield and Kan's language:, is "closely related") to $\pi_{s+t} \operatorname{holim}_{\overleftarrow{J}} X$. Especially, if J has finite cohomological dimension, the spectral sequence converges. If $J = \mathbf{N}$ it collapses to the exact sequence

$$0 \to \lim_{\overleftarrow{\mathbf{N}}}{}^{(1)} \pi_{t+1} X \to \pi_t \operatorname{holim}_{\overleftarrow{\mathbf{N}}} X \to \lim_{\overleftarrow{\mathbf{N}}} \pi_t X \to 0.$$

If h is some connected reduced homology theory satisfying the wedge axiom then there is a convergent spectral sequence

$$E^2_{s,t} = \lim_{\overrightarrow{J}}{}_{(s)} h_t X \to h_{s+t} \operatorname{holim}_{\overrightarrow{J}} X.$$ ☺

The homotopy limits in abelian groups coincide with what we get if we forget down to $\mathcal{S}_*$, but generally the homotopy colimit will differ. However, if $F\colon I \to Ab$ is a functor and $U\colon Ab \to Ens_*$ is the forgetful functor, there is a natural map

$$\operatorname{holim}_{\overrightarrow{I}} UF \longrightarrow U\operatorname{holim}_{\overrightarrow{I}} F = U\operatorname{diag}^*\left\{[q] \mapsto \bigoplus_{i_0 \leftarrow i_1 \leftarrow \cdots \leftarrow i_q} X(i_q)\right\}$$

given by sending wedges to sums. We leave the proof of the following lemma as an exercises (use that homotopy groups commute with filtered colimit, Lemma A.2.5.3 and the Blakers–Massey Theorem A.8.2.2)

Lemma A.7.4.6 *Let $F\colon I \to A$ be a functor such that $F(i)$ is n-connected for all $i \in \operatorname{ob} I$. Then*

$$\operatorname{holim}_{\overrightarrow{I}} UF \to U \operatorname{holim}_{\overrightarrow{I}} F$$

is $(2n+1)$-connected. ☺

A.7.4.2 Spectra

The category of spectra has two useful notions of fibrations and weak equivalences, the stable Sect. A.4.1(6) and the levelwise Sect. A.4.1(5). For the levelwise case there is no difference from the space case, and so we concentrate on the stable structure. Any spectrum is levelwise equivalent to a cofibrant spectrum (i.e., one for which all the structure maps $S^1 \wedge X^k \to X^{k+1}$ are cofibrations, see Lemma A.3.3.1), so it is no surprise that the levelwise homotopy colimit has good properties also with respect to the stable structure. For homotopy limits we need as usual a bit of preparations. We choose a fibrant replacement functor $X \mapsto QX$ as in Sect. A.3.2.1. Let $X\colon J \to \mathcal{S}pt$ be a functor from a small category to spectra. Then

$$\operatorname*{holim}_{\overrightarrow{J}} X = \left\{k \mapsto \operatorname{diag}^* \coprod{}_* X^k\right\}$$

which is just $\operatorname{holim}_{\to} J$ applied levelwise, and

$$\operatorname*{holim}_{\overleftarrow{J}} X = \left\{k \mapsto Tot^* \prod{}^* Q^k X\right\}$$

which is equivalent to $k \mapsto \operatorname{holim}_{\overleftarrow{J}} Q^k X$ (we just have skipped the extra application of $\sin|-|$).

Lemma A.7.4.7 *Levelwise homotopy limits and colimits preserve stable equivalences of spectra.*

Proof For the homotopy limit this is immediate from the construction since all stable equivalences are transformed into levelwise equivalences of levelwise fibrant spectra by Q. For the homotopy colimit, notice that we just have to prove that for a diagram of spectra X, the canonical map $X \to QX$ induces a stable equivalence of homotopy colimits. By Lemma A.3.3.1 we may assume that all spectra in X are $-n$ connected, and then Freudenthal's suspension Theorem A.8.2.3 gives that the maps in $X^k \to Q^k X$ are $2k-n$ connected. Since homotopy colimits preserve connectivity (Lemma A.7.3.1) this means that the map of levelwise homotopy colimits is a weak equivalence. □

A.7.5 Enriched Homotopy (Co)Limits

There are "enriched" versions of homotopy (co)limits. In fact, Hochschild homology itself is a close relative of a homotopy colimit with an *Ab*-enrichment (see Remark A.7.5.2 below). We will spell out the details in the case of enrichment in simplicial sets.

Let I be a small $\mathcal{S}$-category, i.e., a category with a set of objects obI and morphism *spaces* $\underline{I}(i,i')$ satisfying the usual axioms for a category, see Sect. A.10.2 for precise definitions and some background on enriched categories. Let $F\colon I \to \mathcal{S}_*$ be an $\mathcal{S}$-functor, i.e., a collection of maps $\underline{I}(i,i') \to \underline{\mathcal{S}_*}(F(i),F(i'))$ satisfying the usual axioms.

Then we may define the $\mathcal{S}$-homotopy (co)limit of F as follows. Given $i_{-1} \in obI$, let

$$N_q^{\mathcal{S}}(I/i_{-1}) = \coprod_{i_0,\dots,i_q} \prod_{0\le k\le q} \underline{I}(i_k, i_{k-1}) \in \mathcal{S},$$

and define the (co)simplicial replacements by

$$\underset{\overleftarrow{I}}{\operatorname{holim}}^{\mathcal{S}_*} F = \left\{[q] \mapsto \int_I \underline{\mathcal{S}}\big(N_q^{\mathcal{S}}(I/-), F\big) \cong \prod_{i_0,\dots,i_q} \underline{\mathcal{S}}\Big(\prod_{1\le k\le q} \underline{I}(i_k, i_{k-1}), F(i_0)\Big)\right\}$$

and

$$\underset{\overrightarrow{I}}{\operatorname{holim}}^{\mathcal{S}_*} F = \left\{[q] \mapsto \int^I N_q^{\mathcal{S}}\big(I^o/-\big)_+ \wedge F \cong \bigvee_{i_0,\dots,i_q} \bigwedge_{1\le k\le q} \underline{I}(i_k, i_{k-1})_+ \wedge F(i_q)\right\}$$

and finally, let

$$\underset{\overleftarrow{I}}{\operatorname{holim}} F = \operatorname{Tot} \underset{\overleftarrow{I}}{\operatorname{holim}}^{\mathcal{S}_*} \sin|F| \text{ and } \underset{\overrightarrow{I}}{\operatorname{holim}} F = \operatorname{diag}^* \underset{\overrightarrow{I}}{\operatorname{holim}}^{\mathcal{S}_*} X.$$

We see that the homotopy limit is a functor of "$\mathcal{S}$-natural comodules", and homotopy colimits are functors of "$\mathcal{S}$-natural modules" (see Sect. A.10.4.1 for terminology).

Example A.7.5.1 As a particularly important example, let G be a simplicial monoid, and X a G-space (i.e., an $\mathcal{S}$-functor $X\colon G \to \mathcal{S}_*$), then the homotopy fixed point and orbit spaces are just $X^{hG} = \operatorname{holim}_{\overleftarrow{G}} X$ and $X_{hG} = \operatorname{holim}_{\overrightarrow{G}} X$. See Sect. A.9 for further details.

Remark A.7.5.2 The notion of homotopy (co)limits carries over almost word by word to other enriched situations but for one small detail. In the definitions above, the top face map in the nerve $N_q^{\mathcal{S}}(I/i_{-1})$ uses the preferred map $\underline{I}(i_q, i_{q-1}) \to *$ coming from the fact that the unit element for the cartesian product is also terminal in the category of spaces. In other words, the functor $I \to \mathcal{S}_*$ sending everything to the one point space is an "I-module". This has no analog in general, so we have to plug in explicit modules on both sides of the construction, or even better, a "bimodule". That is, if $V = (V, \otimes, e)$ is a closed symmetric monoidal category and $\mathcal{C}$ is a cotensored V-category with all coproducts, let $M\colon I^o \otimes I \to \mathcal{C}$ be a V-functor. Then one may define

$$HH(I, M) = \left\{[q] \mapsto \coprod_{i_0,\dots,i_q} M(i_0, i_q) \otimes \bigotimes_{j=1}^{q} \underline{I}(i_j, i_{j-1})\right\},$$

with Hochschild-style face and degeneracy maps.

If there is a sensible projection $pr\colon I^o \otimes I \to I$ one may define the homotopy colimit of a V-functor $F\colon I \to \mathcal{C}$ as $\operatorname{diag}^* HH(I, F\,pr)$. Likewise we define a Hochschild cohomology which may give rise to a homotopy limit.

All the usual results for homotopy (co)limits generalize, for instance

Lemma A.7.5.3 (Homotopy Lemma) *Let I be a small $\mathcal{S}$-category, let $X, Y\colon I \to \mathcal{S}_*$ be $\mathcal{S}$-functors, and let $\eta\colon X \to Y$ be an $\mathcal{S}$-natural equivalence* (*see Sect.* A.10.2.1).

Then η *induces weak equivalences* $\operatorname{holim}_{\overleftarrow{I}} X \xrightarrow{\sim} \operatorname{holim}_{\overleftarrow{I}} Y$ *and* $\operatorname{holim}_{\overrightarrow{I}} X \xrightarrow{\sim} \operatorname{holim}_{\overrightarrow{I}} Y$.

Proof The homotopy colimit statement is clear since by Theorem A.6.0.1 a map of simplicial spaces which induces an equivalence in each degree induce an equivalence on the diagonal. For the homotopy limit case, the proof proceeds just as the one sketched in [40, p. 303]: first one shows that

$$\left\{[q] \mapsto \prod_{i_0,\dots,i_q} \underline{\mathcal{S}_*}\left(\bigwedge_{1\le k\le q} \underline{I}(i_k, i_{k-1})_+, \sin\left|X(i_0)\right|\right)\right\}$$

is a fibrant cosimplicial space (this uses the "matching spaces" of [40, p. 274], essentially you fix an i_0 and use that the degeneracy map

$$\sum_j s_j\colon \bigvee_{0\le j\le q} \bigvee_{i_1,\dots,i_{q-1}} \bigwedge_{1\le k\le q-1} \underline{I}(i_k, i_{k-1})_+ \to \bigvee_{i_1,\dots,i_q} \bigwedge_{1\le k\le q} \underline{I}(i_k, i_{k-1})_+$$

is an inclusion). Then one uses that a map of fibrant cosimplicial spaces that is a pointwise equivalence, induces an equivalence on Tot. □

The (co)finality statements carry over from the discrete case. Since we use the following version in the text we spell it out in all detail

Lemma A.7.5.4 (Cofinality Lemma) *Let* $f\colon I \to J$ *be an* $\mathcal{S}$*-functor. Then*

$$\operatorname{holim}_{\overleftarrow{J}} F \xrightarrow{f^*} \operatorname{holim}_{\overleftarrow{I}} Ff$$

is an equivalence for all $\mathcal{S}$*-functors* $F\colon J \to \mathcal{S}_*$ *if and only if* f *is "left cofinal" in the sense that for all* $j \in \operatorname{ob} J$ $N^{\mathcal{S}}(f/j)$ *is contractible.*

Proof Assume $* \simeq N^{\mathcal{S}}(f/j)$, and let $X = \sin|F|$. Consider the bicosimplicial space C which in bidegree p, q is given by

$$C^{pq} = \prod_{\substack{i_0,\dots,i_p\in I\\ j_0,\dots,j_q\in J}} \underline{\mathcal{S}_*}\Bigg(\underline{J}\big(f(i_0), j_q\big)_+ \wedge \bigwedge_{1\le k\le p} \underline{I}(i_k, i_{k-1})_+ \wedge \bigwedge_{1\le l\le q} \underline{J}(j_l, j_{l-1})_+, X(j_0)\Bigg).$$

Fixing q, we get a cosimplicial space

$$\prod_{j_0,\dots,j_q\in J} \underline{\mathcal{S}_*}\left(\bigwedge_{1\le l\le q} \underline{J}(j_l, j_{l-1})_+, \underline{\mathcal{S}_*}\big(N^{\mathcal{S}}(f/j_q)_+, X(j_0)\big)\right)$$

which by hypothesis is equivalent to

$$\prod_{j_0,\dots,j_q\in J} \underline{\mathcal{S}_*}\Big(\bigwedge_{1\le l\le q} \underline{J}(j_l, j_{l-1})_+, X(j_0)\Big)$$

which, when varying q and taking the total, is $\operatorname{holim}_{\overleftarrow{J}} X$.

Fixing p we get a cosimplicial space

$$[q]\mapsto \prod_{i_0,\dots,i_p\in I} \underline{\mathcal{S}_*}\Big(\bigwedge_{1\le k\le p} \underline{I}(i_k, i_{k-1})_+,$$
$$\prod_{j_0,\dots,j_q\in J} \underline{\mathcal{S}_*}\Big(\underline{J}(f(i_0), j_q)_+\wedge \bigwedge_{1\le l\le q} \underline{J}(j_l, j_{l-1})_+, X(j_0)\Big)\Big).$$

Note that $X(f(i_0))\to \{[q]\mapsto \prod_{j_0,\dots,j_q\in J} \underline{\mathcal{S}_*}(\underline{J}(f(i_0), j_q)_+\wedge \bigwedge_{1\le l\le q} \underline{J}(j_l, j_{l-1})_+, X(j_0))\}$ is an equivalence (the right hand side has an extra codegeneracy), and so, when varying p again, we get $\operatorname{holim}_{\overleftarrow{I}} Ff$. One also has to show compatibility with the map in the statement.

In the opposite direction, let $F(j)=\underline{\mathcal{S}_*}(\underline{J}(j, j'), Z)$ for some $j'\in obJ$ and fibrant space Z. Writing out the cosimplicial replacements for $\operatorname{holim}_{\overleftarrow{J}} F$ and $\operatorname{holim}_{\overleftarrow{J}} Ff$ we get that the first is $\underline{\mathcal{S}_*}(N(J/j')_+, Z)\simeq Z$, whereas the latter is $\underline{\mathcal{S}_*}(N(f/j')_+, Z)$, and so $N(f/j')$ must be contractible. □

Note that in the proof, for a given $F\colon J\to \mathcal{S}_*$, the crucial point was that for all $j, j'\in obJ$, we had an equivalence $\underline{\mathcal{S}_*}(N^{\mathcal{S}}(f/j')_+, X(j))\simeq X(j)$. This gives the corollary

Corollary A.7.5.5 *Given $\mathcal{S}$-functors*

$$I \xrightarrow{f} J \xrightarrow{F} \mathcal{S}_*,$$

then

$$\operatorname{holim}_{\overleftarrow{J}} F \xrightarrow{f^*} \operatorname{holim}_{\overleftarrow{I}} Ff$$

is an equivalence if for all $j, j'\in obJ$, the projection $N^{\mathcal{S}}(f/j')_+\to S^0$ induces an equivalence $Map_*(N^{\mathcal{S}}(f/j')_+, F(j))\simeq F(j)$. ☺

A.7.6 Completions and Localizations

We review the basic facts about the completion and localization functors. The authoritative references are Bousfield and Kan's book [40], and Bousfield's papers [37] and [38]. Let R be either the field $\mathbf{F}_p$ with p elements for a prime p or a subring of the rationals $\mathbf{Q}$. The free/forgetful adjoint pair

$$sR\text{-mod} \overset{\tilde{R}}{\leftrightarrows} \mathcal{S}_*$$

gives rise to a cosimplicial functor on spaces which in dimension q takes $X \in \mathcal{S}_*$ to the simplicial R-module $\tilde{R}^{q+1}(X)$ considered as a space. In favourable circumstances the total, or homotopy limit $R_\infty X$, of this cosimplicial space has the right properties of an *R-completion*. We say that X is *good* (with respect to R) if $X \to R_\infty X$ induces an isomorphism in R-homology, and $R_\infty X \to R_\infty R_\infty X$ is an equivalence.

Especially, Bousfield and Kan [40] prove that simply connected spaces and loop spaces are good, and so R-completion of connective spectra is well behaved (this is a homotopy limit construction, so we should be prepared to make our spectra Ω-spectra before applying R_∞ to each space).

Explicitly, for a spectrum X, let J be a set of primes, I the set of primes not in J and p any prime, we let

$$X_{(J)} = \{k \mapsto X^k_{(J)} = (\mathbf{Z}[I^{-1}])_\infty X^k\}$$

and

$$X\hat{_p} = \{k \mapsto (Q^k X)\hat{_p} = (\mathbf{Z}/p\mathbf{Z})_\infty (Q^k X)\}.$$

For many purposes it is advantageous to use Bousfield's model [38]

$$X\hat{_p} = \underline{\mathcal{S}pt}(\Sigma^{-1} M\mathbf{Z}/p^\infty, X)$$

for the completion, where $M\mathbf{Z}/p^\infty$ is the Moore spectrum Example A.3.3.2 associated to the abelian group $\mathbf{Z}/p^\infty = \mathbf{Z}[1/p]/\mathbf{Z}$. For instance, this way one clearly sees that profinite completion commutes with homotopy limits. Using that $\mathbf{Q}/\mathbf{Z} \cong \bigoplus_{p \text{ prime}} \mathbf{Z}[1/p]/\mathbf{Z}$ we may choose $M\mathbf{Q}/\mathbf{Z} = \bigvee_{p \text{ prime}} M\mathbf{Z}[1/p]/\mathbf{Z}$, and the *profinite completion* $X\hat{}$ may then be written as

$$X\hat{} = \underline{\mathcal{S}pt}(\Sigma^{-1} M\mathbf{Q}/\mathbf{Z}, X). \tag{A.2}$$

When J is empty, $X_{(J)}$ is the rationalization of X, which is more customarily denoted $X_{(0)}$ or $X_\mathbf{Q}$. We say that X is *rational* if $X \to X_\mathbf{Q}$ is an equivalence, which is equivalent to asserting that $\pi_* X$ is a graded rational vector space. Generally, $X_{(J)}$ is a *localization*, in the sense that $X \to X_{(J)}$ induces an equivalence in spectrum homology with coefficients in $\mathbf{Z}[I^{-1}]$, and $\pi_* X_{(J)} \cong \pi_* X \otimes \mathbf{Z}[I^{-1}]$.

Also, $X\hat{_p}$ is a *p-completion* in the sense that $X \to X\hat{_p}$ induces an equivalence in spectrum homology with coefficients in $\mathbf{Z}/p\mathbf{Z}$, and there is a natural short exact (non-naturally splittable) sequence

$$0 \to \mathrm{Ext}^1(C_{p^\infty}, \pi_* X) \to \pi_* X\hat{_p} \to \mathrm{Hom}(C_{p^\infty}, \pi_{*-1} X) \to 0$$

where $C_{p^\infty} = \mathbf{Z}[1/p]/\mathbf{Z}$.

There is an "arithmetic square" [38]:

Theorem A.7.6.1 *Let X be any spectrum, then*

$$\begin{array}{ccc} X & \longrightarrow & X_{\mathbf{Q}} \\ \downarrow & & \downarrow \\ \prod_{p\text{prime}} X\hat{_p} & \longrightarrow & (\prod_{p\text{prime}} X\hat{_p})_{\mathbf{Q}} \end{array}$$

is homotopy cartesian. ☺

Also, from the description of completion and localization in Bousfield we get that p-completion commutes with arbitrary homotopy limits and J-localization with arbitrary homotopy colimits.

One says that an abelian group M is Ext-p-*complete* if $M \to \mathrm{Ext}(C_{p^\infty}, M)$ is an isomorphism and $\mathrm{Hom}(C_{p^\infty}, M) = 0$. A spectrum X is p-complete (i.e., $X \to X\hat{_p}$ is an equivalence) if and only if $\pi_* X$ is Ext-p-complete.

Lemma A.7.6.2 *Any simplicial space satisfying the π_*-Kan condition and which is "good" in every degree (and in particular, any simplicial spectrum) may be p-completed or localized degreewise.*

Proof We prove the less obvious completion part. Let Y be the simplicial space $\{q \mapsto (X_q)\hat{_p}\}$. We must show that the map $\mathrm{diag}^* X \to \mathrm{diag}^* Y$ is a p-completion. Use the spectral sequence of Theorem A.6.0.5 for the simplicial space Y, and that Ext-p completeness is closed under extension to see that $\mathrm{diag}^*(Y)$ is p-complete. Then use the spectral sequence for the simplicial space $\mathbf{F}_p Y$ to see that $H_*(\mathrm{diag}^* X, \mathbf{F}_p) \to H_*(\mathrm{diag}^* Y, \mathbf{F}_p)$ is an isomorphism. □

We end this section with two results that are needed in the text.

Lemma A.7.6.3 *Given $\mathcal{S}$-functors*

$$I \xrightarrow{f} J \xrightarrow{F} \mathcal{S}_*$$

such that F has p-complete values and such that $(N^{\mathcal{S}}(f/j))\hat{_p}$ is contractible for every $j \in ob J$, then

$$\underset{\overleftarrow{J}}{\mathrm{holim}} F \xrightarrow{f^*} \underset{\overleftarrow{I}}{\mathrm{holim}} Ff$$

is an equivalence.

Proof This follows from Corollary A.7.5.5 of the cofinality lemma and the fact that if A is a space and B is a p-complete space, then $Map(A\hat{_p}, B) \xrightarrow{\sim} Map(A, B)$. □

Corollary A.7.6.4 *The inclusion $C_{p^\infty} = \lim_{r\to\infty} C_{p^r} \subseteq \mathbf{S^1} = \sin|S^1|$ induces a weak equivalence*

$$BC_{p^\infty} \xrightarrow{\sim} B\mathbf{S^1} \simeq K(\mathbf{Z}, 2)$$

after p-completion. Thus we get that for any p-complete space or spectrum X with $\mathbf{S^1}$*-action, that the map*

$$X^{h\mathbf{S^1}} \to X^{hC_{p^\infty}}$$

is an equivalence.

Proof Given Lemma A.7.6.3 we only need to see that $BC_{p^\infty}\widehat{{}_p} \xrightarrow{\sim} B\mathbf{S^1}\widehat{{}_p}$: We have a short exact sequence $C_{p^\infty} \subseteq \mathbf{S^1} \to \lim_{\overrightarrow{p}} \mathbf{S^1}$, and so it is enough to show that $B(\lim_{\overrightarrow{p}} \mathbf{S^1})\widehat{{}_p} \simeq *$. But this is clear, since the homotopy groups of $B(\lim_{\overrightarrow{p}} \mathbf{S^1}) \simeq K(\mathbf{Z}[1/p], 2)$ are uniquely p-divisible. □

A.7.6.1 Completions and Localizations of Simplicial Abelian Groups

If M is a simplicial abelian group, then we can complete or localize the Eilenberg–Mac Lane spectrum HM. The point here is that this gives new Eilenberg–Mac Lane spectra which can be described explicitly. The proofs of the statements below follow from the fact that Eilenberg–Mac Lane spectra, and completion and localization are determined by their homotopy groups.

Let $M \in obA = sAb$ be a simplicial abelian group. Then $H(M \otimes_{\mathbf{Z}} \mathbf{Q})$ is clearly a model for $HM_{(0)}$. The map $HM \to HM_{(0)}$ is given by $M \cong M \otimes_{\mathbf{Z}} \mathbf{Z} \to M \otimes_{\mathbf{Z}} \mathbf{Q}$.

Choose a free resolution $R \overset{\sim}{\twoheadrightarrow} \mathbf{Z}[1/p]/\mathbf{Z}$. Then we may define the p-completion as

$$M\widehat{{}_p} = \underline{A}\big(R, \tilde{\mathbf{Z}}[S^1] \otimes_{\mathbf{Z}} M\big)$$

(internal function object in A, see Sect. A.3.1.1) which is a simplicial abelian group whose Eilenberg–Mac Lane spectrum $H(M\widehat{{}_p})$ is equivalent to $(HM)\widehat{{}_p}$ (note the similarity with the up to homotopy definition commonly used for spectra). The homotopy groups are given by considering the second quadrant spectral sequence (of the bicomplex associated with the simplicial direction of the morphism spaces and the cosimplicial direction of the resolution R—just as explained in Sect. A.3.1.4 for simplicial abelian groups, cosimplicial abelian groups give rise to chain complexes)

$$E^2_{s,t} = \operatorname{Ext}^{-s}_{\mathbf{Z}}\big(\mathbf{Z}[1/p]/\mathbf{Z}, \pi_{t-1}M\big) \Rightarrow \pi_{s+t}M\widehat{{}_p}$$

whose only nonvanishing columns are in degree 0 and -1. The map $M \to M\widehat{{}_p}$ is given as follows. Let $Q = R \times_{\mathbf{Z}[1/p]/\mathbf{Z}} \mathbf{Z}[1/p]$, and consider the short exact sequence

$$0 \to \mathbf{Z} \to Q \to R \to 0$$

giving rise to the exact sequence

$$0 \to A\big(R, \tilde{\mathbf{Z}}[S^1] \otimes_{\mathbf{Z}} M\big) \to A\big(Q, \tilde{\mathbf{Z}}[S^1] \otimes_{\mathbf{Z}} M\big) \to \tilde{\mathbf{Z}}[S^1] \otimes_{\mathbf{Z}} M \to 0$$

which gives the desired map $M \to M\widehat{{}_p}$.

A.8 Cubical Diagrams

Cubical diagrams are used in many places in the text. We collect some useful facts here for reference.

Definition A.8.0.1 If S is a set, let $\mathcal{P}S$ be the category of subsets of S and inclusions. We introduce the shorthand $\mathcal{P} = \mathcal{P}\{1, 2, \dots\}$ and $\mathcal{P}n = \mathcal{P}\{1, \dots, n\}$. We let $\mathcal{P}^{\neq\emptyset}S \subseteq \mathcal{P}S$, $\mathcal{P}^{\neq\emptyset} \subseteq \mathcal{P}$ and $\mathcal{P}^{\neq\emptyset}n \subseteq \mathcal{P}n$ be the subcategories of nonempty subsets.

An S-*cube* is a functor $\mathcal{X}$ from the category $\mathcal{P}S$. A *cubical diagram* is a functor from $\mathcal{P}$. An n-*cube* is an S-cube for some S of cardinality $|S| = n$.

A d-*subcube* of an S-cube $\mathcal{X}$ is a d-cube resulting as the precomposite of $\mathcal{X}$ along an injection $F\colon \mathcal{P}T \to \mathcal{P}S$ satisfying that if $U, V \subseteq T$, then $F(U \cap V) = F(U) \cap F(V)$ and $F(U \cup V) = F(U) \cup F(V)$. A d-*face* is a d-subcube induced by an F given by $F(V) = U \cup f(V)$ where $f\colon \{1, \dots, d\} \to S$ is an injection and $U \subseteq S$ is disjoint from the image of f.

So, a 0-cube is an object $\mathcal{X}_\emptyset$, a 1-cube is a map $\mathcal{X}_\emptyset \to \mathcal{X}_{\{1\}}$, and a 2-cube is a (commuting) square

$$\begin{array}{ccc} \mathcal{X}_\emptyset & \longrightarrow & \mathcal{X}_{\{1\}} \\ \downarrow & & \downarrow \\ \mathcal{X}_{\{2\}} & \longrightarrow & \mathcal{X}_{\{1,2\}} \end{array}$$

and so on. Note that if $\mathcal{X}$ is a 47-cube, then

$$\begin{array}{ccc} \mathcal{X}_{\{3\}} & \longrightarrow & \mathcal{X}_{\{1,3,4\}} \\ \downarrow & & \downarrow \\ \mathcal{X}_{\{2,3\}} & \longrightarrow & \mathcal{X}_{\{1,2,3,4\}} \end{array}$$

is a 2-subcube, but not a 2-face.

We will regard a natural transformation of n-cubes $\mathcal{X} \to \mathcal{Y}$ as an $n+1$-cube. In particular, if $\mathcal{X}$ is an n-cube and $F \to G$ is a natural transformation of functors from the target category of $\mathcal{X}$, then we get an $n+1$-cube $F\mathcal{X} \to G\mathcal{X}$.

Definition A.8.0.2 Let $\mathcal{X}$ be an S-cube with values in any one of the categories where we have defined homotopy (co)limits (see Sect. A.7). We say that $\mathcal{X}$ is k-*cartesian* if the canonical map

$$\mathcal{X}_\emptyset \to \underset{\overleftarrow{S \neq \emptyset}}{\operatorname{holim}}\, \mathcal{X}_S$$

is k-connected, and k-*cocartesian* if the canonical map

$$\underset{\overrightarrow{S \neq \{1,\dots,n\}}}{\operatorname{holim}}\, \mathcal{X}_S \to \mathcal{X}_{\{1,\dots,n\}}$$

is k-connected. It is *homotopy cartesian* if it is k-cartesian for all k, and *homotopy cocartesian* if it is k-cocartesian for all k.

A 0-cube is k-cartesian (resp. k-cocartesian) if its single value is $(k-1)$-connected (resp. k-connected). A 1-cube is k-(co)cartesian if it is k-connected as a map.

When there is no possibility of confusion with the categorical notions, we drop the "homotopy" and write just cartesian and cocartesian. Homotopy (co)cartesian cubes are also called *(homotopy) pullback cubes* (resp. *(homotopy) pushout cubes*), and the initial (resp. final) vertex is then called the *(homotopy) pullback* (resp. *(homotopy) pushout*).

Definition A.8.0.3 Let $\mathcal{X}$ be an S-cube with values in any one of the categories where we have defined homotopy (co)limits (see Sect. A.7). The *iterated fiber* (resp. *iterated cofiber*) of $\mathcal{X}$ is the homotopy fiber (resp. cofiber) of the canonical map

$$\mathcal{X}_\emptyset \to \underset{\overleftarrow{S\neq\emptyset}}{\operatorname{holim}} \mathcal{X}_S \qquad \Big(\text{resp.} \quad \underset{\overrightarrow{S\neq\{1,\dots,n\}}}{\operatorname{holim}} \mathcal{X}_S \to \mathcal{X}_{\{1,\dots,n\}}\Big).$$

The reason for the term "iterated" is that one may obtain these homotopy types by iteratively taking homotopy (co)limits in one direction at a time.

A.8.1 Cubes and (Co)Simplicial Spaces

Simplicial and cosimplicial spaces can be approximated through cubes, by "writing out" all the (co)simplicial relations. Since we will need the cosimplicial version, we write this out in some detail, and leave the simplicial statement to the reader.

Lemma A.8.1.1 *Let Ord_n be the category of ordered non-empty sets of cardinality less than or equal $n+1$. The inclusion $f\colon \mathcal{P}^{\neq\emptyset}n \subseteq Ord_n$ is left cofinal. Hence, by the cofinality Lemma* A.7.2.1 *the total may be calculated by a pullback: if X is a cosimplicial space (in the form of a functor from the category of ordered nonempty finite sets), spectrum or abelian group, then the map*

$$\underset{\overleftarrow{S\in Ord_n}}{\operatorname{holim}} X^S \to \underset{\overleftarrow{S\in\mathcal{P}^{\neq\emptyset}n}}{\operatorname{holim}} X^S$$

is an equivalence.

Proof We must show that for $t\leq n$ the over categories $f/[t]$ are all contractible. Since two objects in $f/[t]$ have maximally one morphism connecting them, it is enough to show that $f/[t]$ is connected. We will produce a path from an arbitrary object $\phi\colon S\to[t]$, where $S\subseteq[n]$, to the inclusion $\{0\}\subseteq[n]$. By restricting, we may assume that S is a one-point set $\{s\}$. If $s=\phi(s)=0$ we are done. If $s\neq 0$, consider

$\{s\} \subseteq \{0, s\} \supseteq \{0\}$ and extend ϕ by sending 0 to 0, and we are again done. If $s = 0$ and $\phi(s) \neq 0$, consider $\{0\} \subseteq \{0, t\} \supseteq \{t\} \subseteq \{0, t\} \supseteq \{0\}$, where we extend ϕ to $\{0, t\}$ by sending t to t, and on the second instance of $\{0, t\}$ the inclusion to $[t]$. □

There is a preferred equivalence $Ord \to \Delta$ to the skeletal subcategory, and this lemma is used to identify homotopy limits over Δ with homotopy limits over $\mathcal{P}^{\neq\emptyset}$. In particular, one may apply this to the cosimplicial replacement of homotopy limits: for any functor $X\colon J \to \mathcal{S}_*$ from a small category J we have a natural equivalence

$$\underset{\overleftarrow{J}}{\text{holim}} X \simeq \underset{\overleftarrow{S \in \mathcal{P}^{\neq\emptyset}}}{\text{holim}} \left(\prod_{j_0 \leftarrow \cdots \leftarrow j_{|S|} \in B_{|S|} J} X(j_0) \right)$$

and dually

$$\underset{\overrightarrow{J}}{\text{holim}} X \simeq \underset{\overrightarrow{S \in (\mathcal{P}^{\neq\emptyset})^o}}{\text{holim}} \left(\bigvee_{j_0 \leftarrow \cdots \leftarrow j_{|S|} \in B_{|S|} J} X(j_{|S|}) \right).$$

This is especially interesting if BJ is a finite space, for then the homotopy limit is a homotopy pullback of a finite cube, and the homotopy colimit is the homotopy pushout of a finite cube. Explicitly, if $B(J) = sk_k B(J)$ (that is, as a functor from Δ^o, it factors through the homotopy equivalent subcategory Δ_k of objects $[q]$ for $q \leq k$), then $\text{holim}_{\overleftarrow{J}} X$ is equivalent to the homotopy pullback of the punctured k-cube which sends $S \in \mathcal{P}^{\neq\emptyset} k$ to $\prod_{j_0 \leftarrow \cdots \leftarrow j_{|S|} \in B_{|S|} J} X(j_0)$, and dually for the homotopy colimit.

A.8.2 The Blakers–Massey Theorem

The discussion above means that statements for homotopy pullbacks and pushouts are especially worthwhile listening to. The Blakers–Massey Theorem A.8.2.2 is an instance of such a statement. It relates homotopy limits and homotopy colimits in a certain range. The ultimate Blakers–Massey theorem is the following. See [106, 2.5 and 2.6] for a proof.

Theorem A.8.2.1 *Let S be a finite set with $|S| = n \geq 1$, and let $k\colon \mathcal{P}S \to \mathbf{Z}$ be a monotone function. Set $M(k)$ to be the minimum of $\sum_\alpha k(T_\alpha)$ over all partitions $\{T_\alpha\}$ of S by nonempty sets. Let $\mathcal{X}$ be an S-cube.*

1. *If $\mathcal{X}|_T$ is $k(T)$-cocartesian for each nonempty $T \subseteq S$, then $\mathcal{X}$ is $1 - n + M(k)$-cartesian.*
2. *If $\mathcal{X}(- \cup (S - T))|_T$ is $k(T)$-cartesian for each nonempty $T \subseteq S$, then $\mathcal{X}$ is $n - 1 + M(k)$-cocartesian.*

☺

The usual Blakers–Massey theorem is a direct corollary of this. We say that a cube is *strongly (co)cartesian* if all subcubes of dimension strictly greater than one are homotopy (co)cartesian (demanding this also for dimension one would be the same as demanding that all maps were equivalences, and would lead to a rather uninteresting theory!).

Corollary A.8.2.2 (Blakers–Massey) *Let $\mathcal{X}$ be a strongly cocartesian n-cube, and suppose that $\mathcal{X}_\emptyset \to \mathcal{X}_{\{s\}}$ is k_s-connected for all $1 \leq s \leq n$. Then $\mathcal{X}$ is $1 - n + \sum_s k_s$-cartesian. Dually, if $\mathcal{X}$ is strongly cartesian, and $\mathcal{X}_{\{1,\dots,n\}-\{s\}} \to \mathcal{X}_{\{1,\dots,n\}}$ is k_s connected for $1 \leq s \leq n$, then $\mathcal{X}$ is $n - 1 + \sum_s k_s$-cocartesian.* ☺

By applying the Blakers–Massey theorem to the cocartesian square

$$\begin{array}{ccc} X & \longrightarrow & \Delta[1]\wedge X \\ \downarrow & & \downarrow \\ * & \longrightarrow & S^1\wedge X \end{array}$$

you get

Corollary A.8.2.3 (Freudenthal) *If X is $(n-1)$-connected, then the natural map $X \to \Omega^1(S^1\wedge X)$ is $(2n-1)$-connected.* ☺

For reference we list the following useful corollary which is the unstable forerunner of the fact that stably products are sums.

Corollary A.8.2.4 *Let X and Y be pointed spaces where X is m-connected and Y is n-connected. Then $X \vee Y \to X \times Y$ is $m + n$-connected.*

Proof This is much easier by using the Whitehead and Künneth theorems, but here goes. Assume for simplicity that $m \geq n$. Consider the cocartesian square

$$\begin{array}{ccc} X \vee Y & \longrightarrow & X \times Y \\ \downarrow & & \downarrow \\ * & \longrightarrow & X\wedge Y. \end{array}$$

Now, $X\wedge Y$ is $m + n + 1$-connected (by e.g., considering the spectral sequence Theorem A.6.0.5 of the associated bisimplicial set), the left vertical map is $n+1$-connected and the top horizontal map is—for trivial reasons—n-connected. Using the Blakers–Massey Theorem A.8.2.1, we get that the diagram is $2n$-cartesian and so the top horizontal map must be at least $2n$-connected (since $m + n \geq 2n$). With this improved connectivity, we can use Blakers–Massey again. Repeating this procedure until we get cartesianness that exceeds $m + n$ we get that the top map is $m + n$-connected (and finally, the diagram is $m + 2n$-cartesian). □

The Blakers–Massey theorem has the usual consequence for spectra:

Corollary A.8.2.5 *Let $\mathcal{X}$ be an n-cube of bounded below spectra. Then $\mathcal{X}$ is homotopy cartesian if and only if it is homotopy cocartesian.* ☺

Lemma A.8.2.6 *Let $X\colon I \times J \to \mathcal{S}pt$ be a functor where BI is finite. Then the canonical maps*

$$\underset{\overrightarrow{I}}{\mathrm{holim}}\,\underset{\overleftarrow{J}}{\mathrm{holim}}\, X \to \underset{\overleftarrow{J}}{\mathrm{holim}}\,\underset{\overrightarrow{I}}{\mathrm{holim}}\, X$$

and

$$\underset{\overrightarrow{J}}{\mathrm{holim}}\,\underset{\overleftarrow{I}}{\mathrm{holim}}\, X \to \underset{\overleftarrow{I}}{\mathrm{holim}}\,\underset{\overrightarrow{J}}{\mathrm{holim}}\, X$$

are equivalences.

Proof The homotopy colimit of X over I is equivalent to the homotopy pushout of a punctured cube with finite wedges of copies of $X(i)$'s on each vertex. But in spectra, finite wedges are equivalent to products, and homotopy pushout cubes are homotopy pullback cubes, and homotopy pullbacks commute with homotopy limits. This proves the first equivalence, the other is dual. □

Corollary A.8.2.7 *Let $X\colon \Delta^o \times J \to \mathcal{S}pt$ be a functor regarded as a functor from J to simplicial spectra. Assume J has finite cohomological dimension (cf. Definition* A.7.4.4) *and* $\mathrm{diag}^* X$ *is bounded below. Then*

$$\mathrm{diag}^*\,\underset{\overleftarrow{J}}{\mathrm{holim}}\, X \to \underset{\overleftarrow{J}}{\mathrm{holim}}\,\mathrm{diag}^* X$$

is an equivalence.

Proof Assume $\lim_{\overleftarrow{J}}{}^{(s)} \equiv 0$ for $s > n$, and $\pi_s \mathrm{diag}^* X = 0$ for $s < m$. Let

$$sk_k X = \underset{\overrightarrow{[q]\in\Delta_k}}{\mathrm{holim}}\, X_q$$

This maps by a $k-m$-connected map to $\mathrm{holim}_{\overrightarrow{\Delta^o}} X \cong \mathrm{diag}^* X$, and let F be the homotopy fiber of this map. Then $E^2_{s,t} = \lim_{\overleftarrow{J}}{}^{(-s)} \pi_t F = 0$ if $s < -n$ or $t < k-m$, so $\pi_q \,\mathrm{holim}_{\overleftarrow{J}} F = 0$ for $q < k-m-n$. All in all, this means that the last map in

$$sk_k \underset{\overleftarrow{J}}{\mathrm{holim}}\, X = \underset{\overrightarrow{[q]\in\Delta_k}}{\mathrm{holim}}\,\underset{\overleftarrow{J}}{\mathrm{holim}}\, X_q \xrightarrow{\sim}$$

$$\underset{\overleftarrow{J}}{\mathrm{holim}}\,\underset{\overrightarrow{[q]\in\Delta_k}}{\mathrm{holim}}\, X_q = \underset{\overleftarrow{J}}{\mathrm{holim}}\, sk_k X \to \underset{\overleftarrow{J}}{\mathrm{holim}}\,\mathrm{diag}^* X$$

is $k-n-m$-connected. Letting k go to infinity we have the desired result. □

As we see in the next section even in the unstable case there is a shadow of these nice properties.

A.8.3 Uniformly Cartesian Cubes

Definition A.8.3.1 If f is some integral function, we say that an S-cube $\mathcal{X}$ is f-cartesian if each d-subcube of $\mathcal{X}$ is $f(d)$-cartesian. Likewise for f-cocartesian.

Lemma A.8.3.2 *Let $k > 0$. An S-cube of spaces is $id + k$-cartesian if and only if it is $2 \cdot id + k - 1$-cocartesian. The implication cartesian to cocartesian holds even if $k = 0$.*

Proof Note that it is trivially true if $|S| \leq 1$. Assume it is proven for all d-cubes with $d < n$.

To prove one implication, let $\mathcal{X}$ be an $id + k$-cartesian $n = |S|$-cube. All strict subcubes are also $id + k$-cartesian, and so $2 \cdot id + k - 1$-cocartesian, and the only thing we need to show is that $\mathcal{X}$ itself is $2n + k - 1$-cocartesian. This follows from the second part of the Blakers–Massey Theorem A.8.2.1: $\mathcal{X}$ is K-cocartesian where

$$K = n - 1 + \min\left(\sum_\alpha (|T_\alpha| + k)\right)$$

where the minimum is taken over all partitions $\{T_\alpha\}$ of S by nonempty sets. But this minimum is clearly attained by the trivial partition, for if we subdivide T into T_1 and T_2 then $|T| + k = |T_1| + |T_2| + k \leq |T_1| + k + |T_2| + k$, and so $K = (n-1) + (n+k) = 2n + k - 1$.

In the opposite direction, let $\mathcal{X}$ be a $2 \cdot id + k - 1$-cocartesian $n = |S|$-cube. This time, all strict subcubes are by assumption $id + k$ cartesian, and so we are left with showing that $\mathcal{X}$ is $n + k$-cartesian. Again this follows from Theorem A.8.2.1: $\mathcal{X}$ is K-cartesian where

$$K = (1 - n) + \min\left(\sum_\alpha (2|T_\alpha| + k - 1)\right)$$

where the minimum is taken over all partitions $\{T_\alpha\}$ of S by nonempty sets. But this minimum is clearly attained by the trivial partition, for if we subdivide T into T_1 and T_2 then $2|T| + k - 1 = 2|T_1| + 2|T_2| + k - 1 \leq 2|T_1| + k - 1 + 2|T_2| + k - 1$, and so $K = (1 - n) + (2n + k - 1) = n + k$. □

Notice that this statement is undisturbed if one replaces all instances of "subcube" by "face" in the definitions.

Homology takes cofiber sequences to long exact sequences. This is a reflection of the well-known statement

Lemma A.8.3.3 *If $\mathcal{X}$ is a homotopy cocartesian cube of spaces, then $\tilde{\mathbf{Z}}\mathcal{X}$ is homotopy cartesian.*

Proof This follows by induction on the dimension d of $\mathcal{X}$. If $d \leq 1$ it follows since homology is a homotopy functor, and if $\mathcal{X}$ has dimension $d > 1$, split $\mathcal{X}$ into two $d-1$ dimensional cubes $\mathcal{X}^i \to \mathcal{X}^f$. Do a functorial replacement so that each map in $\mathcal{X}^i \rightarrowtail \mathcal{X}^f$ is a cofibration and take the cofiber $\mathcal{X}^c$. As $\mathcal{X}$ was cocartesian, so is $\mathcal{X}^c$, and by assumption $\tilde{\mathbf{Z}}\mathcal{X}^c$ is cartesian, and $\tilde{\mathbf{Z}}\mathcal{X}^i \to \tilde{\mathbf{Z}}\mathcal{X}^f \to \tilde{\mathbf{Z}}\mathcal{X}^c$ is a short exact sequence of cubes of simplicial abelian groups, and so $\tilde{\mathbf{Z}}\mathcal{X}$ must be cartesian. □

We will need a generalization of the Hurewicz theorem. Recall that the Hurewicz theorem states that if X is $k-1>0$ connected, then $\pi_k X \to H_k(X)$ is an isomorphism and $\pi_{k+1} X \to H_{k+1} X$ is a surjection, or in other words that

$$X \xrightarrow{h_X} \tilde{\mathbf{Z}}X$$

is $k+1$-connected.

Using the transformation $h\colon 1 \to \tilde{\mathbf{Z}}$ on $h_X\colon X \to \tilde{\mathbf{Z}}X$ we get a square

$$\begin{array}{ccc} X & \xrightarrow{h_X} & \tilde{\mathbf{Z}}X \\ {\scriptstyle h_X}\downarrow & & \downarrow{\scriptstyle h_{\tilde{\mathbf{Z}}X}} \\ \tilde{\mathbf{Z}}X & \xrightarrow{\tilde{\mathbf{Z}}h_X} & \tilde{\mathbf{Z}}\tilde{\mathbf{Z}}X. \end{array}$$

One may check by brute force that this square is $k+2$-cartesian if X is $k-1>0$ connected. We may continue this process to obtain arbitrarily high dimensional cubes by repeatedly applying h and the generalized Hurewicz theorem states that the result gets linearly closer to being cartesian with the dimension.

Theorem A.8.3.4 (The Hurewicz Theorem (Generalized Form)) *Let $k > 1$. If $\mathcal{X}$ is an $id + k$-cartesian cube of spaces, then so is $\mathcal{X} \to \tilde{\mathbf{Z}}\mathcal{X}$.*

Proof To fix notation, let $\mathcal{X}$ be an $n = |S|$-cube with iterated fiber F and iterated cofiber C. Let $\mathcal{C}$ be the S-cube which sends S to C, and all strict subsets to $*$. Then the $|S|+1$-cube $\mathcal{X} \to \mathcal{C}$ is cocartesian.

As $\mathcal{X}$ is $id+k$-cartesian, it is $2\cdot id+k-1$-cocartesian, and in particular C is $2n+k-1$ connected. Furthermore, if $\mathcal{X}|T$ is some d-subcube of $\mathcal{X}$ with T not containing the terminal set S, then $\mathcal{X}|T$ is $2d+k-1$-cocartesian, and so $\mathcal{X}|T \to \mathcal{C}|T = *$ is $2d+k$-cocartesian. Also, if $\mathcal{X}|T$ is some strict subcube with T containing the terminal set S, then $\mathcal{X}|T \to \mathcal{C}|T$ is still $2d+k$-cocartesian because C is $2n+k-1$ connected, and $d<n$. Thus $\mathcal{X} \to \mathcal{C}$ is $2\cdot id+k-2$-cocartesian, and cocartesian. Using the Blakers–Massey Theorem A.8.2.1 again, we see that $\mathcal{X} \to \mathcal{C}$ is $1-n+2(n+1+k-2) = n+2k-1$-cartesian as the minimal partition is obtained by partitioning $S \cup \{n+1\}$ in two.

This implies that the map of iterated fibers $F \to \Omega^n C$ is $n+2k-1$ connected. We note that $n+2k-1 \geq n+k+1$ as $k>1$.

Furthermore, as C is $2n+k-1$ connected, $\Omega^n C \to \Omega^n \tilde{\mathbf{Z}}C$ is $n+k+1$ connected. But Lemma A.8.3.3 implies that $\tilde{\mathbf{Z}}\mathcal{X} \to \tilde{\mathbf{Z}}\mathcal{C}$ is cartesian. Hence the iterated fiber of $\tilde{\mathbf{Z}}\mathcal{X}$ is $\Omega^n\tilde{\mathbf{Z}}C$, and we have shown that the map from the iterated fiber of $\mathcal{X}$ is $n+1+k$ connected. Doing this also on all subcubes gives the result. □

In particular

Corollary A.8.3.5 *Let X be a $k-1>0$-connected space. Then the cube you get by applying the Hurewicz map n times to X is $id+k$-cartesian.* ☺

Lastly, the equivalence between $(id+k)$-cartesian and $(2id+k-1)$-cocartesian implies good behavior of the plus construction.

Lemma A.8.3.6 *If $\mathcal{X}$ is an $(id+k)$-cartesian n-cube of spaces for $k\geq 1$, then so is $\mathcal{X}^+$.*

Proof Since all the maps in $\mathcal{X}$ are $1+k\geq 2$-connected, they induce isomorphisms on fundamental groups. Let $\pi=\pi_1\mathcal{X}_\emptyset$, and let P be the maximal perfect subgroup. From the comment following immediately after the proof of Lemma 3.1.1.1, we have that for each $S\subseteq\{1,\dots,n\}$ the map $(q_{\mathcal{X}_S})_*\colon H_*(\mathcal{X}_S;(q_{\mathcal{X}_S})^*\mathbf{Z}[\pi/P])\to H_*(\mathcal{X}_S^+;\mathbf{Z}[\pi/P])$ is an isomorphism. Hence, the spectral sequence of the homotopy colimit Theorem A.7.4.5 over the proper subsets S of $\{1,\dots,n\}$,

$$\underset{\rightarrow}{\lim}{}_{(s)}H_t\big(\mathcal{X}_S;(q_{\mathcal{X}_S})^*\mathbf{Z}[\pi/P]\big)\Rightarrow H_{s+t}\big(\underset{\rightarrow}{\operatorname{holim}}\,\mathcal{X}_S;(q_{\operatorname{holim}_\rightarrow\mathcal{X}_S})^*\mathbf{Z}[\pi/P]\big),$$

and the corresponding spectral sequence for $\mathcal{X}^+$ are isomorphic. Again by the comment following Lemma 3.1.1.1, this gives that $\operatorname{holim}_\rightarrow\mathcal{X}_S\to\operatorname{holim}_\rightarrow\mathcal{X}_S^+$ is acyclic and kills the maximal perfect subgroup, and so by the uniqueness of the plus construction, Theorem 3.1.1.7, $\operatorname{holim}_\rightarrow(\mathcal{X}_S^+)$ is equivalent to $(\operatorname{holim}_\rightarrow\mathcal{X}_S)^+$.

Now, since $\mathcal{X}$ is $(id+k)$-cartesian, $\mathcal{X}$ is $2id+k-1$-cocartesian, and in particular $\operatorname{holim}_\rightarrow\mathcal{X}_S\to\mathcal{X}_{\{1,\dots,n\}}$ is $2n+k-1$-connected, which implies that $(\operatorname{holim}_\rightarrow\mathcal{X}_S)^+\to\mathcal{X}^+_{\{1,\dots,n\}}$ is $2n+k-1$-connected, where again the homotopy colimit is over the proper subsets S of $\{1,\dots,n\}$. Consequently $\mathcal{X}^+$ is $2n+k-1$-cocartesian.

Repeating this argument for all subcubes of $\mathcal{X}$ gives that $\mathcal{X}^+$ is $(2id+k-1)$-cocartesian, and so $(id+k)$-cartesian. □

A.9 G-Spaces

In this section we collect some useful facts on G-spaces used in Chap. 6. We will not strive for the maximal generality, and there is nothing here which can not be found elsewhere in some form.

Let G be a simplicial monoid. A *G-space* X is a space X together with a pointed G-action $\mu\colon G_+\wedge X\to X$ such that the expected diagrams commute. Differently put: it is an $\mathcal{S}$-functor (see Sect. A.10.2.1, or alternatively: a map between simplicial categories with discrete object classes)

$$G \xrightarrow{X} \mathcal{S}_*$$

with G considered as an $\mathcal{S}$-category with one object and morphism space G. We let X denote both the functor and the image of the object in G. That the functor is enriched over $\mathcal{S}$ asserts that the map $G \to \underline{\mathcal{S}_*}(X, X)$ is simplicial, and by adjointness it gives rise to μ (the "plus" in $G_+ \wedge X \to X$ comes from the fact that $G \to \underline{\mathcal{S}_*}(X, X)$ is not basepoint preserving as it must send the identity to the identity). Then the functoriality encodes the desired commuting diagrams. Note that a morphism $X \to Y$ in $G\mathcal{S}_*$ is a G-equivariant map of pointed G-spaces.

According to our general convention of writing $C\mathcal{S}_*$ for the category of functors from a category C to $\mathcal{S}_*$ (blatantly violated in our notation $\Gamma\mathcal{S}_*$ for functors from Γ^o to spaces), we write $G\mathcal{S}_*$ for the category of G-spaces. The category of G-spaces is a pointed $\mathcal{S}$-category with function spaces

$$\underline{G\mathcal{S}_*}(X, Y) = \bigl\{[q] \mapsto G\mathcal{S}_*\bigl(X \wedge \Delta[q]_+, Y\bigr)\bigr\}$$

where G acts trivially on $\Delta[q]_+$.

If X is a G-space and Y is a G^o-space (a right G-space), we let their smash product be the space

$$Y \wedge_G X = Y \wedge X/(yg \wedge x \sim y \wedge gx).$$

The forgetful map $G\mathcal{S}_* \to \mathcal{S}_*$ has a left adjoint, namely $X \mapsto G_+ \wedge X$, the *free G-space on the space X*.

If G is a simplicial group we say that a G-space X is *free* if for all non-base points $x \in X$ the *isotropy groups* $I_x = \{g \in G \mid gx = x\}$ are trivial, whereas $I_{\text{base point}} = G$ ("free away from the basepoint"). A *finite free G-space* is a G-space Y with only finitely many non-degenerate *G-cells* (you adjoin a "G-cell" of dimension n to Y_j by taking a pushout of maps of G spaces

$$\begin{array}{ccc} \partial\Delta[n]_+ \wedge G_+ & \xrightarrow{\text{incl.} \wedge id} & \Delta[n]_+ \wedge G_+ \\ \downarrow & & \downarrow \\ Y_j & \longrightarrow & Y_{j+1} \end{array}$$

where G acts trivially on $\partial\Delta[n]$ and $\Delta[n]$).

A.9.1 The Orbit and Fixed Point Spaces

Let $f: M \to G$ be a map of monoids. Precomposition with f gives a functor

$$f^*: [G, Ens_*] \to [M, Ens_*],$$

and since all (co)limits exist this functor has both a right and a left adjoint. If f is surjective and G a group, let $H \subset M$ be the submonoid of elements mapping to the identity. Then the right adjoint of f^* is

$$X \mapsto X^H = \lim_{\overleftarrow{H}} X = \{x \in X | h \cdot x = x \text{ for all } h \in H\}$$

the set of *fixed points*, and the left adjoint is

$$X \mapsto X_H = \lim_{\overrightarrow{H}} X = X/(h \cdot x \sim x)$$

the set of *orbits*. The same considerations and definitions holds in the simplicial case, and we even get simplicial adjoints:

$$M\mathcal{S}_* \quad \begin{matrix} \xrightarrow{X \mapsto X^H} \\ \xleftarrow{f^*} \\ \xrightarrow{X \mapsto X_H} \end{matrix} \quad G\mathcal{S}_*$$

$$\underline{G\mathcal{S}_*}(X_H, Y) \cong \underline{M\mathcal{S}_*}\big(X, f^*Y\big), \quad \text{and} \quad \underline{G\mathcal{S}_*}\big(Y, X^H\big) \cong \underline{M\mathcal{S}_*}\big(f^*Y, X\big)$$

If G is a simplicial group, the homomorphism $G \to G^o \times G$ sending g to (g^{-1}, g) makes it possible to describe $-\wedge_G-$ and $\underline{G\mathcal{S}_*}(-,-)$ in terms of orbit and fixed point spaces. If $X, Y \in obG\mathcal{S}_*$ and $Z \in obG^o\mathcal{S}_*$ then $Z \wedge X$ and $\underline{\mathcal{S}_*}(X, Y)$ are naturally $G^o \times G$-spaces, and since G is a group also G-spaces, and we get that

$$Z \wedge_G X \cong (Z \wedge X)_G, \quad \text{and} \quad \underline{G\mathcal{S}_*}(X, Y) \cong \underline{\mathcal{S}_*}(X, Y)^G.$$

A.9.2 The Homotopy Orbit and Homotopy Fixed Point Spaces

Let G be a simplicial monoid. When regarded as a simplicial category, with only one object $*$, we can form the over (resp. under) categories, and the nerve $B(G/*)_+$ (resp. $B(*/G)_+$) is a contractible free G-space (resp. contractible free G^o-space), and the G-orbit space is BG. For G a group, $B(G/*) \cong B(G, G, *)$ (resp. $B(*/G) \cong B(*, G, G)$) is the one sided bar construction Sect. A.5.2, and we note that in this case the left and right distinction is inessential.

Recalling the notion of $\mathcal{S}$-homotopy (co)limits from Sect. A.7.5 (if G is discrete this is nothing but the usual homotopy (co)limit), we get as in Example A.7.5.1.

Definition A.9.2.1 Let G be a simplicial monoid and X a G-space. Then the *homotopy fixed point space* is

$$X^{hG} = \operatorname{holim}_{\overleftarrow{G}} X = \underline{G\mathcal{S}_*}\big(B(G/*)_+, \sin|X|\big)$$

and the *homotopy orbit space* is

$$X_{hG} = \underset{\overrightarrow{G}}{\operatorname{holim}}\, X = B(*/G)_+ \wedge_G X.$$

A nice thing about homotopy fixed point and orbit spaces is that they preserve weak equivalences (since homotopy (co)limits do):

Lemma A.9.2.2 *Let $X \to Y$ be a map of G-spaces which is a weak equivalence of underlying spaces. Then the induced maps $X^{hG} \to Y^{hG}$ and $X^{hG} \to Y^{hG}$ are weak equivalences.*

We have maps $X^G \to X^{hG}$ and $X_{hG} \to X_G$, and a central problem in homotopy theory is to know when they are equivalences.

Lemma A.9.2.3 *Let U be a free G-space, and X any fibrant G-space (i.e., a G space which is fibrant as a space). Then*

$$\underline{G\mathcal{S}_*}(U, X) \xrightarrow{\sim} \underline{G\mathcal{S}_*}\big(U \wedge B(G/*)_+, X\big),$$

and so if G is a group $\underline{\mathcal{S}_*}(U, X)^G \simeq \underline{\mathcal{S}_*}(U, X)^{hG}$. Furthermore, if U is d-dimensional, then $\underline{G\mathcal{S}_*}(U, -)$ sends n-connected maps of fibrant spaces to $(n-d)$-connected maps.

Proof By induction on the G-cells, it is enough to prove the claim for $U = S^k_+ \wedge G_+$. But then the map is the composite from top left to top right in

$$\begin{array}{ccccc}
\underline{G\mathcal{S}_*}(S^k_+ \wedge G_+, X) & \longrightarrow & \underline{G\mathcal{S}_*}(S^k_+ \wedge G_+ \wedge B(G/*)_+, X) & \xrightarrow[\cong]{i^*} & \underline{G\mathcal{S}_*}(S^k_+ \wedge G_+ \wedge B(G/*)_+, X) \\
\cong\downarrow & & \cong\downarrow & & \\
\underline{\mathcal{S}_*}(S^k_+, X) & \xrightarrow{\sim} & \underline{\mathcal{S}_*}(S^k_+ \wedge B(G/*)_+, X) & &
\end{array}$$

where i_* is the G-isomorphism from $S^k_+ \wedge G_+ \wedge B(G/*)_+$ (with trivial action on $B(G/*)_+$) to $S^k_+ \wedge G_+ \wedge B(G/*)_+$ (with the diagonal action) given by the *shear map* $(s, g, e) \mapsto (s, g, ge)$. The last statement follows from induction on the skeleta, and the fact that $\underline{G\mathcal{S}_*}(S^k_+ \wedge G_+, -) \cong \underline{\mathcal{S}_*}(S^k_+, -)$ sends n-connected maps of fibrant spaces to $(n-k)$-connected maps. □

Corollary A.9.2.4 *Let G be a simplicial group and E a free contractible G-space. If X is a G-space there are natural equivalences*

$$X^{hG} \simeq Map_*(E_+, X)^G, \quad \text{and} \quad X_{hG} \simeq (E_+ \wedge X)_G$$

(where the G-action is by conjugation and diagonal).

This gives us considerable freedom; in particular, if $H \subseteq G$ is a subgroup and EG is a free contractible G-space, EG will serve as a free contractible H-space as well. Hence there may be some ambiguity as to what EG will mean, but the one-sided bar construction gives a functorial choice.

A.10 A Quick Review on Enriched Categories

To remind the reader, and set notation, we give a short presentation of enriched categories (see e.g., [62, 107, 162] or [32]), together with some relevant examples. Our guiding example will be *Ab*-categories, also known as linear categories. These are categories where the morphism sets are actually abelian groups, and composition is bilinear. That is: in the definition of "category", sets are replaced by abelian groups, Cartesian product by tensor product and the one point set by the group of integers. Besides *Ab*-categories, the most important example will be the $\Gamma\mathcal{S}_*$-categories, which are used frequently from Chap. II on, and we go out of our way to point out some relevant details for this case. Note however, that scary things like limits and ends are after all not that scary since limits (and colimits for that matter) are calculated pointwise.

A.10.1 Closed Categories

Recall the definition of a symmetric monoidal closed category $(V, \square, e)$, see e.g., [191]. For convenience we repeat the definition below, but the important thing to remember is that it behaves as $(Ab, \otimes_{\mathbf{Z}}, \mathbf{Z})$.

Definition A.10.1.1 A *monoidal category* is a tuple $(\mathcal{C}, \square, e, \alpha, \lambda, \rho)$ where $\mathcal{C}$ is a category, $\square$ is a functor $\mathcal{C} \times \mathcal{C} \to \mathcal{C}$, and α, λ and γ are natural ("structure") isomorphisms

$$\alpha_{a,b,c}\colon a\square(b\square c) \xrightarrow{\cong} (a\square b)\square c, \qquad \lambda_a\colon e\square a \xrightarrow{\cong} a, \quad \text{and}$$

$$\rho_a\colon a\square e \xrightarrow{\cong} a$$

with $\lambda_e = \rho_e : e \Box e \to e$, satisfying the coherence laws given by requiring that the following diagrams commute:

$$
\begin{array}{ccccc}
a\big(\Box b \Box (c \Box d)\big) & & \xrightarrow{\quad\alpha\quad} & & (a \Box b) \Box (c \Box d)\ , \\
\downarrow{\scriptstyle 1\Box\alpha} & & & & \downarrow{\scriptstyle \alpha} \\
a \Box \big((b \Box c) \Box d\big) & & & & \big((a \Box b) \Box c\big) \Box d \\
& \searrow{\scriptstyle \alpha} & & \nearrow{\scriptstyle \alpha\Box 1} & \\
& & \big(a \Box (b \Box c)\big) \Box d & &
\end{array}
$$

$$
\begin{array}{ccc}
a \Box (e \Box c) & & \\
& \searrow{\scriptstyle q\Box\lambda} & \\
\downarrow{\scriptstyle \alpha} & & a \Box c \\
& \nearrow{\scriptstyle \rho\Box 1} & \\
(a \Box e) \Box c & &
\end{array}
$$

A monoidal category is *symmetric* when it is equipped with a natural isomorphism

$$\gamma_{a,b} : a \Box b \xrightarrow{\cong} b \Box a$$

such that the following diagrams commute

$$
\begin{array}{ccc}
a \Box b & = & a \Box b \\
& \searrow{\scriptstyle\gamma} \quad \nearrow{\scriptstyle\gamma} & \\
& b \Box a &
\end{array}
\qquad
\begin{array}{ccc}
a \Box e & \xrightarrow{\quad\rho\quad} & a \\
& \searrow{\scriptstyle\gamma} \quad \nearrow{\scriptstyle\lambda} & \\
& e \Box a &
\end{array}
$$

$$
\begin{array}{ccccc}
a \Box (b \Box c) & & \xrightarrow{\quad\alpha\quad} & & (a \Box b) \Box c \\
\downarrow{\scriptstyle 1\Box\gamma} & & & & \downarrow{\scriptstyle \gamma} \\
a \Box (c \Box b) & & & & (c \Box a) \Box b \\
& \searrow{\scriptstyle \alpha} & & \nearrow{\scriptstyle \gamma\Box 1} & \\
& & (a \Box c) \Box b & &
\end{array}
$$

A *symmetric monoidal closed category* (often just called a *closed category*) is a symmetric monoidal category such that

$$-\Box b\colon \mathcal{C}\to\mathcal{C}$$

has a right adjoint $\underline{\mathcal{C}}(b,-)\colon \mathcal{C}\to\mathcal{C}$ (which is considered to be part of the data).

If $\mathcal{C}$ is a closed category, we will refer to $\underline{\mathcal{C}}(b,c)$ as *the internal morphism objects*.
If there is no ambiguity we typically will omit mentioning α, λ, ρ, γ and $\underline{\mathcal{C}}$ explicitly when declaring that "$(\mathcal{C},\Box,e)$ is a closed category" (or variants thereof).

Example A.10.1.2 The prime example of a closed category is the category of sets with the cartesian product as monoidal pairing. The unit element is the one-point set $*$, and for sets X, Y and Z, the structure isomorphisms are given by $\alpha_{X,Y,Z}(x,(y,z))=((x,y),z)$, $\lambda_X(*,x)=\rho_X(x,*)=x$ and $\gamma_{X,Y}(x,y)=(y,x)$. Letting $\underline{Ens}(X,Y)$ simply **be** the set $Ens(X,Y)$ of functions from X to Y, the structure of a closed category follows from the one-to-one correspondence between $Ens(X\times Y,Z)$ and $Ens(Y,Ens(X,Z))$ sending f to $y\mapsto\{x\mapsto f(x,y)\}$. This is an example of a cartesian closed category: a closed category with monoidal structure arising from a (choice of) products.

In the pointed case, the cartesian product is replaced by the smash product $X\wedge Y=(X\times Y)/(X\vee Y)$ where $X\vee Y$ is the coproduct of the pointed sets X and Y over the base point, and the unit is the two-point set S^0. The same formula as above give us a closed structure: $Ens_*(X\wedge Y,Z)\cong Ens_*(Y,Ens_*(X,Z))$, $f\mapsto\{y\mapsto\{x\mapsto f(x\wedge y)\}\}$.

To us, the most important example is $(\mathcal{S}_*,\wedge,S^0)$, the closed category of pointed spaces (simplicial sets). The symmetric monoidal structure is given by taking the smash product in every degree. The mapping spaces $\underline{\mathcal{S}_*}(X,Y)$ is given in degree q by the set $\mathcal{S}_*(X\wedge\Delta[q]_+,Y)$ of pointed simplicial maps from $X\wedge\Delta[q]_+$ to Y with structure maps given by the cosimplicial structure on $[q]\mapsto\Delta[q]$. The structure isomorphism $\mathcal{S}_*(X\wedge Y,X)\cong\mathcal{S}_*(Y,\underline{\mathcal{S}_*}(X,Z))$ sends a pointed simplicial map $f\colon X\wedge Y\to Z$ to $Y\to\underline{\mathcal{S}_*}(X,Z)$ via $y\mapsto\{x\wedge\sigma\mapsto f(x\wedge\sigma^*y)\}$ for $y\in Y_q$, $\sigma\in\Delta([p],[q])$ and $x\in X_p$.

Definition A.10.1.3 Let $(\mathcal{C},\Box,e)$ and $(\mathcal{D},\boxtimes,f)$ be monoidal categories. A *monoidal functor* from $\mathcal{C}$ to $\mathcal{D}$ consists of a functor $F\colon\mathcal{C}\to\mathcal{D}$ together with a natural transformation $F_{c,c'}\colon F(c)\boxtimes F(c')\to F(c\Box c')$ (of functors $\mathcal{C}\times\mathcal{C}\to\mathcal{D}$) and a morphism

$F_1\colon f \to F(e)$ such that for all objects c_0, c_1, c_2 in $\mathcal{C}$ the diagrams

$$\begin{array}{ccc}
\big(F(c_0)\boxtimes F(c_1)\big)\boxtimes F(c_2) & \xrightarrow{\alpha_{\mathcal{D}}} & F(c_0)\boxtimes\big(F(c_1)\boxtimes F(c_2)\big) \\
\downarrow{\scriptstyle F_{c_0,c_1}\boxtimes 1} & & \downarrow{\scriptstyle 1\boxtimes F_{c_1,c_2}} \\
F(c_0\,\Box\, c_1)\boxtimes F(c_2) & & F(c_0)\boxtimes F(c_1\,\Box\, c_2) \\
\downarrow{\scriptstyle F_{c_0\Box c_1,c_2}} & & \downarrow{\scriptstyle F_{c_0,c_1\Box c_2}} \\
F\big(c_0\,\Box\,(c_1\,\Box\, c_2)\big) & \xrightarrow{F\alpha_{\mathcal{C}}} & F\big((c_0\,\Box\, c_1)\,\Box\, c_2\big),
\end{array}$$

$$\begin{array}{ccc}
F(c)\boxtimes f & \xrightarrow{1\boxtimes F_1} & F(c)\boxtimes F(e) \\
\downarrow{\scriptstyle \rho_{\mathcal{D}}} & & \downarrow{\scriptstyle F_{c,e}} \\
F(c) & \xleftarrow[F\rho_{\mathcal{C}}]{} & F(c\,\Box\, e)
\end{array}
\quad\text{and}\quad
\begin{array}{ccc}
f\boxtimes F(c) & \xrightarrow{F_1\boxtimes 1} & F(e)\boxtimes F(c) \\
\downarrow{\scriptstyle \lambda_{\mathcal{D}}} & & \downarrow{\scriptstyle F_{e,c}} \\
F(c) & \xleftarrow[F\lambda_{\mathcal{C}}]{} & F(e\,\Box\, c)
\end{array}$$

commute. The monoidal functor is *strong* (resp. *strict*) if the F_1 and F_{c_0,c_1} are isomorphisms (resp. identities) for all $c_0, c_1 \in ob\mathcal{C}$.

If $\mathcal{C}$ and $\mathcal{D}$ are both symmetric monoidal, a monoidal functor F from $\mathcal{C}$ to $\mathcal{D}$ is a *symmetric monoidal functor* if for all objects c, c' in $\mathcal{C}$ the square

$$\begin{array}{ccc}
F(c)\boxtimes F(c') & \xrightarrow{\gamma_{\mathcal{D}}} & F(c')\boxtimes F(c) \\
\downarrow{\scriptstyle F_{c,c'}} & & \downarrow{\scriptstyle F_{c',c}} \\
F(c\,\Box\, c') & \xrightarrow{\gamma_{\mathcal{C}}} & F(c'\,\Box\, c)
\end{array}$$

commutes.

Some authors reserve the term “monoidal functor” for what we call a strong monoidal functor and would use the term “lax monoidal” for the weaker version.

A.10.1.1 Monoids

Definition A.10.1.4 Let $(\mathcal{C}, \Box, e)$ be a monoidal category. A *monoid* in $\mathcal{C}$ is an object M in $\mathcal{C}$ together with two morphisms

$$\eta\colon e \to M, \text{ and } \mu\colon M\,\Box\, M \to M$$

satisfying unitality and associativity, that is the diagrams

$$\begin{array}{ccccc} M \Box e & \xrightarrow{1 \Box \eta} & M \Box M & \xleftarrow{\eta \Box 1} & e \Box M, \\ & {\scriptstyle \rho_M} \searrow & \downarrow {\scriptstyle \mu} & \swarrow {\scriptstyle \lambda_M} & \\ & & M & & \end{array}$$

$$\begin{array}{ccccc} M \Box (M \Box M) & & \xrightarrow{\alpha} & & (M \Box M) \Box M \\ \downarrow {\scriptstyle 1 \Box \mu} & & & & \downarrow {\scriptstyle \mu \Box 1} \\ M \Box M & \xrightarrow{\mu} & M & \xleftarrow{\mu} & M \Box M \end{array}$$

commute. If $\mathcal{C}$ is a symmetric monoidal category, M is a *symmetric monoid* if the diagram

$$\begin{array}{ccc} M \Box M & \xrightarrow{\gamma_{M,M}} & M \Box M \\ & {\scriptstyle \mu} \searrow \quad \swarrow {\scriptstyle \mu} & \\ & M & \end{array}$$

commutes.

A.10.2 Enriched Categories

Let $(V, \Box, e)$ be any closed symmetric monoidal category.

Definition A.10.2.1 A V-category $\mathcal{C}$ is a class of objects, $ob\mathcal{C}$, and for objects $c_0, c_1, c_2 \in ob\mathcal{C}$ objects in V, $\underline{\mathcal{C}}(c_i, c_j)$, and a "composition"

$$\underline{\mathcal{C}}(c_1, c_0) \Box \underline{\mathcal{C}}(c_2, c_1) \to \underline{\mathcal{C}}(c_2, c_0)$$

and a "unit"

$$e \to \underline{\mathcal{C}}(c, c)$$

in V subject to the usual unit and associativity axioms: given objects $a, b, c, d \in ob\mathcal{C}$ then the following diagrams in V commute

$$\begin{array}{ccc}
\underline{\mathcal{C}}(c,d)\Box\big(\underline{\mathcal{C}}(b,c)\Box\underline{\mathcal{C}}(a,b)\big) & \xrightarrow{\cong} & \big(\underline{\mathcal{C}}(c,d)\Box\underline{\mathcal{C}}(b,c)\big)\Box\underline{\mathcal{C}}(a,b) \\
\downarrow & & \downarrow \\
\underline{\mathcal{C}}(c,d)\Box\underline{\mathcal{C}}(a,c) & & \underline{\mathcal{C}}(b,d)\Box\underline{\mathcal{C}}(a,b) \\
& \searrow \quad \swarrow & \\
& \underline{\mathcal{C}}(a,d) &
\end{array}$$

$$\begin{array}{ccccc}
\underline{\mathcal{C}}(a,b)\Box e & \xrightarrow{\cong} & \underline{\mathcal{C}}(a,b) & \xleftarrow{\cong} & e\Box\underline{\mathcal{C}}(a,b) \\
\downarrow & & {=}\downarrow & & \downarrow \\
\underline{\mathcal{C}}(a,b)\Box\underline{\mathcal{C}}(a,a) & \longrightarrow & \underline{\mathcal{C}}(a,b) & \longleftarrow & \underline{\mathcal{C}}(b,b)\Box\underline{\mathcal{C}}(a,b).
\end{array}$$

We see that $\mathcal{C}$ gives rise to an ordinary category (an "*Ens*-category") too, which we will call $\mathcal{C}$ also, or $U_0\mathcal{C}$ if we need to be precise, with the same objects and with morphism sets $U_0\mathcal{C}(c,d) = V(e, \underline{\mathcal{C}}(c,d))$.

We see that $\underline{\mathcal{C}}$ can be viewed as a functor $U_0\mathcal{C}^o \times U_0\mathcal{C} \to V$: if $f \in \mathcal{C}(c', c) = V(e, \underline{\mathcal{C}}(c', c))$ and $g \in \mathcal{C}(d, d') = V(e, \underline{\mathcal{C}}(d, d'))$ then $f^* g_* = g_* f^* = \underline{\mathcal{C}}(f, g) : \underline{\mathcal{C}}(c,d) \to \underline{\mathcal{C}}(c', d') \in V$ is defined as the composite

$$\underline{\mathcal{C}}(c,d) \cong e\Box\underline{\mathcal{C}}(c,d)\Box e \xrightarrow{g\Box id\Box f} \underline{\mathcal{C}}(d,d')\Box\underline{\mathcal{C}}(c,d)\Box\underline{\mathcal{C}}(c,c') \to \underline{\mathcal{C}}(c',d').$$

Definition A.10.2.2 Let $\mathcal{C}$ be a V-category. We say that $\mathcal{C}$ is a *tensored* V-category if it comes equipped with a functor $\mathcal{C} \times V \to \mathcal{C}$ sending (c, v) to $c \otimes v$ and a natural isomorphism

$$\underline{\mathcal{C}}(c \otimes v, d) \cong \underline{V}\big(v, \underline{\mathcal{C}}(c,d)\big),$$

and *cotensored* if it comes equipped with a functor $\mathcal{C} \times V^o \to \mathcal{C}$ sending (c, v) to c^v and a natural isomorphism

$$\underline{\mathcal{C}}\big(c, d^v\big) \cong \underline{V}\big(v, \underline{\mathcal{C}}(c,d)\big).$$

We notice that the closed structure of V makes V into a tensored and cotensored V-category.

A.10.2.1 Some Further Definitions

If $\mathcal{C}$ and $\mathcal{D}$ are two V-categories, we define their *tensor product* (or whatever the operator in V is called) $\mathcal{C}\Box\mathcal{D}$ to be the V-category given by $ob(\mathcal{C}\Box\mathcal{D}) = ob\mathcal{C} \times ob\mathcal{D}$, and $\underline{\mathcal{C}\Box\mathcal{D}}((c,d),(c',d')) = \underline{\mathcal{C}}(c,c')\Box\underline{\mathcal{D}}(d,d')$.

Let $\mathcal{C}$ be a V-category where V has finite products. If $U_0\mathcal{C}$ is a category with sum (i.e. it has an initial object $*$, and categorical coproducts), then we say that $\mathcal{C}$ is a *V-category with sum* if the canonical map $\underline{\mathcal{C}}(c \vee c', d) \to \underline{\mathcal{C}}(c,d) \times \underline{\mathcal{C}}(c',d)$ is an isomorphism.

A *V-functor* F from $\mathcal{C}$ to $\mathcal{D}$ is an assignment $ob\mathcal{C} \to ob\mathcal{D}$ together with maps

$$\underline{\mathcal{C}}(c,c') \to \underline{\mathcal{D}}(F(c),F(c'))$$

preserving unit and composition.

A V-functor $F\colon \mathcal{C} \to \mathcal{D}$ is *V-full* (resp. *V-faithful*) if $\underline{\mathcal{C}}(c,d) \to \underline{\mathcal{D}}(F(c),F(d))$ is epic (resp. monic).

A *V-natural transformation* between two V-functors $F,G\colon \mathcal{C} \to \mathcal{D}$ is a map $\eta_c\colon F(c) \to G(c) \in U_0\mathcal{D}$ for every $c \in ob\mathcal{C}$ such that all the diagrams

$$\begin{array}{ccc} \underline{\mathcal{C}}(c,c') & \longrightarrow & \underline{\mathcal{D}}(F(c),F(c')) \\ \big\downarrow & & \big\downarrow{\scriptstyle (\eta_{c'})_*} \\ \underline{\mathcal{D}}(G(c),G(c')) & \xrightarrow{(\eta_c)^*} & \underline{\mathcal{D}}(F(c),G(c')) \end{array}$$

commute. The V-natural transformation η is a *V-natural isomorphism* if each η_c is an isomorphism.

A V-functor $F\colon \mathcal{C} \to \mathcal{D}$ is a *V-natural equivalence* if there is a V-functor $G\colon \mathcal{D} \to \mathcal{C}$ and V-natural isomorphisms $GF \cong 1$ and $FG \cong 1$.

If

$$\mathcal{D} \underset{U}{\overset{F}{\rightleftarrows}} \mathcal{C}$$

is a pair of V-functors, we say that F is *V-left adjoint* to U (and U is *V-right adjoint* to F) if there are V-natural transformations $FU \xrightarrow{\epsilon} 1_{\mathcal{C}}$ (the *counit*) and $UF \xleftarrow{\eta} 1_{\mathcal{D}}$ (the *unit*) such that the following diagrams commute:

$$\begin{array}{ccc} U & \xrightarrow{\eta U} & UFU \\ & \searrow^{=} & \big\downarrow{\scriptstyle U\epsilon} \\ & & U, \end{array} \qquad \begin{array}{ccc} F & \xrightarrow{F\eta} & FUF \\ & \searrow^{=} & \big\downarrow{\scriptstyle \epsilon F} \\ & & U. \end{array}$$

A.10.2.2 Examples of Enriched Categories

1. Any closed symmetric monoidal closed category $(V, \square, e)$ is enriched in itself due to the internal morphism objects.
2. A *linear* category is nothing but an *Ab*-category, that is a category enriched in $(Ab, \otimes, \mathbf{Z})$. Note that an additive category is *something else* (it is a linear category with a zero object and all finite sums). "Linear functor" is another name for *Ab*-functor.
3. Just as a ring is an *Ab*-category with one object, or a k-algebra is a (k-mod)-category with only one object, an $\mathbf{S}$-algebra is a $\Gamma\mathcal{S}_*$-category with only one object. This is equivalent to saying that it is a monoid in $(\Gamma\mathcal{S}_*, \wedge, \mathbf{S})$, which is another way of saying that an $\mathbf{S}$-algebra is something which satisfies all the axioms of a ring, if you replace $(Ab, \otimes, \mathbf{Z})$ by $(\Gamma\mathcal{S}_*, \wedge, \mathbf{S})$.
4. "Function spaces" appear in many applications, mirroring an enrichment in spaces. In particular, the category $s\mathcal{C}$ of simplicial objects in some category $\mathcal{C}$ can be given an enrichment in $\mathcal{S}$ as in [235, II.1.7]. The structure is easiest to describe if $\mathcal{C}$ has finite colimits. Then one may define $c \otimes K$ for any $c \in s\mathcal{C}$ and $K \in \mathcal{S}$ to be the simplicial object $[n] \mapsto \coprod_{K_n} c_n$ (the coproduct of c_n with itself indexed over K_n), and the function space becomes
$$\underline{s\mathcal{C}}(c, d) = \big\{[q] \mapsto s\mathcal{C}\big(c \otimes \Delta[q], d\big)\big\},$$
which is a simplicial set since $[q] \mapsto \Delta[q]$ is cosimplicial.
5. Let $\mathcal{C}$ be a category with sum (and so is "tensored over Γ^0" by the formula $c \square k_+ = \bigvee_k c$). This defines a (discrete) $\Gamma\mathcal{S}_*$-category $\mathcal{C}^\vee$ by setting $\underline{\mathcal{C}}^\vee(c, c')(X) = \mathcal{C}(c, c' \square X)$ for $X \in ob\Gamma^o$ and $c, c' \in ob\mathcal{C}$, and with composition given by
$$\begin{aligned}
\underline{\mathcal{C}}^\vee(c,d)(X) \wedge \underline{\mathcal{C}}^\vee(b,c)(Y) &= \mathcal{C}(c, d \square X) \wedge \mathcal{C}(b, c \square Y) \\
&\xrightarrow{(-\square Y)\wedge id} \mathcal{C}\big(c \square Y, (d \square X) \square Y\big) \wedge \mathcal{C}(b, c \square Y) \\
&\longrightarrow \mathcal{C}\big(b, (d \square X) \square Y\big) \\
\cong \mathcal{C}\big(b, d \square (X \square Y)\big) &= \underline{\mathcal{C}}^\vee(b, d)(X \wedge Y).
\end{aligned}$$
Slightly more generally, we could have allowed $\mathcal{C}$ to be an $\mathcal{S}_*$-category with sum.

A.10.3 Monoidal V-Categories

There is nothing hindering us from adding a second layer of complexity to this. Given a closed category $(V, \boxtimes, \epsilon)$, a *(symmetric) monoidal (closed) V-category* is a (symmetric) monoidal (closed) category $(\mathcal{C}, \square, e)$ in the sense that you use Definition A.10.1.1, but do it in the V-enriched world (i.e., $\mathcal{C}$ is a V-category, $\square\colon \mathcal{C} \boxtimes \mathcal{C} \to \mathcal{C}$ a V-functor, the required natural transformations are V-natural (and $\underline{\mathcal{C}}(b, -)$ is V-right adjoint to $- \square b$)).

A.10.3.1 Important Convention

All categories are considered to be enriched over $(\mathcal{S}_*, \wedge, S^0)$ without further mention. In particular, $(V, \Box, e)$ is a closed $\mathcal{S}_*$-category, and any V-category $\mathcal{C}$ is also an $\mathcal{S}_*$-category which is sometimes also called $\mathcal{C}$, with morphism spaces $\mathcal{C}(b, c) = V(e, \underline{\mathcal{C}}(b, c)) \in ob\mathcal{S}_*$. This fits with the convention of not underlining function spaces. Of course, it also defines a set-based category $U_0\mathcal{C}$ too by considering zero-simplices only.

A.10.4 Modules

A *left $\mathcal{C}$-module* P is an assignment $ob\mathcal{C} \to obV$, and a morphism $P(c)\Box\underline{\mathcal{C}}(c, b) \to P(b)$ in V such that the obvious diagrams commute; or in other words, a $\mathcal{C}$-module is a V-functor $P: \mathcal{C} \to V$. Right modules and bimodules are defined similarly as V-functors $\mathcal{C}^o \to V$ and $\mathcal{C}^o\Box\mathcal{C} \to V$. If V has finite products and $\mathcal{C}$ is a V-category with sum, a $\mathcal{C}^o$-module M is said to be *additive* if the canonical map $M(c \vee c') \to M(c) \times M(c')$ is an isomorphism, and a bimodule is additive if $P(c \vee c', d) \to P(c, d) \times P(c', d)$ is an isomorphism.

Example A.10.4.1 If a ring A is considered to be an Ab-category with just one object, one sees that a left A-module M in the ordinary sense is nothing but a left A-module in the sense above: consider the functor $A \to Ab$ with M as value, and sending the morphism $a \in A$ to multiplication on $M \xrightarrow{m \mapsto am} M$. Similarly for right modules and bimodules.

Likewise, if A is an $\mathbf{S}$-algebra, then an A-module is a $\Gamma\mathcal{S}_*$ functor $A \to \Gamma\mathcal{S}_*$. Again, this another way of saying that an A-module is an "$-\wedge A$"-algebra, which is to say that it satisfies all the usual axioms for a module, mutatis mutandem.

A.10.4.1 *V*-Natural Modules

A *V-natural bimodule* is a pair $(\mathcal{C}, P)$ where $\mathcal{C}$ is a V-category and P is a $\mathcal{C}$-bimodule. A map of V-natural bimodules $(\mathcal{C}, P) \to (\mathcal{D}, Q)$ is a V-functor $F: \mathcal{C} \to \mathcal{D}$ and a V-natural transformation $P \to F^*Q$ where F^*Q is the $\mathcal{C}$-bimodule given by the composite

$$\mathcal{C}^o\Box\mathcal{C} \xrightarrow{F\times F} \mathcal{D}^o\Box\mathcal{D} \xrightarrow{Q} V.$$

Similarly one defines V-natural modules as pairs $(\mathcal{C}, P)$ where $\mathcal{C}$ is a V-category and P a $\mathcal{C}$-module. A map of V-natural modules $(\mathcal{C}, P) \to (\mathcal{D}, Q)$ is a V-functor $F: \mathcal{C} \to \mathcal{D}$ and a V-natural transformation $P \to F^*Q$ where F^*Q is the $\mathcal{C}$-bimodule given by the composite

$$\mathcal{C} \xrightarrow{F} \mathcal{D} \xrightarrow{Q} V.$$

The V-natural (bi)modules form a 2-category: the maps between two V-natural (bi)modules $(\mathcal{C}, P)$ and $(\mathcal{D}, Q)$ are themselves objects of a category. The morphisms in this category are (naturally) called natural transformations; a natural transformation $\eta\colon F \to G$ where F, G are two maps of V-natural bimodules $(\mathcal{C}, P) \to (\mathcal{D}, Q)$ is a V-natural transformation $\eta\colon F \to G$ of V-functors $\mathcal{C} \to \mathcal{D}$ such that the diagram

$$\begin{array}{ccc} P(c,c') & \longrightarrow & Q(F(c),F(c')) \\ \downarrow & & \downarrow{\scriptstyle (\eta_{c'})_*} \\ Q(G(c),G(c')) & \xrightarrow{(\eta_c)^*} & Q(F(c),G(c')) \end{array}$$

commutes. A natural isomorphism is a natural transformation such that all the η_c are isomorphisms. Likewise one defines the notion of a natural transformation/isomorphism for maps of V-natural modules.

For cohomology considerations, the dual notion of V-natural co(bi)modules is useful. The objects are the same as above, but a morphism $f\colon (\mathcal{C}, P) \to (\mathcal{D}, Q)$ is a functor $f\colon \mathcal{D} \to \mathcal{C}$ together with a natural transformation $f^*P \to Q$, and so on.

Example A.10.4.2 Let $\mathcal{C}$ be a category with sum, and let P be an additive $\mathcal{C}$-bimodule (i.e., $P(c \vee c', d) \xrightarrow{\cong} P(c,d) \times P(c',d)$). Recall from Sect. 2.1.6.1 the definition of $\mathcal{C}^\vee$. We define a $\mathcal{C}^\vee$-bimodule $P^\vee$ by the formula $P^\vee(c,d)(X) = P(c, d \square X)$. Note that since P is additive we have a canonical map $P(c,d) \to P(c \square X, d \square X)$, and the right module action uses this. Then $(\mathcal{C}^\vee, P^\vee)$ is a natural module, and $(\mathcal{C}^\vee, P^\vee) \to ((- \square X)^*\mathcal{C}^\vee, (- \square X)^* P^\vee)$ is a map of natural modules.

A.10.5 Ends and Coends

Ends and coends are universal concepts as good as limits and colimits, but in the set-based world you can always express them in terms of limits and colimits, and hence they are less often used. The important thing to note is that this is the way we construct natural transformations: given two (set-based) functors $F, G\colon \mathcal{C} \to \mathcal{D}$, a natural transformation η from F to G is a collection of maps $\eta_c\colon F(c) \to G(c)$ satisfying the usual condition. Another way to say the same thing is that the set of natural transformations is a set $\mathcal{D}^\mathcal{C}(F, G)$ together with a family of functions

$$\mathcal{D}^\mathcal{C}(F,G) \xrightarrow{\eta \mapsto p_c(\eta)=\eta_c} \mathcal{D}(F(c), G(c))$$

such that for every $f: c_1 \to c_0$

$$\begin{array}{ccc} \mathcal{D}^{\mathcal{C}}(F,G) & \xrightarrow{p_{c_1}} & \mathcal{D}\big(F(c_1),G(c_1)\big) \\ {\scriptstyle p_{c_0}}\big\downarrow & & \big\downarrow{\scriptstyle G(f)_*} \\ \mathcal{D}\big(F(c_0),G(c_0)\big) & \xrightarrow{F(f)^*} & \mathcal{D}\big(F(c_1),G(c_0)\big) \end{array}$$

commutes. Furthermore, $\mathcal{D}^{\mathcal{C}}(F,G)$ is universal among sets with this property: It is "the end of the functor $\mathcal{D}(F(-),G(-)) \colon \mathcal{C}^o \times \mathcal{C} \to \mathcal{D}$". This example is the only important thing to remember about ends. What follows is just for reference.

Definition A.10.5.1 Let $\mathcal{C}$ and $\mathcal{D}$ be V-categories and $T: \mathcal{C}^o \square \mathcal{C} \to \mathcal{D}$ a V-functor. A *V-natural family* is an object $d \in ob\mathcal{D}$, and for every object $c \in ob\mathcal{C}$ a map $f_c \colon d \to T(c,c)$ such that the diagram

$$\begin{array}{ccc} \underline{\mathcal{C}}(c_1,c_0) & \xrightarrow{T(c_1,-)} & \underline{\mathcal{D}}\big(T(c_1,c_1),T(c_1,c_0)\big) \\ {\scriptstyle T(-,c_0)}\big\downarrow & & \big\downarrow{\scriptstyle f_{c_1}^*} \\ \underline{\mathcal{D}}\big(T(c_0,c_0),T(c_1,c_0)\big) & \xrightarrow{f_{c_0}^*} & \underline{\mathcal{D}}\big(d,T(c_1,c_0)\big) \end{array}$$

commutes.

Definition A.10.5.2 Let $\mathcal{C}$ be a V-category. The *end* of a bimodule $T: \mathcal{C}^o \square \mathcal{C} \to V$ is a V-natural family

$$\int_c T(c,c) \xrightarrow{p_x} T(x,x)$$

such that for any other V-natural family $f_x \colon v \to T(x,x)$, there exists a unique morphism $v \to \int_c T(c,c)$ making the following diagram commute:

$$\begin{array}{ccc} v & \longrightarrow & \int_c T(c,c) \\ & {\scriptstyle f_x}\searrow \quad \swarrow{\scriptstyle p_x} & \\ & T(x,x) & \end{array}$$

Definition A.10.5.3 Let $T: \mathcal{C}^o \square \mathcal{C} \to \mathcal{D}$ be a V-functor. The *end* of T is a V-natural family

$$\int_c T(c,c) \xrightarrow{p_x} T(x,x)$$

such that for every $d \in ob\mathcal{D}$

$$\underline{\mathcal{D}}\Big(d, \int_c T(c,c)\Big) \xrightarrow{p_{x*}} \underline{\mathcal{D}}\big(d,T(x,x)\big)$$

is the end of

$$\mathcal{C}^o \Box \mathcal{C} \xrightarrow{\underline{\mathcal{D}}(d, T(-,-))} V.$$

With mild assumptions, this can be expressed as a limit in $\mathcal{D}$ (see [62, p. 39]). The dual of the end is the *coend*. The most basic is the tensor product: considering a ring A as an Ab-category with one object (called A), a left module $M\colon A \to Ab$ and a right module $N\colon A^o \to Ab$, the tensor product $N \otimes_A M$ is nothing but the coend $\int^A N \otimes M$.

A.10.6 Functor Categories

Assume that V has all limits. If I is a small category, we define the V-category $\int_I \mathcal{C}$ of "functors from I to $\mathcal{C}$" as follows. The objects are just the functors from I to $U_0\mathcal{C}$ (the underlying category of $\mathcal{C}$), but the morphisms $\int_I \underline{\mathcal{C}}(F, G)$ is set to be the end $\int_I \underline{\mathcal{C}}(F, G) = \int_{i\in I} \mathcal{C}(F(i), G(i))$ of

$$I^o \times I \xrightarrow{(F,G)} U_0\mathcal{C}^o \times U_0\mathcal{C} \xrightarrow{\mathcal{C}} V.$$

We check that this defines a functor $[I, U_0\mathcal{C}]^o \times [I, U_0\mathcal{C}] \to V$. The composition is defined by the map

$$\begin{aligned}&\left(\int_I \underline{\mathcal{C}}(G, H)\right) \Box \left(\int_I \underline{\mathcal{C}}(F, G)\right)\\ &\quad \to \int_I \int_I \underline{\mathcal{C}}(G, H) \Box \underline{\mathcal{C}}(F, G) \xrightarrow{diag^*} \int_I \underline{\mathcal{C}}(G, H) \Box \underline{\mathcal{C}}(F, G)\\ &\quad \to \int_I \underline{\mathcal{C}}(F, H).\end{aligned}$$

Note that I is here an ordinary category, and the end here is an end of set-based categories. In the case where the forgetful map $V \xrightarrow{N \mapsto V(e,N)} Ens$ has a left adjoint, say $X \mapsto e \Box X$, then there is a left adjoint functor from categories to V-categories, sending a category I to a "free" V-category $e \Box I$, and the functor category we have defined is the usual V-category of V-functors from $e \Box I$ to $\mathcal{C}$ (see [62, 162] or [107]).

Also, a $\mathcal{C}$-bimodule P gives rise to a $\int_I \mathcal{C}$-bimodule $\int_I P$ with $\int_I P(F, G)$ defined as the end. The bimodule structure is defined as

$$\int_I \underline{\mathcal{C}} \Box \int_I P \Box \int_I \underline{\mathcal{C}} \to \int_{I^{\times 3}} \underline{\mathcal{C}} \Box P \Box \underline{\mathcal{C}} \to \int_I \underline{\mathcal{C}} \Box P \Box \underline{\mathcal{C}} \to \int_I P.$$

As an example, one has the fact that if $\mathcal{C}$ is any category and $\mathcal{D}$ is an Ab-category, the free functor from sets to abelian groups $\mathbf{Z}\colon Ens_* \to Ab$ induces an equivalence between the Ab-category of Ab-functors $\mathbf{Z}\mathcal{C} \to \mathcal{D}$ and the Ab-category of functors $\mathcal{C} \to \mathcal{D}$. See [107] for a discussion on the effect of change of base-category.

Example A.10.6.1 (Modules over an **S**-Algebra) Let A be an **S**-algebra. The category $\mathcal{M}_A$ of A-modules is again a $\Gamma\mathcal{S}_*$-category . Explicitly, if M and N are A-modules, then

$$\underline{\mathcal{M}}_A(M, N) = \int_A \underline{\Gamma\mathcal{S}_*}(M, N) \cong \lim_{\leftarrow}\{\underline{\Gamma\mathcal{S}_*}(M, N) \rightrightarrows \underline{\Gamma\mathcal{S}_*}(A \wedge M, N)\}$$

with the obvious maps.

We refer to [253] for a more thorough discussion of **S**-algebras and A-modules and their homotopy properties. See also Chap. 2.

References

1. Théorie des Intersections et Théorème de Riemann-Roch, volume 225 of Lecture Notes in Mathematics. Springer, Berlin, 1971. Séminaire de Géométrie Algébrique du Bois-Marie 1966–1967 (SGA 6), Dirigé par P. Berthelot, A. Grothendieck et L. Illusie. Avec la collaboration de D. Ferrand, J.P. Jouanolou, O. Jussila, S. Kleiman, M. Raynaud et J.P. Serre.
2. Guido's book of conjectures. Enseign. Math. (2), 54(1–2):3–189, 2008. A gift to Guido Mislin on the occasion of his retirement from ETHZ, June 2006. Collected by Indira Chatterji.
3. J.F. Adams. On the groups $J(X)$, IV. Topology, 5:21–71, 1966.
4. J.F. Adams. Stable Homotopy and Generalised Homology, Chicago Lectures in Mathematics. University of Chicago Press, Chicago, 1974.
5. J.F. Adams. Infinite Loop Spaces, volume 90 of Annals of Mathematics Studies. Princeton University Press, Princeton, 1978.
6. V. Angeltveit. On the algebraic K-theory of $\mathbf{Z}/p^n$. arXiv:1101.1866, 2011.
7. V. Angeltveit and J. Rognes. Hopf algebra structure on topological Hochschild homology. Algebr. Geom. Topol., 5:1223–1290, 2005 (electronic).
8. M. Artin and B. Mazur. On the van Kampen theorem. Topology, 5:179–189, 1966.
9. M.F. Atiyah. K-Theory, 2nd edition, Advanced Book Classics. Addison-Wesley, Redwood City, 1989. Notes by D.W. Anderson.
10. C. Ausoni. On the algebraic K-theory of the complex K-theory spectrum. Invent. Math., 180(3):611–668, 2010.
11. C. Ausoni and J. Rognes. Algebraic K-theory of topological K-theory. Acta Math., 188(1):1–39, 2002.
12. M. Barratt and S. Priddy. On the homology of non-connected monoids and their associated groups. Comment. Math. Helv., 47:1–14, 1972.
13. H. Bass. Algebraic K-Theory. Benjamin, New York, 1968.
14. H. Bass. Personal reminiscences of the birth of algebraic K-theory. K-Theory, 30(3):203–209, 2003. Special issue in honor of Hyman Bass on his seventieth birthday, Part III.
15. A.J. Berrick and L. Hesselholt. Topological Hochschild homology and the Bass trace conjecture. Available from http://www-math.mit.edu/~larsh/papers/029/, 2008.
16. A.J. Berrick. An Approach to Algebraic K-Theory, volume 56 of Research Notes in Mathematics. Pitman, Boston, 1982.
17. S. Bloch. On the tangent space to Quillen K-theory. In Algebraic K-Theory, I: Higher K-Theories, Proc. Conf., Battelle Memorial Inst., Seattle, WA., 1972, volume 341 of Lecture Notes in Mathematics, pages 205–210. Springer, Berlin, 1973.
18. S. Bloch. Algebraic K-theory and crystalline cohomology. Inst. Hautes Études Sci. Publ. Math., 47(1978):187–268, 1977.
19. A.J. Blumberg. A discrete model of S^1-homotopy theory. J. Pure Appl. Algebra, 210(1):29–41, 2007.

B.I. Dundas et al., *The Local Structure of Algebraic K-Theory*,
Algebra and Applications 18,
DOI 10.1007/978-1-4471-4393-2,

20. A.J. Blumberg, R.L. Cohen, and C. Schlichtkrull. Topological Hochschild homology of Thom spectra and the free loop space. Geom. Topol., 14(2):1165–1242, 2010.
21. A.J. Blumberg, D. Gepner and G. Tabuada. Uniqueness of the multiplicative cyclotomic trace, March 21 2011.
22. A.J. Blumberg and M.A. Mandell. The localization sequence for the algebraic K-theory of topological K-theory. Acta Math., 200(2):155–179, 2008.
23. A.J. Blumberg and M.A. Mandell. Localization theorems in topological Hochschild homology and topological cyclic homology. arXiv:0802.3938v3, 2008.
24. A.J. Blumberg and M.A. Mandell. Localization for $THH(ku)$ and the topological Hochschild and cyclic homology of Waldhausen categories. arXiv:1111.4003, 2011.
25. J.M. Boardman. Conditionally convergent spectral sequences. In Homotopy Invariant Algebraic Structures, Baltimore, MD, 1998, volume 239 of Contemporary Mathematics, pages 49–84. Am. Math. Soc., Providence, 1999.
26. M. Bökstedt, G. Carlsson, R. Cohen, T. Goodwillie, W.C. Hsiang, and I. Madsen. On the algebraic K-theory of simply connected spaces. Duke Math. J., 84(3):541–563, 1996.
27. M. Bökstedt, W.C. Hsiang, and I. Madsen. The cyclotomic trace and algebraic K-theory of spaces. Invent. Math., 111(3):465–539, 1993.
28. M. Bökstedt and I. Madsen. Topological cyclic homology of the integers. Astérisque, 226(7–8):57–143, 1994. K-theory (Strasbourg, 1992).
29. M. Bökstedt and I. Madsen. Algebraic K-theory of local number fields: the unramified case. In Prospects in Topology, Princeton, NJ, 1994, volume 138 of Ann. of Math. Stud., pages 28–57. Princeton University Press, Princeton, NJ, 1995.
30. M. Bökstedt. Topological Hochschild homology. Preprint, Bielefeld, 1986.
31. F. Borceux. Handbook of Categorical Algebra, 1, volume 50 of Encyclopedia of Mathematics and Its Applications. Cambridge University Press, Cambridge, 1994. Basic category theory.
32. F. Borceux. Handbook of Categorical Algebra, 2, volume 51 of Encyclopedia of Mathematics and Its Applications. Cambridge University Press, Cambridge, 1994. Categories and structures.
33. F. Borceux. Handbook of Categorical Algebra, 3, volume 52 of Encyclopedia of Mathematics and Its Applications. Cambridge University Press, Cambridge, 1994. Categories of sheaves.
34. A. Borel. Stable real cohomology of arithmetic groups. Ann. Sci. École Norm. Sup. (4), 7(1975):235–272, 1974.
35. A. Borel and J.-P. Serre. Le théorème de Riemann-Roch. Bull. Soc. Math. Fr., 86:97–136, 1958.
36. R. Bott. The periodicity theorem for the classical groups and some of its applications. Adv. Math., 4(1970):353–411, 1970.
37. A.K. Bousfield. The localization of spaces with respect to homology. Topology, 14:133–150, 1975.
38. A.K. Bousfield. The localization of spectra with respect to homology. Topology, 18(4):257–281, 1979.
39. A.K. Bousfield and E.M. Friedlander. Homotopy theory of Γ-spaces, spectra, and bisimplicial sets. In Geometric Applications of Homotopy Theory II, Proc. Conf., Evanston, IL, 1977, volume 658 of Lecture Notes in Mathematics, pages 80–130. Springer, Berlin, 1978.
40. A.K. Bousfield and D.M. Kan. Homotopy Limits, Completions and Localizations, volume 304 of Lecture Notes in Mathematics Springer, Berlin, 1972.
41. L. Breen. Extensions du groupe additif. Inst. Hautes Études Sci. Publ. Math., 48:39–125, 1978.
42. M. Brun. Topological Hochschild homology of $\mathbf{Z}/p^n$. J. Pure Appl. Algebra, 148(1):29–76, 2000.
43. M. Brun. Filtered topological cyclic homology and relative K-theory of nilpotent ideals. Algebr. Geom. Topol., 1:201–230, 2001 (electronic).
44. M. Brun, G. Carlsson, and B. Dundas. Covering homology. Adv. Math., 225:3166–3213, 2010.

45. M. Brun, Z. Fiedorowicz, and R.M. Vogt. On the multiplicative structure of topological Hochschild homology. Algebr. Geom. Topol., 7:1633–1650, 2007.
46. R.R. Bruner, J.P. May, J.E. McClure, and M. Steinberger. H_∞ Ring Spectra and Their Applications, volume 1176 of Lecture Notes in Mathematics. Springer, Berlin, 1986.
47. D. Burghelea and Z. Fiedorowicz. Cyclic homology and algebraic K-theory of spaces, II. Topology, 25(3):303–317, 1986.
48. D. Burghelea, Z. Fiedorowicz, and W. Gajda. Power maps and epicyclic spaces. J. Pure Appl. Algebra, 96(1):1–14, 1994.
49. D. Burghelea. Some rational computations of the Waldhausen algebraic K-theory. Comment. Math. Helv., 54(2):185–198, 1979.
50. D. Burghelea. Cyclic homology and the algebraic K-theory of spaces, I. In Applications of Algebraic K-Theory to Algebraic Geometry and Number Theory, Part I, II, Boulder, CO, 1983, volume 55 of Contemporary Mathematics, pages 89–115. Am. Math. Soc., Providence, 1986.
51. G.E. Carlsson, R.L. Cohen, T. Goodwillie, and W.C. Hsiang. The free loop space and the algebraic K-theory of spaces. K-Theory, 1(1):53–82, 1987.
52. G. Carlsson. Equivariant stable homotopy and Segal's Burnside ring conjecture. Ann. Math. (2), 120(2):189–224, 1984.
53. G. Carlsson, L.D. Cristopher, and B. Dundas. Higher topological cyclic homology and the Segal conjecture for tori. Adv. Math., 226:1823–1874, 2011.
54. P.M. Cohn. Some remarks on the invariant basis property. Topology, 5:215–228, 1966.
55. A. Connes. Cohomologie cyclique et foncteurs Ext^n. C. R. Acad. Sci. Paris Sér. I Math., 296(23):953–958, 1983.
56. A. Connes. Noncommutative differential geometry. Inst. Hautes Études Sci. Publ. Math., 62:257–360, 1985.
57. G. Cortiñas. The obstruction to excision in K-theory and in cyclic homology. Invent. Math., 164(1):143–173, 2006.
58. J. Cuntz and D. Quillen. Excision in bivariant periodic cyclic cohomology. Invent. Math., 127(1):67–98, 1997.
59. B. Day. On closed categories of functors. In Reports of the Midwest Category Seminar, IV, volume 137 of Lecture Notes in Mathematics, pages 1–38. Springer, Berlin, 1970.
60. R.K. Dennis and M.R. Stein. K_2 of discrete valuation rings. Adv. Math., 18(2):182–238, 1975.
61. V. Drinfeld. On the notion of geometric realization. Mosc. Math. J., 4(3):619–626, 782, 2004.
62. E.J. Dubuc. Kan Extensions in Enriched Category Theory, volume 145 of Lecture Notes in Mathematics. Springer, Berlin, 1970.
63. B.I. Dundas. Relative K-theory and topological cyclic homology. Acta Math., 179(2):223–242, 1997.
64. B.I. Dundas. On K-theory of simplicial rings and degreewise constructions. K-Theory, 18(1):77–92, 1999.
65. B.I. Dundas. The cyclotomic trace for symmetric monoidal categories. In Geometry and Topology, Aarhus, 1998, volume 258 of Contemporary Mathematics, pages 121–143. Am. Math. Soc., Providence, 2000.
66. B.I. Dundas. Localization of V-categories. Theory Appl. Categ., 8(10):284–312, 2001 (electronic).
67. B.I. Dundas and H.Ø. Kittang. Excision for K-theory of connective ring spectra. Homol. Homotopy Appl., 10(1):29–39, 2008.
68. B.I. Dundas and H.Ø. Kittang. Integral excision for K-theory. arXiv:1009.3044, 2010.
69. B.I. Dundas and R. McCarthy. Stable K-theory and topological Hochschild homology. Ann. Math. (2), 140(3):685–701, 1994.
70. B.I. Dundas and R. McCarthy. Topological Hochschild homology of ring functors and exact categories. J. Pure Appl. Algebra, 109(3):231–294, 1996.

71. B.I. Dundas, O. Röndigs, and P.A. Østvær. Enriched functors and stable homotopy theory. Doc. Math., 8:409–488, 2003 (electronic).
72. W. Dwyer, W.C. Hsiang, and R. Staffeldt. Pseudo-isotopy and invariant theory. Topology, 19(4):367–385, 1980.
73. W. Dwyer, W.C. Hsiang, and R.E. Staffeldt. Pseudo-isotopy and invariant theory, II: rational algebraic K-theory of a space with finite fundamental group. In Topology Symposium, Proc. Sympos., Univ. Siegen, Siegen, 1979, volume 788 of Lecture Notes in Mathematics, pages 418–441. Springer, Berlin, 1980.
74. W.G. Dwyer, M.J. Hopkins, and D.M. Kan. The homotopy theory of cyclic sets. Trans. Am. Math. Soc., 291(1):281–289, 1985.
75. W.G. Dwyer and D.M. Kan. Function complexes in homotopical algebra. Topology, 19(4):427–440, 1980.
76. W. Dwyer, E. Friedlander, V. Snaith, and R. Thomason. Algebraic K-theory eventually surjects onto topological K-theory. Invent. Math., 66(3):481–491, 1982.
77. W.G. Dwyer and E.M. Friedlander. Algebraic and étale K-theory. Trans. Am. Math. Soc., 292(1):247–280, 1985.
78. W.G. Dwyer, P.S. Hirschhorn, D.M. Kan, and J.H. Smith. Homotopy limit functors on model categories and homotopical categories, volume 113 of Mathematical Surveys and Monographs. Am. Math. Soc., Providence, 2004.
79. P. Elbaz-Vincent, H. Gangl, and C. Soulé. Quelques calculs de la cohomologie de $\mathrm{GL}_N(\mathbb{Z})$ et de la K-théorie de $\mathbb{Z}$. C. R. Math. Acad. Sci. Paris, 335(4):321–324, 2002.
80. A.D. Elmendorf, I. Kriz, M.A. Mandell, and J.P. May. Rings, modules, and algebras in stable homotopy theory, volume 47 of Mathematical Surveys and Monographs. Am. Math. Soc., Providence, 1997. With an appendix by M. Cole.
81. F.T. Farrell and L.E. Jones. Rigidity in geometry and topology. In Proceedings of the International Congress of Mathematicians, Vol. I, II, Kyoto, 1990, pages 653–663. Math. Soc. Japan, Tokyo, 1991.
82. S.C. Ferry, A. Ranicki, and J. Rosenberg. A history and survey of the Novikov conjecture. In Novikov Conjectures, Index Theorems and Rigidity, Vol. 1, Oberwolfach, 1993, volume 226 of London Mathematical Society Lecture Note Series, pages 7–66. Cambridge University Press, Cambridge, 1995.
83. Z. Fiedorowicz, C. Ogle, and R.M. Vogt. Volodin K-theory of A_∞-ring spaces. Topology, 32(2):329–352, 1993.
84. V. Franjou, J. Lannes, and L. Schwartz. Autour de la cohomologie de Mac Lane des corps finis. Invent. Math., 115(3):513–538, 1994.
85. V. Franjou and T. Pirashvili. On the Mac Lane cohomology for the ring of integers. Topology, 37(1):109–114, 1998.
86. E.M. Friedlander and D.R. Grayson, editors. Handbook of K-Theory, Vols. 1, 2. Springer, Berlin, 2005.
87. E.M. Friedlander and B. Mazur. Filtrations on the homology of algebraic varieties. Mem. Am. Math. Soc., 110(529):110, 1994. With an appendix by Daniel Quillen.
88. O. Gabber. K-theory of Henselian local rings and Henselian pairs. In Algebraic K-Theory, Commutative Algebra, and Algebraic Geometry, Santa Margherita Ligure, 1989, volume 126 of Contemporary Mathematics, pages 59–70. Am. Math. Soc., Providence, 1992.
89. P. Gabriel and M. Zisman. Calculus of Fractions and Homotopy Theory, volume 35 of Ergebnisse der Mathematik und ihrer Grenzgebiete. Springer, New York, 1967.
90. W. Gajda. On $K_*(\mathbb{Z})$ and classical conjectures in the arithmetic of cyclotomic fields. In Homotopy Theory: Relations with Algebraic Geometry, Group Cohomology, and Algebraic K-Theory, volume 346 of Contemporary Mathematics, pages 217–237. Am. Math. Soc., Providence, 2004.
91. S. Galatius, U. Tillmann, I. Madsen, and M. Weiss. The homotopy type of the cobordism category. Acta Math., 202(2):195–239, 2009.
92. T. Geisser and L. Hesselholt. Topological cyclic homology of schemes. In Algebraic K-Theory, Seattle, WA, 1997, volume 67 of Proceedings of Symposia in Pure Mathematics,

pages 41–87. Am. Math. Soc., Providence, 1999.
93. T. Geisser and L. Hesselholt. Bi-relative algebraic K-theory and topological cyclic homology. Invent. Math., 166(2):359–395, 2006.
94. T. Geisser and L. Hesselholt. On the K-theory and topological cyclic homology of smooth schemes over a discrete valuation ring. Trans. Am. Math. Soc., 358(1):131–145, 2006 (electronic).
95. T. Geisser and M. Levine. The Bloch-Kato conjecture and a theorem of Suslin-Voevodsky. J. Reine Angew. Math., 530:55–103, 2001.
96. S.M. Gersten. K_3 of a ring is H_3 of the Steinberg group. Proc. Am. Math. Soc., 37:366–368, 1973.
97. S.M. Gersten. K-theory of free rings. Commun. Algebra, 1:39–64, 1974.
98. H. Gillet and D.R. Grayson. The loop space of the Q-construction. Ill. J. Math., 31(4):574–597, 1987.
99. P.G. Goerss and J.F. Jardine. Simplicial homotopy theory, volume 174 of Progress in Mathematics. Birkhäuser, Basel, 1999.
100. T.G. Goodwillie. Cyclic homology, derivations, and the free loopspace. Topology, 24(2):187–215, 1985.
101. T.G. Goodwillie. On the general linear group and Hochschild homology. Ann. Math. (2), 121(2):383–407, 1985.
102. T.G. Goodwillie. Relative algebraic K-theory and cyclic homology. Ann. Math. (2), 124(2):347–402, 1986.
103. T.G. Goodwillie. Letter to F. Waldhausen, August 10 1987.
104. T.G. Goodwillie. The differential calculus of homotopy functors. In Proceedings of the International Congress of Mathematicians, Vols. I, II, Kyoto, 1991, pages 621–630. Math. Soc. Japan, Tokyo, 1990.
105. T.G. Goodwillie. Notes on the cyclotomic trace. Lecture notes for a series of seminar talks at MSRI, Spring 1990, December 1991.
106. T.G. Goodwillie. Calculus, II: analytic functors. K-Theory, 5(4):295–332, 1991/92.
107. J.W. Gray. Formal Category Theory: Adjointness for 2-Categories, volume 391 of Lecture Notes in Mathematics. Springer, Berlin, 1974.
108. D. Grayson. Higher algebraic K-theory. II (after Daniel Quillen). In Algebraic K-Theory, Proc. Conf., Northwestern Univ., Evanston, IL, 1976, volume 551 of Lecture Notes in Mathematics, pages 217–240. Springer, Berlin, 1976.
109. D.R. Grayson. The K-theory of endomorphisms. J. Algebra, 48(2):439–446, 1977.
110. D.R. Grayson. Exact sequences in algebraic K-theory. Ill. J. Math., 31(4):598–617, 1987.
111. D.R. Grayson. Exterior power operations on higher K-theory. K-Theory, 3(3):247–260, 1989.
112. D.R. Grayson. On the K-theory of fields. In Algebraic K-Theory and Algebraic Number Theory, Honolulu, HI, 1987, volume 83 of Contemporary Mathematics, pages 31–55. Amer. Math. Soc., Providence, 1989.
113. D.R. Grayson. Adams operations on higher K-theory. K-Theory, 6(2):97–111, 1992.
114. D.R. Grayson. Weight filtrations via commuting automorphisms. K-Theory, 9(2):139–172, 1995.
115. J.P.C. Greenlees. Representing Tate cohomology of G-spaces. Proc. Edinb. Math. Soc. (2), 30(3):435–443, 1987.
116. J.P.C. Greenlees and J.P. May. Generalized Tate cohomology. Mem. Am. Math. Soc., 113(543):178, 1995.
117. T. Gunnarsson. Algebraic K-theory of spaces as K-theory of monads. Preprint, Aarhus University, 1982.
118. T. Gunnarsson, R. Schwänzl, R.M. Vogt, and F. Waldhausen. An un-delooped version of algebraic K-theory. J. Pure Appl. Algebra, 79(3):255–270, 1992.
119. B. Harris. Bott periodicity via simplicial spaces. J. Algebra, 62(2):450–454, 1980.
120. A. Hatcher. Algebraic Topology. Cambridge University Press, Cambridge, 2002.
121. A. Hattori. Rank element of a projective module. Nagoya Math. J., 25:113–120, 1965.

122. J.-C. Hausmann and D. Husemoller. Acyclic maps. Enseign. Math. (2), 25(1–2):53–75, 1979.
123. L. Hesselholt. On the p-typical curves in Quillen's K-theory. Acta Math., 177(1):1–53, 1996.
124. L. Hesselholt. Witt vectors of non-commutative rings and topological cyclic homology. Acta Math., 178(1):109–141, 1997.
125. L. Hesselholt. Correction to: "Witt vectors of non-commutative rings and topological cyclic homology" [Acta Math. **178** (1997), no. 1, 109–141]. Acta Math., 195:55–60, 2005.
126. L. Hesselholt. On the K-theory of the coordinate axes in the plane. Nagoya Math. J., 185:93–109, 2007.
127. L. Hesselholt. The big de Rham–Witt complex. arXiv:1006.3125, 2010.
128. L. Hesselholt and I. Madsen. Cyclic polytopes and the K-theory of truncated polynomial algebras. Invent. Math., 130(1):73–97, 1997.
129. L. Hesselholt and I. Madsen. On the K-theory of finite algebras over Witt vectors of perfect fields. Topology, 36(1):29–101, 1997.
130. L. Hesselholt and I. Madsen. On the K-theory of nilpotent endomorphisms. In Homotopy Methods in Algebraic Topology, Boulder, CO, 1999, volume 271 of Contemporary Mathematics, pages 127–140. Amer. Math. Soc., Providence, 2001.
131. L. Hesselholt and I. Madsen. On the K-theory of local fields. Ann. Math. (2), 158(1):1–113, 2003.
132. L. Hesselholt and I. Madsen. On the De Rham-Witt complex in mixed characteristic. Ann. Sci. École Norm. Sup. (4), 37(1):1–43, 2004.
133. G. Higman. The units of group-rings. Proc. Lond. Math. Soc. (2), 46:231–248, 1940.
134. H.L. Hiller. λ-rings and algebraic K-theory. J. Pure Appl. Algebra, 20(3):241–266, 1981.
135. P.S. Hirschhorn. Model Categories and Their Localizations, volume 99 of Mathematical Surveys and Monographs. Am. Math. Soc., Providence, 2003.
136. G. Hochschild, B. Kostant, and A. Rosenberg. Differential forms on regular affine algebras. Trans. Am. Math. Soc., 102:383–408, 1962.
137. L. Hodgkin and P.A. Østvær. The homotopy type of two-regular K-theory. In Categorical Decomposition Techniques in Algebraic Topology, Isle of Skye, 2001, volume 215 of Progress of Mathematics, pages 167–178. Birkhäuser, Basel, 2004.
138. M.J. Hopkins. Notes on E_∞ ring spectra. Typed Notes, 1993.
139. M. Hovey. Model Categories, volume 63 of Mathematical Surveys and Monographs. Am. Math. Soc., Providence, 1999.
140. M. Hovey. Spectra and symmetric spectra in general model categories. J. Pure Appl. Algebra, 165(1):63–127, 2001.
141. M. Hovey, B. Shipley, and J. Smith. Symmetric spectra. J. Am. Math. Soc., 13(1):149–208, 2000.
142. W.C. Hsiang and R.E. Staffeldt. A model for computing rational algebraic K-theory of simply connected spaces. Invent. Math., 68(2):227–239, 1982.
143. W.C. Hsiang and R.E. Staffeldt. Rational algebraic K-theory of a product of Eilenberg-Mac Lane spaces. In Proceedings of the Northwestern Homotopy Theory Conference, Evanston, IL, 1982, volume 19 of Contemporary Mathematics, pages 95–114. Amer. Math. Soc., Providence, 1983.
144. W.C. Hsiang. Geometric applications of algebraic K-theory. In Proceedings of the International Congress of Mathematicians, Vols. 1, 2, Warsaw, 1983, pages 99–118. PWN, Warsaw, 1984.
145. L. Illusie. Complexe Cotangent et Déformations, II, volume 283 of Lecture Notes in Mathematics. Springer, Berlin, 1972.
146. L. Illusie. Complexe de de Rham-Witt et cohomologie cristalline. Ann. Sci. École Norm. Sup. (4), 12(4):501–661, 1979.
147. H. Inassaridze. Algebraic K-Theory, volume 311 of Mathematics and Its Applications. Kluwer Academic, Dordrecht, 1995.

148. J.F. Jardine. The homotopical foundations of algebraic K-theory. In Algebraic K-Theory and Algebraic Number Theory, Honolulu, HI, 1987, volume 83 of Contemporary Mathematics, pages 57–82. Amer. Math. Soc., Providence, 1989.
149. J.F. Jardine. The K-theory presheaf of spectra. In New Topological Contexts for Galois Theory and Algebraic Geometry, BIRS, 2008, volume 16 of Geometry & Topology Monographs, pages 151–178. Geom. Topol. Publ., Coventry, 2009.
150. M. Jibladze and T. Pirashvili. Cohomology of algebraic theories. J. Algebra, 137(2):253–296, 1991.
151. J.D.S. Jones. Cyclic homology and equivariant homology. Invent. Math., 87(2):403–423, 1987.
152. B. Kahn. The Quillen-Lichtenbaum conjecture at the prime 2. Preprint, available at http://www.math.uiuc.edu/K-theory/0208/.
153. B. Kahn. Algebraic K-theory, algebraic cycles and arithmetic geometry. In Handbook of K-Theory, Vols. 1, 2, pages 351–428. Springer, Berlin, 2005.
154. D.M. Kan and W.P. Thurston. Every connected space has the homology of a $K(\pi, 1)$. Topology, 15(3):253–258, 1976.
155. D.M. Kan. A combinatorial definition of homotopy groups. Ann. Math. (2), 67:282–312, 1958.
156. M. Karoubi and T. Lambre. Quelques classes caractéristiques en théorie des nombres. J. Reine Angew. Math., 543:169–186, 2002.
157. M. Karoubi and O. Villamayor. K-théorie algébrique et K-théorie topologique, I. Math. Scand., 28(1972):265–307, 1971.
158. M. Karoubi and O. Villamayor. K-théorie algébrique et K-théorie topologique, II. Math. Scand., 32:57–86, 1973.
159. C. Kassel. K-théorie relative d'un idéal bilatère de carré nul: étude homologique en basse dimension. In Algebraic K-Theory, Proc. Conf., Northwestern Univ., Evanston, IL, 1980, volume 854 of Lecture Notes in Mathematics, pages 249–261. Springer, Berlin, 1981.
160. C. Kassel. La K-théorie stable. Bull. Soc. Math. Fr., 110(4):381–416, 1982.
161. G.M. Kelly. Coherence theorems for lax algebras and for distributive laws. In Category Seminar, Proc. Sem., Sydney, 1972/1973, volume 420 of Lecture Notes in Mathematics, pages 281–375. Springer, Berlin, 1974.
162. G.M. Kelly. Basic Concepts of Enriched Category Theory, volume 64 of London Mathematical Society Lecture Note Series. Cambridge University Press, Cambridge, 1982.
163. M.A. Kervaire. Le théorème de Barden-Mazur-Stallings. Comment. Math. Helv., 40:31–42, 1965.
164. M.A. Kervaire. Smooth homology spheres and their fundamental groups. Trans. Am. Math. Soc., 144:67–72, 1969.
165. M.A. Kervaire. Multiplicateurs de Schur et K-théorie. In Essays on Topology and Related Topics (Mémoires dédiés à Georges de Rham), pages 212–225. Springer, New York, 1970.
166. F. Keune. The relativization of K_2. J. Algebra, 54(1):159–177, 1978.
167. Ch. Kratzer. λ-structure en K-théorie algébrique. Comment. Math. Helv., 55(2):233–254, 1980.
168. M. Larsen and A. Lindenstrauss. Topological Hochschild homology of algebras in characteristic p. J. Pure Appl. Algebra, 145(1):45–58, 2000.
169. M. Larsen and A. Lindenstrauss. Topological Hochschild homology and the condition of Hochschild-Kostant-Rosenberg. Commun. Algebra, 29(4):1627–1638, 2001.
170. T. Lawson. Commutative Γ-rings do not model all commutative ring spectra. Homol. Homotopy Appl., 11(2):189–194, 2009.
171. W.G. Leavitt. Modules without invariant basis number. Proc. Am. Math. Soc., 8:322–328, 1957.
172. R. Lee and R.H. Szczarba. The group $K_3(Z)$ is cyclic of order forty-eight. Ann. Math. (2), 104(1):31–60, 1976.
173. L.G. Lewis Jr., J.P. May, M. Steinberger, and J.E. McClure. Equivariant Stable Homotopy Theory, volume 1213 of Lecture Notes in Mathematics. Springer, Berlin, 1986. With contri-

butions by J.E. McClure.
174. L.G. Lewis Jr. Is there a convenient category of spectra? J. Pure Appl. Algebra, 73(3):233–246, 1991.
175. S. Lichtenbaum. Values of zeta-functions, étale cohomology, and algebraic K-theory. In Algebraic K-Theory, II: "Classical" Algebraic K-Theory and Connections with Arithmetic, Proc. Conf., Battelle Memorial Inst., Seattle, WA, 1972, volume 342 of Lecture Notes in Mathematics, pages 489–501. Springer, Berlin, 1973.
176. A. Lindenstrauss. A relative spectral sequence for topological Hochschild homology of spectra. J. Pure Appl. Algebra, 148(1):77–88, 2000.
177. A. Lindenstrauss and I. Madsen. Topological Hochschild homology of number rings. Trans. Am. Math. Soc., 352(5):2179–2204, 2000.
178. J.-L. Loday. Applications algébriques du tore dans la sphère et de $S^p \times S^q$ dans S^{p+q}. In Algebraic K-Theory, II: "Classical" Algebraic K-Theory and Connections with Arithmetic, Proc. Conf., Seattle Res. Center, Battelle Memorial Inst., 1972, volume 342 of Lecture Notes in Mathematics, pages 79–91. Springer, Berlin, 1973.
179. J.-L. Loday. K-théorie algébrique et représentations de groupes. Ann. Sci. École Norm. Sup. (4), 9(3):309–377, 1976.
180. J.-L. Loday. Cohomologie et groupe de Steinberg relatifs. J. Algebra, 54(1):178–202, 1978.
181. J.-L. Loday. Cyclic Homology, 2nd edition, volume 301 of Grundlehren der Mathematischen Wissenschaften [Fundamental Principles of Mathematical Sciences]. Springer, Berlin, 1998. Appendix E by María O. Ronco, Chapter 13 by the author in collaboration with Teimuraz Pirashvili.
182. J.-L. Loday and D. Quillen. Cyclic homology and the Lie algebra homology of matrices. Comment. Math. Helv., 59(4):569–591, 1984.
183. W. Lück and H. Reich. Detecting K-theory by cyclic homology. Proc. Lond. Math. Soc. (3), 93(3):593–634, 2006.
184. S. Lunøe-Nielsen and J. Rognes. The topological Singer construction. arXiv:1010.5633, 2010.
185. S. Lunøe-Nielsen and J. Rognes. The Segal conjecture for topological Hochschild homology of complex cobordism. J. Topol., 4(3):591–622, 2011.
186. J. Lurie. Higher Topos Theory, volume 170 of Annals of Mathematics Studies. Princeton University Press, Princeton, 2009.
187. M. Lydakis. Simplicial functors and stable homotopy theory. Preprint 98-049, SFB 343, Bielefeld, June 1998.
188. M. Lydakis. Smash products and Γ-spaces. Math. Proc. Camb. Philos. Soc., 126(2):311–328, 1999.
189. H. Maazen and J. Stienstra. A presentation for K_2 of split radical pairs. J. Pure Appl. Algebra, 10(3):271–294, 1977.
190. S. Mac Lane. Homologie des anneaux et des modules. In Colloque de Topologie Algébrique, Louvain, 1956, pages 55–80. Georges Thone, Liège, 1957.
191. S. Mac Lane. Categories for the Working Mathematician, 2nd edition, volume 5 of Graduate Texts in Mathematics. Springer, New York, 1998.
192. I. Madsen. Algebraic K-theory and traces. In Current Developments in Mathematics, Cambridge, MA, 1995, pages 191–321. International Press, Cambridge, 1994.
193. I. Madsen and M. Weiss. The stable mapping class group and stable homotopy theory. In European Congress of Mathematics, pages 283–307. Eur. Math. Soc., Zürich, 2005.
194. B.A. Magurn, editor. Reviews in K-Theory, 1940–1984. Am. Math. Soc., Providence, 1985. Reviews reprinted from Mathematical Reviews.
195. M.A. Mandell and J.P. May. Equivariant orthogonal spectra and S-modules. Mem. Am. Math. Soc., 159(755):108, 2002.
196. M.A. Mandell, J.P. May, S. Schwede, and B. Shipley. Model categories of diagram spectra. Proc. Lond. Math. Soc. (3), 82(2):441–512, 2001.
197. M.A. Mandell. Equivariant symmetric spectra. In Homotopy Theory: Relations with Algebraic Geometry, Group Cohomology, and Algebraic K-Theory, volume 346 of Contempo-

rary Mathematics, pages 399–452. Amer. Math. Soc., Providence, 2004.
198. J.P. May. E_∞ spaces, group completions, and permutative categories. In New Developments in Topology, Proc. Sympos. Algebraic Topology, Oxford, 1972, volume 11 of London Mathematical Society Lecture Note Series, pages 61–93. Cambridge University Press, London, 1974.
199. J.P. May. A Concise Course in Algebraic Topology. Chicago Lectures in Mathematics. University of Chicago Press, Chicago, 1999.
200. J.P. May. E_∞ Ring Spaces and E_∞ Ring Spectra, volume 577 of Lecture Notes in Mathematics. Springer, Berlin, 1977. With contributions by Frank Quinn, Nigel Ray, and Jørgen Tornehave.
201. J.P. May. Simplicial Objects in Algebraic Topology, Chicago Lectures in Mathematics. University of Chicago Press, Chicago, 1992. Reprint of the 1967 original.
202. B. Mazur. Relative neighborhoods and the theorems of Smale. Ann. Math. (2), 77:232–249, 1963.
203. R. McCarthy. On fundamental theorems of algebraic K-theory. Topology, 32(2):325–328, 1993.
204. R. McCarthy. The cyclic homology of an exact category. J. Pure Appl. Algebra, 93(3):251–296, 1994.
205. R. McCarthy. Relative algebraic K-theory and topological cyclic homology. Acta Math., 179(2):197–222, 1997.
206. J.E. McClure and R.E. Staffeldt. The chromatic convergence theorem and a tower in algebraic K-theory. Proc. Am. Math. Soc., 118(3):1005–1012, 1993.
207. J.E. McClure and R.E. Staffeldt. On the topological Hochschild homology of bu. I. Am. J. Math., 115(1):1–45, 1993.
208. J.E. McClure and J.H. Smith. A solution of Deligne's Hochschild cohomology conjecture. In Recent Progress in Homotopy Theory, Baltimore, MD, 2000, volume 293 of Contemporary Mathematics, pages 153–193. Amer. Math. Soc., Providence, 2002.
209. D. McDuff and G. Segal. Homology fibrations and the "group-completion" theorem. Invent. Math., 31(3):279–284, 1975.
210. J. Milnor. Whitehead torsion. Bull. Am. Math. Soc., 72:358–426, 1966.
211. J. Milnor. The Steenrod algebra and its dual. Ann. Math. (2), 67:150–171, 1958.
212. J. Milnor. Two complexes which are homeomorphic but combinatorially distinct. Ann. Math. (2), 74:575–590, 1961.
213. J. Milnor. Introduction to Algebraic K-Theory, volume 72 of Annals of Mathematics Studies. Princeton University Press, Princeton, 1971.
214. J.W. Milnor and J.D. Stasheff. Characteristic Classes, volume 76 of Annals of Mathematics Studies. Princeton University Press, Princeton, 1974.
215. B. Mitchell. Rings with several objects. Adv. Math., 8:1–161, 1972.
216. S.A. Mitchell. On the Lichtenbaum-Quillen conjectures from a stable homotopy-theoretic viewpoint. In Algebraic Topology and Its Applications, volume 27 of Mathematical Sciences Research Institute Publications, pages 163–240. Springer, New York, 1994.
217. I. Moerdijk. Bisimplicial sets and the group-completion theorem. In Algebraic K-Theory: Connections with Geometry and Topology, Lake Louise, AB, 1987, volume 279 of NATO Advanced Study Institutes Series. Series C, Mathematical and Physical Sciences, pages 225–240. Kluwer Academic, Dordrecht, 1989.
218. D. Mumford. Towards an enumerative geometry of the moduli space of curves. In Arithmetic and Geometry, Vol. II, volume 36 of Progress of Mathematics, pages 271–328. Birkhäuser, Boston, 1983.
219. S. Oka. Multiplicative structure of finite ring spectra and stable homotopy of spheres. In Algebraic Topology, Aarhus, 1982, volume 1051 of Lecture Notes in Mathematics, pages 418–441. Springer, Berlin, 1984.
220. R. Oliver. Whitehead Groups of Finite Groups, volume 132 of London Mathematical Society Lecture Note Series. Cambridge University Press, Cambridge, 1988.

221. I.A. Panin. The Hurewicz theorem and K-theory of complete discrete valuation rings. Izv. Akad. Nauk SSSR, Ser. Mat., 50(4):763–775, 878, 1986.
222. T. Pirashvili. On the topological Hochschild homology of $\mathbf{Z}/p^k\mathbf{Z}$. Commun. Algebra, 23(4):1545–1549, 1995.
223. T. Pirashvili. Polynomial approximation of Ext and Tor groups in functor categories. Commun. Algebra, 21(5):1705–1719, 1993.
224. T. Pirashvili. Spectral sequence for Mac Lane homology. J. Algebra, 170(2):422–428, 1994.
225. T. Pirashvili and F. Waldhausen. Mac Lane homology and topological Hochschild homology. J. Pure Appl. Algebra, 82(1):81–98, 1992.
226. D. Popescu. Letter to the editor: "General Néron desingularization and approximation" [Nagoya Math. J. **104** (1986), 85–115]. Nagoya Math. J., 118:45–53, 1990.
227. D.G. Quillen. Spectral sequences of a double semi-simplicial group. Topology, 5:155–157, 1966.
228. D. Quillen. Cohomology of groups. In Actes du Congrès International des Mathématiciens, Tome 2, Nice, 1970, pages 47–51. Gauthier-Villars, Paris, 1971.
229. D. Quillen. Letter to Graeme Segal, July 25 1972.
230. D. Quillen. On the cohomology and K-theory of the general linear groups over a finite field. Ann. Math. (2), 96:552–586, 1972.
231. D. Quillen. Finite generation of the groups K_i of rings of algebraic integers. In Algebraic K-Theory, I: Higher K-Theories, Proc. Conf., Battelle Memorial Inst., Seattle, WA, 1972, volume 341 of Lecture Notes in Mathematics, pages 179–198. Springer, Berlin, 1973.
232. D. Quillen. Higher algebraic K-theory. I. In Algebraic K-Theory, I: Higher K-Theories, Proc. Conf., Battelle Memorial Inst., Seattle, WA, 1972, volume 341 of Lecture Notes in Mathematics, pages 85–147. Springer, Berlin, 1973.
233. D. Quillen. Higher algebraic K-theory. In Proceedings of the International Congress of Mathematicians, Vol. 1, Vancouver, BC, 1974, pages 171–176. Can. Math. Congress, Montreal, 1975.
234. D. Quillen. Projective modules over polynomial rings. Invent. Math., 36:167–171, 1976.
235. D.G. Quillen. Homotopical Algebra, volume 43 of Lecture Notes in Mathematics. Springer, Berlin, 1967.
236. A.A. Ranicki, A.J. Casson, D.P. Sullivan, M.A. Armstrong, C.P. Rourke, and G.E. Cooke. The Hauptvermutung Book, volume 1 of K-Monographs in Mathematics. Kluwer Academic, Dordrecht, 1996. A collection of papers of the topology of manifolds.
237. D.C. Ravenel. Nilpotence and periodicity in stable homotopy theory, volume 128 of Annals of Mathematics Studies. Princeton University Press, Princeton, 1992. Appendix C by Jeff Smith.
238. D.C. Ravenel. Life after the telescope conjecture. In Algebraic K-Theory and Algebraic Topology, Lake Louise, AB, 1991, volume 407 of NATO Advanced Study Institutes Series. Series C, Mathematical and Physical Sciences, pages 205–222. Kluwer Academic, Dordrecht, 1993.
239. J. Rognes and C. Weibel. Two-primary algebraic K-theory of rings of integers in number fields. J. Am. Math. Soc., 13(1):1–54, 2000. Appendix A by Manfred Kolster.
240. J. Rognes. Algebraic K-theory of the two-adic integers. J. Pure Appl. Algebra, 134(3):287–326, 1999.
241. J. Rognes. $K_4(\mathbf{Z})$ is the trivial group. Topology, 39(2):267–281, 2000.
242. J. Rognes. Two-primary algebraic K-theory of pointed spaces. Topology, 41(5):873–926, 2002.
243. J. Rognes. The smooth Whitehead spectrum of a point at odd regular primes. Geom. Topol., 7:155–184, 2003.
244. J. Rosenberg. Algebraic K-Theory and Its Applications, volume 147 of Graduate Texts in Mathematics. Springer, New York, 1994.
245. M. Schlichting. Negative K-theory of derived categories. Math. Z., 253(1):97–134, 2006.
246. C. Schlichtkrull. The transfer map in topological Hochschild homology. J. Pure Appl. Algebra, 133(3):289–316, 1998.

247. C. Schlichtkrull. Units of ring spectra and their traces in algebraic K-theory. Geom. Topol., 8:645–673, 2004 (electronic).
248. C. Schlichtkrull. The cyclotomic trace for symmetric ring spectra. In New Topological Contexts for Galois Theory and Algebraic Geometry, BIRS, 2008, volume 16 of Geometry & Topology Monographs, pages 545–592. Geom. Topol. Publ., Coventry, 2009.
249. R. Schwänzl, R.M. Vogt, and F. Waldhausen. Topological Hochschild homology. J. Lond. Math. Soc. (2), 62(2):345–356, 2000.
250. R. Schwänzl, R. Staffeldt, and F. Waldhausen. Stable K-theory and topological Hochschild homology of A_∞ rings. In Algebraic K-Theory, Poznań, 1995, volume 199 of Contemporary Mathematics, pages 161–173. Amer. Math. Soc., Providence, 1996.
251. S. Schwede. An untitled book project about symmetric spectra. Preliminary version of a book project in progress. Available from the author's home page.
252. S. Schwede. Spectra in model categories and applications to the algebraic cotangent complex. J. Pure Appl. Algebra, 120(1):77–104, 1997.
253. S. Schwede. Stable homotopical algebra and Γ-spaces. Math. Proc. Camb. Philos. Soc., 126(2):329–356, 1999.
254. S. Schwede and B. Shipley. Equivalences of monoidal model categories. Algebr. Geom. Topol., 3:287–334, 2003.
255. S. Schwede and B.E. Shipley. Algebras and modules in monoidal model categories. Proc. Lond. Math. Soc. (3), 80(2):491–511, 2000.
256. G.B. Segal. Equivariant stable homotopy theory. In Actes du Congrès International des Mathématiciens, Tome 2, Nice, 1970, pages 59–63. Gauthier-Villars, Paris, 1971.
257. G. Segal. Categories and cohomology theories. Topology, 13:293–312, 1974.
258. J.-P. Serre. Faisceaux algébriques cohérents. Ann. Math. (2), 61:197–278, 1955.
259. J.-P. Serre. Corps locaux. In Publications de l'Institut de Mathématique de l'Université de Nancago, VIII, volume 1296 of Actualités Scientifiques Et Industrielles. Hermann, Paris, 1962.
260. N. Shimada and K. Shimakawa. Delooping symmetric monoidal categories. Hiroshima Math. J., 9(3):627–645, 1979.
261. B. Shipley. Symmetric spectra and topological Hochschild homology. K-Theory, 19(2):155–183, 2000.
262. B. Shipley. A convenient model category for commutative ring spectra. In Homotopy Theory: Relations with Algebraic Geometry, Group Cohomology, and Algebraic K-Theory, volume 346 of Contemporary Mathematics, pages 473–483. Amer. Math. Soc., Providence, 2004.
263. U. Shukla. A relative cohomology for associative algebras. Proc. Am. Math. Soc., 15:461–469, 1964.
264. S. Smale. On the structure of manifolds. Am. J. Math., 84:387–399, 1962.
265. V.P. Snaith. Towards the Lichtenbaum-Quillen conjecture concerning the algebraic K-theory of schemes. In Algebraic K-Theory, Number Theory, Geometry and Analysis, Bielefeld, 1982, volume 1046 of Lecture Notes in Mathematics, pages 349–356. Springer, Berlin, 1984.
266. C. Soulé. Rational K-theory of the dual numbers of a ring of algebraic integers. In Algebraic K-Theory, Proc. Conf., Northwestern Univ., Evanston, IL, 1980, volume 854 of Lecture Notes in Mathematics, pages 402–408. Springer, Berlin, 1981.
267. C. Soulé. On higher p-adic regulators. In Algebraic K-Theory, Proc. Conf., Northwestern Univ., Evanston, IL, 1980, volume 854 of Lecture Notes in Mathematics, pages 372–401. Springer, Berlin, 1981.
268. J. Spaliński. Strong homotopy theory of cyclic sets. J. Pure Appl. Algebra, 99(1):35–52, 1995.
269. E.H. Spanier. Algebraic Topology. McGraw-Hill, New York, 1966.
270. V. Srinivas. Algebraic K-Theory, 2nd edition, volume 90 of Progress in Mathematics. Birkhäuser, Boston, 1996.
271. R.E. Staffeldt. On fundamental theorems of algebraic K-theory. K-Theory, 2(4):511–532, 1989.

272. J.R. Stallings. On infinite processes leading to differentiability in the complement of a point. In Differential and Combinatorial Topology (A Symposium in Honor of Marston Morse), pages 245–254. Princeton University Press, Princeton, 1965.
273. A. Suslin. On the K-theory of algebraically closed fields. Invent. Math., 73(2):241–245, 1983.
274. A. Suslin. On the Grayson spectral sequence. Tr. Mat. Inst. Steklova, 241:218–253, 2003.
275. A.A. Suslin. Projective modules over polynomial rings are free. Dokl. Akad. Nauk SSSR, 229(5):1063–1066, 1976.
276. A.A. Suslin. On the equivalence of K-theories. Commun. Algebra, 9(15):1559–1566, 1981.
277. A.A. Suslin. Algebraic K-theory of fields. In Proceedings of the International Congress of Mathematicians, Vols. 1, 2, Berkeley, CA, 1986, pages 222–244. Amer. Math. Soc., Providence, 1987.
278. A.A. Suslin. Excision in integer algebraic K-theory. Tr. Mat. Inst. Steklova 208:290–317, 1995. Dedicated to Academician Igor' Rostislavovich Shafarevich on the occasion of his seventieth birthday (in Russian).
279. A.A. Suslin and M. Wodzicki. Excision in algebraic K-theory. Ann. Math. (2), 136(1):51–122, 1992.
280. R.G. Swan. Vector bundles and projective modules. Trans. Am. Math. Soc., 105:264–277, 1962.
281. R.G. Swan. Excision in algebraic K-theory. J. Pure Appl. Algebra, 1(3):221–252, 1971.
282. R.W. Thomason. Homotopy colimits in the category of small categories. Math. Proc. Camb. Philos. Soc., 85(1):91–109, 1979.
283. R.W. Thomason. Algebraic K-theory and étale cohomology. Ann. Sci. École Norm. Sup. (4), 18(3):437–552, 1985.
284. R.W. Thomason and T. Trobaugh. Higher algebraic K-theory of schemes and of derived categories. In The Grothendieck Festschrift, Vol. III, volume 88 of Progress of Mathematics, pages 247–435. Birkhäuser, Boston, 1990.
285. U. Tillmann. On the homotopy of the stable mapping class group. Invent. Math., 130(2):257–275, 1997.
286. T.T. Dieck. Orbittypen und äquivariante Homologie, I. Arch. Math. (Basel), 23:307–317, 1972.
287. T.T. Dieck. Orbittypen und äquivariante Homologie, II. Arch. Math. (Basel), 26(6):650–662, 1975.
288. S. Tsalidis. On the topological cyclic homology of the integers. Am. J. Math., 119(1):103–125, 1997.
289. S. Tsalidis. Topological Hochschild homology and the homotopy descent problem. Topology, 37(4):913–934, 1998.
290. B.L. Tsygan. Homology of matrix Lie algebras over rings and the Hochschild homology. Usp. Mat. Nauk, 38(2(230)):217–218, 1983.
291. W. van der Kallen. Descent for the K-theory of polynomial rings. Math. Z., 191(3):405–415, 1986.
292. W. van der Kallen, H. Maazen, and J. Stienstra. A presentation for some $K_2(n, R)$. Bull. Am. Math. Soc., 81(5):934–936, 1975.
293. V. Voevodsky. Motivic cohomology with $\mathbf{Z}/2$-coefficients. Publ. Math. Inst. Hautes Études Sci., 98:59–104, 2003.
294. V. Voevodsky. On motivic cohomology with $\mathbf{Z}/l$-coefficients. Ann. Math. (2), 174(1):401–438, 2011.
295. R. Vogt. Boardman's Stable Homotopy Category, volume 21 of Lecture Notes Series. Matematisk Institut, Aarhus Universitet, Aarhus, 1970.
296. F. Waldhausen, B. Jahren, and J. Rognes. Spaces of PL manifolds and categories of simple maps. Preprint available at http://www.mn.uio.no/math/personer/vit/rognes/index.html, 2008.
297. F. Waldhausen. Algebraic K-theory of generalized free products, I, II. Ann. Math. (2), 108(1):135–204, 1978.

298. F. Waldhausen. Algebraic K-theory of topological spaces, I. In Algebraic and Geometric Topology, Part 1, Proc. Sympos. Pure Math., Stanford Univ., Stanford, CA, 1976, volume XXXII of Proceedings of Symposia in Pure Mathematics, pages 35–60. Amer. Math. Soc., Providence, 1978.
299. F. Waldhausen. Algebraic K-theory of topological spaces, II. In Algebraic Topology, Proc. Sympos., Univ. Aarhus, Aarhus, 1978, volume 763 of Lecture Notes in Mathematics, pages 356–394. Springer, Berlin, 1979.
300. F. Waldhausen. Algebraic K-theory of spaces, localization, and the chromatic filtration of stable homotopy. In Algebraic Topology, Aarhus, 1982, volume 1051 of Lecture Notes in Mathematics, pages 173–195. Springer, Berlin, 1984.
301. F. Waldhausen. Algebraic K-theory of spaces. In Algebraic and Geometric Topology, New Brunswick, NJ, 1983, volume 1126 of Lecture Notes in Mathematics, pages 318–419. Springer, Berlin, 1985.
302. F. Waldhausen. Algebraic K-theory of spaces, concordance, and stable homotopy theory. In Algebraic Topology and Algebraic K-Theory, Princeton, NJ, 1983, volume 113 of Annals of Mathematics Studies, pages 392–417. Princeton University Press, Princeton, 1987.
303. C. Weibel. The norm residue isomorphism theorem. J. Topol., 2(2):346–372, 2009.
304. C.A. Weibel. Mayer-Vietoris sequences and module structures on NK_*. In Algebraic K-Theory, Proc. Conf., Northwestern Univ., Evanston, IL, 1980, volume 854 of Lecture Notes in Mathematics, pages 466–493. Springer, Berlin, 1981.
305. C.A. Weibel. Mayer-Vietoris sequences and mod p K-theory. In Algebraic K-Theory, Part I, Oberwolfach, 1980, volume 966 of Lecture Notes in Mathematics, pages 390–407. Springer, Berlin, 1982.
306. C. Weibel. An introduction to algebraic K-theory. In progress, some of it is on the web, http://math.rutgers.edu/~weibel/Kbook.html.
307. C. Weibel. K-théorie algébrique homotopique. C. R. Acad. Sci. Paris Sér. I Math., 305(18):793–796, 1987.
308. C. Weibel. The 2-torsion in the K-theory of the integers. C. R. Acad. Sci. Paris Sér. I Math., 324(6):615–620, 1997.
309. C. Weibel. Algebraic K-theory of rings of integers in local and global fields. In Handbook of K-Theory, Vols. 1, 2, pages 139–190. Springer, Berlin, 2005.
310. C.A. Weibel. The development of algebraic K-theory before 1980. In Algebra, K-Theory, Groups, and Education, New York, 1997, volume 243 of Contemporary Mathematics, pages 211–238. Am. Math. Soc., Providence, 1999.
311. M. Weiss and B. Williams. Automorphisms of manifolds and algebraic K-theory, II. J. Pure Appl. Algebra, 62(1):47–107, 1989.
312. H. Weyl. Philosophy of Mathematics and Natural Science. Revised and Augmented English Edition Based on a Translation by Olaf Helmer. Princeton University Press, Princeton, 1949.
313. G.W. Whitehead. Elements of Homotopy Theory, volume 61 of Graduate Texts in Mathematics. Springer, New York, 1978.
314. J.H.C. Whitehead. Simplicial spaces, nuclei and m-groups. Proc. Lond. Math. Soc. (2), 45:243–327, 1939.
315. J.H.C. Whitehead. On incidence matrices, nuclei and homotopy types. Ann. Math. (2), 42:1197–1239, 1941.
316. J.H.C. Whitehead. Simple homotopy types. Am. J. Math., 72:1–57, 1950.
317. D. Yao. A devissage theorem in Waldhausen K-theory. J. Algebra, 176(3):755–761, 1995.
318. D. Yau. Lambda-Rings. World Scientific, Hackensack, 2010.

Index

B.I. Dundas et al., *The Local Structure of Algebraic K-Theory*,
Algebra and Applications 18,
DOI 10.1007/978-1-4471-4393-2, © Springer-Verlag London 2013

MIX
Papier aus verantwortungsvollen Quellen
Paper from responsible sources
FSC® C105338

If you have any concerns about our products,
you can contact us on
ProductSafety@springernature.com

In case Publisher is established outside the EU,
the EU authorized representative is:
Springer Nature Customer Service Center GmbH
Europaplatz 3, 69115 Heidelberg, Germany

Printed by Libri Plureos GmbH
in Hamburg, Germany